Seasonal Adaptations of Insects

Seasonal Adaptations of Insects

MAURICE J. TAUBER
CATHERINE A. TAUBER
Cornell University

AND

SINZO MASAKI
Hirosaki University

New York Oxford
OXFORD UNIVERSITY PRESS
1986

Oxford University Press

Oxford New York Toronto
Delhi Bombay Calcutta Madras Karachi
Petaling Jaya Singapore Hong Kong Tokyo
Nairobi Dar es Salaam Cape Town
Melbourne Auckland

and associated companies in
Beirut Berlin Ibadan Nicosia

Published by Oxford University Press, Inc.,
200 Madison Avenue, New York, New York 10016

Library of Congress Cataloging in Publication Data
Tauber, Maurice J.
Seasonal adaptations of insects.
Bibliography: p.
Includes index.
1. Insects—Adaptation. 2. Dormancy in insects.
3. Insects—Wintering. 4. Insects—Migration. 5. Diapause.
I. Tauber, Catherine A. II. Masaki, Shinzo, 1927–
III. Title.
QL495.T37 1985 595.7′0543 85-4917
ISBN 0-19-503635-2

Printing (last digit): 9 8 7 6 5 4 3 2 1
Printed in the United States of America

To Our Parents

Foreword

Thirty years ago the physiological and ecological problems associated with insect diapause appeared to be well defined and indeed readily accessible to experimental solutions. A number of significant environmental factors, particularly photoperiod and temperature, had been found to control the onset and termination of dormancy in many insects, particularly those from temperate latitudes. In the case of photoperiod, there was evidently a photoreceptor and a clock that measured day or night length, and this was somehow coupled to the insect's endocrine system, which, in turn, regulated its growth patterns. Moreover, from laboratory evidence tentative inferences could be drawn regarding the probable responses of the insect in its natural environment. Authors such as Andrewartha and Danilevsky had pointed out that diapause must be of prime adaptive significance in adjusting the insect's entire life cycle to seasonal change and should not be regarded as merely a means of avoiding adverse climatic conditions or seasonal food shortages. Nevertheless, the construction of predictive models was seriously hampered by the relative scarcity of combined field and laboratory data on the seasonal progress of diapause.

Since then much water has flowed under the bridge. Although physiological studies of diapause have multiplied, many fundamental problems are still unsolved. The photoperiodic clock mechanisms remain tantalizingly elusive. The mode of action of temperature, with its unexpected subtleties, is still obscure. And the endocrinology has taken on the complexities associated with this discipline in other animal groups. No longer is it possible to predict that larval, pupal, and adult diapauses will be caused by simple hormone deficiencies. In other words, this field remains a fascinating challenge to the physiologist.

Perhaps because of these difficulties, as well as its inherent attractions, interest has recently veered toward another facet of diapause—namely, the elucidation of those diapause-mediated seasonal strategies that ensure the survival of the insect in a fluctuating and hostile environment. Interest has been stimulated by the relevance of diapause and quiescence mechanisms in population biology, a realm in which slight imperfections in the system may have more significance than predators or parasites. The discovery of many intraspecific races, both geographical and local, has also directed attention toward the genetic and evolutionary aspects of diapause, in which areas the present authors have themselves made many notable contributions. There has also been an increasing appreciation of the significance of diapause in relation to the practical problems of pest management. Work with this as a primary or secondary objective has, in turn, uncovered many new types of response to such environmental variables as photoperiod, temperature, and food.

From this viewpoint environmental physiology is a necessary adjunct for interpreting diapause in all its seasonal manifestations. Such a balance has been preserved in this volume. In the past diapause has often been regarded in rather narrow physiological terms—simply as an arrest or slowing of growth. The broader concept of a "diapause syndrome," advocated by de Wilde, is more appropriate in the context of seasonal adaptation and embraces many types of pre- and postdiapause responses, such as those affecting behavior (e.g., migration), cold hardiness, and so forth. In this respect the subject has assumed a much wider scope.

The voluminous literature on diapause and seasonal adaptation has long been in need of summary and critical review. The present volume will therefore be most welcome. Nevertheless, the implications of this survey are indeed overwhelming. If a human modeler was given the task of devising a set of optimum responses for controlling winter and summer dormancies in an irregularly fluctuating environment, he would surely be hard put to invent the staggering assay of strategies adopted by insects. As in their physiology, diversity rather than uniformity is the rule, no doubt an eloquent testimony to the powers of natural selection.

A. D. LEES
Professor Emeritus and Senior Research Fellow
Department of Pure and Applied Biology
Imperial College at Silwood Park
Ascot, U.K.

Preface

We contemplate our cultural inheritance largely through historical, archaeological, and comparative anthropological investigations of human and other anthropoid species. By contrast, perceptions of our biological inheritance and adaptations have received their primary influence from observations of a much broader range of organisms—prokaryotes and eukaryotes, plants and animals, invertebrates and vertebrates.

Our interest in both cultural and biological inheritance often includes a metaphysical component which has its origin in humanity's profound concern with mortality, the origin and cyclic nature of life, and the flow of time. Although we cannot achieve immortality by abolishing time, we can seemingly slow time's progress by decelerating biological processes, thus prolonging the life cycle. Within this context, insect dormancy is an especially intriguing phenomenon because it can, in some instances, extend the lifespan a hundredfold or more.

If we view an organism's life history as a chink of light within an eternity of darkness, we may partially see why penumbral states such as dormancy and sleep captivate our attention. It is as if our deeply seated curiosity concerning the prenatal and postmortem darkness could be satisfied vicariously by peering into the shadowy world of suspended life as we view it from the perspective of dormant insects.

Thus far, with each peek at insect seasonality, investigators have come face to face with a series of views representing facets of the life cycle—inactivity, covert activity, and overt activity (including movement over vast distances that span continents). Despite many notable insights derived from ecological, physiological, and biochemical studies, our present concept of insect seasonal adaptations, that is, dormancy, migration, and polyphenism, remains highly circumscribed. We are like spectators attempting to view a dynamic panorama through an opaque barrier perforated by a series of small, unevenly spaced peepholes. Although the present state of our knowledge of seasonality finds us largely on the periphery of a fascinating biological phenomenon, the broadly based investigations by scientists from round the world provide great hope for significant enlightenment in the near future.

Ithaca, N.Y.
January 1985

M.J.T.
C.A.T.

Acknowledgments

We owe much to many individuals and institutions. We particularly recognize the New York State College of Agriculture and Life Sciences and its Albert R. Mann Library, Cornell University (MJT, CAT), as well as Hirosaki University (SM).

Part of the research and writing was accomplished during a sabbatical leave (MJT), while many of the ideas expressed herein resulted from work generously funded by the National Science Foundation, the U.S. Department of Agriculture, and Cornell University's College of Agriculture and Life Sciences (MJT, CAT) and by Hirosaki University (SM).

A number of people have contributed at various times and in significant ways to our project, notably Anthony D. Lees, Imperial College, Silwood Park, United Kingdom; Carl B. Huffaker, University of California, Berkeley; Robert L. Rabb, North Carolina State University; Yoshikazu Ando, Hirosaki University; Sadaya Katsuno, Hokkaido University; Lloyd V. Knutson, SEA, USDA; Richard L. Ridgway, SEA, USDA; Donald W. Kaufman, Kansas State University; David L. Call, Cornell University; Samuel B. Tauber, Winnipeg, Canada; Paul, Michael, and Agatha Tauber, Ithaca, N.Y.; John J. Obrycki, Iowa State University; James R. Nechols, Kansas State University; Brian Gollands, Cornell University. To them we offer our deep-felt thanks.

Contents

1. Introduction 3

1.1 Objective 3

1.2 Historical Perspective 4

2. Insect Adaptations to Environmental Changes 7

2.1 Adaptations to Aseasonal Exigencies 10

2.2 Seasonal Adaptations 20

3. The Course of Diapause 38

3.1 Diapausing Stage 38

3.2 Stages Sensitive to Diapause-Inducing Stimuli 43

3.3 Diapause Induction Period 47

3.4 Diapause Completion 52

3.5 Postdiapause Transitional Period 59

3.6 Postdormancy Effects of Diapause 62

4. The Diapause Syndrome 67

4.1 Endocrine Mediation of Diapause 67

4.2 Behavioral Expression of Diapause 81

4.3 Physiological Expression of Diapause 91

4.4 Morphological Expression of Diapause 106

5. Environmental Regulation of Seasonal Cycles 111

5.1 Photoperiod 112
5.2 Temperature 135
5.3 Other Environmental Factors 151

6. Seasonal Adaptations—Special Cases 161

6.1 Parasitoids 162
6.2 Social Insects 171
6.3 Tropical Insects 178
6.4 Arctic and Desert Insects 185

7. Variability and Genetics of Seasonal Adaptations 192

7.1 Variability in Seasonal Cycles 192
7.2 Genetics of Seasonal Cycles 208

8. Evolution of Seasonal Cycles 218

8.1 Origin of Insect Diapause 219
8.2 Evolutionary Changes in Seasonal Cycles 229

9. Seasonality, the Evolution of Life History, and Speciation 256

9.1 Seasonality as a Life-History Trait 257
9.2 Seasonality in the Evolution of Life History 265
9.3 Seasonal Cycles and Speciation 280

10. Seasonality and Insect Pest Management 287

10.1 Predictive Capability as the Key to Pest Management 288

10.2 Seasonal Considerations in Implementing Management Tactics 298

10.3 Biological Control and Seasonality 305

Bibliography 310

Author Index 385

Species Index 402

Subject Index 408

Seasonal Adaptations of Insects

Chapter 1

Introduction

1.1 Objective

We wrote this book primarily for teachers, researchers, and university students in evolutionary biology and applied ecology. Our objective is twofold. First, we seek to give biologists a much fuller understanding of the broad array of insect seasonal adaptations than they generally find in current texts. In doing so we attempted to give a critical interpretation to the voluminous scientific literature that now exists on the subject, clarify various ideas relating to dormancy, migration, and polyphenism, and simplify the terminology and concepts dealing with seasonality. Second, we want to provide a strong basis from which to proceed in the evolutionary and ecological investigation of seasonal adaptations. Although insect seasonal adaptations do show patterns, knowledge of the mechanisms underlying seasonal cycles, their diversity, and their interrelationships is, at present, much too meager for solving specific problems in either evolutionary biology or applied ecology. As a result we have emphasized a comparative approach to seasonal cycles with the hope of providing a stimulus and a guide for future research.

Our book is organized into five main parts. Initially, we deal with the long- and short-term environmental changes that insects encounter and how insects adapt to these changes (Chapter 2). Then we focus on the types of seasonal changes that insects undergo, and we discuss diapause as the major physiological adaptation underlying the seasonal expressions of dormancy, migration, and polyphenism (Chapters 3 and 4). The diverse mechanisms through which environmental and genetic factors influence seasonal cycles form the topic of the next part of the book (Chapters 5, 6, and 7). Subsequently, we delve into the evolutionary implications of seasonal cycles (Chapters 8 and 9), and we consider such topics as the evolution of diapause, the influence of seasonality on the evolution of life histories, and the importance of seasonal cycles in speciation. Finally, the last chapter (Chapter 10) deals with the use of phenological knowledge in applied ecology, specifically in insect pest management.

1.2 Historical Perspective

In the 1940s the study of insect seasonal cycles began a course that advanced it through a series of phases—the first of which culminated with the publication of Andrewartha's review, "Diapause in relation to the ecology of insects" (Andrewartha 1952). This extensive paper summarized previous research and placed diapause within an ecological setting. It showed how diapause is essential to the survival of insects in seasonally inhospitable environments; it defined and clarified terms; and it provided a comprehensive perspective for the analysis and interpretation of data. Publication of *The Distribution and Abundance of Animals* by Andrewartha and Birch (1954) reinforced the impact of this review.

Since the 1950s, phenology has assumed a central role in insect ecology. Not only has research into the ecological aspects of insect seasonality maintained its initial impetus, but it has expanded into the areas of population and community ecology (e.g., see Root & Chaplin 1976; Thompson & Price 1977; Wood & Olmstead 1984). On the practical side, the effects of this expansion are particularly evident in modern insect pest management, an approach that often requires the development of accurate phenological models (Chapter 10).

The next major milestone in the study of insect seasonality was A. D. Lees' (1955) monograph, *The Physiology of Diapause in Arthropods*, in which he analyzed diapause from both a physiological and an ecophysiological standpoint. This classic admirably reviewed and summarized the current state of the art in a very active field of study, and it served as a stimulus and reference for both laboratory and field-oriented physiologists around the world for many years. Although several fine monographs have reviewed aspects of seasonal cycles in insects, there has been no other major review concerned with the comprehensive analysis of insect seasonal cycles since Lees' (see also Lees 1956; Tauber *et al.* 1984).

Also during the 1950's, the study of insect photoperiodism began to expand enormously. The pioneering studies of Marcovitch (1923) and Kogure (1933) formed the basis for a myriad of studies that established photoperiod as the primary seasonal cue for insects. The summarization and analysis of these advances in important reviews and books by de Wilde (1962a), Danilevsky (1965), Beck (1980), Lees (1968), Danilevsky *et al.* (1970), Saunders (1982), and Tyshchenko (1977) provided great insight into the role of anticipatory (token) cues in the regulation of seasonal cycles in insects. They also provided the necessary background for clearing the next hurdle—an understanding of the endocrine control of diapause.

The fields of insect physiology and endocrinology flowered during the late 1950s and the early 1960s and this blossoming had important implications for the analysis of the hormonal regulation of seasonal cycles. The laboratories of V. B. Wigglesworth in the United Kingdom, C. M. Williams and G. S. Fraenkel in the United States, and S. Fukuda in Japan led the way, and subsequently the laboratories of J. de Wilde and S. D. Beck contributed enormously by examining the environmental influences on the endocrine functions that regulate diapause (see Chapter 4). Concomitant with this came the work of A. D. Lees and H. J. Müller in unraveling the complexities of the environmental and hormonal control of seasonal changes in morphology and color. Recent work has built on these fine foundations so that at present insect seasonal responses can be used as model systems for analyzing circadian rhythms, time measurement, and the "black box" that translates environmental stimuli into physiological function (see Chapter 4).

On the evolutionary front, variability in seasonal life cycles, and the underlying physiological and genetic bases for the variability, initially received major attention from researchers in the Soviet Union. Soviet workers, particularly A. S. Danilevsky, N. I. Geyspits, and E. B. Vinogradova, significantly advanced the understanding of geographical and local variation in seasonal cycles. Nonetheless, the variability and genetics of seasonal cycles were not ignored elsewhere. In North America, Beck and Apple (1961) unraveled the subtleties of the geographical variability in the European corn borer populations that had adapted to the diverse climatic conditions of their introduced homeland in North America (see Chapter 8). Similarly, the genetic perspective developed very early in Japan, particularly in work dealing with voltinism in the commercial silkworm (see summary by Tazima 1964).

Researchers during this time considered the evolutionary implications of their findings. However, before phenological phenomena could be incorporated into the mainstream of evolutionary biology, data and ideas concerning the ecophysiological and genetic mechanisms underlying seasonal cycles had to accumulate and modern evolutionary theory had to develop. To date, the evolutionary approach to seasonality studies has passed two major milestones. First was the recognition of the physiological and ecological relationship between dormancy and migration in the adaptation of insects to variable environments. Studies and reviews by J. S. Kennedy, C. G. Johnson, and others in the 1960s laid the groundwork, and this was followed and expanded by later workers (see, e.g., Dingle 1978b). From this basis, analysis of the evolution of life-history adaptations focused on a second major topic during the 1970s and early 1980s. This work centered in North America, western and northern Europe, and Japan, and it emphasized the incorporation of a seasonal perspective into life-history theory. Noteworthy in promoting scientific interaction in this area have been the International Entomological Congresses (see Dingle 1978b; Brown & Hodek 1983).

It is our view that the evolutionary analysis of insect seasonal cycles is now at the threshold of a new era of synthesis involving a broad array of disciplines—that is, life-history evolution, developmental biology, and quantitative genetics (e.g., see Dingle & Hegmann 1982; Stearns 1983). But before such a broad-ranging integration can be successful it is essential to bring to the fore and analyze comprehensively the diversity of insect seasonal adaptations, their environmental, physiological, and genetic control, and their function in the adaptation of life cycles. Only with a clear understanding of the full array of developmental, genetic, and evolutionary pathways that insect seasonal cycles have, can such an integrative approach reach its full potential. This is true whether research seeks to incorporate genetic and physiological approaches into applied ecology or to integrate physiological, genetic, and ecological approaches into the evolutionary analysis of life histories.

Our impetus for writing this book stems from a recognition of the broad diversity in responses and mechanisms underlying the seasonal adaptations and life histories of insects. We emphasize that this diversity must be examined with greater precision and its significance must be considered more fully than it has been in the past if physiological, ecological, and systematic studies are to broaden their evolutionary bases in a comprehensive fashion.

Chapter 2

Insect Adaptations to Environmental Changes

To various degrees, all environments on earth, whether terrestrial or aquatic, change over time. These changes involve alterations in both biotic and abiotic elements. They are long or short term, cyclic or acyclic, severe or mild, as well as widespread or localized. Their effects on organisms vary enormously, but because of their pervasiveness, environmental changes constitute a major selective force shaping plant and animal life cycles. As such, they act as a major factor in the evolution of the vastly diverse groups of plants and animals, including the class Insecta. The flora and fauna of our world would certainly be grossly different if physical and biotic conditions remained constant!

In approaching the main topic of our book—adaptations of insects to seasonal fluctuations in their environment—we start with an overview of the types of environmental alterations that insects face as individuals, as populations, and as species. We consider that environmental changes have three general characteristics that are relevant to organisms—magnitude, predictability, and duration. Of these three, magnitude is the most difficult to categorize in relation to its effects on organisms. Few studies have related the magnitude of environmental changes to the adaptations of insects, but some investigations show that insects perceive and respond to what appear to be very slight or subtle environmental alterations. Typically, tropical areas seem to be least subject to changes over time. Nevertheless, even in these areas, variation in rainfall, human disturbances, and other factors have influences ranging from almost undetectable to grossly obvious. As one moves from the equator into temperate regions, daily and seasonal fluctuations in physical factors become more and more pronounced. Seasonal changes finally reach their extreme at the poles.

The predictability and duration of environmental changes are more easily characterized in relation to their effects on organisms (Table 2.1). Some changes (e.g., diurnal and seasonal) recur on a regular, cyclic basis; thus, environmental cues and the inherent ability of insects to measure time provide a degree of predictability to such changes. Although the duration of these cyclic changes varies along a continuum, we can classify changes as short term or long term. For example, we consider such phenomena as diur-

Table 2.1 Categories of environmental changes and the nature of insect adaptations to these changes.

DURATION	PREDICTABILITY	
	Acyclic	Cyclic
Short-term or localized	*Examples:* Sudden temperature change Localized deterioration of food Lack of mates	*Examples:* Diurnal, lunar, and tidal cycles in temperature, humidity, food, enemies, competitors
	Adaptations: Aseasonal quiescence Aseasonal migration Aseasonal polyphenism	*Adaptations:* Diurnal, lunar, and tidal rhythms in physiological and behavioral processes
	Primary control of adaptation: NERVOUS → physiological & behavioral adjustments	*Primary control of adaptation:* BIOLOGICAL CLOCK → behavioral and physiological rhythms
Long-term or widespread	*Examples:* Drought Forest and prairie fires Habitat modification (logging, agriculture, urbanization)	*Examples:* Seasonal cycles in temperature, humidity, food, enemies, competitors
	Adaptations: Aseasonal quiescence Genetic polymorphisms Colonization and succession Evolutionary change	*Adaptations:* Dormancy Seasonal migration Seasonal polyphenism
	Primary control of adaptation: GENETIC → physiological and genetic alterations	*Primary control of adaptation:* NEUROHORMONAL → diapause

nal cycles, lunar cycles, and tidal cycles to be short-term cyclic fluctuations, and we categorize seasonal cycles as long term because most of the short-term cyclic fluctuations themselves have annual cycles.

In contrast to cyclic events, acyclic changes occur irregularly and thus do not provide advance warning that allows organisms to prepare physiologically and behaviorally for the changes. They also vary in duration but, again, the short and long term form a continuum—from quickly reversible changes such as periods of cold or drought in midsummer, to medium and long-term alterations in the habitat, such as those produced by forest and prairie fires, large-scale logging, horticulture, and urbanization (see Wellington & Trimble 1984 for a discussion of the effects of weather).

The ability to tolerate environmental changes involves more or less highly evolved adaptations (see Table 2.1). Although these adaptations ultimately originate in the genetic makeup of the organism, it is useful to categorize the dominant proximal systems that govern the species' responses to the various types of environmental changes. Adaptations to cyclic fluctuations, whether long or short term, generally involve biological clocks or other time-measuring mechanisms and rhythmic (diurnal, lunar, tidal, annual) responses. To a greater or lesser extent, the expression of such rhythmic responses is mediated through the neural and endocrine systems.

Compared with responses to cyclic environmental fluctuations, the responses to acyclic factors show a very different array of dominant adaptive mechanisms. Because acyclic changes are largely unpredictable, time-measuring capabilities and the neuroendocrine system generally do not play leading roles in these adaptations. Rather, the nervous system and various physiological, behavioral, and genetic mechanisms are dominant. Perhaps the simplest response to short-term acyclic exigencies is for insects to stop or decelerate growth and reproduction and to become quiescent while the unfavorable conditions remain. However, if unfavorable conditions persist for a long period, such a reaction is rarely successful unless it involves highly evolved physiological adaptations that subserve survival. We discuss these adaptations in Section 2.1.

Insects may also respond to a lack of essential requisites in a locality by moving to another area, thus increasing the likelihood of survival and reproduction. Again, such migrations are not likely to be successful unless they are based on highly evolved behavioral and physiological adaptations.

Some insect species have evolved genetic polymorphisms for some aspect of their lives; these can provide protection against acyclic fluctuations in environmental conditions. Part of the population may remain in a protected state of dormancy even when conditions are favorable for growth and reproduction. Similarly, genetic polymorphisms for migratory behavior and wing length serve as means whereby part of the population escapes when local habitats become unsuitable (see Section 9.2.2.1; also Dingle 1980 for a review).

Finally, extreme or long-term changes in the environment may lead to local extinction of species, species replacement, or both. In some cases (e.g., after forest or prairie fires), a regular, gradual change in species composition occurs (succession), and the original species may ultimately return to the area (e.g., Southwood *et al.* 1983). In other cases, the environment may be maintained in a perturbed state, usually through human action (agriculture or urbanization). If this occurs, the original species may be permanently replaced by a new complex of introduced or opportunistic species, any of which may undergo genetic modifications through selection in the perturbed environment. Many adaptations to acyclic environmental changes (genetic polymorphisms and evolutionary adaptation) involve the species' seasonal cycle; we discuss these in later sections.

In summary, insects adapt to changing environments in a number of highly characteristic ways. Acyclic, unpredictable changes are largely overcome by (1) physiological and behavioral adaptations whose primary function is survival during the adverse periods and (2) genetic polymorphisms that protect populations against extinction. In contrast, highly regular, predictable fluctuations in environmental conditions lead to the evolution of behavioral and physiological cycles based on (3) elaborate time-measuring mechanisms and (4) associated neuroendocrine control of growth, development, reproduction, and behavior.

2.1 Adaptations to Aseasonal Exigencies

Among the diverse adaptations of insects to environmental changes, certain types can serve as protection against both relatively short-term, acyclic exigencies and seasonal changes in the environment. Some forms of dormancy or migration can have this dual role. However, seasonal adaptations are usually based on physiological mechanisms that are different from those controlling aseasonal responses. In addition, their ecological function and consequences are generally different from those of seasonal adaptations. Thus, although they may overlap somewhat in controlling mechanisms and functions, it is important to differentiate between seasonal and aseasonal adaptations and to provide some examples that typify each category.

As we designate them here, aseasonal exigencies are irregularly occurring, usually short-term or localized stresses such as extremely high or low temperature, drought, depletion of food, or lack of mates. Because the occurrence of these exigencies is unpredictable, survival is dependent on the insects' immediate reaction (Fig. 2.1). Insects typically respond to these unpredictable changes by becoming torpid or quiescent or by moving away from the unfavorable area (aseasonal migration). Also, some species have evolved the ability to undergo morphological or color changes (aseasonal polyphenism) that may increase tolerance to unpredictable adversities. These adaptations generally do not entail the long-term preparatory physiological changes that characterize most seasonal adaptations. Some involve responses triggered by sensory stimuli immediately preceding the adversity itself. Most important, from an ecological view, is that these adaptations allow growth, development, and reproduction to resume quickly after favorable conditions are encountered.

Insects vary considerably in their abilities to withstand aseasonal adversities. Sometimes their activity is simply hampered by adverse conditions such as low temperature. In this case, the immobilized state itself cannot be considered an adaptation any more than death caused by heat or cold is an adaptation. However, the ability to tolerate a period of torpor (quiescence) and the ability to move to a more suitable site or area (migration) or to assume a protective phenotype (polyphenism) are clearly adaptive traits, and insect species show a wide range of variation in their expression of these

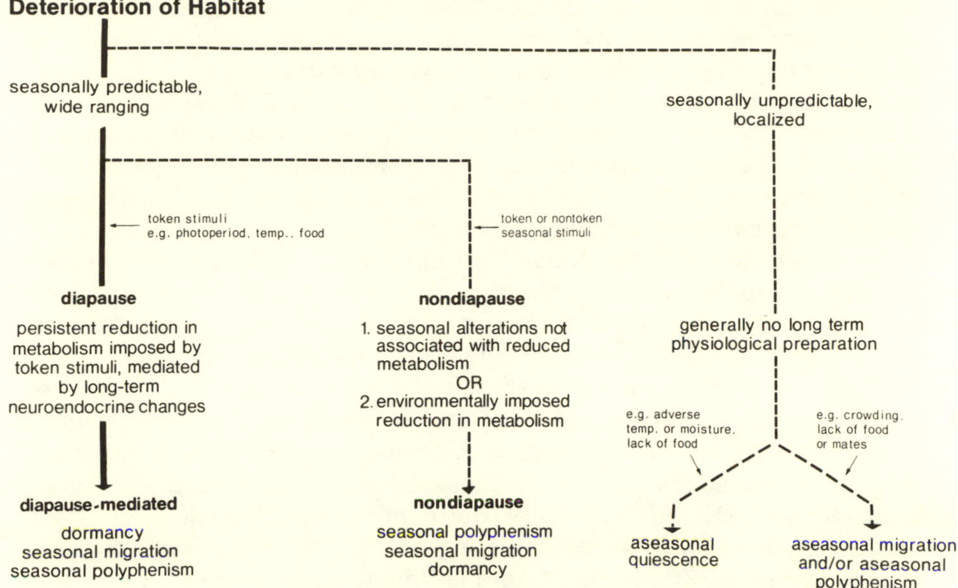

Fig. 2.1 Responses of insects to environmental deterioration: diapause-mediated seasonal adaptations, nondiapause seasonal adaptations, and aseasonal adaptations. Usually, but not always, when changes are regular and widespread, diapause-mediated responses control dormancy, migration, and polyphenism. Seasonally unpredictable or localized exigencies elicit the immediate responses of quiescence, aseasonal migration, and aseasonal polyphenism.

adaptations (see, e.g., Cohet *et al.* 1980 for a discussion of the responses of tropical and temperate-zone *Drosophila*).

2.1.1 Aseasonal quiescence

Quiescence is a reversible state of suppressed metabolism imposed by conditions beyond certain thresholds in temperature, moisture, and nutrition. It is an adaptation that allows animals to tolerate periods of cold or heat, drought, or lack of food in the immediate environment. Although it is probably the most common means whereby insects withstand aseasonal periods of stress, there are no comparative studies that illustrate how quiescence, as an adaptation to aseasonal exigencies, varies among insect groups. Yet, it is clear that there is considerable range and variety in the adaptations comprising aseasonal quiescence. The essential feature that all types of aseasonal quiescence share is the immediate resumption of development or reproduction at any time if favorable conditions return.

In general, most terrestrial, temperate-zone species in a growing or breeding state do not withstand prolonged periods of cold torpor without the occurrence of pathological changes. However, there are rare cases of exceptionally long periods of cold torpor; eggs of a nondiapause strain of the migratory grasshopper, *Melanoplus sanguinipes*, can be held at 4°C for two to three months and yet yield 55 to 67% hatch (Brelje & Blickenstaff 1974).

In comparison to thermally induced quiescence, aseasonal quiescence imposed by drought conditions has received a little more attention. It may involve any stage of the life cycle, including eggs, larvae, or adults. *Anopheles maculipennis messeae* females typically oviposit during the whole summer in most areas of South Russia, but in some places breeding sites occasionally dry up. When this happens, the mosquitoes cease ovipositing and only take occasional blood meals (Eichler 1951). Return of water to the breeding sites results in the immediate resumption of oviposition.

One of the most spectacular examples of a highly evolved form of quiescence occurs in the aquatic larvae of the chironomid *Polypedilum vanderplanki*. This species breeds in shallow, temporary rock pools in northern Nigeria. When the pools dry up, the larvae become dehydrated and development ceases (Hinton 1951). These dormant larvae are in a quiescent state and extremely resistant to both heat and desiccation. Under these circumstances larvae can survive after repeated dehydration down to 8% or less water content, and they can recover after exposure to temperatures of over 100°C for up to one hour. Some survive storage in warm, dry conditions for 7 to 10 years (Hinton 1960). Development resumes within hours of return to moist conditions.

A similar, moisture-regulated quiescence occurs in the eggs of *Hippelates* eye gnats (Spielman 1962; Legner *et al.* 1966). Under relatively dry conditions, eggs collapse but remain viable for up to 40 days. Upon exposure to moisture, hatch occurs within two to 24 hours. When this phenomenon occurs in nature has not been determined, but it may have a role in aestivation (Spielman 1962).

Another example of quiescence that occurs in response to unpredictable moisture conditions is found in the eggs of certain aedine mosquitoes. Generally aedine eggs require both conditioning (i.e., exposure to high humidity or contact moisture) and a hatching stimulus (deoxygenated water) before they will hatch (see Clements 1963). When these requisites are not fulfilled, hatching is delayed at any time of the year, and the unhatched eggs withstand long periods of quiescence. With the arrival of favorable conditions, quiescence ends and hatching occurs quickly. Such an adaptation allows the species to inhabit areas where water abundance and quality vary unpredictably.

Abundance and quality of food are other environmental factors that fluctuate irregularly, and insects vary in their ability to withstand these unpredictable periods of starvation. For example, larvae of the mosquitoes *Culex nigripalpus* and *Aedes taeniorhynchus* develop within five to eight days under optimal conditions and are unable to prolong their developmental period for more than 21 days under suboptimal nutritional conditions. In contrast, some larvae of *Wyeomyia vanduzeei, W. medioalbipes, W. smithii,* and *Toxorhynchites rutilus rutilus* may survive many weeks and even months with suboptimal nutrition (Istock *et al.* 1975; Frank & Curtis 1977; Moeur & Istock 1980; Lounibos *et al.* 1982). Addition of food results in immediate resumption of development.

2.1.2 Aseasonal migration

In contrast with aseasonal quiescence, aseasonal migration allows insects to take advantage of environmental heterogeneity (variability over space) to offset the detrimental effects of variability in time.[1] Rather than staying in place and postponing growth and reproduction until conditions improve, migrants may attain the requisites for continued growth and reproduction by moving out of the deteriorating area. Thus, unlike aseasonal quiescence, aseasonal migration may result in large changes in population density and in the spatial distribution of the population. However, despite the differences between aseasonal quiescence and aseasonal migration, we expect that studies ultimately will show that they share some important characteristics. Although the two phenomena may be based on different response patterns, both involve highly evolved behavioral and physiological adaptations (see reviews by Kennedy 1961, 1975; Johnson 1963, 1969; Dixon 1971b, 1977; Dingle 1972, 1980; Rainey 1974; Natuhara 1983; Rankin & Singer 1984).

Movement out of one area and into another is rarely successful unless the insect possesses the ability to predict the quality of its environment and to make the necessary physiological, and sometimes morphological, changes that facilitate migration. Moreover, migration itself is a highly evolved form of behavior involving reduction in sensitivity to cues from the host plant and a simultaneous increase in sensitivity to factors that promote movement away from the feeding or breeding site (Kennedy 1961, 1975). In view of these characteristics, it is not surprising that most migration occurs in response to relatively long-term, cyclic (i.e., seasonal) fluctuations in the environment. However, as Kennedy (1961, 1975) has pointed out, this pattern is not universal; some insects migrate in response to irregularly arising changes in local conditions and some undergo migration as a regular manifestation of metamorphosis.

Among the species that regularly undergo aseasonal migration is the frit fly, *Oscinella frit*. A large proportion of newly emerged *O. frit* adults actively migrate from the oat fields where they developed (Southwood *et al.* 1961; Johnson *et al.* 1962). Year-round seasonal trapping demonstrated that these migrations occur during all seasons in which adults emerge (Southwood 1964) and that females have a greater tendency to migrate than males (Ade-

1. We exclude from consideration here short-distance foraging movements (e.g., see Rankin & Singer 1984) and within-season movements between feeding and breeding sites (= Class II of Johnson 1969). These occur primarily as a result of the spatial distribution of resources and they are not primarily concerned with temporal changes in the environment.

siyun & Southwood 1979). Migration by this species and by others, such as the salt marsh-inhabiting planthopper, *Prokelisia marginata*, and the goldenrod leaf beetles, *Trirhabda virgata* and *T. borealis*, is associated with the utilization of ephemeral food plants or habitats (Southwood 1962, 1977; Denno & Grissell 1979; Messina 1982) and involves dispersal by females with undeveloped ovaries (Johnson 1969; Dingle 1978a).

Numerous other examples of aseasonal migratory flights are given by Johnson (1969) and Rabb and Kennedy (1979). In some cases, e.g., certain economically important noctuid moths, a seasonal increase in population size is correlated with large migrations during a particular season. Many of these species originate in the tropics, and they exhibit what appear to be one-way northward, seasonal migrations (the "Pied Piper" migrations of Rabb & Stinner 1978). One-way migrations present many interesting problems for evolutionary and applied biologists, largely because so little is known about them (see Dingle 1982; Iwao 1983; and various authors in Rabb & Kennedy 1979). They are often windborne and very long range, and upon completion, they are followed immediately by reproduction and the buildup of large populations in the new areas. Although migrations continue all summer, no return migrations are known and northern populations suffer complete or almost complete mortality during the winter. Species undergoing such migrations include important crop pests, such as the Oriental armyworm, *Mythimna separata*, the grass leafroller, *Cnaphalocrocis medinalis*, and the planthoppers *Nilaparvata lugens* and *Sogatella furcifera* in Japan, and the armyworms *Spodoptera frugiperda* and *S. exigua* and the potato leafhopper, *Empoasca fabae*, in North America.

Dingle (1982) categorized these movements with seasonal migrations; however, we agree with Rabb and Stinner (1978) that they could be considered aseasonal migrations. Other than the observation that growth and development in the southern, permanently occupied sites may be slowed down during cold months (various authors in Rabb & Kennedy 1979), no aspect of the life cycles of these species, including the migrations, has been shown to be controlled by seasonal factors. Thus, it is very likely that these species exhibit general adaptive dispersion in all directions at all times of the year when local conditions are sufficiently warm to allow growth and reproduction. If this is so, then the annual patterns of northward range extension are largely population phenomena, determined not by the seasonal stimulation of northward movement but by the annual availability of suitable northbound winds and hospitable conditions at the northern site. Clearly, these migratory species need additional behavioral and ecophysiological studies, especially in those localities where populations are permanent and where, presumably, natural selection is acting to maintain the migratory behavior patterns (see Rabb & Kennedy 1979). A similar situation could be occurring in the butterfly *Nathalis iole* that builds up large populations during the summer in midwestern North America (Douglas & Grula 1978). On a local level, aseasonal or opportunistic migration could also account for seasonal differences in population levels of the planthopper,

Prokelisia marginata in two habitats that vary seasonally in their relative suitability (Denno & Grissell 1979).

Aphids, too, exhibit aseasonal migration by flight, in which parthenogenetic females are transported between summer hosts (see Bonnemaison 1951; Lees 1966; Hille Ris Lambers 1966; Dixon 1971b; Blackman 1974a). Aphid migration is especially interesting because it often involves both migration and polyphenism for wing length, and because the aphids' perception of deteriorating environmental conditions can occur well in advance of the migratory stage. In polyphenic species, environmental conditions of crowding and food quality (Lees 1966; Dixon *et al.* 1968; Dixon 1971b, 1977; Dixon & Glen 1971; Mittler 1973; Harrewijn 1978) function as stimuli for production of winged forms (alatae). These same stimuli, with a greater or lesser degree of independence (Dixon 1971b), also favor migratory behavior. In certain tree-dwelling species, the parthenogenetic females are all alates, and in these cases, flight behavior itself clearly depends on environmental cues (Dixon 1969).

Once the migrating aphid is in motion, external factors, such as wind, largely determine the direction of movement, but the duration of the movement is under the control of the aphid. Two examples illustrate this in very different ways. In the black bean aphid, *Aphis fabae*, a few hours of flight usually result in a weakening of flight activity and an increase in responsiveness to growth- and reproduction-promoting stimuli (Kennedy & Booth 1963a, b). If the aphid is successful in finding a suitable host plant, feeding and embryonic development begin. In another aphid, *Pemphigus treherni*, summer migrations are accomplished by first-instar larvae floating on seawater (Foster & Treherne 1978). Nonmigrant individuals, which feed on the roots of the summer host (saltmarsh aster), exhibit strong negative phototaxis, whereas individuals destined to undergo migration do not respond to food and are positively phototactic (Foster 1978). They climb up on the stems and leaves of the host plant from which they are swept by the tides. Approximately 30 minutes of floating on seawater results in the migrants' behavior reversing back to negative phototaxis, and if a suitable host plant is encountered, the aphid crawls down to the roots and begins feeding.

In analyzing within-season migrations by summer aphids (as opposed to the regular seasonal movements between primary and secondary hosts by spring and autumn migrants), it is often difficult to categorize them as seasonal or aseasonal. The difficulty stems from the fact that there are both seasonal and aseasonal components to many important aspects of the summer migrations. First, the two primary stimuli that underlie the summer migrations, crowding and food quality, separately and in combination, are indicative of two types of interacting environmental changes—population buildup and deterioration of the host plant. Each of these environmental changes can occur on a seasonal and an aseasonal basis. Second, in response to these cues, aphids disperse from the host plant and, if successful, arrive at a new, suitable host. Summer hosts can include a series of plants of one species that is available throughout summer, or they can include several different species of plants that are available on a regular, sequential basis. For example, the sycamore aphid, *Drepanosiphum platanoides*, is autoecious and may colonize a succession of sycamore trees (Dixon 1969), whereas the bean aphid, *Aphis fabae*, may colonize a series of different summer hosts (Kennedy & Booth 1954). Finally, the stimuli that induce wing formation and migratory behavior patterns can occur seasonally and are perceived, not only during the lifetime of the aphid itself, but often during the parental generation as well (see Dixon 1969). Thus, there are both immediate, aseasonal and long-term seasonal aspects to the buildup of responsiveness to flight-promoting stimuli.

While recognizing the above problems, we include most examples of summer migrations by aphids in the aseasonal category for two reasons. First, migrating summer aphids are generally less discriminatory in their range of acceptable hosts than are autumnal migrants (see Kennedy & Booth 1954, 1961; Dixon 1971b, c). Thus, seasonal sequences of summer hosts are usually secondarily imposed by the seasonal availability of suitable plants and are not determined primarily by physiological or behavioral alterations, such as a seasonally induced change in host preference by the aphids. Similarly, seasonal changes in host plant substrate by the drosophilid *Scaptomyza pallida* do not involve a change in host preference (Toda & Kimura 1978). These patterns are in contrast to the seasonal changes in host-seeking behavior recently reported for certain butterflies (see Section 2.2.2.1). Second, the neuroendocrinological mechanisms involved in the control of wing length determination in autumnal migrants appear to differ significantly from those in summer forms (Hardie 1981a; Lees 1978, 1980). The autumnal wing-length changes represent an integral part of the neuroendocrinologically mediated adaptation to the seasons whereas the alate summer form is induced as a specific response to immediate conditions, usually crowding.

Like aphids, locusts undergo migratory flights that are preceded by morph changes and that have both seasonal and aseasonal components. Several species of locusts in Africa, Asia Minor, southern Europe, India, Australia, and tropical America (see Nolte 1974; Harvey 1983) form swarms that travel thousands of kilometers between breeding sites. Reproduction is suppressed during migration, and the flights occur largely during the rainy season, when the directional wind patterns are favorable (Morant 1947; Davies 1952; Fortescue-Foulkes 1953; Rainey 1963, 1974, 1978; Waloff 1966; Batten 1967). However, locust migrations mainly involve movement between breeding sites that are subject to considerable within- and between-year variation in moisture and food availability. The flights, therefore, appear to serve mainly as means whereby the locusts utilize patchily distributed resources (including moist soil for the development of the desiccation-susceptible eggs), rather than as adaptations to seasonal changes *per se*. Obviously, the resources have seasonal cycles, as well as spatial heterogeneity in their availability, and thus it is not surprising that the migratory patterns of locusts are not clearly seasonal or aseasonal. However, because of the apparent dominance of the spatial distribution of resources in the determination of migration, we consider locust flights to be primarily aseasonal.

Locusts are polymorphic species of gregarious Acrididae with a range of morphological and physiological characters varying between two extremes, the *solitaria* and the *gregaria* (see Uvarov 1966; Nolte 1974). The change from *solitaria* to *gregaria* is important in the formation of swarms and the initiation of migration (e.g., see Kennedy 1956). Both gregarious and solitary phases are found in migrating swarms, but the gregarious phase is far more active and prone to flight. Thus the environmentally induced, neuroendocrinologically mediated (e.g., Pener *et al.* 1972; Pener 1974, 1976, 1983; Joly *et al.* 1977; Tobe & Chapman 1979; Injeyan & Tobe 1981) changes in morphology and behavior that determine locust phases are of direct importance to the migration of locust swarms. Clearly, population density plays a major role in this function; both tactile and olfactory stimuli associated with crowding contribute to phase changes (see Uvarov 1966 for a summary; also Norris 1970). Photoperiod, temperature, food, parental age, and humidity also are involved (Albrecht 1973a, b, c; Cassier 1966, 1967c, 1968; Albrecht & Lauga 1979; Gillett 1978; Laugé & Launois 1980). The influence of these environmental factors on phase expressions is cumulative over generations, and there is considerable variation among individuals, even those from a single egg mass, in the ability to respond to the stimuli (Cassier 1967a, b; see Nolte 1974). The immediate causes of flight are largely related to meteorological conditions (Rainey 1974; Farrow 1979) and to daily changes in photoperiod and temperature (Albrecht *et al.* 1978).

Numerous other species of insects (e.g., gerrids, fulgorids, planthoppers, crickets) exhibit environmentally controlled, aseasonal variation in wing length and migration (see Dingle 1980). However, in contrast to studies of aphids and locusts in which both migration and polyphenism have been analyzed, most investigations of these species have emphasized the wing polyphenism that facilitates migration, rather than migration *per se*. Thus, we discuss them in the next section.

Aseasonal migration is also expressed in the periodic movements of army ant colonies, for example, *Eciton hamatum* and *E. burchelli* (Schneirla 1945, 1957; Rettenmeyer 1963). Apparently, maturation of new brood in the colony stimulates mature workers to undergo changes in behavior that result in movement of the whole colony—queen, brood, and all—to a new site, where growth, development, and reproduction continue uninterrupted.

Another example is the aggregation and marching behavior of immature desert locusts and lepidopteran larvae (armyworms) (Ellis 1964a, b; see Uvarov 1966; Iwao 1983). In large part, these movements are determined by nervous stimulation associated with crowded conditions; solitary individuals are much less active than those with a history of crowding. These migratory behaviors can also have a seasonal component, and we discuss them more fully later.

Both temperate zone and tropical bugs undergo aseasonal migration. In the seed bug, *Neacoryphus bicrucis*, from temperate-zone North America, only adults that have fed and mated reproduce without migrating; lack of either food or mates induces flight to another area where feeding and reproduction occur (Solbreck 1978; Solbreck & Pehrson 1979). Starvation of adults of the Nearctic species *Oncopeltus fasciatus* also increases flight activity but only for a short period, and it is likely that this form of migration is different both physiologically and ecologically from the photoperiodically induced migration that occurs each autumn. The environmental stimuli, the duration, and the neuroendocrine involvement differ between the two (Rankin 1980). Similar types of aseasonal migrations, but apparently no seasonal migrations, occur in the African cotton stainer bugs, *Dysdercus fasciatus, D. nigrofasciatus*, and *D. superstitiosus*. All three species utilize seasonally transient, patchily distributed sources of food; deterioration of one food source stimulates migration to another. Food is both the proximate and ultimate cause of these migrations (Dingle & Arora 1973; Fuseini & Kumar 1975; Derr *et al.* 1981).

From the above discussion of aseasonal migration, it becomes apparent that various environmental factors can provide cues that indicate unsuitable conditions or the imminent deterioration of the environment. Migratory insects respond to these cues by switching their behavioral sensitivity from stimuli that signal food, mates, or oviposition sites to stimuli that promote movement. This change in sensitivity is again reversed during migration, and if a suitable host or habitat is found, growth, development, and reproduction resume immediately. Thus, from an ecological viewpoint, aseasonal migration shares some important characteristics with aseasonal quiescence—the immediacy of its control by environmental conditions and some of its demographic consequences, that is, short-term postponement of reproduction in favor of survival. However, there are some clear differences between the phenomena. There is a considerable difference between aseasonal quiescence and aseasonal migration in the time taken for their expression. For example, when wing or other morphological changes are involved in migration, the morphogenetic processes must proceed prior to the migratory stage. No such changes must precede aseasonal quiescence. Moreover, the demographic consequences of the two phenomena, other than the short-term postponement of reproduction, may differ considerably. Migration results in a drastic change in population density or at least in the spatial pattern of distribution, while quiescence does not, as long as the environmental conditions do not cause high mortality.

2.1.3 Aseasonal polyphenism

As stated above, insects sometimes undergo changes in color and/or structure that provide crypsis or other forms of adaptation to nonseasonal environmental variables in their environment. This type of polyphenism does not involve a delay in either growth or reproduction, and it is not usually associated with a seasonal adaptation. It is largely triggered by the immediate conditions of the environment.

Notable examples of aseasonal polyphenism involving background coloration are found in pupae of papilionid and other butterflies, acridoid grasshoppers, and tenebrionid beetles (see Rowell 1971; Wigglesworth 1972; Nijhout & Wheeler 1982 for examples). Nondiapausing *Papilio* summer pupae vary in color—either green or brown—depending on optical, textural, and chemical cues they perceive from their substrate (Hidaka 1956; Ohtaki 1960; West *et al.* 1972; Hazel & West 1979; Honda 1981). These color changes are not related ecologically to diapause, and they are controlled by different neuroendocrine mechanisms from those that control diapause (Awiti & Hidaka 1982). Similarly, although diapausing pupae of the same species lack the ability to form green pupae, background color determines the depth of their brown coloration.

Nymphs and adults of several species of grasshoppers also show relatively fast homochromic responses to their substrate. Responses can occur at moults after only several days' exposure to new backgrounds, and they are limited to individuals in the solitary phase (Faure 1932; Rowell 1971). The grasshopper *Paulinia acuminata*, which is darkly colored when it occurs on brown leaves of its host plant, *Salvinia*, and light green on fresh leaves, was thought to have a homochromic response. However, studies showed that the darker color is the result of melanic deposition associated with unfavorable diet and increased activity, not with the color of the substrate (Meyer 1979). Color change (i.e., the assumption of black and orange or yellow in bold patterns) associated with the gregarious phase of locusts is also unrelated to substrate but occurs in response to crowding and other factors that control phase changes (Section 2.1.2). A similar type of melanization and phase change occurs in response to crowding in many lepidopterous larvae, particularly armyworms (Iwao 1968; Ogura 1975).

Another interesting example of aseasonal polyphenism is found in the desert tenebrionid beetle, *Cryptoglossa verrucosa*, which exhibits distinct color phases that range from light blue to jet black, depending on humidity conditions in the local environment. The color phases result from wax secretion and may be important in regulating water balance and body temperature in the desert environment (Hadley 1979).

As discussed in the section above, some aspects of the wing-length polyphenisms in aphids, gerrids, crickets, and other insects are aseasonal, whereas others are seasonal and associated with a relatively long period of suppressed development. For aphids, katydids, crickets, and planthoppers, summer conditions of crowding and poor food quality may provide cues that stimulate production of winged forms (see above; Kisimoto 1956, 1957; Denno & Grissell 1979; Ando & Hartley 1982; Strong & Stiling 1983). In polyphenic gerrids, temperature determines whether nondiapausing summer adults will be short or long winged (Vepsäläinen 1978). Low temperatures, characteristic of aquatic habitats that are not likely to dry up, result in short-winged forms that undergo reproduction soon after emergence. By contrast, high temperatures signal the danger of habitat deterioration and promote the production of long-winged forms that remain in a teneral state for some time. If these tenerals experience crowding, food shortage, or high temperature—stimuli that signal a deteriorating habitat—migration precedes reproduction. Under favorable environmental conditions, long-winged tenerals become reproductive without migration.

2.1.4 Summary

Insects cope with short-term, acyclic (aseasonal) fluctuations in their environment primarily through physiological and behavioral adaptations that specifically subserve survival during environmental exigencies. In general, responses to aseasonal changes are elicited by unfavorable conditions; growth, development, and reproduction resume immediately when suitable conditions are experienced.

Such adaptations involve (1) various physiological mechanisms that allow insects to survive long periods under an environmentally imposed state of suppressed metabolism (aseasonal quiescence), (2) highly evolved behavioral patterns that result in movement out of a deteriorating area, with the opportunity of finding a new, favorable area (aseasonal migration), and (3) environmentally induced changes in color or structure that subserve protection or movement (aseasonal polyphenism).

2.2 Seasonal Adaptations

Most biotic and abiotic variables that organisms can utilize or must cope with in their environment are either directly or indirectly related to seasonal changes. Thus, one of the most basic and unifying aspects of any species' ecological adaptation is its *phenology*, that is, the set of adaptations that leads to the appropriate seasonal timing of recurring biological events—growth, development, reproduction, dormancy, and short- or long-distance migration—in relation to annual cycles of both biotic and abiotic factors in the environment.

In contrast to the irregularly occurring and/or localized environmental changes discussed above, seasonal changes are characterized as being cyclic, persistent, and geographically widespread. Most insects have evolved the ability to perceive environmental cues that signal the upcoming seasonal changes, and they respond to these cues by undergoing specific physiological, behavioral, and morphological modifications that prepare them for approaching adverse conditions (Fig. 2.1). These modifications occur in a species-specific sequence, and they comprise the *diapause syndrome.*

It is very difficult to define diapause as a specific physiological phenomenon because it varies enormously from species to species, it is polyphyletic in origin, and it serves as a timing device for diversified phenological adaptations—including nearly all cases of dormancy and most cases of seasonal migration and seasonal polyphenism. The problems in defining diapause are similar to those encountered in defining "dormancy" and "quiescence" in seeds and "hibernation" and "torpor" for mammals and birds; there are several, sometimes conflicting, definitions, the diversity of which stems from the numerous approaches used in research on the phenomena (see Bewley & Black 1982; Lyman *et al.* 1982; Dawe 1983). However, given these problems, we define *diapause as a neurohormonally mediated, dynamic state of low metabolic activity. Associated with this are reduced morphogenesis, increased resistance to environmental extremes, and altered or reduced behavioral activity. Diapause occurs during a genetically determined stage(s) of metamorphosis, and its full expression develops in a species-specific manner, usually in response to a number of environmental stimuli that precede unfavorable conditions. Once diapause has begun, metabolic activity is suppressed even if conditions favorable for development prevail.* As we will illustrate later, therefore, diapause serves as the primary means whereby insect life cycles are kept in phase with the changing seasons.

The environmental cues that regulate diapause are called *token stimuli* (after Fraenkel & Gunn 1940) because they are not, in themselves, favorable or unfavorable for growth, development, or reproduction, but they herald a change in environmental conditions. In this regard, the most common token stimulus is photoperiod, although temperature, moisture, and biotic factors can also provide reliable cues to future seasonal changes. These stimuli function during periods favorable for growth, development, and reproduction and are often, but not exclusively, perceived by earlier developmental stages than those that undergo diapause (Section 3.2, below).

Originally, W. M. Wheeler (1893) applied the term "diapause" to a resting stage during embryogenesis, and subsequently its use was altered to refer to suppressed growth, development, or reproduction at any stage of insect development (see Andrewartha 1952, and Lees 1955 for the early history of the term). Later, de Wilde (1962a, b, 1970) and Hodek (1973) expanded the definition of diapause to include a variety of physiological and behavioral changes that subserve dormancy; they referred to the many changes as the diapause syndrome. Our definition of diapause is similar to that proposed by de Wilde and Hodek in that it emphasizes the *syndrome* of physiological and behavioral changes that adapt insects to approaching seasonal changes. However, our concept of the diapause syndrome is broader than previous ones in that we emphasize the dynamic aspects of diapause—that is, the full development, maintenance, and completion of all related symptoms, whether they are under the control of the primary diapause-inducing token stimulus or other environmental factors that precede seasonal changes. Thus, we include as expressions of the diapause syndrome most forms of seasonal migration and seasonal polyphenism, in addition to dormancy, with its associated cold-, heat-, and drought-hardiness because these phenomena are inextricable from, and vital to, the successful accomplishment of diapause in nature.

Some authors have used the term "diapause" to denote dormancy; with this usage the term has ecological as well as physiological and behavioral connotations. However, we restrict the term to its physiological and behavioral meanings for two reasons. First, dormancy, which is an ecological phenomenon, can occur without diapause; as will be illustrated later, it can involve seasonal periods of quiescence. Second, the physiological and behavioral changes that comprise the diapause syndrome can form the basis for ecological adaptations other than dormancy. Seasonal migration and seasonal polyphenism are often intimately and inextricably associated with diapause, and dormancy is sometimes not the predominant expression of diapause. Thus, we find it useful to restrict the term diapause to its physiological and behavioral meanings. In doing so, we also establish a basis for broadening its usage by recognizing that the physiological and behavioral changes involved in diapause can underlie several interrelated ecological manifestations (dormancy, seasonal migration, seasonal polyphenism).

In the past, the various phenological expressions of the diapause syndrome have been considered independently. In fact, dormancy (= diapause of various authors) and migration have often been contrasted as evolutionary alternatives, that is, as "escape in time" versus "escape in space." However, the intimate association between seasonal migration, seasonal polyphenism, and dormancy (see Müller 1960b; Kennedy 1961; Southwood 1962; de Wilde 1962a; Shapiro 1976; Dingle 1979, 1982; various authors in Dingle 1978b) speaks against such a polarity. In many cases the physiological bases and ecological functions of dormancy, seasonal migration, and seasonal polyphenism appear to be closely related. All three may involve a delay in reproduction and, for all three, diapause is often the pivotal adaptation.

Below we discuss the three major ecological expressions of diapause—dormancy, seasonal migration, and seasonal polyphenism. We illustrate how the three phenomena vary in their roles as components of seasonal cycles, and we analyze their relationship to diapause.

2.2.1 Dormancy

Dormancy is a general term that refers to a seasonally recurring period (*phenophase*) in the life cycle of a plant or animal during which growth, development, and reproduction are suppressed. Insects show considerable variability in their expressions of dormancy. At the physiological level, it can involve either diapause or quiescence (or both), provided they occur on a regular *seasonal* basis. At the ecological level, dormancy can involve suppression of growth or reproduction, and it can extend anywhere from a short period of time within one season to a much longer span of time over several seasons, even years.

Dormancy occurs during summer, fall, winter, or spring; these periods are termed *aestivation, autumnal dormancy, hibernation*, and *vernal dormancy*, respectively. We stress that these terms do *not* specify the organism's physiological state during dormancy; rather, they define only the seasonal timing of dormancy. The life cycles of insects may include a period of dormancy during one season or one or more periods of dormancy spanning almost any combination of seasons. But as we define it, dormancy is a recurring event, the beginning and end of which are associated with specific seasons.

Although dormancy is the means by which most plants and animals withstand seasonal periods of environmental extremes, not all species undergo dormancy. Some species of tropical cotton stainers in the genus *Dysdercus* are reported not to have a resting stage. These species feed preferentially on the ripening seeds of Malvales, and they withstand seasonal periods when the preferred food is not available by moving onto less preferred, but available, alternate hosts (Southwood 1962; Dingle & Arora 1973; Edmunds 1978). Apparently dormancy does not exist in some tropical and subtropical species of *Oncopeltus* milkweed bugs. Their food plants are more or less continuously available, and they can withstand relatively long periods of starvation during which they migrate to new sites (Root & Chaplin 1976; Dingle 1978a; Miller & Dingle 1982). Similarly, in some populations of the carabid beetles *Pogonus chalceus* and *Bembidion andreae* from North Africa, the Australian grasshopper, *Austracris proxima*, and the Queensland fruit fly, *Dacus tryoni*, ovarian development can occur all year round (Paarmann 1975, 1976b, d, 1977; Meats 1976; Farrow & O'Neill 1978). All of the above species occur in tropical areas or areas of relatively mild climates that favor year-round activity.

Temperate-zone insects may also lack dormancy, but these species usually occur in protective habitats. In the crickets *Gryllodes sigillatus* and *Myrmecophilus sapporoensis*, which are confined to habitats such as houses and ant nests, development appears to proceed year-round without interruption by dormancy (Masaki 1978a). Apparently, reproduction, growth, and development continue all year in the fruit fly *Drosophila subobscura* from northern England (Begon 1976). However, the rates of these processes are reduced during winter and there is an increase in fat body deposition during winter. Despite the above examples, it is important to stress that examples of insects, even tropical insects, that totally lack dormancy are not common. Dormancy in tropical species and in species with protective habitats can be very subtle and often goes undetected (Tauber & Tauber 1981a). We will discuss this more fully below.

Because dormancy in insects and mites usually involves specific physiological and behavioral modifications that prepare the animal for adverse seasonal conditions, it is reasonable to conclude that in most cases dormancy is diapause mediated. Nevertheless, there are some examples in which insects undergo dormancy in a state of environmentally imposed quiescence. We refer to this type of dormancy as nondiapause dormancy. Each of the two types of dormancy, nondiapause and diapause mediated, has specific characteristics and different modes of environmental regulation whereby they may be differentiated. However, when testing insects for diapause, it is important to keep in mind that many cases of dormancy in insects involve a period of diapause followed by a period of postdiapause quiescence. Thus, the seasonal timing of studies with dormant insects is crucial in determining the type of dormancy and the controlling factors.

2.2.1.1 Nondiapause dormancy

We define nondiapause dormancy as a state of suppressed metabolism imposed by seasonally recurring environmental conditions that are unfavorable for growth and reproduction. Thus, nondiapause dormancy has two distinguishing features: (1) a seasonal component that separates it from aseasonal quiescence that occurs on an irregular basis in response to unpredictable, localized exigencies (see Section 2.1.1 above) and (2) a high degree of responsiveness to the immediate conditions of the environment, which sets it apart from diapause-mediated dormancy, which is under the control of token stimuli occurring in advance of seasonal changes.

Cases of nondiapause dormancy are uncommon—the few relatively well-known examples occur in certain strains of aphids (see Hille Ris Lambers 1966; Blackman 1971, 1974a; Müller 1971; Landis *et al.* 1972; Müller & Möller 1973; Bevan & Carter 1980; Annis *et al.* 1981; Tamaki *et al.* 1982a; Hand 1983; Komazaki 1983), the leafhopper *Euscelis incisus* (Müller 1981), and the Antarctic terrestrial mite *Alaskozetes antarcticus* (Young & Block 1980). The pemphigine aphid, *Thecabius affinis*, studied by Sutherland (1968), provides a particularly good example. In this species, as in the other aphid species with overwintering quiescence, both holocyclic and anholocyclic strains are known. During autumn, the holocyclic strains develop winged, parthenogenetic morphs that move to the winter host and produce sexual forms. Mating occurs and females oviposit diapausing eggs that overwinter. In contrast, the anholocyclic strains do not leave the summer host and do not produce either sexual morphs or diapausing eggs. In these strains, both larvae and adult females overwinter on the summer host in a state of quiescence that is maintained by low temperature. This quiescence has a preparatory quality in that low temperatures of autumn not only impose quiescence but they also induce certain physiological changes that enhance cold-hardiness (acclimation). Individuals reared under low temperatures accumulate large lipid reserves that are not found in summer aphids. Postquiescence growth by aphids that have accumulated large lipid reserves is slower than that by summer aphids that do not have high amounts of lipids, and resumed reproduction by quiescent aphids requires time for egg maturation. However, the dormancy is presumed to be the result of quiescence, because overwintering individuals commence feeding and wax production within hours of transfer to laboratory conditions. Thus, the exact mechanisms controlling this type of dormancy are not known, but certainly they are worthy of further study.

The Antarctic terrestrial mite, *A. antarcticus*, appears to hibernate in a nondiapause state of quiescence. All life stages overwinter, and photoperiod does not appear to affect the rate of growth and development or the development of cold-hardiness. This species is freezing-susceptible but it is capable of supercooling to −26.5°C without an increase in glycerol. This adaptation is important because subzero temperatures can occur unpredictably at any time of the year in this species' habitat. Further suppression of the supercooling point requires the accumulation of glycerol; this occurs in response to low temperature (Young & Block 1980). Suppression of feeding is also necessary for full development of cold tolerance in this freezing-susceptible species, so that gut contents do not contain efficient ice nucleating agents. It is not known what factors regulate feeding.

Laboratory studies with the psocid, *Peripsocus quadrifasciatus*, suggest that at southern latitudes, quiescence may act as an alternative to the photoperiodically induced egg diapause that occurs in northern populations (Eertmoed 1978). Nondiapause eggs of this species are relatively cold-hardy, but field data are needed to substantiate the claim that eggs overwinter in a state of quiescence in southern areas.

One method through which nondiapause dormancy (or, indeed, diapause-mediated dormancy) could function in a thermally predictable environment, for example, certain aquatic environments, is for the overwintering stage to have a relatively high thermal threshold for development. Thus, growth would be prevented during the cold period and posthibernal development would be synchronized when environmental temperatures exceed the thermal threshold. Such a mechanism has not been demonstrated experimentally, but field data suggest the occurrence of a similar mechanism in some Odonata (Corbet 1963). Here, the last few larval instars preceding emergence appear to have progressively higher thermal thresholds and/or coefficients for development. Thus, metamorphosis during the early spring and emergence in summer are synchronized among individuals by the successively higher thermal requirements for each succeeding stage. However, in this case, diapause could be involved in changing the thermal thresholds from one stage to the next.

2.2.1.2 Diapause-mediated dormancy

By far the majority of cases of dormancy in insects are diapause mediated. The distinguishing features of diapause-mediated dormancy—its anticipatory nature and its persistence—are based on the fact that diapause occurs in response to token stimuli. First, diapause-mediated dormancy involves specific physiological and behavioral preparations before the arrival of unfavorable conditions. That is, the insect perceives specific seasonal cues that herald the approach of seasonal changes, it responds by embarking on a course of neurohormonally mediated changes, and thus it prepares in advance for the arrival of unfavorable conditions. Second, diapause-mediated dormancy is persistent; it does not terminate until after certain physiological processes have occurred. Thus, when insects in diapause experience favorable conditions for growth and reproduction, they do not resume development. Diapause prevents them from undergoing premature termination of dormancy and untimely growth and reproduction.

The expression of diapause-controlled dormancy varies considerably among species. Usually insects stop growing and feeding during diapause; however, dormant embryos of some grasshoppers may undergo morphometric development, but only to a certain stage of embryogenesis (Steele 1941; Birch 1942; Andrewartha 1943; Visscher 1976). Moreover, caterpillars of *Cirphus unipunctata* and *Laphygma exigua* (Saulich 1975) and *Dasychira pudibunda* (Geyspits & Zarankina 1963) and immature Neuropterans in the genera *Myrmeleon* (Furunishi & Masaki 1981) and *Chrysopa* (Tauber & Tauber unpublished) continue to feed and develop during diapause, but the rate and extent to which they do so are reduced and regulated, at least in part, by token stimuli.

2.2.2 *Seasonal migration*

The life cycles of many insect species involve specific, seasonally restricted periods of migration. These migrations are also seasonally cued and are based on characteristic behavioral patterns that subserve well-defined and directional movement; in some cases they require environmentally influenced changes in wing musculature and structure (e.g., Kennedy 1961, 1975; Johnson 1963, 1969; Dixon 1971b, 1977; Dingle 1979, 1980, 1982; Rankin & Singer 1984). Seasonal migration differs from aseasonal migration, discussed above, in that it is associated with regular and predictable seasonal changes in the environment; in addition, it may involve neuroendocrinological mechanisms that are distinct from those controlling aseasonal migration (Kennedy 1961; Lees 1966; Lees & Hardie 1981; Johnson 1969; Dingle 1972, 1978a, 1982; Rankin 1978; Herman 1981).

Seasonal migration often involves movement to and from dormancy sites. In such cases, it typically occurs as part of the diapause syndrome, and we refer to this type of seasonal movement as diapause-mediated migration. By contrast, *nondiapause* seasonal movements are not associated with dormancy; they are usually followed by resumed growth and development immediately after arrival at the new host or habitat. These movements serve as a means for insects to utilize a seasonally distributed series of hosts and/or habitats. In some instances, nondiapause alterations in foraging behavior may also serve this function.

2.2.2.1 Nondiapause seasonal migration and changes in foraging behavior

We define nondiapause seasonal migration as persistent, directional movement that (a) is induced by seasonal cues, (b) subserves transportation between seasonally available hosts or habitats, and (c) is immediately followed by growth and reproduction. Thus, nondiapause seasonal migration has two essential features. First, it occurs each year on a regular, seasonal basis, often with an anticipatory component, and sometimes in response to token stimuli. Because of this, it is distinguishable from aseasonal migration that occurs on an irregular basis and as an immediate response to unpredictable, localized exigencies (see Fig. 2.1). Second, nondiapause seasonal migration serves primarily as a means of moving from one seasonally available site of growth and reproduction to another, and it does not precede a long period of delayed growth, reproduction, or development. Thus, it is usually clearly distinguishable from diapause-mediated seasonal migration between the site of growth and reproduction and the site of dormancy.

Documented examples of nondiapause seasonal migration are few; we discuss two special cases here. The life cycles of dioecious aphids involve movements from the winter to the summer host; this migration is seasonal, but probably not diapause mediated. Overwintered eggs hatch and the resulting females (fundatrices) remain on the winter host to reproduce and give rise to one or more spring generations of parthenogenetic, usually apterous females (fundatrigenae). Sometime during spring, alates are produced and migration to the summer host(s) takes place. Alate production and spring migration can be influenced by a number of factors—mainly crowding and host quality (Lees 1961, 1966; Hille Ris Lambers 1966; Dixon 1971b, 1977; Dixon & Glen 1971; Mittler 1973). And, at least in some species, it involves a shift in host plant preference from the winter to the summer host (Kennedy & Booth 1954; Dixon 1971c; Hardie 1980b).

Nondiapause migration by the bivoltine leafhoppers *Lindbergina aurovittata* and *Edwardsiana rosae* is also associated with distinct seasonal changes in host preference. For example, in South Wales, *L. aurovittata* overwinters as eggs on various *Rubus* species; nymphal feeding and growth continues on *Rubus* during the spring. During the summer adult leafhoppers move to *Quercus* and other trees, where another generation is produced. The leafhoppers return in the fall to *Rubus*, where they oviposit overwintering eggs on the evergreen leaves. First- and second-generation females show distinct differences in their reactions to the host plants as oviposition sites and, if given a choice, will move from an unpreferred host to a preferred host (Claridge & Wilson 1978). The movement by first-generation adults from the spring host for oviposition on the summer host is considered a nondiapause seasonal migration. By contrast, the autumnal return migrations of both the aphids and the leafhoppers to their winter hosts are probably diapause mediated because they ultimately result in the oviposition of diapausing eggs on the winter host.

The rice skipper, *Parnara guttata guttata,* also migrates and changes its host plant seasonally; it switches from its summer host of wet lowland grasses (e.g., rice) to dry upland grasses (e.g., cogon) in winter. Host-plant alteration is associated with photoperiodically controlled increases in egg and larval size, but it does not appear to be related to diapause (Nakasuji & Kimura 1984).

Recent studies with papilionid and pierid butterflies show seasonal variation in foraging behavior that is probably not related to diapause. In *Colias philodice eriphyle* second-brood females show less restriction in their selection of oviposition sites than do first-brood females. This phenomenon could be the result of one of three factors: environmental changes between the first and second broods in host plant quality or the time available for oviposition, genetic differences between first- and second-brood females, and developmental differences between the broods (Stanton & Cook 1983). Females of another butterfly, *Battus philenor*, also exhibit a seasonal shift in the behavior involved in searching for an oviposition site. First-brood females seek a broad leaf host, which is the more abundant host early in the season. Second brood females seek a narrow-leaved host, which is the more favorable host during the late season (Rausher 1980; Papaj & Rausher 1983). Neither the papilionid nor the pierid enter diapause in the adult stage, and the seasonal changes in their foraging behavior do not appear to be related to diapause.

2.2.2.2 Diapause-mediated seasonal migration

Two features characterize diapause-mediated migration (= Type III migration of Johnson 1969)—its anticipatory nature and its association with an endocrinologically mediated state of reduced metabolism and reduced response to vegetative and reproductive stimuli. Insects that are to undergo diapause-mediated migration perceive token stimuli, often far in advance of the seasonal change, and they respond by undergoing the endocrinologically mediated physiological and behavioral changes associated with diapause (Chapter 4). Diapause-mediated migration occurs during a very specific phase in the course of diapause. Thus, movements to and from the dormancy site are timed, not by the conditions of the immediate environment, but by the interaction between the insects' physiological state and the occurrence of environmental factors that precede seasonal change.[2]

Although almost all insects exhibit some movement to sheltered sites as a symptom of the diapause syndrome, the degree to which it is expressed, and its timing in relation to dormancy, varies considerably. For example, prediapause coccids do not migrate but remain in place on twigs to undergo dormancy. Other insects, such as the European corn borer (*Ostrinia nubilalis*) and the alfalfa weevil (*Hypera postica*), move several centimeters or meters to sheltered sites for dormancy. Still others undergo spectacular, long-distance migrations to dormancy sites. In some species seasonal migration is accomplished by the prediapause stage immediately before dormancy; in other species, the prediapause adult moves to the dormancy site and oviposits eggs that undergo diapause. In aphids, this may occur as many

2. Like Johnson (1969) we do not include in this category the postdiapause migrations of insects, like the gypsy moth (*Porthetria dispar*), that lack prediapause migration. Although they occur on a regular seasonal basis, such migrations do not appear to occur in response to, or as an adaptation to seasonal changes in the environment. Their primary relationship appears to be with the spatial, rather than temporal, distribution of requisites, and they more readily fall into the category of nondiapause seasonal migration.

as two generations prior to the oviposition of diapausing eggs (see Blackman 1974a). We discuss all of these types of movements as part of the diapause syndrome in a later section.

Although, on an ecological basis, most cases of nondiapause migration are relatively easy to distinguish from diapause-mediated migration, such distinctions do not provide information regarding the evolutionary origin of migration. Diapause itself is polyphyletic. Moreover, the diapause syndrome involves numerous symptoms, any of which can evolve to dominate the life cycle at the "expense" of the others. Thus, what appear to be cases of "primitive" nondiapause migration may actually have their origins in diapause.

An interesting example that illustrates this issue lies in the black cutworm, *Agrotis ipsilon*, which migrates between the northern parts of Japan, and the southern, subtropical islands where its occurrence is restricted to the period between September and May (Oku & Kobayashi 1978; Sugimoto & Kobayashi 1978). This species breeds continuously throughout the year by migrating between these seasonally favorable localities. Thus, its migration appears to fall into the nondiapause category. However, when larvae are reared under short day length (LD 11:13), the ovaries of the resulting females mature within a day after adult emergence, whereas under long day conditions (LD 16:8) ovarian maturation is delayed by a relatively short period (four days at 21°C) (Oku & Kobayashi 1978). Although there is no evidence for a persistent period of dormancy in Japan, it is possible that this extremely abbreviated, but seasonally recurring, photoperiodically controlled delay in ovarian maturation may actually represent a relic form of diapause that has persisted as the species' migratory strategy evolved. This species is reported to overwinter as larvae, pupae, or adults in other parts of the world (Odiyo 1975; Druzhelyubova 1976; von Kaster & Showers 1982; Troester *et al.* 1982).

The difficulties in categorizing migration in insects are illustrated further by the locusts. Migration by most locust swarms is primarily an aseasonal phenomenon because it occurs as an adaptation to spatial heterogeneity in habitat or host suitability (Section 2.1.2). However, there often is a strong seasonal component to the formation and movement of swarms. In such species as *Nomadacris septemfasciata* and *Schistocerca gregaria*, which migrate and enter diapause as adults (see Pener & Lazarovici 1979), the relationship between migration and diapause is not clear because the factors controlling diapause and migration may overlap but may not be identical (see Norris 1958, 1962, 1965a, b). In other species, such as *Locusta migratoria migratoria*, *Locusta pardalina*, and *Chortoicetes terminifera*, which diapause in the egg stage, the two phenomena appear to be separate, but the distinction is not always clear because there can be a correlation between the phase of the mother and diapause induction in the egg (Matthée 1951; Nolte 1974). However, in all of these species, the factors controlling diapause and migration apparently are distinct at least in part (see Wardhaugh 1980a, b; Hunter 1980).

Diapause may have an interesting secondary effect on phase changes in some locust species that undergo diapause as eggs. The accumulation of diapausing eggs in the soil, especially during periods of high rainfall, followed by drought in mid to late summer, may contribute substantially to crowding conditions after the first rain in spring (Matthée 1951). The crowded conditions, in turn, influence the induction of the gregarious phase (see Section 2.1.2).

From the examples above, it is clear that our categories must not be viewed as rigid or without intergradation. Their ecological meaningfulness may vary with the groups of insects under consideration, and they certainly are not meant to represent evolutionary trends. Clearly, migration in *A. ipsilon*, the locusts, and other species without dormancy deserves further in-depth, physiological, and genetic study in order to elucidate the relationship between diapause, dormancy, and migration. In Chapter 8 we discuss the origin of diapause and the evolution of diapause-mediated dormancy and migration.

2.2.3 Seasonal polyphenism

Many insects exhibit seasonal patterns of change in the color and/or structure of their bodies and wings. These alterations subserve such seasonal functions as crypsis during dormancy, thermoregulation during active periods of growth and reproduction, and other forms of adaptation to seasonal factors in the environment. Together, these changes comprise *seasonal polyphenism*, which is defined here as an annually repeating pattern of changing phenotypic ratios under some kind of control by seasonally recurring environmental factor(s). This definition is the same as that given in a recent review of seasonal polyphenism by Shapiro (1976), with two exceptions. First, Shapiro restricted his definition to changes that occur over successive generations, thus eliminating seasonal changes in color and form that are reversible in the same individual. Although examples of such reversible seasonal changes in insects and acari are uncommon [e.g., they are reported from adults of the green lacewing *Chrysopa carnea* (MacLeod 1967; Tauber *et al.* 1970b) and the bug *Nexara viridula* (Michieli 1968; Michieli & Žener 1968), adult females of the spider mite, *Tetranychus urticae* (= *T. telarius*) (Lees 1953; Parr & Hussey 1966), and caterpillars of the butterfly *Lethe* sp. (Cardé *et al.* 1970)], we include them because of their regular seasonal occurrence and their dependence on diapause. Second, unlike ours, Shapiro's definition did not restrict the environmental control of seasonal polyphenism to seasonally recurring factors. We make this stipulation in order to separate seasonal polyphenism more clearly from the phenomenon of aseasonal polyphenism that we discussed earlier (Section 2.1.3).

Seasonal polyphenism has been examined for many insects, especially for butterflies, moths, leafhoppers, bugs, aphids, and the Orthoptera, and this subject was thoroughly reviewed by Shapiro (1976). Therefore, in this section we use selected examples to illustrate the types of seasonal polyphenism found in insects. Other examples occur in later sections, when we discuss the environmental and genetic control of diapause (Chapters 5, 6, and 7) and the evolution of seasonal cycles and life histories (Chapters 8 and 9). At this point, it is interesting to note that seasonal polyphenism in butterflies and other insects is controlled by the same sort of environmental factors that control dormancy. In most cases, the primary controlling factor is photoperiod. Photoperiod and temperature interact in many cases, and in a few cases, especially near the equator, temperature and/or moisture appear to be dominant. Responses underlying both dormancy and seasonal polyphenism also show similar ranges and patterns of intra- and interpopulation variation. In many insect species, seasonal polyphenism thus appears to be an integral part of the diapause syndrome.

Seasonal polyphenism, however, has a full complement of expressions that range from phenotypic changes that are unrelated to diapause but that are associated with growth, development, and reproduction during the active season, to those involved in the diapause syndrome and expressed during the prediapause, diapause, or postdiapause periods. Thus, we divide seasonal polyphenism into two types: nondiapause and diapause mediated. In doing so, we stress that very few studies have addressed the problem of the physiological, endocrinological, and genetic relationships between diapause and seasonal polyphenism, and thus our assignment of most cases to categories must be considered tentative. Moreover, as in the case of seasonal migration, our categories generally are ecologically meaningful, but they are *not* meant to reflect evolutionary relationships. Both nondiapause and diapause-mediated polyphenism have evolved independently numerous times in the Insecta, and trends in the evolution of seasonal polyphenism will be uncovered only by close examination of related taxa (e.g., see work of Shapiro 1978, 1980a, c), and not by designating examples to broad categories.

2.2.3.1 Nondiapause seasonal polyphenism

We include in this category those cases in which seasonal changes in body form and/or color are associated with continuous growth and reproduction year round. Such changes are controlled by a variety of factors, including photoperiod, food quality, and crowding.

Nondiapause seasonal polyphenism in body color in the leafhopper *Stirellus bicolor* is under photoperiodic control, but growth and reproduction continue all year. This species has two color forms (an ornate summer form and a drab winter form) that are photoperiodically induced and that are associated with seasonal alteration in host plant. Aestival generations are produced primarily on broomsedge (*Andropogon virginicus*), whereas hibernal generations are produced on hardy grasses that persist throughout winter (Whitcomb *et al.* 1972) Dormancy is not reported for this species, and reproductive adults have been collected year round, at least in the southern

United States. Nondiapause seasonal polyphenism in wing length may also come under photoperiodic control. In the cricket *Scapsipedus asperus* long day length and crowding during the nymphal stages favor the development of long wings (Saeki 1966a, b). However, adults do not enter diapause; the egg is the diapause stage. Both macropterous and brachypterous females lay only diapausing eggs.

In contrast to the photoperiodic control of polyphenism, nondiapause seasonal polyphenism in the acridid *Zonocerus variegatus* is controlled by biotic conditions. This tropical pest exists in two forms—a long-winged form that is prevalent during the dry season, and a short-winged form that is typical of the wet season. Crowding during the last nymphal instar and low or high food quality promote the formation of long-winged adults (McCaffery & Page 1978). Diapause is probably not involved in this seasonal polyphenism. Temperature and crowding also induce long wings in the polyphenic mirid *Leptopterna dolobrata*; diapause occurs in the egg stage and is unrelated to wing length (Braune 1983).

Diapause is also probably not associated with the seasonal color changes exhibited by summer populations of two species of aphids, the sycamore aphid, *Drepanosiphum platanoides*, and the lime aphid, *Eucallipterus tilliae* (Dixon 1972a, 1973; Kidd 1979). The sycamore aphid, which is active from early spring to late autumn, develops melanic bands in response to low temperatures during spring and fall. The presence or absence of bands appears to be important in thermoregulation during the long growing season that this species encounters (Dixon 1972a). Bands form as an immediate response to environmental conditions, in that the temperatures during nymphal and early adult stages, when melanic pigment is deposited, determine the degree of melanism. By contrast, control of the red color that some summer aphids assume, is more remote in that high temperatures induce red coloration only after two or more generations. In the second example, nymphs of the lime aphid acquire black pigmentation in response to the conditions of crowding and food quality experienced by their mother and first instar larvae (Kidd 1979). This color change apparently does not function in thermoregulation (Dixon 1973).

Seasonal color change occurs in the adults of many psyllid species that overwinter as diapausing eggs. The adults characteristically are bright green upon emergence in early summer and undergo a rapid color change to dark reddish brown in mid August. In at least one species, *Psylla peregrina*, the timing of color change coincides with the timing of sexual maturation and apparently functions to maintain crypsis in a seasonal environment (Sutton 1983).

Our final examples of nondiapause seasonal polyphenism concern photoperiodically controlled thermoregulatory wing melanization in a pierid butterfly and in the rice-plant skipper. Many species of diurnal Lepidoptera are polyphenic for thermoregulatory characteristics. Typically, cool weather forms are small and have dark (melanized) scales on their wings, whereas midsummer forms are larger and have few dark scales. For most species of pierids (e.g., *Pieris napi* and *P. virginiensis*), the seasonal changes in body size and wing form are under some sort of photoperiodic control and are associated very closely with diapause (Shapiro 1976). Thus, they are included in the category of diapause-mediated polyphenism. However, in *Nathalis iole*, the photoperiodic regulation of wing melanization occurs independently of diapause. Short day lengths induce dark adult forms, and long day lengths produce pale adults. There is no evidence for winter diapause in any part of this species' range in North America, and individuals do not survive the winter under northern conditions. Each spring and summer, adults migrate northward in North America. The melanic forms occur under the cool, short-day conditions of spring and fall; whereas the pale form appears during summer (Douglas & Grula 1978). Thus, although it is probably closely related to diapause-mediated polyphenism in the other pierid species, we consider *N. iole*'s polyphenism an example of nondiapause seasonal polyphenism.

In the rice-plant skipper, *Parnara guttata guttata*, long day lengths experienced by larvae and pupae also induce pale wing coloration, small adult size, and the production of small eggs, but these characteristics are controlled independently of each other (Ishii & Hidaka 1979; Nakasuji & Kimura 1984). Temperature can affect body size, wing melanization, and wing coloration, with low temperature enhancing the influence of short day lengths and high temperatures intensifying the effects of long day lengths. Diapause in this species occurs in the larval stage and does not appear to affect adult polyphenism.

2.2.3.2 Diapause mediated seasonal polyphenism

We include as diapause mediated those cases of seasonal polyphenism that in some way or another are associated with the prediapause, diapause, or postdiapause phases of the seasonal cycle. Although their adaptive value in most cases has not been investigated, they appear to serve three main functions associated with dormancy. First, they may provide some sort of protection during dormancy (e.g., Kimura & Masaki 1977; Hazel & West 1983). Examples include tightly spun as opposed to coarsely spun cocoons, increased wax secretion on pupal cuticle, variation in egg shape and shell surface, variation in prediapause wing length, and color changes that contribute to crypsis during dormancy. Second, they may subserve the efficient allocation of energy between wing development and seasonal migration, on the one hand, and growth and reproduction, on the other. Examples include seasonal variations in wing length and body size that are commonly associated with migratory species. Third, they may contribute to pre- or postdiapause development during periods of seasonal transition. Examples include variation in body and wing size and thermoregulatory wing melanization. All of these changes are expressed and induced at specific times during the course of diapause; we will discuss further details later (Chapter 3 and Section 4.4).

In assigning examples of seasonal polyphenism to the diapause-mediated category, it would be ideal to have data showing a direct genetic, endocrinological, or physiological link between diapause and the seasonal change in color or form. Some excellent studies have demonstrated a hormonal linkage between seasonal polyphenism and diapause (Lees 1961; Pammer 1966; Nopp-Pammer & Nopp 1967; Endo 1970, 1972; Pener 1970, 1974; Yin & Chippendale 1974). Unfortunately, such data are restricted to few species and, in most cases, our designations must be made from indirect evidence.

In many cases, seasonal polyphenism has been shown to be controlled by the environmental cues—token stimuli—that influence diapause (see Pammer 1966; Müller 1958a, 1960b, 1974; Stross & Hill 1968; Tauber *et al.* 1970b; Shapiro 1976, 1977, 1978, 1980a, c; Kimura & Masaki 1977; Wardhaugh 1977, 1980a; Hazel & West 1983). From these data, we conclude that the color and/or morph changes occur as part of, or in association with, the diapause syndrome. However, the relationship between diapause and seasonal polyphenism appears to vary considerably among species. Below we illustrate some of the variability.

A very close relationship between diapause and seasonal polyphenism exists in the green lacewing, *Chrysopa carnea*. The degree of color change from aestival green to autumnal brown is quantitatively related to the quality of the diapause-inducing and diapause-intensifying stimuli. Just as there is an inverse relationship between day length below the critical photoperiod and the intensity (duration) of diapause, there is also an inverse relationship between day length and the intensity of the diapause color (Tauber *et al.* 1970b; Tauber & Tauber unpublished).

The association between color change and diapause in the bug *Euryeloma oleracea* is interesting in that diapause acts as a delayer of age-related color changes. Newly molted adults are typically white, and as they age, they proceed through a series of color changes that are not dependent on photoperiod or sexual maturation. Photoperiodically controlled diapause can intervene when the insects are at any stage of color change, and when it does, the pigmentation processes are halted until diapause is ended. Pigmentation resumes and is completed after the end of diapause (Fasulati 1979).

In some species, especially those in which the polyphenic stage precedes the diapausing stage, the stimuli that control the polyphenism can be separated from those that induce diapause. In the cricket *Pteronemobius taprobanensis*, wing form is seasonal and is controlled by population density and photoperiod, but the response curve for wing determination is clearly different from that for induction of egg diapause (Masaki 1979). Each of the short- and long-winged adults can produce both diapause and nondiapause eggs (Masaki 1979), although egg production is considerably delayed and decreased in the long-winged form (Tanaka 1976). Similar conditions occur in the subtropical strain of *Pteronemobius fascipes* and in the swallowtail *Papilio polyxenes*, in which the diapause-inducing stimuli can be separated from those controlling wing polyphenism (Masaki unpublished; West & Hazel 1983). In these cases, a series of photoperiods may act in the full induction of diapause (see Section 3.3.2). The thermally and photoperiodically controlled color change in *Hyphantria cunea* larvae shows some association with diapause, but, again, the relationship is not strict (Masaki 1977). Finally, in the saturniid *Philosamia cynthia* the sensitive stages for diapause induction and for winter morph induction overlap but are not identical (Pammer 1966). Shapiro (1976, 1980c) gives other examples in which seasonal polyphenism and diapause may be decoupled.

Water striders in the genus *Gerris* present a large array of wing length variation that has been well studied by Vepsäläinen (1974a, 1978). Both intrinsic and environmental factors are involved in the determination of wing length patterns, which range from monomorphic winglessness to monomorphic long wingedness and from seasonal polyphenism to genetic polymorphism. Among the environmental determinants of the seasonal polyphenism are photoperiod (incremental vs. decremental change of day length), which provides relatively long-term predictions of seasonal conditions, and temperature, which provides a more immediate signal of habitat endangerment. Reproductive state (diapause) in these species is also determined by day length changes but it is not strictly associated with wing length. Both wing length variation and variation in diapause are related to spatial and temporal heterogeneity in the occurrence of suitable habitats (Vepsäläinen 1978). A similar type of relationship between diapause and wing length may exist in the small, wing-polymorphic *Ptinella* beetles; however, diapause has not yet been reported for these species. The incidence of long-winged *Ptinella* varies seasonally, but both spatial and temporal heterogeneity of the habitat are involved (V. Taylor 1981).

Aphids provide some of the best examples in which seasonal polyphenism for prediapause morphs (alate sexuparae and sexual morphs) appear to be distantly connected to diapause. Aphid seasonal cycles are typified by the seasonal progression of an elegant series of morphs and by an equally elegant series of terms to describe the morphs. We refer readers to Blackman (1974a) for a thorough and clear description of aphid annual cycles and to Hille Ris Lambers (1966) and Lees (1966) for details. Here we discuss only that period of the life cycle that occurs at the end of the summer breeding period and that leads to the oviposition of the sexually produced, diapausing eggs.

Two aphid species serve as examples—the autoecious vetch aphid, *Megoura viciae*, studied by Lees (1959, 1960, 1963) and the heteroecious bird cherry-oat aphid, *Rhopalosiphum padi*, studied by Dixon (1971c) and Dixon and Glen (1971). The autumnal morph changes in the vetch aphid are relatively simple in that there is only one generation of autumnal morphs, that of the sexual females and males. Under the long day conditions of summer, reproduction is by parthenogenesis with virginoparae giving rise to virginoparous offspring. Under the short days of autumn, virginoparae produce oviparae—that is, females that have shorter legs, are sexual, and can lay fertilized diapausing eggs after mating (see Lees 1966).

In the heteroecious aphids, like *R. padi*, summer-breeding virginoparae give rise to winged forms (parthenogenetic females = gynoparae, and males) that make the autumnal migration from the summer host to the winter host. These autumnal females, gynoparae, are very similar to summer migrants (alate virginoparae) except that (1) they are induced by short day lengths (Dixon & Glen 1971), (2) they show a stronger preference for the winter host plant than summer migrants do (Kennedy & Booth 1954; Dixon 1971c; Hardie 1980b), and (3) unlike summer migrants, which produce only parthenogenetic females, they give rise to oviparae—the wingless, sexual females that mate and oviposit diapausing eggs.

Photoperiod appears to be the primary factor inducing autumnal morph changes in aphids; however, temperature, crowding, food quality, and perhaps the photoperiodic experience of the host plant also have a role in some species (Bonnemaison 1951; Lees 1966, 1967; Hille Ris Lambers 1966; Young 1972; Dixon 1971b, 1972a, 1977; Blackman 1975; Matsuka & Mittler 1979; Searle & Mittler 1981; Takada 1982a, b; Hardie & Lees 1983). In all cases, sexual reproduction leads to the oviposition of diapausing eggs that overwinter. Those species or strains that do not reproduce sexually either overwinter as quiescent adults, without diapause, or they die off during the winter (Sutherland 1968; Blackman 1971, 1974a; Müller 1971; Müller & Möller 1973, Tamaki *et al.* 1982a). Thus, the factors that control autumnal morph determination and sexuality in the aphids are somehow related to diapause, but the connection is far from clear. Very little is known about the environmental and endocrinological control of aphid egg diapause.

2.2.4 Summary

In contrast to the physiological and behavioral patterns that insects have evolved to cope with acyclic phenomena in their environments, most seasonal adaptations are not directed primarily at survival during particular types of unfavorable conditions. Rather, they are more concerned with the overall timing of the whole life cycle (the periods of growth, reproduction, dormancy, and migration) with seasonal changes in the environment. In general, they are based on the longer-term, hormonally mediated, physiological and behavioral changes associated with diapause. They are less responsive to the immediate conditions of the environment, and they serve to harmonize the entire life cycle with the seasonal cycle of the environment.

In our treatise we restrict application of the term "diapause" to the physiological and behavioral changes brought about by token stimuli that herald seasonal changes in the environment. But, in doing so, we simultaneously establish a basis for broadening its usage by recognizing that diapause can have several interrelated ecological manifestations—dormancy, seasonal migration, and seasonal polyphenism.

In the past, the various symptoms of the diapause syndrome generally have been considered in isolation, and dormancy (= diapause of various authors) and seasonal migration have often been contrasted with each other vis à vis their ecological significance and evolutionary origins. Dormancy and migration were viewed as evolutionary alternatives, that is, as "escape in time" versus "escape in space." This contrast can be of value in reference to some species, particularly those whose seasonal cycles strongly emphasize *either* dormancy (e.g., European corn borer) *or* migration (e.g., the black cutworm in Japan). However, the intimate association between seasonal migration and dormancy in most species limits the usefulness of this approach, and we conclude that in most cases dormancy, seasonal migration, and seasonal polyphenism are interrelated through the phenomenon of diapause.

It is important to note that not all seasonal adaptations are associated with diapause, and we discussed examples of nondiapause dormancy, seasonal migration, and seasonal polyphenism. These seasonal adaptations may or may not be characterized by long periods of suppressed growth and reproduction; however, they lack the seasonal synchronizing function of the diapause-mediated seasonal adaptations.

Chapter 3

The Course of Diapause

In the previous chapter, we defined diapause and discussed various adaptations associated with it. Here we provide an overview of specific characteristics of diapause and the various changes that an insect experiences as it enters, undergoes, and leaves the state of diapause. We concentrate on the insect's sensitivity and reactions to environmental stimuli during the various phases of diapause, and, as a setting for later chapters, we illustrate that the insect's physiological state changes as diapause progresses. This chapter stresses the concept that diapause is a dynamic state with multifaceted symptoms, the full expression of which generally follows a predictable course.

3.1 Diapausing Stage

Central to each species' life cycle is the metamorphic stage that undergoes diapause, and thus we begin to discuss the course of diapause by considering the diapausing stage. The utilization of a particular developmental stage for diapause not only influences the degree of resistance to environmental extremes during dormancy, but perhaps more important, it provides the pivot around which growth, development, and reproduction are timed. Later (Section 9.1), we will discuss the various factors that impinge on the evolution of the diapausing stage; here we consider the interspecific and intraspecific variability that insects express in their characteristic diapausing stage.

3.1.1 Species specificity

Diapause typically occurs at a specific stage and during a specific season in an insect's life history. Although closely related species often differ in their diapausing stages, there are no records in which this characteristic of diapause has been altered either by artificial selection in the laboratory or by natural selection acting on recently introduced or colonizing species.

Diapause has been recorded for all times of the year and in all stages of metamorphosis, but for each species the diapausing stage(s) is(are) genetically determined. Examples are known for various stages of embryogenesis, all forms of immatures, pharate adults, and in both sexes of adults. Danilevsky (1965), Saunders (1982), and Beck (1980) present comprehensive lists of species that undergo diapause in the various metamorphic stages (see also Hartley & Warne 1972; Kimura 1975; Falkovich 1979; Hartman & Hynes 1980).

Diapause in the adult stage has been studied most thoroughly in females, but males have also been examined (e.g., Norris 1964; MacLeod 1967; Hodek & Landa 1971; Ferenz 1975b; Barnes 1976; Pener 1977; Thiele 1977a; Ascerno *et al.* 1978; Boiteau *et al.* 1979; Orshan & Pener 1979a, b; Wellso & Hoxie 1981; Hayes & Dingle 1983; Pener & Orshan 1983; Sims & Munstermann 1983). In some species with imaginal diapause, only mated females enter diapause; males die off. Examples are found in the culicine and anopheline mosquitoes (Danilevsky 1965), certain bees and wasps (Section 6.2.1), and phytoseiid, eriophyoid, and tetranychid mites (Lees 1953; Parr & Hussey 1966; Hoy & Flaherty 1970; Jeppson *et al.* 1975).

In species with long life cycles, more than one stage may enter diapause. The alfalfa snout beetle, *Otiorhynchus ligustici*, which has a two-year life cycle, spends the first winter as a dormant, late-stage larva and the second winter as a dormant adult (Lincoln & Palm 1941). The cockroach *Ectobius lapponicus*, which also has a two-year life cycle, spends the first winter as a diapausing egg and the second winter as a quiescent second- or third-instar larva or diapausing fourth instar (Brown 1973). For most species with very long life cycles [e.g., cicadas with 13- and 17-year life cycles (White & Lloyd 1975), Alaskan *Chironomus* spp. with seven-year life cycles (Butler 1982), and the dragonfly *Cordulia aenea amurensis* with a five-year life cycle (Ubukata 1980)], it is not known how many diapause periods occur or how long they last.

Other examples of insects with diapause in two stages include species that enter diapause at one stage in summer and at another stage in winter (Masaki 1980). The western tree-hole mosquito, *Aedes sierrensis*, has a one-year life cycle with diapause occurring in two stages (an aestival egg diapause and an hibernal larval diapause) (Jordan 1980a, b). Diapause intervenes in two stages of the life cycle of other mosquitoes too, for example, *Wyeomyia smithii, Aedes triseriatus, Aedes togoi*, and *Aedes geniculatus* (Lounibos & Bradshaw 1975; Bradshaw & Lounibos 1977; Holzapfel & Bradshaw 1981; Mogi 1981; Sims & Munstermann 1983). Eggs of katydids, such as *Ephippiger cruciger*, also exhibit two periods of diapause; both periods intervene during embryogenesis. The first occurs at the unsegmented embryonic primordium stage and the second in the late embryonic stage (Dean & Hartley 1977a, b).

In some cases the occurrence of diapause in two stages is not readily demonstrated and may vary even among very closely related species with similar life cycles. For example, one species of swallowtail butterflies, *Leuhdorfia puziloi inexpecta*, spends summer, autumn, and winter within the chrysalis, and during this time, two distinct stages undergo diapause. The first diapause occurs during summer soon after pupation. In autumn, adult development is almost completed and the pharate adult enters the second diapause, which persists throughout winter (Kimura 1975). In a congeneric species, *L. japonica*, summer, autumn, and winter are also spent within the chrysalis, during which time the pupa undergoes diapause twice. The first, an aestival diapause, is terminated by short day lengths but little or no development commences afterwards and the pupae enter a second diapause. Low temperatures hasten the termination of the second diapause (Hidaka *et al.* 1971; Ishii & Hidaka 1982, 1983). Concealed variation also occurs in the megachilid bees, *Osmia* spp., that overwinter in their nests. Populations of these bees are mixed; some individuals have a one-year life cycle, while others have a two-year life cycle. Univoltine individuals enter diapause only as adults; the semivoltine individuals spend the first winter as diapausing larvae and the second winter as diapausing adults (Torchio & Tepedino 1982). Torchio and Tepedino term this phenomenon "parsivoltinism."

3.1.2 Intraspecific variability

Although cessation of growth at a particular stage is most often the rule, some species have considerable variability in their diapausing stage. There is a general tendency for those species entering diapause as midinstar larvae to express variation in the diapausing stage. For example, an overwintering population of *Pteronemobius nitidus* in the field contains more than five different instars (Tanaka 1983). Even under constant photoperiod in the laboratory, growth is retarded in several instars (Tanaka 1979). Similar variation characterizes the larval diapause of the arctiid moth *Spilarctia imparilis* (Sugiki & Masaki 1972) and the tortricid moths *Adoxophyes orana* and *Platynota idaeusalis* (Homma 1966; Shaffer & Rock 1983). In the dragonfly, *Tetragoneuria cynosura*, most individuals are univoltine and overwinter in the final larval instar. However, a few slow-growing larvae require two years for development. These individuals spend their first winter as diapausing young larvae, their first summer in the penultimate larval stage, and their second winter in the final larval stage (Lutz & Jenner 1964). Thus, overwintering populations are composed of young larvae and mature larvae of two different ages.

In some instances environmental conditions prior to diapause influence the diapausing stage. Hibernal diapause in two species of tettigoniids may occur in three different embryonic stages depending on the maternal photoperiod and the temperature during early embryonic development (Helfert 1980). The acridid *Tetrix undulata* can enter diapause in the fourth or fifth larval instar depending on when, during the sensitive period, the diapause-inducing photoperiodic stimulus is perceived (Poras 1981). A more extreme example occurs in the mosquito *Aedes togoi*, which undergoes diapause during either of two widely separated stages of development, depending on larval photoperiod within the diapause-inducing range. If the larval photoperiod is LD 12:12, adults emerge and oviposit eggs that enter diapause; however, at LD 10:14, adult emergence is delayed by diapause in the larval stage (Mogi 1981).

A somewhat similar form of variation related to prediapause conditions is found in the nymphs of the tick *Hyalomma anatolicum* in central Asia. Although diapause in ticks usually occurs in unfed nymphs, both unfed and engorged nymphs of *H. anatolicum* enter a photoperiodically induced hibernal diapause. In this species the critical photoperiod for engorged nymphs is somewhat shorter than that for unfed nymphs. Thus, diapause induction by engorged nymphs is several weeks later than that for unfed nymphs (Belozerov & Galyal'murad 1977).

Some species continue growth at a reduced rate during diapause, and for these species there may be considerable variation in the stages undergoing diapause. Development proceeds very slowly, but measurably, in diapausing eggs of the grasshoppers *Austroicetes cruciata* and *Locustana pardilina* (Steele 1941; Birch 1942; Andrewartha 1943; Matthée 1951). Eggs of another species of grasshopper, *Aulocara elliotti*, enter diapause at several morphological stages before blastokinesis; they also undergo some morphometric development during diapause (Visscher 1976). In the odonates, diapause in the final instar is characteristic of many species that emerge in spring. However, species that emerge in the summer lack diapause in the final instar and any of the last three, four, or five larval instars may undergo diapause (Corbet 1963). In some cases (e.g., *Lestes eurinus*) larvae continue to grow during winter, but the rate of growth is modulated by photoperiod (Norling 1971, 1975; Lutz 1968). We consider this to be a form of diapause, because of the similarities in controlling mechanisms (photoperiodic retardation of growth) and in function (timing of subsequent growth and reproduction) it has with diapause. Similar variation in the overwintering stage occurs in the aquatic larvae of the mayfly, *Stenacron interpunctatum* (McCafferty & Huff 1978), but it has not been established if these overwintering larvae are in diapause or in a thermally controlled state of quiescence.

In the univoltine antlion, *Myrmeleon formicarius*, all three larval instars may undergo photoperiodically regulated diapause, although the first instar is the most common overwintering stage. Successively longer day lengths are required for larval development to the cocoon stage. This type of response to a series of increasing critical photoperiods allows larvae within the population to become synchronized in their development. Synchronous adult emergence and the maintenance of a univoltine life cycle are thus achieved (Furunishi & Masaki 1981, 1982). Photoperiodically regulated growth rates in several instars may also regulate the life cycles of another myrmeleontid, *Hagenomyia micans* (Furunishi & Masaki 1983), the acridid *Acrotylus insubricus* (Abou-Elela & Hilmy 1977), and two green lacewing species, *Chrysopa flavifrons* and *Mallada perfecta* (Principi *et al.* 1977; Tauber & Tauber unpublished).

Two species of mosquitoes exhibit geographical variability in the stage that undergoes overwintering diapause. In northern North America, *Wyeo-*

mia smithii enters diapause as a third instar larva, whereas in the southern United States, the overwintering stage is the fourth larval instar (Bradshaw & Lounibos 1977). In *Aedes triseriatus*, the diapausing embryo is apparently the only stage that can withstand freezing, and thus populations from northern United States and southern Canada overwinter exclusively in this stage. In southern United States, overwintering is accomplished by both diapausing eggs and diapausing larvae (Holzapfel & Bradshaw 1981; Sims 1982). It is interesting to note that northern and southern populations of each of these species retain the potential to enter both types of diapause under certain conditions.

Finally, some species show variability in their overwintering stage, even though only one stage enters diapause. All larval instars of the carrot fly, *Psila rosae*, can overwinter, but only the pupae enter hibernal diapause. Overwintering larvae, which are cold-resistant, continue to feed and develop at a slow rate during autumn and winter (Städler 1970). The satyrid butterfly *Pararge aegeria* in Great Britain exhibits a similar type of variability in its overwintering stage. Some individuals stop growth and enter diapause at the second or third larval instar, whereas others continue to develop and pupate in December. Such flexibility presumably allows the species great latitude in extending its activity late into fall (Lees & Tilley 1980). In these cases, there is less pressure to reach one particular stage of development for overwintering; however, emergence of adults is prevented until spring.

From the numerous examples above, it is clear that insects show great variability in the stage that undergoes diapause and in the degree of suppressed development during diapause. Such variability serves many functions and will be the subject of later discussion (Chapters 7, 8, 9).

3.1.3 Summary

Diapause typically occurs during a specific stage of metamorphosis. For each species, the diapausing stage is genetically determined and difficult to alter through artificial selection. Although diapause usually occurs in only one stage, insects with life cycles of a year or longer may enter diapause in two or more stages. But, in all of these cases, the diapausing stages are species-specific.

Various forms of intraspecific variation underlie the expression of the diapausing stage. Some are largely a function of the genetic makeup of the species; others are related to variability in environmental conditions prior to or during diapause.

For most insect species studied, diapause is the primary factor synchronizing the life cycle with seasonal changes in the environment. The diapausing stage is central to this synchronization and thus provides the pivotal point for the timing of growth, development, and reproduction, both before and after dormancy.

3.2 Stages Sensitive to Diapause-Inducing Stimuli

The primary feature of diapause is its anticipatory nature. Environmental cues that signal future environmental changes are perceived by the insect, often long in advance of the diapausing stage itself. This information is stored and later translated into neuroendocrine functions in the form of diapause induction. The duration over which the information is stored may span a considerable period, encompassing a number of developmental stages and even generations. Therefore, insect diapause offers a unique and valuable tool for research into the biochemical, physiological, and genetic mechanisms involved in information storage and translation in animals (Section 4.1.5).

It is not surprising, therefore, that in addition to their ecological significance, analyses of the insect stages that perceive diapause-inducing stimuli are of special interest to insect physiologists. Ecologically, the sensitive stages determine if and when development is to proceed on a reproduction-destined or diapause-destined course. Physiologically, the sensitive stages are involved in the perception of environmental cues and the storage of information within the insect for later translation into neuroendocrine function. Two major features of the sensitive periods have been examined: their proximity to the diapausing stage and their duration over the life cycle and over time. We discuss both aspects below and give some methods for analyzing them in the laboratory and applying this knowledge to analysis of diapause induction in the field.

3.2.1 Proximity to diapausing stage

Not only is diapause expressed in a specific, genetically determined stage of metamorphosis, but diapause-inducing stimuli are also perceived only during specific, genetically determined stages. In some species, the sensitive stages and the diapausing stage are widely separated within the same generation or even between generations, whereas in other species, especially those that diapause as adults, they overlap (see de Wilde *et al.* 1959; Tauber & Tauber 1970b; Kono 1980). Thus, in the case of egg diapause, the sensitive stage ranges from various periods within the parental generation, as in the silkworm *Bombyx mori*, the lepidopterans *Orgyia antiqua* and *O. thyellina*, and the cricket *Pteronemobius fascipes*, to the embryo itself, as in another cricket *Teleogryllus* (= *Acheta*) *commodus* and the mosquito *Aedes triseriatus* (Kogure 1933; Hogan 1960a; Kappus & Venard 1967; Kind 1969; Kimura & Masaki 1977; Kidokoro & Masaki 1978; Shroyer & Craig 1980) (Fig. 3.1). In the aphids *Megoura viciae* and *Myzus persicae* (Lees 1959; Blackman 1975), the sensitive stage for egg diapause apparently extends back to the grandparental generation; however, photoperiod may determine not diapause itself but the autumnal appearance of sexual morphs (males and oviparae) that mate and lay diapausing eggs.

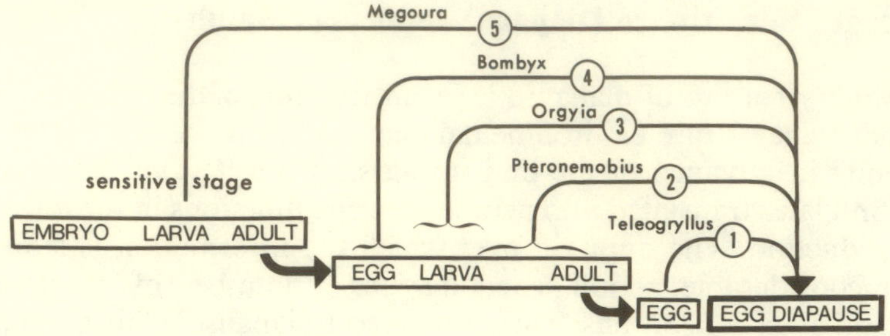

Fig. 3.1 Examples of insects that undergo diapause in a similar stage (the egg) but that differ in the stages that perceive diapause-inducing stimuli. 1. In the cricket *Teleogryllus,* the young embryo is sensitive and diapause ensues during a later stage of embryonic development. 2–4. In *Pteronemobius, Orgyia,* and *Bombyx,* diapause-inducing cues are perceived during various stages of maternal development. 5. Finally, in the aphids *Megoura* and *Myzus,* the sensitive stage (for sexual morph determination and subsequently diapause in overwintering eggs) occurs during the grandparental generation. See text for references.

Species in which the sensitive stage occurs in the female parent are often referred to as having a "maternal effect." In addition to the examples given above, others are reported from parasitic Hymenoptera (e.g., Ryan 1965; Griffiths 1969a; Saunders 1965; McNeil & Rabb 1973; Anderson & Kaya 1974), sarcophagid flies (Depner 1962; Ring 1967; Vinogradova & Zinovyeva 1972d), and mosquitoes (Khelevin 1958; Anderson 1968; McHaffey & Harwood 1970; McHaffey 1972; Beach 1978; see Pinger & Eldridge 1977 for a review). In some cases, such as several larvapositing flesh flies, the "maternal effect" has been distinguished from prenatal embryonic sensitivity (Denlinger 1971; Vinogradova 1976b).

3.2.2 Duration of sensitivity

The extent of the period during which insects are sensitive to diapause-inducing stimuli varies considerably within and among species, ranging from a few days to almost an entire year. In *Teleogryllus* sp., the period of maximum sensitivity for induction of egg diapause extends over two days (days 5 and 6 after oviposition) (Hogan 1960a), and among geographical populations of the mosquito, *Aedes atropalpus,* the duration of the photosensitive period for induction of egg diapause varies from four to nine days during the fourth instar and pupal stages (Beach 1978). By contrast, in the aphid *Megoura* almost the entire grandparental generation is sensitive to photoperiodic stimuli (Lees 1959). In the antlion, *Myrmeleon formicarius,* sensitivity extends over several immature stages and persists for almost an entire year (Furunishi & Masaki 1981). Similarly, the tufted apple bud moth, *Platynota idaeusalis,* is sensitive to photoperiod over the whole larval period. A maximum of 39% diapause occurs when four of the five larval instars experience short day lengths, whereas 100% diapause occurs when all five instars are exposed to short days (Rock *et al.* 1983).

During an extended period of sensitivity, the degree of responsiveness to diapause-inducing stimuli may change over time. For example, the same diapause-inducing stimulus may cause different responses depending on the time during the sensitive period that it is perceived; species in which this occurs include the tobacco hornworm *Manduca sexta* (Rabb 1966), the green lacewing *Chrysopa carnea* (Tauber & Tauber 1970b), the mite, *Neoseiulus fallacis* (Rock *et al.* 1971), and the blowfly, *Calliphora vicina* (Vinogradova 1974). And, in some cases, environmental factors may influence the duration of the sensitive stage. For example, when larvae of the silkworm, *Bombyx mori*, are reared on mulberry leaves diapause is induced by the temperature and photoperiod during the egg stage of the previous generation. Temperature and light conditions during the larval stage have no role. However, larvae become responsive to diapause-averting photoperiods if they have been reared on an artificial diet (see Shimizu 1982; Tsuchida & Yoshitake 1983a, b). In a comparable situation, reduced amounts of food prolong the sensitive stage in the coccinellid *Henosepilachna vigintioctopunctata* (Kono 1982). The problem, therefore, is both to determine the position and extent of the sensitive period in the life cycle and to analyze relative sensitivity throughout the sensitive period. All of these factors appear to have intrinsic as well as environmental determinants.

The choice of methods for analyzing the sensitive stage(s) depends on the stage that enters diapause, the degree of sensitivity of the diapausing stage, and the techniques for culturing and handling the particular species in the laboratory (Tauber & Tauber 1973b). Within the Chrysopidae, experiments have involved several methods. For species that undergo diapause in the prepupal stage within the cocoon (*Chrysopa oculata, C. nigricornis, C. coloradensis*, and *Meleoma emuncta*), tests included exposing the immature stadia, both individually and in combination, to diapause-inducing and diapause-averting day lengths. Differences in *percentage* of individuals entering diapause revealed which stages are sensitive (Propp *et al.* 1969; Tauber & Tauber 1972b; Tauber & Tauber unpublished). Studies using similar techniques have been conducted for other species, including the mite, *Metaseiulus occidentalis*, the bertha armyworm, *Mamestra configurata*, and the sunflower moth, *Homoeosoma electellum* (Hoy 1975a; Hegdekar 1977; Chippendale & Kikukawa 1983).

In contrast to species that diapause as immatures, *Chrysopa carnea*, which undergoes diapause in the adult stage, presented considerable problems in determining the sensitive stage because this species remains sensitive to diapause-inducing and diapause-averting day lengths during the adult stage. In this case, preimaginal sensitivity tended to be masked by adult responses to photoperiod. This problem was overcome by using the depth, or duration, of diapause as a measure of the relative sensitivity of preimaginal stages to diapause-inducing stimuli (Tauber & Tauber 1970b). All those individuals reared under constant short day lengths (LD 10:14) entered imaginal diapause, but the diapause was of rather short duration. A decrease in photoperiod from long day length (LD 16:8) to short day length increased the duration of diapause, if it was experienced during late larval instars or the pupal or adult stages, but not if it was experienced during the egg or first instar. Thus, in addition to the adult stage, the late instars and pupal stage are very sensitive to diapause-inducing stimuli.

The above methods are not equally applicable to all species, and experiments with each species require an appropriate, sometimes unique approach and bioassay. For example, tests for larval sensitivity to photoperiods controlling adult diapause cannot use diapause duration as an assay in species for which diapause duration is relatively invariable. This problem can be overcome, in part, by the use of "neutral" photoperiods. By this method, insects are reared under diapause-inducing conditions until a particular stage, at which time they are transferred to the "neutral" photoperiod (either constant dark or a day length between long day and short day). The incidence of diapause is then used as a measure of the relative sensitivity of the various stages (see Hodek 1971b).

In applying the knowledge of sensitive stages to analyses of the timing of diapause induction in the field, it is important to be aware that during diapause induction, both qualitative and quantitative changes in the effectiveness of diapause-inducing stimuli may occur. For example, as late-summer day lengths approach, cross, and depart from the critical photoperiod, their effect on induction of autumnal diapause may change. For some species, very short day lengths are capable of inducing a more intense diapause than day lengths just below the critical photoperiod. In other species, the response is reversed (see Section 5.1.2).

In nature, the percentage of a population entering diapause and the depth of diapause are influenced (l) by the degree of coincidence between the sensitive stages and the diapause-inducing stimuli and (2) by the effectiveness of the diapause-inducing stimuli. The degree of coincidence, in turn, is affected by the age-class structure and rate of growth in the population (see Taylor 1980a, b); the effectiveness of the stimuli may or may not vary within and among years. Therefore, all conditions that influence these population parameters interact with the diapause-inducing stimuli and also with the genetic characteristics of the population to affect the percentage and depth of diapause at any given time. Environmental conditions during the prediapause and diapause induction periods also influence diapause induction and intensification; we will discuss these later.

3.2.3 Summary

Insects perceive environmental cues that signal future deterioration of the habitat often long in advance of the diapausing stage. The sensitive stage therefore has a vital function in the life cycle because it determines when development will change from a developmental pathway to a diapause-destined course. Physiologically, the sensitive stages of insects are of great interest because they not only perceive the environmental cues but also store the information for later translation into neuroendocrine function in the form of diapause induction.

Insects vary considerably in the proximity of the sensitive period to the diapausing stage. In some species, both sensitivity and diapause occur in the same stage; in others they are separated within the same generation or even between generations. Similarly, the duration of the sensitive period, as well as the degree of sensitivity over time, varies among species and sometimes with environmental conditions.

Identification of the sensitive stages requires a suitable bioassay for measuring sensitivity to diapause-inducing factors. Percentage diapause and the depth of diapause (as measured by its duration) have been useful measures.

For very few species is sufficient information available to allow the prediction of diapause induction in the field. It requires knowledge of sensitive stages and analysis of the population parameters that influence (1) the degree of coincidence between the sensitive stage and diapause-inducing stimuli and (2) the effectiveness of diapause-inducing stimuli over time.

3.3 Diapause Induction Period

After the sensitive stage perceives the diapause-inducing stimuli, changes begin to occur within the animal. These changes are recognizable at four general levels: neuroendocrine, metabolic, behavioral, and morphological, and together they constitute the diapause syndrome (see Chapter 4). Within each of these four categories diapausing individuals undergo a characteristic sequence of events that can generally be described by a U-shaped curve that is subdivided into four parts—the (1) prediapause, (2) diapause induction and intensification, (3) diapause maintenance and termination, and (4) postdiapause transitional periods of dormancy (Fig. 3.2).

Some aspects of the diapause syndrome are shared by all insects undergoing diapause (e.g., reduced metabolism). Others are common to a particular stage (e.g., suppressed reproductive function during adult diapause), and still others are highly species-specific (e.g., seasonal polyphenism). But, in all cases the various symptoms of the diapause syndrome develop in a species-specific pattern.

3.3.1 Prediapause and early diapause changes

The prediapause and early diapause period ensures that individuals reach the proper stage for diapause induction at the correct time of the year and that they reach a suitable place in which to undergo dormancy. It involves three essential features: (1) the regulation of prediapause growth and reproduction so that the diapausing stage is reached prior to seasonal exigencies

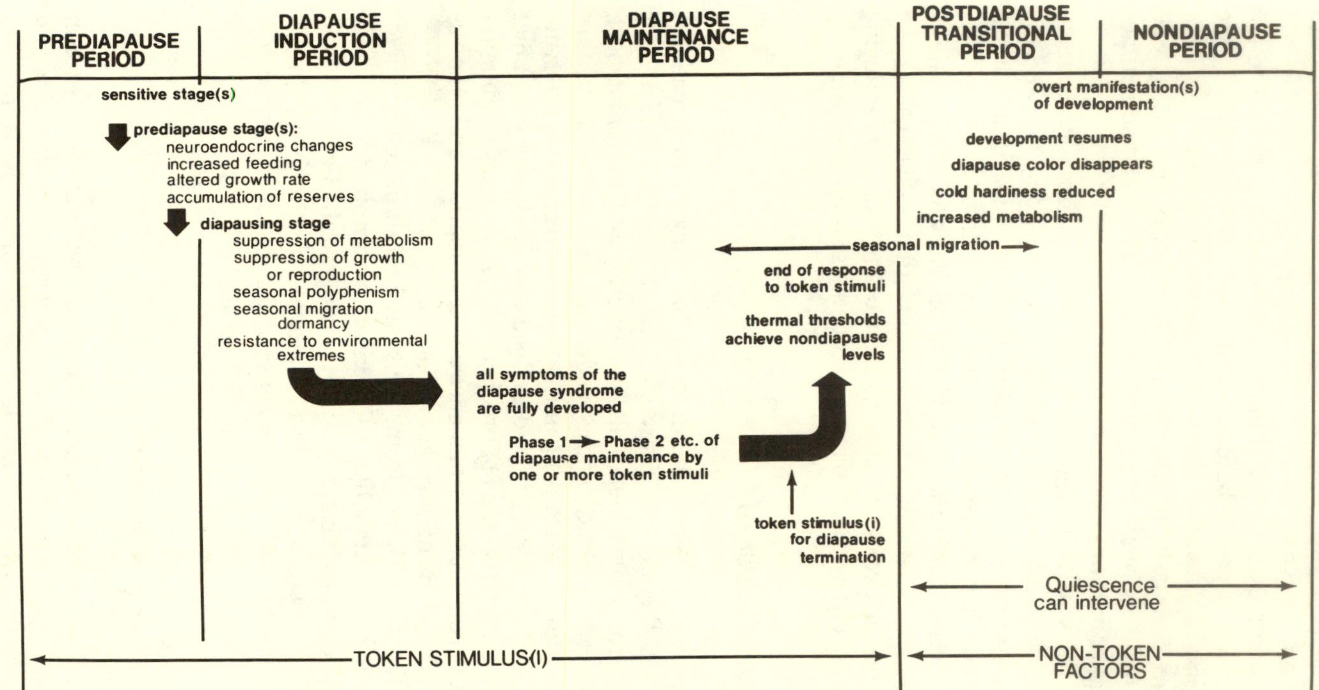

Fig. 3.2 Insect seasonal cycles comprise five periods, each characterized by species-specific sets of responses. Token stimuli largely govern the first three periods, which thus constitute the diapause-regulated periods of the seasonal cycle. During the last two periods, the immediate conditions of the environment (not token stimuli) regulate growth and development. Dormancy commences during the diapause induction period and ends at a species-specific time during the postdiapause transitional period. Migration to the site of dormancy usually occurs during the early part of diapause; return migration occurs near the end of diapause maintenance or after diapause has ended.

and in coincidence with any prediapause requisites, (2) the accumulation of metabolic reserves and other substances that are essential for survival during and after dormancy, and (3) the development of behavioral and morphological changes that subserve movement or migration to the dormancy site and various forms of protection during dormancy.

One of the important characteristics of diapause is that it usually occurs in only one developmental stage. Thus, the successful entry into dormancy is influenced significantly by the insect's ability to reach the stage that is sensitive to diapause inducing stimuli at the appropriate time of the year. If it reaches that stage too soon, diapause will not be induced and an entire life cycle must take place before diapause can be induced. This could be disastrous, especially for insects with a relatively long generation time. By comparison, if development is too slow, the insect may not reach the diapausing stage soon enough to avoid the effects of harsh conditions. Thus, the regulation of prediapause growth and development can be essential to the timely and successful induction of diapause and therefore to survival.

Several factors influence prediapause growth and reproduction. In many species, especially multivoltine species with short generation times, development continues unaffected by token stimuli until it is halted at the diapause stage. In other species token stimuli regulate the rate of prediapause development. For example, photoperiod controls the rate of nymphal development and the number of nymphal moults in several species of crickets. Because morphogenesis is accelerated by short days and decelerated by long days (Masaki 1978b; Tanaka 1979), the primary function of this response pattern is probably to ensure that the insects reach the diapausing stage before conditions become too cold. In other insects, the effect of photoperiod on development is the opposite, and short day lengths decelerate prediapause growth (see Section 5.1.4). In this case, delayed development ensures the induction of diapause before the initiation of another generation.

In many of the examples above, deceleration of development occurs during the actively feeding prediapause stages. These stages may also exhibit accelerated feeding rates (see Ready & Croset 1980; Kono 1980), and presumably the energy thus derived is devoted in large part to the buildup of reserves for dormancy. As diapause begins, the accumulation of metabolic reserves becomes evident through the enlargement of fat body and other storage tissues and through an increase in the chemicals involved in cold-hardiness, heat tolerance, and resistance to desiccation. In some cases, diapause-associated cold-hardiness may develop in diapause-destined individuals before they reach the diapausing stage (Adedokun & Denlinger 1984). We will discuss these physiological changes later (Sections 4.3.2 and 4.3.3).

In addition to accelerated feeding, described above, insects express other prediapause behavioral changes that are associated with migration to, and the selection of, appropriate sites for dormancy. These changes include not only the movements of prediapause insects to dormancy sites, but also prediapause alterations in oviposition behavior that result in diapausing eggs being laid in suitable situations. Very often these behavioral changes are associated with diapause-mediated morphological modifications (e.g., in wing length) that subserve movement or that provide protection during dormancy (Sections 2.2.3.2 and 4.2.3). It is important to stress that the timing of these prediapause changes ranges from several generations before the diapausing stage (as in the aphids, with their autumnal migrations to their primary host and their production of morphs that oviposit diapausing eggs) to the diapausing stage itself. But, we emphasize that no matter how remote they are from the diapause stage, the behavioral and morphological changes that precede diapause constitute an essential aspect of the diapause syndrome. They are induced by token stimuli that herald seasonal change, they develop at a specific time during the course of the life cycle, and they are vital to the successful induction of diapause.

3.3.2 Diapause intensification

An early sign of diapause initiation is a noticeable reduction in feeding and growth rates. In association with this external manifestation are numerous internal changes that occur in a regular sequence as diapause commences and intensifies (Fig. 3.2). These alterations can be monitored and they provide a good measure of the intensification of diapause and a good indication of the degree of metabolic suppression during diapause. A notable example is the often-measured depression in oxygen consumption that occurs with the reduction of metabolic activity typical of diapause (Schneiderman & Williams 1953; Davey 1956; Tombes 1964a; Siew 1966; others). As diapause begins and intensifies, the rate of O_2 consumption decreases; the rate continues to fall until diapause reaches its full depth. Thus, O_2 consumption provides a good indicator of diapause depth.

Several other criteria can also illustrate the intensification of diapause, and again, most of the factors measured show a characteristic U-shaped pattern of change as diapause proceeds. These criteria include reduced RNA synthesis (Wigglesworth 1972), increased muscular resistance to electrical stimulation (Ilyinskaya 1969), altered sensitivity to applied or injected hormones (Bodnaryk 1977; Bradfield & Denlinger 1980), increased resistance to temperature extremes (see Section 4.3.2.4), increased resistance to irradiation (Brower 1980), increase in the time to terminate diapause (Bodnaryk 1978; Tauber *et al.* 1970b), alteration in thermal relations (Ando 1978; Nechols *et al.* 1980), intensification of diapause color (Tauber *et al.* 1970b; Honek 1973) and possibly decrease in the critical photoperiod of long-day insects (see Saunders 1983; Section 5.1.1).

Many of the above symptoms develop simultaneously and thus a measure of one can sometimes be used as a measure of another. For example, O_2 consumption by the bug *Nexara viridula* decreases during diapause induction in conjunction with the loss of green summer color and the assumption of brown winter color. Return of the spring green color is correlated with an increase in O_2 consumption (Michieli 1968; Michieli & Žener 1968). However, it is important to stress that the development of each of the symptoms of diapause occurs at its own time during diapause, at its own rate, and under the influence of specific external and internal factors. Thus, the diapause syndrome is multifaceted both in its expression and in its control.

The multifaceted control of diapause is well illustrated by the fact that environmental conditions during the early period of diapause often have profound effects on the successful induction and full intensification of diapause. Numerous examples show that the temperature conditions prevailing immediately before or after the sensitive stage has perceived the diapause-inducing photoperiods play a very important role in determining the full expression of diapause. In some cases a period of unusually high or low temperatures during these times can completely reverse the diapause-inducing effects of photoperiod (see Kimura & Minami 1980). Similarly, temperature and other conditions during diapause induction have a major effect on the full development of cold-hardiness (Section 4.3.2) and other diapause symptoms (see Kono 1979, 1980). A good example occurs in the papilionid butterfly, *Papilio polyxenes.* In this species, diapause and pupal color are highly correlated if day lengths are either very long (nondiapause, green pupae) or very short (diapausing, brown pupae). At intermediate day lengths there is a decoupling of responses (Hazel & West 1983), perhaps indicating that the development of each of these diapause symptoms is controlled by a slightly different critical photoperiod. Diapause-mediated seasonal polyphenism in other cases can also be decoupled from diapause under artificial conditions in the laboratory (Shapiro 1976). But, presumably, in the field where the insects experience a natural sequence of changing day lengths and temperatures, the full expression of the diapause syndrome would be induced.

With the above thoughts in mind, researchers who examine aspects of dormancy should carefully control both the prediapause environmental conditions and the age of diapausing animals in their experiments. A comparison of the data obtained by Browning (1952) and Hogan (1960b) provides an illustration of the pitfalls. There is a conspicuous difference between the two sets of data in the time required for diapause in *Teleogryllus commodus* to terminate under low temperature. The difference can be accounted for by the fact that Browning used eggs that had not yet reached the diapause stage, and Hogan used those that had. Similarly ambiguous results could be obtained with most other species, if they were tested at different times during diapause induction and intensification or if prediapause conditions were allowed to vary.

3.3.3 Summary

The first symptoms of the diapause syndrome express themselves as developmental or behavioral alterations during prediapause stages. Whether they occur long before or immediately before diapause, their primary function is to ensure that the insect reaches the proper stage for diapause at an appropriate time during the season and in the appropriate place. Thus, the prediapause period involves three essential features: (1) the regulation of prediapause growth and reproduction so that the diapausing stage is reached prior to seasonal exigencies and in coincidence with any prediapause requisites, (2) the accumulation of metabolic reserves that are necessary for survival during and after dormancy, and (3) the development of behavioral and morphological changes that result in prediapause movement to the dormancy site and that provide various forms of protection during dormancy.

Prediapause alterations in development and behavior may occur far in advance of the diapausing stage—sometimes as many as two generations prior to the diapausing stage. But, no matter how remote they are from the diapausing stage, these changes are essential aspects of the diapause syndrome. They are induced by token stimuli that precede seasonal change, they develop during a specific time during metamorphosis, and they are vital to the successful induction of diapause.

Once begun, diapause follows a regular, species-specific pattern of behavioral, physiological, and phenotypic changes that leads to its full induction. Many measures of metabolic function can provide evidence for the intensification of diapause, and in most cases, there is a relatively rapid decline in metabolic and behavioral activity as diapause intensifies.

Environmental factors vary in their effects during the period of diapause intensification, and the various symptoms of diapause develop to their full expression in response to a variety of external and internal factors. In some cases, environmental conditions during the diapause intensification period can substantially alter or even reverse the effects of diapause-inducing stimuli. Thus, diapause should be viewed as multifaceted, not only in its expression, but in its development and control.

3.4 Diapause Completion

Diapause induction and intensification are followed by a period of diapause maintenance during which all or most of the species-specific symptoms of diapause are present. At this time the animals are sensitive to token stimuli and/or their thermal reactions are altered from the nondiapause pattern so as to prevent growth and development even if environmental conditions are favorable. However, the diapausing animals concurrently are undergoing changes that lead to the timely completion of diapause. These changes constitute what is commonly called *diapause development*, which, although considered by some to be an ambiguous term (see Sheldon & MacLeod 1974), is useful in referring to the progress made toward the completion of diapause. Other terms (e.g., diapause processes, diapause-terminating processes, diapause-ending processes, reactivation, conditioning) have been used to refer to the processes during diapause (see Sheldon & MacLeod 1974; Waldbauer 1978), but it is well recognized that knowledge of the ecological and physiological processes occurring during this time is too limited to warrant precise definition or rigid application of specific terms to events during diapause. Therefore, we use the term "diapause development" because it implies, correctly, that diapause normally undergoes a regular course of events that eventually lead to diapause completion and because it is consistent with the fact that diapause involves numerous, simultaneous and/or serial processes—some of which promote the maintenance of diapause, others of which hasten its termination.

3.4.1 Diapause duration

The duration of diapause in nature is a species-specific and/or strain-specific characteristic—typically ranging from several weeks to several years. During this period, the depth of diapause as well as the responses of diapausing insects to external factors change, in a species-specific or strain-specific pattern. Many species that undergo autumnal-hibernal diapause exhibit a gradual loss in sensitivity to diapause-maintaining factors and a concomitant decrease in diapause depth, until diapause terminates spontaneously sometime during late autumn or early winter. Examples of the spontaneous termination of diapause are found in numerous species including the lacewing, *Chrysopa carnea* and the parasitoid, *Tetrastichus julis* (Tauber & Tauber 1973d; Nechols *et al.* 1980; see Tauber & Tauber 1976a for a review). In other species with autumnal-hibernal diapause, especially those that require a diapause-terminating stimulus, diapause depth and responses to environmental conditions may change slightly at the beginning of diapause and then remain relatively stable throughout fall and winter; subsequently, they decrease rapidly sometime in spring or early summer.

In many species a proportion of individuals undergo an unusually long period of diapause lasting more than one year (= prolonged diapause). Powell (1974) recognized that there are two types of prolonged diapause. In some cases, whole populations may respond to adverse environmental conditions by remaining in diapause for periods longer than a year (see Section 6.4.2); in other instances, prolonged diapause regularly occurs in a certain proportion of the population as a bet-hedging tactic (see Section 9.2.2.1). In both types of prolonged diapause, genetic and environmental factors may be important.

Generally, the inherent length of diapause is a characteristic with considerable genetic variability, as illustrated by the high heritability values it exhibits in selection experiments (e.g., Dingle 1974; Tauber & Tauber 1978, 1979). In nature, however, the ultimate duration of diapause is determined by the interaction between (a) the genetic characteristics of the species, strain, or individual, (b) the environmental factors that determine diapause depth (i.e., the effectiveness of diapause-inducing and -intensifying stimuli), (c) the sensitivity of the stage(s) perceiving the stimuli, and (d) the environmental conditions that determine the rate of diapause development (i.e., the physical conditions during diapause and the need for, and seasonal occurrence of, a diapause-terminating stimulus)(Fig. 3.2).

The degree to which diapause duration depends on any one of the above factors varies with species. In some species, diapause duration is largely determined at or near the time the diapause-inducing stimuli are perceived, but obviously is not expressed until much later. For example, temperature conditions during diapause induction play a significant role in determining diapause depth in larvae of the codling moth, *Laspeyresia pomonella* (Sieber & Benz 1980a). In other species, photoperiodic information determining diapause depth is passed between stages (Bell & Adkisson 1964; Tauber & Tauber 1970b; Kimura *et al.* 1982) and even between generations (Rock *et al.* 1971; Vinogradova 1974, 1975b).

In a very fine piece of work, Denlinger and Bradfield (1981) showed that the brain is the main point of storage for photoperiodic information determining diapause depth in the tobacco hornworm, *Manduca sexta.* First, they demonstrated that photoperiodic conditions during diapause induction influence diapause depth. If eggs and larvae are maintained under constant short-day conditions, most individuals enter a diapause of relatively short duration. However, when larvae experience a decrease in day length from long to short days, a longer diapause ensues. Subsequently, they showed that the brain stores photoperiodic information that programs diapause duration and that this information can be transferred between individuals through brain transplantation.

In nature, insects may or may not rely heavily on prediapause conditions to determine diapause duration. Nevertheless, it has been well demonstrated that environmental factors during diapause strongly influence the course and duration of diapause. This is the subject of the next two sections.

3.4.2 Diapause maintenance

In nature, one or more mechanisms may maintain diapause; sensitivity to day length and altered thermal thresholds for diapause development are the most common. As diapause progresses there usually is a gradual alteration in response to diapause-maintaining stimuli (Tauber & Tauber 1976a). During photoperiodically maintained diapause, insects usually undergo a gradual diminution of sensitivity to day lengths. Similar changes occur in insects' responses to temperature during diapause (e.g., Ando 1978; Tauber *et al.* 1982).

In some species more than one factor maintains diapause. In such cases, the diapausing animal may show, simultaneously, both sensitivity to photoperiod and a lowered thermal range for diapause development. Such is the case in the parasitoid *Tetrastichus julis* (Nechols *et al.* 1980). In other species, different physiological processes controlling diapause may act sequentially, and when they do it is often possible to distinguish two or more phases of diapause. For example, during early autumn, embryonic diapause in the damselflies *Lestes disjunctus* and *L. unguiculatus* is maintained by lowered thermal thresholds for diapause development. By the end of October the thermal control of diapause ends, and the insects enter a second phase of diapause that is maintained through December by the action of photoperiod (Sawchyn & Church 1973). In another aquatic arthropod, *Daphnia pulex*, photoperiod largely controls the autumnal phase of diapause, but in late spring, although light is necessary for diapause termination, a stimulus associated with population density is almost completely effective in terminating diapause (Stross 1969b).

In the *mohave* strain of *Chrysopa carnea*, the autumnal phase of diapause is maintained by day length and the hibernal-early vernal phase is maintained by the absence of prey (Tauber & Tauber 1973c). Thus, diapause is maintained until prey become available, even though abiotic factors may be favorable for breeding. Similarly, the European corn borer has a two-phase diapause. Photoperiod maintains the first phase of diapause; termination of the second phase requires water intake (Beck 1967).

Different levels of the same diapause-maintaining factor may function in different phases of diapause. In diapausing eggs of the muscid fly, *Leptohylemyia coarctata*, the early phase of diapause ends more readily at high temperatures than the second phase (Way 1960). In the cutworm, *Heliothis armigera*, the reverse is true. Here, the first phase of pupal diapause proceeds at low temperatures, whereas the second phase requires temperatures that are higher than those for nondiapause pupal development (Wilson *et al.* 1979). Photoperiod alone can also function in maintaining two phases of diapause. In the green lacewing, *Chrysopa downesi*, short days of autumn terminate the first phase of an aestival-autumnal-hibernal diapause and long vernal day lengths terminate the second phase (Tauber & Tauber 1976b).

In the carabid beetle, *Pterostichus nigrita*, the recognition of a two-phase diapause was aided by morphological and endocrinological changes. During the first phase the ovaries remain almost completely undifferentiated, and during the second phase, development of the ovaries is arrested at the previtellogenesis stage (Thiele 1966; Hoffmann 1970). Photoperiod regulates both phases of this diapause.

It is generally not easy to make the distinction between the transition from one phase of diapause to another and the actual termination of diapause; such analyses often require considerable experimentation because the changes in responses are sometimes very subtle. Such a situation is found in the cricket, *Teleogryllus commodus*, in which diapause persists after a long period (60-80 days) of exposure to cold conditions. Termination of a second phase of diapause requires only brief (as few as three days) exposure to temperatures above 25°C (Masaki *et al.* 1979).

The tree-hole mosquito, *Aedes triseriatus*, illustrates further, but in a very different way, that the termination of diapause is not always easily delineated. These mosquitoes generally enter the overwintering period as diapausing embryos; apparently after a sufficiently long period of low temperature, the eggs hatch in response to warmth (Holzapfel & Bradshaw 1981).

Development of the resulting larvae is then dependent on photoperiod and temperature. If hatching occurs in midwinter, development proceeds to the fourth instar, at which time short day length and low temperature may induce a second diapause. If hatching occurs in spring, the long day lengths do not cause diapause induction, but they may interact with temperature to influence the rate of larval development. Under the relatively short days of early spring, larval development is slow and unrelated to temperature; under the long day lengths of late spring, it is directly correlated with temperature. Because of the interaction of photoperiod and temperature, there is considerable geographic and annual variation in the timing of diapause termination among and within *A. triseriatus* populations—a situation that necessitates the acquisition of considerable data before the end of diapause and the initiation of postdiapause emergence can be predicted accurately.

From the above, it is clear that diapause is a dynamic state characterized by physiological change. During the course of diapause, there is generally a decrease (to a greater or lesser extent) in diapause intensity, as well as continual alterations in the animals' responses to environmental stimuli. These changes can occur even if the animals are held under constant conditions (e.g., McLeod & Beck 1963; Tauber & Tauber 1973d; Ando 1978; see also Zaslavsky 1972, 1975). However, the rates of these processes are largely determined by temperature and light, as well as by the innate characteristics of the species or biotype. All together, the physiological changes within the diapausing animals, in addition to the seasonal progression in photoperiod and temperature, contribute to the dynamism of the diapause state. Therefore, to be ecologically meaningful, investigations of diapause development should consider numerous interacting, constantly changing variables—the altering reactions of the organisms and the seasonally changing environmental factors.

3.4.3 Diapause termination

The general statement, "The termination of diapause is dependent upon stimuli received from outside the insect" (Andrewartha & Birch 1973) is sound, but it is necessary to use caution in applying it to specific cases because insects vary greatly in the degree to which they depend on external cues to time the end of diapause and in the nature of the environmental factors involved in diapause termination. On the one hand, long day lengths often can terminate both laboratory-induced and naturally induced autumnal diapause. On the other hand, the photoperiodic responses of diapausing insects usually diminish as diapause progresses (see Section 5.1.1.6). It is therefore presumptuous to conclude that because day length maintains diapause at the beginning of dormancy it actively terminates diapause after the overwintering period in the field. Similarly, it is often assumed because chilling hastens diapause development in some species that low temperatures function to terminate diapause. However, the evidence supporting this assumption is tenuous because there have been few studies testing the effect of temperature on field populations throughout the season. Therefore, it should *not* be presumed (a) that because chilling hastens the end of diapause in the laboratory it also acts to terminate diapause in nature, (b) that because chilling precedes the end of diapause in nature it is the causal factor for diapause termination, or (c) that because chilling can act directly on the brain to terminate diapause in the laboratory, as was demonstrated for *Hyalophora cecropia* (Williams 1956), diapause is terminated in nature by this mechanism.

In fact, for most temperate-zone species that undergo an overwintering diapause and in which the conditions influencing diapause maintenance have been examined in natural populations throughout the winter and spring, no specific diapause-terminating stimulus has been identified. Rather, sometime during winter, the insects cease to respond to diapause-maintaining factors, and diapause ends gradually and spontaneously sometime during late fall or early winter. The rate at which these insects lose their responses to diapause-maintaining factors depends on both external and internal conditions (see Section 5.1.5), and because no cue synchronizes diapause termination, diapause usually ends over a broad span of time (e.g., see Tauber & Tauber 1973d). Subsequently, postdiapause loss of the overt symptoms of diapause (e.g., cold-hardiness) and the resumption of development are prevented, sometimes for a considerable period, until temperatures rise above the lower thermal threshold for development. Demonstration of this type of spontaneous diapause termination requires experimentation because the overt symptoms of diapause remain until temperatures rise above the lower threshold for development. Such studies have shown or reasonably implied that it does occur in a number of species, including the weevil, *Sitona cylindricollis* (Hans 1961); the lady beetle, *Coccinella septempunctata* (Hodek 1962); the European corn borer, *Ostrinia nubilalis* (McLeod & Beck 1963); the bugs *Lygus hesperus* (Beards & Strong 1966) and *Pyrrhocoris apterus* (Hodek 1971, 1974); the tetranychid and phytoseiid mites *Tetranychus urticae, Metaseiulus occidentalis*, and *Neoseiulus fallacis* (Rock *et al.* 1971; Hoy & Flaherty 1975; Veerman 1977b), the fly *Sarcophaga bullata* (Denlinger 1972b); the thrips, *Anaphothrips obscurus* (Kamm 1972); the lacewings *Chrysopa carnea* and *Chrysopa harrisii* (Tauber & Tauber 1973, 1975); and the tobacco budworm, *Heliothis virescens* (Lopez *et al.* 1983). In these and other species, the loss of the overt symptoms of diapause and the timing of vernal development and reproduction are synchronized throughout the population by the gradual accumulation of heat units above the thermal threshold for postdiapause development, or by the occurrence of other developmental requirements, such as food, moisture, or hatching stimuli (e.g., Clements 1963; Krysan 1978).

In the relatively few species in which a specific factor is needed to terminate autumnal-hibernal diapause, four stimuli have been identified: photoperiod, food, moisture, and an internal stimulus from insect host to parasitoid or from insect parasitoid to host (see Sections 5.1.1.6 and 6.1). In the majority of these cases, diapause appears to have two phases: an initial phase in which considerable diapause development occurs (i.e., as measured by a progressive reduction in diapause intensity), and a second phase that is somewhat static until the terminating stimulus occurs.

The data suggest that in contrast to hibernal diapause, aestival diapause often is terminated in nature by a specific stimulus (Tauber & Tauber 1976; Masaki 1980; see Section 5.1.6). Except for a few species, however, the factors involved in maintenance and termination of summer diapause in the field have not been well established.

In contrasting the gradual or spontaneous termination of diapause with that stimulated by a specific cue, it is interesting to note the specific circumstances to which each type is adapted. The requirement of a terminating stimulus ensures the maintenance of diapause, with its associated cold-, drought-, and heat-hardiness, until the occurrence of a development-inducing cue, even if other conditions become favorable for development. In areas where winter is mild and conditions are highly variable, such an adaptation reduces the risks of premature termination of diapause. As a result, the responses controlling diapause termination may come under selection pressure. However, in situations in which winter temperatures remain relatively low, postdiapause insects in a state of quiescence can retain all the overt benefits of diapause (e.g., cold-hardiness, suppressed metabolism) without having to maintain diapause. Synchrony with developmental requirements is based on responses occurring after diapause termination. In these cases, postdiapause responses may be subject to more selection pressure than the responses controlling diapause termination.

3.4.4 Summary

Diapause depth is a species-specific characteristic that is subject to modification by environmental conditions during the prediapause, diapause-inducing, and diapause-intensifying periods. Once diapause is induced, its duration is determined by the depth of diapause, by diapause-maintaining conditions, and in some cases by the occurrence of a specific stimulus for diapause termination.

In nature, diapause is maintained by one or more mechanisms: the most common ones being response to photoperiod and altered thermal responses. During diapause, sensitivity to diapause-maintaining factors decreases either gradually or abruptly. In both cases, the progress made toward completion of diapause is termed *diapause development*. In many species, a single mechanism serves to maintain diapause. However, there are instances when more than one mechanism is involved, and in these cases it is sometimes possible to distinguish two phases of diapause. Response to two levels of a single diapause-maintaining factor may also indicate a two-phase diapause.

Either diapause can end gradually without the intervention of a specific stimulus or it requires the occurrence of a specific diapause-terminating stimulus. The requirement or lack of a requirement for diapause termination is a species-specific characteristic with considerable phenological importance. Each mechanism, under different circumstances, can be adaptive. Diapause-terminating stimuli may act in very subtle ways, and when they do, they require experiments with precisely controlled conditions to reveal themselves.

3.5 Postdiapause Transitional Period

In general, we consider that diapause has ended, and the postdiapause transition phase of dormancy has begun, when token stimuli no longer prevent growth and development and when thermal (developmental thresholds) and other responses have returned to the nondiapause level. Thus, postdiapause insects are capable of immediate resumption of development upon return of favorable conditions. For example, thermally maintained diapause in eggs of the grasshopper *Austroicetes cruciata* ends when developmental thresholds and response curves reach normal nondiapause levels; this occurs in Australia between mid-May and late June (Birch 1942).

If conditions after diapause are not favorable for growth and development, the diapause symptoms persist for varying periods of time. Thus, the *postdiapause transitional period* may include a period of *postdiapause quiescence*, during which the diapause symptoms are retained and development is suppressed, and a period of *postdiapause morphogenesis* that results in the loss of diapause symptoms and the attainment of the next recognizable stage of morphogenesis. The disappearance of diapause symptoms and development to the next stage are invoked and controlled by the immediate environmental conditions of temperature, moisture, and food and not by token stimuli. Thus, included in this phase are embryonic development and egg hatching that have moisture or other requirements, if both nondiapause and postdiapause eggs have the same requirements (see Section 5.3.3).

3.5.1 Postdiapause quiescence

Because development during the postdiapause transitional period is directly dependent on prevailing environmental conditions, it may be delayed or interrupted at any time by a period of unfavorable conditions; if this occurs the insects remain dormant in a state of postdiapause quiescence. Postdiapause quiescence is especially notable in insects that end diapause during late autumn or early winter, after which environmental conditions are such that development is prevented and dormancy continues. In such cases, the postdiapause insects retain most of their characteristic symptoms of diapause, such as enlarged fat body, cold-hardiness, and diapause color. However, at this time their reactions to environmental factors are those of nondiapausing animals, and development can begin immediately upon the arrival of favorable conditions. Continuous exposure to low temperatures during the postdiapause period also aids in maintaining cold-hardiness by retarding both postdiapause development and the conversion of protective cryosubstances (Asahina 1969).

Because token stimuli do not regulate quiescence, the arrival of favorable conditions ends postdiapause quiescence and initiates postdiapause development without delay. In the green lacewing, *Chrysopa harrisii*, photoperiodically controlled reproductive diapause ends in December. Subsequently, postdiapause development is inhibited by low temperatures for approximately three months until spring, but it begins immediately when vernal temperatures exceed a threshold of approximately 11°C (Tauber & Tauber 1974, unpublished). Numerous other examples of postdiapause quiescence are given by Tauber and Tauber (1976).

Not all insects that end diapause before or during winter exhibit a period of postdiapause quiescence. The alfalfa weevil, *Hypera postica*, undergoes an aestival diapause in response to long day lengths; this diapause ends in autumn. Subsequently, ovarian development occurs and, if autumnal conditions are mild, some females oviposit immediately. In Wisconsin, fall oviposition does not exceed 2% of the yearly egg production, and fewer than 37% of all females lay autumnal eggs (Litsinger & Apple 1973a). However, all females undergo ovariole development in autumn (Litsinger & Apple 1973b). Farther south, in South Carolina, where winter conditions are milder, both ovarian development and oviposition occur in most of the population during autumn and winter (Tombes 1964b).

3.5.2 Postdiapause development

Once a diapausing animal's ability to respond to environmental cues has returned to the nondiapause level, the diapause symptoms (reduced metabolism, cold-hardiness, diapause color, etc.) begin to disappear when environmental conditions become favorable for development. Not all of the symptoms of diapause disappear at the same rate, and during the postdiapause transitional period there is a species-specific, usually temperature-dependent, progression in the loss of the symptoms (see Fig. 3.2). During this period, development leading to the next stage of metamorphosis or to reproduction is also largely dependent on temperature. However, food and other factors also play important roles (Sections 5.3.1.3 and 5.3.3).

The thermal requirements for postdiapause morphogenesis often differ from those for the comparable nondiapause stage. For example, postdiapause *Hyphantria cunea* pupae require a slightly, but significantly higher number of heat units for emergence than nondiapause pupae. Similar differences are found in the development of nondiapause and postdiapause *Chrysopa oculata* and *C. nigricornis* prepupae (Propp *et al.* 1969; Tauber & Tauber 1972b; Tauber & Tauber unpublished). These dissimilarities may be accounted for by the distinct forms of metabolic resources that are available to diapausing and nondiapausing animals. In addition, mobilization of resources for nondiapause and postdiapause development may occur by different pathways. In contrast to the above, postdiapause alfalfa blotch leafminer, *Agromyza frontella*, pupae require fewer heat units for emergence than nondiapause pupae (Nechols *et al.* 1983). And, in other cases, variation in postdiapause rates of development is correlated with variation in diapause duration. For example, postembryonic development by the dragonfly, *Aeshna mixta*, is faster and requires fewer instars if embryonic diapause is short (approximately 100 days) than if it is long (approximately 200 days)(Schaller 1972). Perhaps a short diapause results in the carry-over of large amounts of metabolic reserves to postdiapause stages.

The term "postdiapause development" has sometimes been broadly used to include nondiapause development occurring long after dormancy has ended and all the diapause symptoms have been lost (e.g., Campbell 1978). However, recognizing the paucity of investigations dealing with the physiological aspects of the diapause, postdiapause, and nondiapause periods, we delineate the end of the postdiapause transitional period in morphological terms. For egg diapause, it encompasses all development leading to the next stage of embryogenesis or to hatching. For larval and pupal diapause, it extends until the insect reaches the next advanced stage of development (i.e., for larvae: molt to a later instar or to a pupa, or formation of a cocoon; for pupae: development of pigmented eyespots, molt to an adult, or emergence). These criteria would exclude stationary or supernumerary molts that occur in some diapausing larvae (Yin & Chippendale 1973; Yagi & Fukaya 1974). Because ovarian development can be reversible, the postdiapause transitional period that follows imaginal diapause in females usually includes all ovarian development leading to the initiation of oviposition. For males, the criteria include transfer of sperm to the seminal vesicles, resumption of mating behavior, and the ability to transfer sperm to a female. For each species, the criteria designating the end of the postdiapause transitional period should be clearly defined.

3.5.3 Summary

The transition from the diapause state to nondiapause growth and reproduction is a gradual process that involves both the loss of the various symptoms of diapause and the resumption of development leading to the next stage of metamorphosis or to reproduction. It begins when diapause ends—that is, when token stimuli no longer prevent growth and development and when thermal and other responses have returned to the level characteristic of the nondiapause period.

During the postdiapause transitional period, the initiation of development is directly dependent on prevailing environmental conditions, and frequently it is preceded by a period of postdiapause quiescence. Such a period of postdiapause quiescence is common among insects that end diapause during late autumn or early winter when temperatures are below the threshold for development. Thus, all the characteristics of diapause (e.g., enlarged fat body, cold-hardiness, diapause color) are retained and development is suppressed, but the insect has regained the capability to respond immediately to growth- and reproduction-promoting conditions.

Once favorable conditions occur, the symptoms of diapause disappear and postdiapause development proceeds in a species-specific pattern. The progression of these events is usually temperature dependent and subject to some diapause-related variability. The postdiapause transitional period ends when the insect reaches the next irreversible stage of metamorphosis.

3.6 Postdormancy Effects of Diapause

Molting or, in the case of adult diapause, the initiation of reproduction generally signals the end of the postdiapause transitional period and thus the end of dormancy. However, this does not always mean that the effects of diapause disappear at this time. A few insects typically display diapause-mediated effects in later stages or in the generation after dormancy has ended. To date, these effects have been shown to include changes in certain behavioral patterns, fecundity, and phenotype, but we believe that more traits will be found to have postdormancy effects when researchers begin to look for them.

Termination of diapause is characterized by the loss of sensitivity to diapause-inducing and diapause-maintaining stimuli. Reduced sensitivity to these stimuli may persist for varying periods of time after dormancy. Thus, to complete our consideration of the course of diapause, we also discuss in this section the return of sensitivity to diapause-inducing stimuli.

3.6.1 Postdormancy behavior

Insects that have undergone diapause-mediated dormancy may exhibit behavior patterns that differ from those of nondiapause insects. Presumably these differences adapt the insects to seasonal variations in their physical and biotic environment. Males of the wild silkmoth, *Antheraea pernyi*, that have been exposed to low temperature during their postdiapause adult development begin flight activity during an earlier part of the night than males that were reared without exposure to cold during the postdiapause period. Females exhibit a similar shift in the diurnal timing of sex pheromone release when they experience cold during adult development (Truman 1973). Another example involves the fall webworm, *Hyphantria cunea*, which also overwinters in the pupal stage. In this species, the diurnal timing of pheromone release differs between postdiapause and nondiapause adults. Neither postdiapause nor nondiapause females assume the courtship posture at constant 25°C in darkness. When the temperature was lowered from 25 to 15°C, however, about two-thirds of nondiapause females assumed courtship posture. Only one out of 18 postdiapause females did so. If a two-hour light period was given prior to the reduction in temperature, courting behavior was elicited in 10 out of 13 postdiapause females (Hidaka 1977).

A somewhat similar effect is seen in postdormancy hatching by the long-horned grasshopper, *Metrioptera hime*, but for this species, nondiapause eggs are not available for comparison. Constant light or dark conditions at either 20 or 25°C retard hatching; the introduction of either a light cycle at constant 20°C or a thermoperiod under constant light or darkness promotes synchronized hatching. A temperature of 25°C consistently retards hatching under all photoperiodic conditions, while 15°C promotes it under all photoperiods (Arai 1977).

Postdormancy behavioral adaptation to the biotic environment may involve food preference or food choice. Some bivoltine species utilize a different host or food in spring from that in summer, and the switch in food preference may constitute a postdiapause effect. For example, the vernal generation of caterpillars in several species of desert-inhabiting moths, such as *Cucullia boryphora, Dyscia malatyana*, and *Casignetella polynella*, feed on leaves, whereas after aestival diapause, the autumnal generation feeds on fruits. A more unusual example involves *Hysterosia subfumida*, in which the autumnal and vernal generations of caterpillars feed on the fruits of host plants in different families (Falkovich 1979). Whether these switches in behavior represent seasonal changes associated with diapause or merely reflect the seasonal availability of food remains to be studied.

3.6.2 Postdormancy life-history traits

It is well established that the termination of dormancy influences the seasonal *timing* of growth and reproduction by the postdiapause and subsequent generations. However, it is less well known, but very important, that there can be significant postdormancy effects on the quantitative expression of life-history traits. Such effects have been well demonstrated for a few species of insects that overwinter as adults and that show reduced fecundity and/or fertility after diapause (Deseö 1973; Fujiie 1980; Denlinger 1981b; Abo-Ghalia & Thibout 1983). However, in some cases, postdiapause fecundity is the same as or higher than that of nondiapause insects (Tauber & Tauber 1969; Tauber *et al.* 1970a; Soni 1976; Poras 1976; Linley *et al.* 1970). In these examples, it is possible that a decrease in prediapause growth rates and/or enhanced prediapause feeding rates result in an overwintering generation of large size or with a high level of nutrient reserves that provide the postdiapause adults with a high reproductive potential (e.g., Sáringer & Szentkirályi 1980).

Postdiapause effects have been examined for relatively few species, and they offer an excellent topic for ecological and evolutionary study of the interactions among life-history traits. We will discuss this aspect in Section 9.2.1.1.

3.6.3 Postdormancy phenotypes

The life cycles of several species of butterflies involve the alternation of a posthibernation spring form with a nondiapause summer form (Shapiro 1976, 1982). The nymphalid *Araschnia levana* provides an excellent example. In this species, an orange and black vernal form (levana) alternates with a white-banded summer form (prorsa). The short day lengths that induce pupal diapause also produce the vernal (levana) adult form. Individuals in which diapause has been prevented by long day lengths produce the aestival (prorsa) form. Although the physiological relationship between diapause, day length, and postdiapause polyphenism has not been determined (Shapiro 1976), it is clear that the token stimuli that induce diapause may have a significant effect on adult form long after diapause and dormancy have ended.

The story is somewhat different in *Hyphantria cunea.* The male moths emerging from postdiapause pupae after hibernation are conspicuously maculated, while those from nondiapause pupae of the summer generation are pure white. Although both short photoperiods during the larval stage and low temperatures during pupal-adult development favor the expression of black spots on the wings, temperature is the more important. At a low temperature, nondiapause pupae produce maculated male moths; conversely, exposure of diapause pupae to high temperature (27.5°C) produces immaculate moths (Umeya & Miyata 1979).

3.6.4 Return of sensitivity to diapause-inducing stimuli

To complete the full course of diapause, we must consider the postdiapause return of sensitivity to diapause-inducing stimuli. For most species, especially those in which the sensitive period is discrete and relatively well separated from the diapause stage, this merely requires development to the life stage that characteristically exhibits sensitivity. For example, *Chrysopa oculata* is sensitive to diapause-inducing photoperiods in the second and free-living third instars, and it diapauses as a mature third-instar larva within the cocoon. Sensitivity to diapause-inducing stimuli returns after diapause has ended, when the next generation of second- and third-instar larvae appear (Propp *et al.* 1969; Tauber & Tauber, unpublished). In some species, such as *Choristoneura fumiferana*, reestablishment of photoperiodic sensitivity involves the induction of a second diapause very soon after termination of the first (Harvey 1957, 1967). Induction of aestival diapause may also rely on the return of photosensitivity after hibernal diapause. In early spring, postdiapause larvae of *Spilarctia imparilis* regain sensitivity to photoperiod. However, the short vernal day lengths do not delay further larval development; they exert their growth-retarding effect much later by inducing an aestival diapause in the pupae (Kimura *et al.* 1982).

Some insects that overwinter as adults regain sensitivity to diapause-inducing stimuli after a period of postdiapause oviposition (e.g., Hodek 1981; Numata & Hidaka 1982). Although interesting from a behavioral and physiological point of view, such a phenomenon probably has little ecological importance because diapause-inducing short day lengths usually do not occur while the postdiapause generation of adults is alive, and thus diapause induction must await a subsequent generation. However, in some species, such as *Aelia acuminata, Coccinella septempunctata*, and *Leptinotarsa decemlineata*, postdiapause adults may live long enough to undergo a second hibernation; in these instances, return of photoperiodic sensitivity may be important for the induction of the second diapause (Hodek 1981). By contrast, *Oedipoda miniata* adults are not long-lived; all die by winter. Apparently, for this species, continued adult sensitivity to photoperiod allows flexibility in the timing of reproduction in a variable environment (Pener & Orshan 1983).

Some multivoltine insects with very short life cycles do not regain full sensitivity to diapause-inducing stimuli for one or more generations after diapause (e.g., Mustafa & Hodgson 1984). Lees (1959, 1960, 1963) demonstrated that a specific interval of time (regardless of the number of generations) must intervene before postdiapause generations of the aphid, *Megoura viciae*, regain sensitivity to the photoperiodic stimuli that induce sexual morphs (diapausing eggs are oviposited only by sexuals). Such a mechanism, termed the "interval timer", also occurs in some other aphid species, and possibly in some strains of the mite *Tetranychus urticae* (Bonnemaison 1951; Dubynina 1965; Azaryan 1966; Geyspits *et al.* 1971, 1974; Dixon 1971a, 1972b), but not in all (Dixon & Glen 1971; Veerman 1977b). It ensures that spring generations that are exposed to short day lengths do not produce sexual forms prematurely (Lees 1960) and that aestival and autumnal generations can take advantage of their hosts late into the season (Dixon 1971a, 1972b). It does not appear to influence the within-season production of alates (MacKay 1977). Geographic variation in the "interval timer" subserves adaptation to differences in the lengths of southern and northern growing seasons (Azaryan 1966). Some sort of cyclic reactions may be associated with the termination of diapause in these and other species (Hussey 1955; Dubynina 1965), but further work on this possibility is clearly needed (see Veerman 1977b).

Unlike the species above that regain sensitivity to day length after a certain period of time, postdiapause *Sarcophaga bullata* require exposure to long day lengths before they can respond to diapause-inducing short day lengths. After pupal diapause, *S. bullata* females that have been reared under short day lengths are unable to produce diapausing progeny even if they are maintained under strongly diapause-inducing conditions (short days). This inability is determined not by diapause itself, but by a short-day effect that is transmitted to progeny by the female parent. Exposure to long day lengths restores the insects' capability to respond to diapause-inducing short day lengths (Henrich & Denlinger 1982a).

European scientists and especially scientists in the U.S.S.R., working with a variety of species such as the mites *Tetranychus urticae, T. crataegi, Panonychus ulmi*, and *Schizotetranychus schizopus*, the fungus flies *Drosophila phalerata* and *D. transversa*, the cabbage moth, *Barathra brassicae*, the knotgrass moth, *Acronycta rumicis*, the flesh fly, *Boettcherisca septentrionalis*, and the carpet beetle, *Anthrenus verbasci*, report cyclic periodicity of approximately a year in the propensity to enter photoperiodically induced diapause (Blake 1958, 1959, 1960, 1963; Geyspits 1960; Dubynina 1965; Geyspits *et al.* 1971, 1974, 1978; Goryshin & Tyshchenko 1976; Stekol'nikov *et al.* 1977; Simonenko 1978; Kuznetsova & Tyshchenko 1979). Annual rhythmicity is also reported for the parasitoid *Trichogramma evanescens* and the fly *Calliphora vicina*, but in these cases it involves reinstating the maternal influence on photoperiodic induction of diapause (Vinogradova & Bogdanova 1980; Zaslavsky & Umarova 1981). Despite attempts to do so, the existence of such annual cycles in photoperiodic sensitivity has not been confirmed for some of the above species (Lees 1953 and Veerman 1977b for *T. urticae*; Masaki unpublished for *A. verbasci*; see also King & Farner 1974 for birds). Moreover, the rhythms described by Geyspits *et al.* (1974, 1978) are much closer to one year than those for other suspected cases of circannual rhythms (e.g., in birds, deer, etc.). Clearly, further studies carefully controlled for inadvertent selection (see Chapter 8), are needed.

3.6.5 Summary

Even after dormancy has ended and postdiapause morphogenesis has been completed, diapause can affect certain aspects of the life cycle. These effects include altered behavioral patterns during the postdormancy period and various effects on fecundity and perhaps other life-history traits, as well as the postdormancy phenotype. In all cases, the exact physiological connection between the postdormancy effect and diapause is not defined. Nevertheless, it is clear that the token stimuli that induce diapause and the physiological changes that occur during diapause have profound effects on subsequent behavior, development, and body form. In general, postdormancy effects appear to serve as adaptations to seasonal alterations in the physical or biotic environment.

Significant among the changes that occur in insects after diapause has ended is the regaining of sensitivity to diapause-inducing and -maintaining stimuli. Most insects do not regain it, and sensitivity does not reappear until the next generation when the characteristic sensitive stage recurs. Other insects (e.g., some species with reproductive diapause, aphids) remain insensitive to diapause-inducing stimuli for a certain period of time after diapause, and several generations may develop before sensitivity returns. Still other species require a specific stimulus to reinstate sensitivity.

Chapter 4

The Diapause Syndrome

Diapause, as we have already illustrated, is not merely the cessation of growth and development. On the contrary, it is a physiological state composed of a species-specific set of traits that are induced in a regular pattern. These traits are expressed in response to a series of stimuli during the prediapause and diapause induction periods, and their development is mediated through specific endocrine changes. Basic to an understanding of insect seasonality, therefore, is an appreciation of the diversity of the behavioral, physiological, and morphological adaptations encompassed by the diapause syndrome and the diversity of ecological and physiological mechanisms that control them. Such an appreciation can be complete only if it includes some understanding of the endocrine and biochemical mechanisms involved.

In the following discussions we try to relate the physiology of diapause to conditions in the field. Thus, when interpreting laboratory data, we have kept in mind the dynamic aspects of diapause occurring under field conditions. That is, as insects enter, maintain, and leave the state of diapause, they undergo characteristic changes that influence not only their lives in the field, but also their responses in the laboratory. These changes are all hormonally mediated, and thus we begin our discussion with a consideration of how the insect endocrine system serves as the mediator of the diverse and dynamic symptoms of diapause. Then, we consider the numerous behavioral, physiological, and morphological symptoms of diapause and how they contribute to seasonal adaptation.

4.1 Endocrine Mediation of Diapause

The insect endocrine system functions as the intermediary between various environmental stimuli and metabolic function. On the one hand, growth, development, reproduction, and behavior are regulated by the neuroendocrine system. On the other, its function (i.e., the synthesis and release of neurohormonal and hormonal material) is influenced by environmental factors such as photoperiod, temperature, food, moisture, and population density (Wigglesworth 1970, 1972; Engelmann 1970; Tombes 1970; de Wilde 1975; Highnam & Hill 1977; Riddiford & Truman 1978; Granger & Bollenbacher 1981). Thus, the neuroendocrine system has a central role in insect diapause. And, it not only mediates the obvious expressions of diapause, such as reduced metabolism (and thus delayed growth, development, and reproduction), but it also functions in the expression of the entire diapause syndrome, including prediapause behavior patterns, migration, cold-hardiness, and diapause-mediated seasonal polyphenism (e.g., de Wilde *et al.* 1968; de Wilde & de Boer 1969; Nijhout & Williams 1974; Chippendale 1978; Shapiro 1976; Rankin 1978; Ivanovic *et al.* 1979; Nijhout & Wheeler 1982).

Despite many recent advances in insect endocrinology, the endocrine control of diapause has been studied in only few species and under laboratory conditions that may not always be ecologically meaningful. Therefore, the connection between naturally occurring environmental stimuli and neuroendocrine function during diapause in the field should be inferred with care. With this word of caution, we present an overview of the neuroendocrine control of diapause and its symptoms in various life-cycle stages. We do not give details of any specific studies, but show the diversity of neuroendocrine mechanisms involved in insect diapause.

In simplest terms, the endocrine system of insects consists of four main parts: neurosecretory cells in the brain, the corpora cardiaca, the corpora allata, and the prothoracic glands (Wigglesworth 1970; Engelmann 1970; Highnam & Hill 1977; Raabe 1982; Downer & Laufer 1983; Gupta 1983). The neurosecretory cells of the brain and the corpora cardiaca together constitute the cerebral neurosecretory system, with the corpora cardiaca functioning, in addition to other ways, as a neurohaemal organ. Hormones (brain hormone or prothoracicotropic hormone) from the neurosecretory cells of the brain are stored in the corpora cardiaca and/or the corpora allata and, when released into the blood, these neurohormones stimulate (a) the corpora allata, (b) the prothoracic glands, and (c) the ovaries. Additionally, the corpora cardiaca not only store and release brain hormone (prothoracicotropic hormone) but they also produce their own hormone.

During the course of development, the cerebral neurosecretory cells function to initiate the molting process, and ultimately ecdysone from the prothoracic glands and juvenile hormone from the corpora allata act together to produce a characteristic molt. The type of molt, whether from larva to larva, larva to pupa, or pupa to adult, is a function of juvenile hormone titres. In general, the prothoracic glands are absent from adult insects; the cerebral neurosecretory cells and the corpora allata, which are reactivated to produce juvenile hormone after adult development, regulate sexual maturation in males and females (e.g., Pener 1977).

On the basis of ablation experiments, which produced diapauselike symptoms, pioneering researchers generally concluded that the diapause state results from a suppression of the hormone secretion necessary for growth, development, molting, and reproduction and that diapause ensues because certain neuroendocrine organs are inactive. This conception was termed the "deficiency syndrome." Although there can be no doubt that during diapause there are relative deficiencies in some hormones that promote growth and reproduction, the deficiency syndrome never satisfactorily described how all the symptoms of diapause are achieved and how their development is timed over the course of diapause. It is certainly an unsatisfactory explanation for egg diapause in which a specific hormone is responsible for diapause. Moreover, it does not describe the neuroendocrine bases for those cases of larval diapause in which the *level* of endocrine function, not its absence, is responsible for diapause (see sections below). A revised view takes into account the positive role played by the neuroendocrine system in controlling diapause—a state that is not merely the lack of growth or reproduction. Diapause is a dynamic condition with highly evolved and delicately controlled component processes.

Summaries of the neuroendocrine basis for insect diapause have also sometimes given an oversimplified impression that diapause in each life stage is mediated by a *single* mechanism. For example, it is frequently stated that diapause is controlled in (a) the embryo, by a diapause hormone from the mother, (b) the larva, by a high juvenile hormone level (or the lack of growth-promoting hormone), (c) the pupa, by the lack of prothoracicotropic hormone, and (d) the adult, by the lack of juvenile hormone. Although these statements may hold true for some species, recent experimentation with a variety of other species provides evidence for wide diversity in the neurohormonal mechanisms involved in diapause regulation (e.g., see Bowen *et al.* 1984).

Perhaps pupal diapause best illustrates the variety of diapause-controlling mechanisms. The breaking of pupal diapause was long considered to be closely associated with low-temperature "reactivation" of the brain; chilling was shown to lead to the release of neuroendocrine material (see Section 4.4.2). However, recent work shows that chilling is not always necessary for termination of pupal diapause and that more than one neurohormonal mechanism subserves the termination of pupal diapause (see below). Similarly, work on egg, larval, and imaginal diapause has shown significant variations in the neuroendocrine mechanisms underlying diapause (see below). These variations carry considerable biological significance because diapause has evolved independently numerous times. The substantial similarities in the neuroendocrine mechanisms controlling diapause in distantly related groups probably resulted from convergent evolution, that is, natural selection acting on common organ systems that are regulated by relatively few reliable environmental stimuli.

4.1.1 Adult

A common feature underlying imaginal diapause is the reduced functioning of the corpora allata and the involvement of the neurosecretory cerebral complex (Pener 1970, 1974; de Wilde & de Loof 1973; Riddiford & Truman 1978; Poras 1978; Kono 1980; de Wilde 1983; Engelmann 1984). In general, diapausing adult insects fail to develop eggs because the corpora allata produce greatly reduced amounts of juvenile hormone, and/or because cerebral neurosecretory cell function is suppressed. For example, in the locust *Anacridium aegyptium*, diapause is associated with short day length, reduced activity of neurosecretory cells of the pars intercerebralis, and small corpora allata (Girardie & Granier 1973a, b). Because production of juvenile hormone by the corpora allata is necessary for sexual development in adult insects, in some cases, diapause can be terminated by the topical application of juvenile hormone or its mimics (e.g., Bowers & Blickenstaff 1966; Kambysellis & Heed 1974; Case *et al.* 1977; Schooneveld *et al.* 1977; Ascerno *et al.* 1981). It should be noted that the neuroendocrine control of adult diapause has been studied only for emerged and hardened adults; diapause in pharate adults (e.g., see Sahota *et al.* 1982) may have different neuroendocrine bases.

Neuroendocrine regulation of adult diapause in the Colorado potato beetle, *Leptinotarsa decemlineata*, has been the subject of intensive investigations by de Wilde and his co-workers at Wageningen. Over the years they have led the way in illustrating the neuroendocrine and biochemical basis for the various changes that comprise the adult diapause syndrome. Much of their work is summarized in recent reviews (e.g., de Kort *et al.* 1980, 1981; de Kort & Granger 1981; de Wilde 1983). Below are a few highlights.

Short day lengths induce diapause in Colorado potato beetle adults, and this diapause is characterized by distinct physiological and behavioral changes, as well as suppressed metabolism. Under long day conditions, all behavioral and developmental symptoms of diapause are achieved with removal of the corpora allata, but diapause termination and sustained postdiapause oocyte growth require the pars intercerebralis, the corpora allata, and long day lengths (Schooneveld *et al.* 1977; de Kort *et al.* 1980; de Wilde 1983). Corpora allata function appears to be controlled by stimulating neurohormones from the brain (de Wilde & de Boer 1969); the brain does not appear to inhibit the corpora allata through neuronal connections (de Kort *et al.* 1980, 1981; de Kort & Granger 1981).

During the prediapause, diapause, and postdiapause periods, corpora allata volume in the Colorado potato beetle undergoes characteristic changes, and these changes are correlated with alterations in juvenile hormone titre (de Wilde *et al.* 1968; de Wilde *et al.* 1971; Schooneveld *et al.* 1977). Although bioassays indicate that titres of juvenile hormone are greatly reduced during diapause (de Kort *et al.* 1982), even very high doses cannot terminate diapause in short-day beetles (de Kort *et al.* 1980). Furthermore, there appears to be a correlation between juvenile hormone titres and various symptoms of diapause in the Colorado potato beetle, including pre- and postdiapause movement and wing muscle histolysis and development (de Kort 1981). Juvenile hormone titre in the hemolymph can be influenced by a number of factors: the rate of biosynthesis and release by the corpora allata, bonding of juvenile hormone to protein during transport, rates of degradation, uptake, and binding to tissues, catabolism within tissues, and excretion. The receptor tissues also may play a significant role in determining juvenile hormone function. All of the above aspects are under study in the Colorado potato beetle (de Kort 1981).

As in the Colorado potato beetle, diapause in the bug *Pyrrhocoris apterus* involves both the corpora allata and the neurosecretory cells (Sláma 1964). In this species, not only do short day lengths exercise control over the corpora allata, through an absence or reduction of neuroendocrine secretion from the brain, but unlike the case in the Colorado potato beetle, nerves from the brain may directly inhibit the corpora allata during diapause (Hodková 1976, 1982). Inhibition of the corpora allata by the brain appears to occur by a similar mechanism in *Tetrix undulata* (Poras 1982).

Diapause is induced in the phytophagous lady beetle *Epilachna vigintioctopunctata* by short day lengths experienced during the first five days of adult life. After the beetles pass their fifth day, they are no longer sensitive to day length, but prediapause feeding continues for approximately another 11 days before diapause ensues. Therefore, the prediapause stage has two phases: a photoperiodically sensitive phase and an insensitive phase. Both the corpora allata and the neurosecretory cells are involved in diapause in *E. vigintioctopunctata* and their activities correlate with the two prediapause phases (Kono 1980). Apparently short day lengths perceived during the first phase cause the secretory material to accumulate in the neurosecretory cells during the second phase. The fat body also develops during the second phase and it may have a role in further decreasing the neurosecretory activity (Kono 1979).

In the carabid, *Pterostichus nigrita*, termination of reproductive diapause occurs in two steps. The first step, previtellogenesis, is induced by short days, and the final step of yolk deposition, vitellogenesis, requires a subsequent period of long days (Thiele 1966). The two steps are correlated with different levels of corpora allata activity (Hoffmann 1970; Ferenz 1977).

In species in which migration or polyphenism occurs as a part of imaginal diapause, there is an association between the hormonal mechanisms controlling the cessation of development and reproduction and those controlling migratory behavior and polyphenism (see reviews by Kennedy 1961; Truman & Riddiford 1974; Shapiro 1976; Rankin 1978; Nijhout & Wheeler 1982; Rankin & Singer 1984; see also Lees 1983). For example, in the milkweed bug, *Oncopeltus fasciatus*, high levels of juvenile hormone lead to reproduction, whereas intermediate levels result in increased flight behavior. Very low levels inhibit both migration and ovarian development (Rankin 1974, 1980; Rankin & Riddiford 1978). A similar endocrine mechanism may also occur in the ladybird beetle, *Hippodamia convergens* (Rankin & Rankin 1980). In the monarch butterfly, *Danaus plexippus*, long day lengths and high juvenile hormone titres are associated with rapid posteclosion reproductive development, whereas short day lengths and decreased production of juvenile hormone result in adult diapause (Herman *et al.* 1981; Herman 1981; Lessman & Herman 1983). The effect on migratory behavior of neither day length nor juvenile hormone is known, but juvenile hormone metabolism is modulated, in part, by flight (Lessman & Herman 1981); an adipokinetic hormone from the corpora cardiaca may be involved (Dallmann & Herman 1978).

An especially interesting example of the hormonal regulation of seasonal polyphenism associated with diapause is found in the aphid *Megoura viciae.* Short day lengths, acting on parthenogenetic viviparous females, stimulate the production of sexual forms that mate and produce diapausing eggs. Photoperiod regulates the release of neurosecretory material from the Group I neurosecretory cells in the brain, and it is possible that this material is delivered by axonal projections to abdominal tissues where it may influence the type of morph produced (Steel & Lees 1977; Steel 1976, 1977; see also Kats 1982). The corpora allata are also involved in determining gamic and parthenogenetic reproduction, and it is possible that the Group I cells act to control juvenile hormone activity.

The role of juvenile hormone in aphid seasonal polyphenism resembles the role it plays in diapause and migration in other insects. Under long days of summer, juvenile hormone levels in *Myzus persicae, Aphis fabae*, and *Megoura viciae* presumably are high. At this time, increased levels of juvenile hormone do not influence the production of either winged or apterous females, and juvenile hormone apparently is not involved in the production of summer alates (Hardie 1980a). Under short day lengths, however, the juvenile hormone level falls and the seasonal sequence of morphs is initiated—that is, in *Megoura viciae* oviparae and males are induced, and in *A. fabae* and *M. persicae* winged gynoparae and males are induced. At this time both long days and juvenile hormone can (a) inhibit the production of oviparae by viviparae (Mittler *et al.* 1976; Lees 1978, 1980; Hardie 1980a, b, 1981b), (b) induce the production of viviparae at the end of a sequence of gynoparae (*A. fabae*) (Hardie 1981a), (c) inhibit male production (*M. persicae*) (Mittler *et al.* 1979), and (d) produce apterization in normally alate gynoparae (*A. fabae*) (Hardie 1981b). Thus, it was proposed that long day lengths can raise the levels of juvenile hormone (Hardie 1981a, b). This effect is analogous to that in *L. decemlineata* and *O. fasciatus* (Hardie 1981b).

4.1.2 Pupa

After the pioneering work on silkworms by Williams and co-workers, it was generally held for many years that pupal diapause is universally caused by a deficiency in prothoracicotropic (brain) hormone. However, studies within the last five years or so have demonstrated a wide variety of neurohormonal mechanisms involved in pupal diapause, and they underscore the need for work with diverse insect species to demonstrate the full range of possible mechanisms.

In the classic example, pupal diapause of the giant silkworm, *Hyalophora* (= *Platysamia*) *cecropia*, results from failure of the neurosecretory cells in the brain, or the corpora cardiaca, to release prothoracicotropic hormone, and thus the prothoracic glands do not release ecdysone, the molting hormone (Williams 1952). Implantation of brains that have been "reactivated" during exposure to low temperature brings diapause to an end in the dormant pupa; hormone from the "reactivated" brain stimulates the prothoracic glands to produce ecdysone, which results in the molt to the adult.

The hormonal mechanism for terminating pupal diapause in *Antheraea pernyi* is similar to that in the giant silkworm. However, in this species both low temperature and photoperiod have a role in bringing diapause to an end; these environmental factors "reactivate" the neurosecretory cells in the brain to produce prothoracicotropic hormone, which in turn acts on the prothoracic glands to release ecdysone (Williams & Adkisson 1964). A somewhat different mechanism occurs in the noctuid *Acronycta rumicis* in which cold apparently acts directly on both the brain and the prothoracic glands during "reactivation" (Kind 1977, 1978). Cold also acts on tissues other than the brain during pupal diapause in the swallowtail *Papilio xuthus* because chilling shortens pupal duration in brainless individuals (Numata & Hidaka 1984).

Working with another species of Lepidoptera, the bollworm, *Heliothis zea*, Meola and Adkisson (1977) showed that pupal diapause is not caused by a deficiency of prothoracicotropic hormone from the brain because the pupal brain secretes prothoracicotropic hormone very soon after the pupal molt, regardless of whether or not the insect will enter diapause. In contrast to silkworm diapause, diapause in the bollworm is maintained by a deficiency in a prohormone (α-ecdysone) that is released from the prothoracic glands. Release of α-ecdysone is prevented when diapausing bollworms are maintained at 21°C, but at 27°C α-ecdysone is released and development proceeds. Recent work indicates that the temperature-sensitive mechanism controls the availability of a humoral factor that is necessary for ecdysone production (Meola & Gray 1984). A similar temperature inhibition of ecdysone production during diapause may occur in *Heliothis punctigera* (Browning 1979, 1981), *Heliothis virescens* (Loeb & Hayes 1980b; Loeb 1982), and the tobacco hornworm, *Manduca sexta* (Bradfield & Denlinger 1980; Bowen *et al.* 1984).

Pupal diapause in *Sarcophaga* flesh flies also differs both ecophysiologically and neuroendocrinologically from pupal diapause in the silkworm. As stated above, diapause development in silkworm pupae proceeds most quickly at low temperatures. However, in flesh flies, as in many other Diptera with pupal diapause, the rate of diapause development is directly related to temperature (see Denlinger 1981a; Tauber *et al.* 1982). Diapause in both flesh fly pupae and silkworm pupae is correlated with a reduction in molting hormone titres (Ohtaki & Takahashi 1972; Walker & Denlinger 1980; see also Giebultowicz & Saunders 1983). However, in contrast to the situation in diapausing silkworm and hornworm pupae (Bradfield & Denlinger 1980; Denlinger *et al.* 1984), juvenile hormone activity in *Sarcophaga crassipalpis* persists during the initial phases of diapause and a sharp increase in juvenile hormone titre precedes the rise in molting hormone that initiates postdiapause adult development. Juvenile hormone apparently does not play a direct role in terminating diapause. Rather it has been proposed to influence the receptivity of tissues to growth promoting hormones or conditions (see work by Flanagan & Hagedorn 1977; Grzelak *et al.* 1981) or to aid in the measurement of time during diapause (Walker & Denlinger 1980).

The prepupal (pharate pupal) diapause of the slug moth, *Monema flavescens*, also is associated with high levels of juvenile hormone in the hemolymph, and it appears that juvenile hormone in this species inhibits the release of neurosecretory materials from the B cells of the pars intercerebralis (Takeda 1972, 1978b). Diapause termination is associated with the release of substances from these cells (Takeda 1977, 1978b).

Juvenile hormone may also be involved in pupal diapause in *Mamestra brassicae.* In this species, as in others, pupal diapause is characterized by deficiencies in prothoracicotropic hormone and ecdysone (molting hormone). However, application of juvenile hormone to last-instar larvae prevents diapause (Hiruma 1979), apparently by activating the prothoracic glands to release ecdysone. This, in turn, may stimulate prothoracicotropic hormone activity in the brain and induce adult development. Thus, although a deficiency of prothoracicotropic hormone occurs with diapause, this deficiency may be a result of, rather than the cause of, the growth-preventing ecdysone deficiency (Hiruma 1979). The above studies and others (Nijhout & Williams 1974; Safranek *et al.* 1980; see larval section later) point to a changing role for juvenile hormone during the prediapause, diapause-induction, and diapause-maintenance periods of dormancy.

Still another neuroendocrine mechanism has been suggested for pupal diapause in *Mamestra* (= *Barathra*) *brassicae*; this one involves ecdysone titres. This species undergoes both aestival and hibernal diapause in the pupal stage. The winter diapause is much more intense (persisting for about eight months at 18°C) than the summer diapause (which lasts for about 1.5 months at the same temperature). Three concentrations of ecdysone typify the three ecophysiological categories: development without diapause, weak aestival diapause, and intense hibernal diapause, and the hormone levels correspond directly to the metabolic level (Maslennikova *et al.* 1976). However, in the experiments with *M. brassicae*, the ecdysone titres were determined for pupae of a uniform age (10 to 13 days old), and differences in rates of adult differentiation alone could account for the differences in ecdysone titres. More studies are needed to establish if there is a relationship between ecdysone titre and diapause.

4.1.3 Larva

It was long considered that larval diapause, like pupal diapause, results from the temporary deficiency of growth-promoting hormones. However, the general concept of larval diapause based solely on hormonal deficiency was questioned in the past and has recently been reevaluated. Reduced activity by the neurosecretory cells of the brain is usually associated with larval diapause (e.g., Schaller *et al.* 1974; Chippendale 1977, 1983; Sieber & Benz 1980b; Bean & Beck 1980). However, there is considerable variability in the function of the other neuroendocrine elements, and evidence indicates that larval diapause may be regulated by the relative levels, and not just the absence (or extremely low level), of growth-promoting hormones.

Currently, there are two general hypotheses explaining the neuroendocrine regulation of larval diapause: (l) the corpora allata remain inactive during larval diapause; diapause occurs because the neuroendocrine complex in the brain is inactive and therefore does not stimulate the ecdysial glands, and (2) diapausing larvae have active corpora allata that secrete juvenile hormone, and diapause occurs because juvenile hormone inhibits the synthesis, transport, and/or release of prothoracicotropic hormone. In this case the neuroendocrine complex of the brain presumably remains active during diapause.

Relatively recent findings provide examples that support both of these hypotheses. In some insects, larval diapause induction and maintenance are correlated with the continuous presence of juvenile hormone (Nair 1974; Yagi & Fukaya 1974; Nijhout & Williams 1974; Yin & Chippendale 1976; Sieber & Benz 1977, 1980a; Scheltes 1978a). In others, juvenile hormone titres are low during diapause (Ismail & Fuzeau-Braesch 1976b; Ismail *et al.* 1984). Evidence for the continuous presence of juvenile hormone during larval diapause first came from the rice stem borer, *Chilo suppressalis*. In this species juvenile hormone is required for the induction and maintenance of diapause, during which time ecdysone levels remain low. Diapause intensity can be estimated from the juvenile hormone titre, and diapause is associated with high corpora allata activity, which gradually decreases near the end of diapause (Fukaya & Mitsuhashi 1961; Mochida & Yoshimeki 1962; Yagi & Fukaya 1974; Yagi 1981). Juvenile hormone and ecdysone also appear to regulate hemolymph concentrations of glycerol and glycogen—two factors that are very important to cold-hardiness and postdiapause growth (Tsumuki & Kanehisa 1981a; see later).

In the southwestern corn borer, *Diatraea grandiosella*, both the brain and the corpora allata are involved in diapause induction and maintenance. The incidence of molting by pre- and early diapause larvae is affected by brain implants from nondiapause larvae, whereas late-diapause larvae, with active brains, are not affected. Removal of the corpora allata from pre- or early diapause larvae results in a high incidence of pupation. Thus, brain activity is reduced and corpora allata activity is maintained during diapause (Yin & Chippendale 1976, 1979a; Chippendale & Yin 1976). During diapause induction, juvenile hormone titre does not decline to the low level required for pupal differentiation. Once diapause is induced, the juvenile hormone level stabilizes at a level slightly lower than that in nondiapause larvae, but sufficiently high so that any molts during this period are "stationary" molts. Juvenile hormone concentrations decrease at the end of diapause to levels low enough to permit pupal development (Yin & Chippendale 1979a). During diapause the corpora allata continue to secrete juvenile hormone and hydrolysis of juvenile hormone by the fat body, a function that is influenced by juvenile hormone esterases, is reduced (Mane & Chippendale 1981). Intact nervous connections between the brain and the corpora allata are necessary for maintenance of juvenile hormone secretion (Yin & Chippendale 1979a). This suggests that the brain stimulates the corpora allata, at least through the middle of diapause. Although the specific sites for the synthesis and storage of juvenile hormone are not known, several intracellular structures of the corpora allata undergo characteristic changes during diapause (Yin & Chippendale 1979b).

Juvenile hormone is also involved in regulating larval behavioral patterns in *D. grandiosella* during the prediapause and diapause periods (Chippendale 1978). And it may function to stimulate the synthesis of a diapause-associated, low molecular weight, protein that is stored in the fat body until middiapause and in the synthesis of a high molecular weight lipoprotein that occurs in the hemolymph and that may have a role in the mobilization of energy reserves during diapause (Turunen & Chippendale 1979, 1980; Chippendale & Turunen 1981; Dillwith & Chippendale 1984).

As in *C. suppressalis* and *D. grandiosella*, both the brain and corpora allata appear to be involved in larval diapause in the codling moth, *Laspeyresia pomonella*, and the European corn borer, *Ostrinia nubilalis*. In these two species, however, diapause occurs in somewhat more mature larvae than in *D. grandiosella* or *C. suppressalis*, and juvenile hormone is not the primary regulator of diapause. Diapause induction in *L. pomonella* requires both short day lengths and high titres of juvenile hormone in the preceding larval stage. Either alone cannot induce diapause, whereas in *C. suppressalis* and *D. grandiosella*, repeated topical application of juvenile hormone or its mimics alone can induce a diapause state. Furthermore, although a high juvenile hormone titre is characteristic of prediapause *L. pomonella*, no juvenile hormone or ecdysteroid can be detected in diapausing larvae, and tissues isolated by ligature from the juvenile hormone source remain in diapause (Sieber & Benz 1980a, b; Friedlander 1982). Thus, although juvenile hormone is necessary to induce diapause in *L. pomonella*, some other factors or a general state of neuroendocrine inactivity maintains it.

Similarly, little or no juvenile hormone can be detected in diapausing *O. nubilalis* larvae, and topical applications of juvenile hormone or its mimics do not induce or maintain diapause (Chippendale & Yin 1979; Bean & Beck 1980). However, in this species, as in *L. pomonella*, juvenile hormone may have a role in prolonging prediapause feeding (Bean & Beck 1980). It is noteworthy that during the prediapause period, both juvenile hormone and juvenile hormone esterase levels are high in *O. nubilalis* (Bean *et al.* 1982), and it appears that diapause in this species intervenes just after juvenile hormone esterases are produced, but before they eliminate juvenile hormone. Recent evidence also suggests that prothoracicotropic hormone release is suppressed during diapause induction and that low levels of ecdysteroids are involved in inducing and maintaining diapause (Bean & Beck 1983; Gelman & Woods 1983).

4.1.4 Egg

Diapause can interrupt development within the egg at a number of stages during embryogenesis. In some species, it ensues very early, before any signs of an embryonic endocrine system are visible. In other species, it occurs just prior to hatching when all larval systems presumably are formed and capable of functioning. Thus, egg diapause may be based on a wide variety of neuroendocrine or other mechanisms.

Two major categories of neuroendocrine control have been associated with egg diapause: one involving a diapause hormone from the mother and the other involving the neuroendocrine system of the embryo itself.

The source and function of maternal diapause hormone have been identified in two lepidopteran species, the commercial silkworm, *Bombyx mori*, and the moth *Orgyia antiqua* (see Yamashita 1983). Both species share a common mechanism in that the hormone produced by the mother's suboesophageal ganglia is transported to eggs in the ovary and causes diapause. A similar mechanism may also occur in the mirid *Adelphocoris lineolatus*, but the cells involved in neurosecretion may be different (Ewen 1966). Silkworms that have been implanted with suboesophagial ganglia from the gypsy moth, *Lymantria dispar*, the giant silk moths, *Antheraea pernyi* and *A. yamamai* (Fukuda 1951; Hasegawa 1952), the noctuid moth, *Phalaenoides glycinae* (Andrewartha *et al.* 1974), the cockroach, *Periplaneta americana* (Takeda 1977), and the armyworm, *Leucania separata* (Ogura & Saito 1973), produce diapause eggs, and it is concluded that the production of "diapause hormone" is not restricted to the silkworm. However, it should be noted that none of the donor species except *L. dispar* have an egg diapause. *L. separata* is a nondiapause species, and the others diapause as pupae. The physiological function of "diapause hormone" in these species has not been determined.

In *B. mori*, diapause occurs in eggs laid in autumn by females that experienced long days (≥LD 14:10) and high temperatures (25°C) when they were eggs and young larvae. Short day lengths act on the brain of the developing female to suppress production of diapause hormone by two large neurosecretory cells in the suboesophageal ganglia (Fukuda & Takeuchi 1967). When eggs within the developing ovaries of female pupae are exposed to high concentrations of diapause hormone, they enter diapause. Low titres lead to the avoidance of diapause. The inhibiting action of the brain on the suboesophageal ganglia is achieved through nervous connection (Hasegawa 1952, 1963; Fukuda 1951, 1952, 1953, 1963; Morohoshi 1959; Morohoshi & Oshiki 1969; see summaries in Wigglesworth 1970; Highnam & Hill 1977; Yamashita 1983). Thus, the activity of the pupal brain and suboesophageal ganglia is determined by the photoperiod experienced during the egg and early larval stage, and the effect (diapause) of this activity is mediated during the pupal stage but not expressed until the following egg stage.

Diapause hormone in *B. mori* has been purified and shown to consist of two kinds of peptides (A and B)(Sonobe & Ohnishi 1971). The hormone is extractable from the suboesophageal ganglia of mature larvae, pupae, and adults, but its effects are restricted to developing oocytes within certain pupal stages (Yamashita & Hasegawa 1970). It appears that the activity of the hormone depends on both the physiological stage of the target organ (the ovaries) (Yamashita & Hasegawa 1970) and the persistence of the hormone *in vivo* (Yamashita *et al.* 1980). Diapause hormone cannot be extracted directly from ovaries, but recent work suggests that diapausing eggs contain the diapause hormone in a bound form (Kai & Kawai 1981).

Further biochemical studies have shown that trehalase in the ovary is a target enzyme for the diapause hormone and that activity by the hormone results in the accumulation of glycogen in eggs. During diapause induction, glycogen is converted to sorbitol and glycerol, but at the end of diapause these polyols are reconverted to glycogen, which serves as an important source of energy for postdiapause embryonic development (Yaginuma & Yamashita 1979; Yamashita *et al.* 1981; Isobe & Goto 1980).

Apparently the diapause hormone persists in its bound form in the diapausing eggs and has a role in suppressing the diapause-terminating effects of esterase A (Kai & Nishi 1976; Kai & Haga 1978). Esterase A activity returns within 30 minutes of treatment with hydrochloric acid (a diapause-terminating stimulus); chilling also increases esterase A activity before any other symptoms of diapause termination become apparent. Thus, it is proposed that diapause termination involves the direct effect of chilling on esterase A activity (Kai & Hasegawa 1973; Kai & Nishi 1976; Kai & Haga 1978; Kai *et al.* 1984). Nucleolar size is also affected directly by chilling, and this too may have a role in diapause termination (Kurata *et al.* 1979a, b). Following the increase in esterase A activity and the increase in the size of egg nucleoli, yolk cell lysis occurs, glycogen begins to reappear, and subsequently ribonucleic acid synthesis begins (Kurata *et al.* 1979a, b). The permeability of the chorion to oxygen may also be involved in diapause termination (Sonobe *et al.* 1979). The biochemical changes associated with these and other functions during diapause induction and termination are an area of active research, and some of the recent advances are summarized by Yamashita *et al.* (1981) and Isobe and Goto (1980).

Diapause induction in *Orgyia antiqua* is similar to that in *B. mori*, but it differs in that short, rather than long, day lengths act on late larval stages, rather than eggs and young larvae, to produce adults that lay diapausing eggs. As in *B. mori*, diapause is caused by a hormone secreted by specific neurosecretory cells in the suboesophageal ganglia, and this hormone acts on the eggs as they develop in the pupae (Kind 1969).

Diapause in advanced embryonic stages is much less well studied than that in either *B. mori* or *O. antiqua*, and much of our information is derived from knowledge of the sensitive stages and other factors rather than from neuroendocrine studies *per se*. In some species of mosquitoes and the cricket *Teleogryllus* (= *Acheta*) *commodus* the early embryo is the sensitive stage for diapause that ensues in a somewhat later embryonic stage (Hogan 1960a; Kappus & Venard 1967; Tauthong & Brust 1977; Pinger & Eldridge 1977), and thus a maternal factor is probably not involved. However, some of these species enter diapause before segmentation, and it is not known if the endocrine system has been formed and is functional in the young embryos. Another type of indirect evidence for the exclusion of a maternal effect on egg diapause comes from crosses between females of *Teleogryllus oceanicus* (nondiapausing species) and males of *T. commodus* (diapausing species). These eggs undergo diapause, while those from the reciprocal cross do not (MacFarlane & Drummond 1970). It appears, therefore, that egg diapause is influenced by the male genotype but not by endocrine conditions during oögenesis.

More direct evidence for the involvement of the embryonic neuroendocrine system in egg diapause comes from studies with the big-headed grasshopper, *Aulocara elliotti*. Diapause in this species occurs prior to blastokinesis and yolk engulfment; it can be terminated by application of a juvenile hormone analogue. Thus it appears that the egg diapause of this species is regulated by an embryonic neuroendocrine system (Visscher 1976).

The discovery of ecdysone and changes of its titre in diapausing eggs of *B. mori* by Ohnishi *et al.* (1971) also suggests that the hormonal system of the very early embryo is functional. They detected a high titre of ecdysone (0.16 to 0.24 μg/g) from eggs about 24 hours after deposition, when the endocrine system has not yet been formed. The titre decreases to 0.02 to 0.097 μg/g during diapause, and increases again to 0.23 μg/g within 72 hours after diapause termination by treatment with HCl. Low ecdysteroid levels are also present in diapausing eggs of the cochineal *Lepidosaphes* (Gharib *et al.* 1981).

And, finally, evidence for the involvement of brain hormone in egg diapause comes from the observation of neurosecretory activity in the brains of chilled, diapausing larvae within eggshells of the gypsy moth, *Lymantria dispar* (Loeb & Hayes 1980a).

4.1.5 Translation of environmental stimuli into neuroendocrine function

Although the neuroendocrine mediation of diapause is well established, many unanswered questions remain. These questions involve not only the diverse functions of the neuroendocrine system but the means whereby photoperiodic and other environmental stimuli are translated into neuroendocrine function at the appropriate time in the life cycle. Equally important are the mechanisms through which the neuroendocrine system affects body tissues to cause the full range of metabolic, behavioral, physiological, and morphological expressions of diapause. These questions have become the target of considerable physiological research. It is beyond the scope of this book to review these investigations. However, we briefly mention the major issues involved and provide pertinent references.

Briefly, translation of environmental stimuli into metabolic function involves three main steps: (a) perception of photoperiodic (or other environmental) stimuli, (b) accumulation and retention (storage) of the information, and (c) use of the stored information. As for the first step, experimental studies have delineated the sites of photoperiodic receptors in some species, and most are located in the brain (e.g., de Wilde *et al.* 1959; Lees, 1964; Williams & Adkisson 1964; Claret 1966; Steel & Lees 1977; Kono *et al.* 1983; see reviews by Beck 1980; Page 1982; Saunders 1982). There are a few reports of photoperiodic receptors located in the compound eyes (Ferenz 1975b; Numata & Hidaka 1983a). Recent work also indicates that carotenoid pigments and vitamin A are involved in photoperiodic reception in some, but not all, insects and mites (see Veerman 1980; Veerman *et al.* 1983).

Accumulation, retention, and use of information, the second and third steps, involve, among other systems, the biological clock, the function of which is also centered in the brain. The role of the biological clock in diapause offers a fascinating area for research, and references to pertinent literature can be found in books and reviews by Bünning (1973), Beck (1980), Saunders (1974, 1981a, 1982), and Tyshchenko (1977, 1980). The use of stored information is also affected through biochemical processes in the brain and elsewhere in the body. In recent years these processes have received considerable attention, and in many cases polymerized biological molecules, such as polypeptides or nucleotides, have been implicated in these functions. We recommend recent reviews by Beck (1980), Tyshchenko (1980), Berry (1981), Chippendale and Turunen (1981), Denlinger (1981a), Granger and Bollenbacher (1981), and de Kort (1981).

4.1.6 Summary

To summarize briefly the environmental-neuroendocrine interaction in diapause regulation: diapause is a distinct physiological state with positive features; it is not simply the cessation of development or reproduction. Thus, diapause is brought about not merely by the absence of hormonal function. On the contrary, full development of the diapause syndrome is mediated through very positive neuroendocrine mechanisms, including (a) a diapause factor (as in the egg diapause of *B. mori* and *O. antiqua*), (b) the regulation of hormone levels (as in *Chilo* and *Diatraea* larval diapause), and (c) suppressed activity of growth-promoting neurohormones (as in *O. nubilalis* and *L. pomonella* larval diapause; *H. cecropia, A. peryni*, and *H. zea* pupal diapause; and *L. decemlineata* adult diapause).

Diversity is the key word characterizing the neuroendocrine control of insect diapause. Even among species of the same order, the numerous patterns observed reflect two very important features of diapause. First, although the diapausing stage is species-specific, diapause can intervene at almost any time within metamorphosis. Among the insects, it not only occurs in all stages, but it can take place at any time within the stage. Because the neuroendocrine system is intimately involved in regulating metamorphosis, its mode of action in diapause varies depending in part on the metamorphic stage during which diapause intervenes. In general, the particular processes involved in diapause regulation appear to coincide with those that control further development.

Second, the diversity of neuroendocrine mechanisms participating in the control of diapause indicates that diapause has evolved independently numerous times among the insects. Similarities in the neuroendocrine control of diapause among distantly related taxa probably result from convergent evolution—the effect of natural selection acting on a relatively limited range of functions by common organ systems.

4.2 Behavioral Expression of Diapause

Insects entering diapause exhibit specialized behavioral patterns that increase their abilities to overcome environmental exigencies during dormancy. These patterns include a decrease in response to feeding and reproductive stimuli, movement to and from dormancy sites, alteration in phototactic and geotactic responses, and so on. The sequence of behavior is related to neuroendocrine changes during diapause induction. For example, both diapause induction and digging behavior in the Colorado potato beetle are correlated with reduced corpora allata activity (e.g., de Wilde *et al.* 1968; de Wilde & de Boer 1969). In the southwestern corn borer, *Diatraea grandiosella*, the specialized prediapause behavior appears to be induced by a relatively high level of juvenile hormone typical of diapause in this species (Chippendale 1978). Many of these behavioral patterns serve to prepare the insect for dormancy prior to the onset of adverse conditions; and thus they are manifested very early in the course of diapause. Consequently, these characteristic behavioral patterns sometimes serve as initial, reliable symptoms for diagnosing the occurrence of diapause.

Some of the alterations may be relatively subtle, such as differences in the angle at which the antennae are held, as in cereal leaf beetles and asparagus beetles (see Wellso *et al.* 1970) or alterations in response to light, moisture, and touch as in the alfalfa weevil, *Hypera postica* (Pienkowski 1976); others are more striking, such as burrowing, cocoon formation, or migratory behavior.

One of the major behavioral changes associated with the diapause syndrome is the reduction in responsiveness to feeding and reproductive stimuli (e.g., stimuli indicating food, mates, oviposition sites, etc.). As noted below, these changes are expressed in insects both entering and undergoing diapause, as well as those preparing for or undergoing migration (Kennedy 1961).

An important, practical aspect of such diapause-related changes concerns reactions to baits and traps used to monitor insect populations. Despite the reduced or altered response to light, food, and pheromones during diapause, light traps, food baits, CO_2 traps, and pheromone traps are used to monitor the seasonal abundance of insects that undergo diapause in the adult stage (e.g., for *Drosophila*: Dobzhansky & Epling 1944; Begon 1976; Lumme *et al.* 1978, 1979; Charlesworth & Shorrocks 1980; for the bark beetle *Ips pini*: Birch 1974; for mosquitoes: Spadoni *et al.* 1974; Meyer 1977; Slaff & Crans 1981); for the boll weevil, *Anthonomus grandis*: Rummel & Bottrell 1976; Wolfenbarger *et al.* 1976; for the rice leaf roller *Cnaphalocrocis medinalis*: Wada *et al.* 1980). These techniques may be adequate to census the active, nondiapausing part of the population, but they may overlook entirely or in part any diapausing, nonreactive individuals. For example, boll weevils captured in traps do not always represent the status of field populations with respect to sex ratio and diapause (Mitchell & Hardee 1974). Nondiapause beetles tend to be overrepresented, presumably because diapause beetles are not attracted to the synthetic sex pheromone used as bait in the traps. This problem is especially acute whenever regular, within-season variability in the incidence of diapause occurs (see Chapter 7).

4.2.1 Changes in responses to feeding stimuli

Insects show a full range of responses to food during the prediapause period. Usually feeding activity increases, apparently to promote the accumulation of metabolic reserves for use during and after diapause. For example, pests such as the Colorado potato beetle, cereal leaf beetle, and alfalfa weevil often cause severe damage to crops during the prediapause feeding stages. In other species, such as *Hyphantria cunea*, feeding rates are slightly lower in prediapause than in nondiapause larvae, but the duration of the feeding stage is longer, resulting in the ingestion of larger amounts of food by prediapause larvae (Masaki 1977).

Sometimes there is considerable intrapopulation variability in prediapause feeding. Three distinct types of prediapause boll weevil (*Anthonomus grandis*) females are recognized (Phillips 1976): (1) Some emerge from the pupal stage with well-developed fat bodies and atrophied ovaries; these individuals probably do not feed or feed very little and subsequently enter diapause without undergoing reproductive development. (2) Some females undergo fat body development and ovarian resorption after emergence; these females require food, but do not reproduce before entering diapause. (3) Still other females emerge, feed, and oviposit before undergoing egg resorption, fat body development, and diapause induction.

During diapause induction, profound changes may occur in insects' responses to food (see review by Stoffolano 1974; also Wellso & Hoxie 1981).

In some species feeding ceases altogether; in others it continues at a reduced rate. Larvae of the rice stem borer, *Chilo suppressalis* (Koidsumi & Makino 1958), the moth *Dasychira pudibunda* (Geyspits & Zarankina 1963), the sugarcane borer *Diatraea saccharalis* (Roe *et al.* 1984), and the lacewing *Mallada perfecta* (Tauber & Tauber, unpublished), as well as adults of the common green lacewing *Chrysopa carnea* (Sheldon & MacLeod 1971) and the chrysomelid *Pyrrhalta humeralis* (Ogata & Sasakawa 1983), feed sporadically or at reduced rates throughout diapause. Alterations in the feeding behavior of diapausing insects may sometimes be overlooked because overt feeding activity continues, but the rate of ingestion diminishes. For example, experimental studies in the laboratory were necessary to show that diapausing aquatic larvae of the pitcher plant midge, *Wyeomyia smithii*, ingest food particles at a much slower rate than do nondiapausing larvae (Evans & Brust 1972).

Feeding during hibernal diapause usually occurs only under relatively mild temperature conditions, and this can give the impression that the insects are in a thermally controlled quiescence, rather than diapause (e.g., see Roe *et al.* 1984). Reduced feeding is only one symptom of diapause, and diagnosis of diapause termination requires examination of a full complement of characters.

Among social insects, many species make and store honey for use during periods of drought or cold (see Section 6.2). This phenomenon has been observed not only among species, like *Apis mellifera*, that remain in large perennial nests during winter, but also in a Texas population of *Polistes annularis* in which females abandon their nests and overwinter in sheltered hibernacula (Strassman 1979). Females return to their nests on warm winter days, feed on the stored honey, and cooperate with sisters in defending their food reserves against use by nonsisters. Honey deprivation reduces winter survival and results in small spring nests.

Blood-sucking insects and ticks exhibit a wide range of variability in their feeding responses before, during, and after diapause. Among the anopheline mosquitoes that overwinter as diapausing adults, some species (*Anopheles maculipennis messeae, A. lyrcanus, A. punctipennis*) are never observed taking a blood meal during diapause (Vinogradova 1960; Washino & Bailey 1970). Diapausing individuals of other species (*Anopheles labranchiae atroparvus, A. sacharovi, A. superpictus, A. freeborni*) may take a full or an occasional blood meal without developing mature eggs (Vinogradova 1960; Clements 1963; Washino 1970, 1977). Blood feeding by these species does not appear to be necessary for successful overwintering (Washino *et al.* 1971; Washino 1977); those that do not feed on blood presumably rely on other food sources (e.g., nectar) for prediapause fat body development. A similar wide range of variability occurs among the ticks that undergo imaginal diapause (Belozerov 1982).

Anopheline mosquitoes that cease reproductive development and enter diapause, despite continued blood feeding, are said to undergo "gonotrophic dissociation," whereas "gonotrophic concordance" may have two different expressions. Both nondiapausing females that feed on blood and subsequently develop mature eggs, and diapausing females that do not feed on blood and do not develop mature eggs, are in a condition of "gonotrophic concordance" (see Washino 1977; also Clements 1963 for criticism of the terms). Several types of conditions intermediate between gonotrophic dissociation and concordance occur (Washino 1977), and some species (e.g., *A. punctipennis*) seem to vary geographically in their ability to undergo gonotrophic dissociation (Washino & Bailey 1970; Magnarelli 1979).

Although less common, gonotrophic dissociation also occurs in culicine mosquitoes (Bellamy & Reeves 1963; Eldridge & Bailey 1979). Some laboratory-reared prediapause females of *Culiseta inornata* from Edmonton, Canada, will take blood meals without undergoing ovarian development (Hudson 1977). Females collected from the field in autumn will feed on blood, but meals are small and no ovarian development occurs (Hudson 1979). However, intraspecific variability in this character occurs; a southern California population of *C. inornata* that undergoes aestival diapause does not blood feed during diapause, nor does it exhibit gonotrophic dissociation (Barnard & Mulla 1977).

Although short day lengths inhibit blood feeding in *Culex tritaeniorhychus, C. tarsalis*, and *C. pipiens* (Eldridge 1963, 1968; Sanburg & Larsen 1973), diapausing *C. pipiens* and *C. tarsalis* have been induced to feed on blood in the laboratory without developing mature eggs (Eldridge & Bailey 1979; Arntfield *et al.* 1982). Host-seeking behavior in these species is altered during diapause and females bite only if they are placed in close proximity to a host (Mitchell 1981, 1983; see also Stoffolano 1974).

On the basis of the above, we generalize that diapause can be associated with both gonotrophic dissociation and gonotrophic concordance. Presumably all individuals undergoing gonotrophic dissociation and those nonfeeding individuals in gonotrophic concordance are in a state of diapause. Thus, gonotrophic dissociation cannot be equated with diapause, but it can be an important aspect of diapause. Certainly, gonotrophic dissociation and reduction in response to feeding stimuli are good indicators of diapause in adult mosquitoes and other blood-feeding arthropods.

It is important to consider here the relevance of blood feeding and gonotrophic dissociation to the study of insect-borne diseases (see Reeves 1974; Spadoni *et al.* 1974; Eldridge & Bailey 1979; Arntfield *et al.* 1982). First, the seasonal inhibition of blood feeding by insect vectors probably represents the major factor influencing the seasonal occurrence of insect-borne diseases. For example, mosquito-borne viral encephalitis and malaria have very distinct seasonal periods of occurrence; decrease in the incidence of disease is markedly associated with decrease in biting by the vectors, although the vectors may still be present around human habitations. Second, it is possible that hibernating female mosquitoes serve as reservoirs for overwintering viruses (Eldridge 1963, 1966, 1968; Bellamy & Reeves 1963; Reeves 1974; Eldridge & Bailey 1979). Some studies have cited the presence of nulliparous females as evidence that blood feeding has not occurred and thus that the mosquitoes could not be harboring virus. Clearly, such evidence is valid only if gonotrophic dissociation does not occur in hibernating females of natural populations (Eldridge 1966).

Changes in feeding behavior can indicate diapause termination for species that undergo diapause as adults or as free-living immatures. For instance, readiness to take a blood meal has been used to show the end of diapause in natural populations of the tick *Dermacentor andersoni* (Wilkinson 1968; see Belozerov 1982 for other examples from the ticks) and the mosquitoes *Culex tarsalis, C. pipiens, Anopheles freeborni*, and *A. punctipennis* (Bellamy & Reeves 1963; Spielman & Wong 1973b; Washino 1970; Washino & Bailey 1970). Other changes in feeding behavior, such as an alteration in tarsal response, as shown in laboratory studies with the face fly, *Musca autumnalis* (Stoffalano 1968), can also indicate diapause termination. Nevertheless, the accuracy of these behavioral changes as indices of diapause termination in nature depends on how closely timed they are to reactivation of the endocrine system.

4.2.2 Changes in responses to reproductive stimuli

Although either mated or unmated adults can undergo diapause, mating itself usually does not occur during diapause. Examples of species with suppressed reproductive behavior during diapause are numerous and include the sweet clover weevil, *Sitona cylindricollis* (Herron 1953), the grasshopper, *Oedipoda miniata* (Orshan & Pener 1979a, b), the drosophilids, *Scaptomyza pallida* and *Drosophila nipponica* (Toda & Kimura 1978), the sunflower beetle, *Zygogramma exclamationis* (Gerber *et al.* 1979), and others. In *C. carnea*, neither diapausing males nor females mate even when paired with receptive, nondiapausing partners (Tauber & Tauber unpublished). In the cereal leaf beetle, *Oulema melanopus*, and the alfalfa weevil, *Hypera postica*, neither prediapause nor diapausing females mate, and although there are reports that diapause-destined males may mate and transfer sperm when presented with a receptive female (Connin & Hoopingarner 1971), there are data indicating that male reproductive behavior and the ability to inseminate females is reduced during diapause (Ascerno *et al.* 1978; Wellso & Hoxie 1981; Barnes 1976). Diapause-destined males and females of the Mexican bean beetle, *Epilachna varivestis*, mate successfully soon after they emerge (Taylor 1984); however, it is not clear if they will mate after diapause is fully induced or if sperm is transferred to the seminal vesicle during diapause. In these cases, care was taken to determine that sperm was or was not transferred during mating, because some diapausing males may mount females even though they do not have mature sperm.

Diapause can interfere with reproduction through its hormonal influence on sex pheromone production and/or release (see Endo 1973). Environmental control of sex pheromone production has been studied in relation to diel periodicity (Shorey 1976), but there are few examples in which seasonal periodicity in pheromone production has been defined experimentally (see Borden 1977 for a review). Notable exceptions are the boll weevil, in which it was shown that diapausing males are not attractive to females (Villavaso & Earle 1974; see also Merkl & McCoy 1978), and in the butterfly *Polygonia c-aureum*, in which sex pheromone production is prevented during diapause (Endo 1973).

The sweetclover weevil, *Sitona cylindricollis*, exhibits an interesting phenomenon that is related to reproduction and that could have importance to the experimental design of studies dealing with postdiapause development. Overwintered (postdiapause) females require continuous or very frequent contact with males for uninterrupted production of oöcytes (Garthe 1970). Similarly, recent work with the milkweed bug, *Oncopeltus fasciatus*, suggests that the presence of a mature, sexually active male may hasten the termination of diapause in females (Hayes & Dingle 1983). Such a response would be of considerable interest and should be tested further in the milkweed bug and other species in which both males and females undergo diapause.

4.2.3 Seasonal migration and selection of dormancy sites

To undergo dormancy, insects generally migrate to an appropriate site and find a suitable hiding or resting place before the onset of adverse conditions. This type of movement, which we term diapause-mediated seasonal migration, is based on the insects' concomitant changes in physiological function and responsiveness to environmental stimuli (Kennedy 1961; Johnson 1963). First, there is a loss of responsiveness to "vegetative" stimuli—cues indicating food, mates, and oviposition sites. Once triggered, this loss persists until after migration has occurred, even if food, mates, and oviposition sites are present. Second, there is an increased responsiveness to factors stimulating migration. This often involves the acquisition of positive phototaxis and negative geotaxis. Environmental (token) stimuli that induce diapause often bring about the above changes in response patterns, and these changes are neurohormonally mediated and expressed as part of the diapause syndrome. Moreover, they occur at a specific time during the course of diapause—usually during the early phases of diapause induction, after the completion of prediapause feeding.

Similarly, return to the site of feeding and reproduction must be seasonally timed to coincide with the return of favorable conditions at that site. When the adult is the stage that undergoes dormancy, return flights are usually made near the end of diapause, as in *Coleomegilla maculata* (Obrycki & Tauber 1979), or during the postdiapause transitional period, as appears to be the case in the monarch butterfly and the convergent lady beetle, *Hippodamia convergens* (Urquhart & Urquhart 1976b, c; Tuskes & Brower 1978; Rankin & Rankin 1980; Davis & Kirkland 1982). However, when it is the larval stage that migrates to the dormancy site, the return may be made at a variety of times. In some cases the larva itself returns during a late phase of diapause or early in the postdiapause transitional period; in other cases the larva pupates and undergoes metamorphosis at the dormancy site, and the adult returns to the host plant or site of reproduction.

Migratory movements to and from the dormancy site vary from long-distance flights of thousands of kilometers to short-distance movements of a few meters or centimeters down a host plant to the soil or from leaves to twigs. In some species, such as the monarch butterfly and the milkweed bug (see below), long-distance migration is associated with a relatively low tolerance to cold, but this is not always the case. The black cutworm, *Agrotis ipsilon*, can overwinter successfully in Iowa, but during the summer, postdiapause moths immigrate from the south and mix with the resident population and their offspring. Thus, localities in Iowa can have five periods of flight for black cutworm moths—two composed of overwintered Iowa residents and their offspring, two of southern immigrants and their progeny, and a fifth that is a mixture of southern and Iowa individuals (von Kaster & Showers 1982; Troester *et al.* 1982). It is important to note here that the flights by the black cutworm moths probably are not diapause-mediated.

The ladybug, *Hippodamia convergens*, apparently migrates to avoid hot, dry, preyless summers in the Great Central Valley of California (Hagen 1962). This species normally produces one annual generation in the valley in coincidence with the occurrence of large populations of aphids in spring. During June most of the adults of this generation enter an aestival-autumnal-hibernal diapause and migrate to mountain canyons in the Sierra Nevada foothills. These canyons provide the diapausing beetles with cool, moist conditions and/or alternative food sources necessary for survival during summer. Hibernation also occurs in the canyons.

A striking and well-studied example of seasonal migration to a dormancy site is exhibited by the monarch butterfly, *Danaus plexippus*. During the summer this species occurs throughout most of the United States and southern Canada, where it produces several successive generations on its hosts, which include various species in the milkweed genus, *Asclepias*. However, monarch butterflies, even those in diapause, are very susceptible to injury by cold (Calvert & Brower 1981; Calvert & Cohen 1983) and apparently are unable to survive cold northern winters. Each autumn large numbers fly southward to hibernate. Populations from the western United States and western Canada congregate in areas along the Pacific coast, south of San Francisco, whereas populations from eastern and midwestern North America move to areas of moderate climate in the Sierra Madre Occidentale mountains of Mexico and the Sierra Madre Mountains of Guatemala and Honduras (Urquhart & Urquhart 1976b, 1977, 1978, 1979b, c; Brower 1977; Brower *et al.* 1977). The overwintering aggregations contain from several thousand to several million individuals clustered on trees; they constitute a spectacular sight.

Monarchs undergoing the southward migration are in a state of reproductive diapause characterized by lack of ovarian development, increased lipid content, and reduced responsiveness to reproductive and vegetative stimuli (Urquhart & Urquhart 1976b; Brower 1977). Short day lengths and low temperatures maintain diapause in the monarch; long days and warm conditions, as well as injection with juvenile hormone isomers, can terminate it and promote oögenesis (Herman 1973). The males remain in reproductive diapause until November, females until December (Herman 1981). Mating occurs in February, and mated adults begin to disperse northward from the dormancy site (Urquhart & Urquhart 1976b, c, 1979a, b; Hill *et al.* 1976; Brower *et al.* 1977; Tuskes & Brower 1978). A relatively high thermal optimum of 28°C for ovarian maturation (Barker & Herman 1976) probably plays a vital role in preventing the conversion of metabolic reserves to ovarian tissue during the northward, postdiapause migration. It is also likely that flight itself plays a role in directing metabolic resources away from ovarian development (Lessman & Herman 1983; c.f. Tanaka 1976).

Not all monarch butterflies undergo autumnal migration. In some areas of coastal California, southern Florida, southern Mexico, and Central America that have mild climates and continual supplies of milkweed, resident populations reproduce year round (Urquhart & Urquhart 1976a, 1977). A similar situation occurs in Australia, where the monarch is a newly introduced species. In this area, the adults cease reproduction and form clusters (James 1982, 1983). They appear to enter a relatively weak diapause, the nature of which needs further study. In Hawaii, reproduction by the monarch occurs under short day lengths and low temperatures, but migratory behavior is unknown (Etchegaray & Nishida 1975). In this regard, it would be significant to determine the differential responses to environmental factors inducing diapause and migration in various populations. It would also be interesting to analyze the intrapopulation variability in responsiveness to these factors; such data would provide valuable insight into the evolutionary relationships between diapause and migration.

Another long-distance migrant in which the relationship between diapause and migration has been examined is the milkweed bug, *Oncopeltus fasciatus* (Dingle 1978a). This is the only American species in the genus that ranges into north temperate areas; however, it is unable to survive cold northern winters. At the end of summer in eastern North America, milkweed bugs respond to short days by entering reproductive diapause and migrating south for the winter. Short days maintain a high level of flight behavior in the diapausing insects, whereas long days terminate both diapause and migration and promote reproduction (Dingle 1978a). To date, the destinations of *O. fasciatus*' southerly flights remain undetermined, and the reproductive condition of the migrants during winter is unknown. The neuroendocrine regulation of both diapause and diapause-mediated migration is discussed in Section 4.1.1.

Migration to dormancy sites sometimes is aided by air movement. For example, when nondiapause summer females of the cabbage whitefly, *Aleyrodes brassicae*, are probed, they take off, but they sink in flight and remain airborne only briefly. By contrast, when diapausing autumnal females are probed, they fly upward and remain in flight for longer periods of time (Iheagwam 1977). This kind of behavior subserves windborne migration and, indeed, cabbage whiteflies have been captured in high aerial nets.

In contrast to the long-distance migrants just discussed, many insects about to undergo diapause move from their feeding and reproductive sites and seek hiding places at the edges of nearby fields and forests. These movements may serve as a means of achieving protection from physical extremes or from predators and parasitoids (see Shiotsu & Arakawa 1982). Colorado potato beetles (*Leptinotarsa decemlineata*) leave the food plant, move to edges of fields, and burrow from 10 to 70 cm in the soil to hibernate (Minder 1966); positive geotaxis and negative phototaxis guide movement during this period (de Wilde 1954). Similarly, in late summer and early autumn, lady beetles in the species *Coleomegilla maculata* move from corn fields and form aggregations in leaf litter near the edges of fields. Apparently the prediapause migration is primarily controlled by photoperiod and postdiapause dispersal by temperature (Solbreck 1974; Obrycki & Tauber 1979). Although the controlling mechanisms are not known, analogous behavior patterns occur in other predaceous species (Hagen 1962; Hodek 1973; Thiele 1977b), alfalfa and sweet clover weevils (e.g., see Prokopy *et al.* 1967), and cereal and rice leaf beetles (Casagrande *et al.* 1977; Kidokoro 1983). Also, diapausing adult mosquitoes sometimes seek shelter under bridges, in caves, or in other structures (Spielman 1971; Bellamy & Reeves 1963; Spielman & Wong 1973b; Buffington 1972; Gallaway & Brust 1982).

Mature lepidopteran larvae, such as the tobacco hornworm, *Manduca sexta*, the woolly bear caterpillar, *Isia isabella*, and the fall webworm, *Hyphantria cunea*, wander for several days prior to digging into the soil or crawling into a protected site to hibernate. Fully grown larvae of the Indian meal moth, *Plodia interpunctella*, when reared under diapause-inducing conditions, leave their food and wander about for a few days, apparently in search of crevices. After finding such sites, they close the entrance with tightly spun silken walls. In contrast, the nondiapausing larvae remain in their food and spin very coarse cocoons for pupation (Tsuji 1958; Kikukawa & Masaki unpublished). The small ermine moth, *Yponomeuta vigintipunctatus*, exhibits an analogous type of behavior. Nondiapause summer larvae pupate primarily on the undersides of the host leaves, whereas the prediapause autumnal larvae leave the host and seek sheltered areas where they spin their cocoons and enter diapause (Veerman & Herrebout 1982).

Many insects form hibernating quarters in the soil or under deep snow. These sites provide relatively stable temperature conditions during winter. Thus, when fungus midges, *Sciara* sp., hibernate in the stems of heart-leaf lilies that are covered with snow to a depth of 84 cm, their body temperatures remain almost constant at 0°C even when the aboveground temperature drops to −20°C (Tanno 1977). Similarly, diapausing pupae of the bertha armyworm, *Mamestra configurata*, which suffer injury if they are exposed for long periods to temperatures of −5 to −20°C, escape cold injury, at least in part, by burrowing 2 to 15 cm in the soil (Turnock *et al.* 1983).

For insects that pass a certain developmental stage in the soil, there apparently is an adaptive advantage in utilizing this developmental stage to overwinter, and this presumably has been a factor in the convergent evolution of the diapausing stage within many taxa. Crickets and grasshoppers, for example, deposit their eggs in the soil, and this is the most common stage of hibernation among these taxa. Also, moths, such as noctuids and sphingids, whose larvae characteristically burrow into the soil for nondiapause pupation, frequently hibernate as prepupae or pupae in the soil.

Selection of a suitable site for dormancy may be made by individuals of the previous generation. For example, when the egg is the overwintering stage, the adult determines the overwintering site. In such cases, behavior patterns associated with the selection of oviposition sites are often altered, and oviposition, rather than occurring on leaves which become senescent and fall, takes place on more permanent sites such as twigs or trunks. Examples of this sort of seasonal alteration in behavior are provided by the red spider mite, *Panonychus* (= *Metatetranychus*) *ulmi*, the wild silkworm, *Bombyx mandarina*, the Australian plague locust, *Chortoicetes terminifera* (Lees 1953; Omura 1950; Wardhaugh 1977), and the leafhoppers *Lindbergina aurovittata* and *Edwardsiana rosae* (Claridge & Wilson 1978).

In some cases, the larvae of the previous generation or even individuals of the grandparental generation determine the dormancy site. In the vapourer, *Orgyia thyellina*, which undergoes diapause in the egg stage, the female larvae determine the oviposition site when they spin their cocoons. Unlike females of the nondiapause generation, females that are destined to lay diapausing eggs are brachypterous, roost on their cocoons, which have been constructed on permanent sites, and lay their eggs on the cocoon's surface (Kimura & Masaki 1977).

Perhaps the most spectacular example of the prediapause determination of dormancy site by a distantly removed generation occurs in dioecious aphids in which the autumnal migration to the overwintering host occurs one, two, or even three generations before the diapausing egg is laid (Hille Ris Lambers 1966; Blackman 1974a; Dixon & Glen 1971). Although far removed from the diapausing stage, these movements constitute part of the diapause syndrome. They are regulated by token stimuli that occur in advance of unfavorable conditions (e.g., see Hardie & Lees 1983), they involve changes in host plant selection (Dixon 1977; Hardie 1980b), and they are essential to the successful induction of diapause.

4.2.4 Summary

Diapause is characterized by a complex repertoire of species-specific, presumably neuroendocrinologically mediated behavioral patterns. Among the most predominant behavioral changes associated with diapause are decreased responsiveness to food and reproductive stimuli and movements to and from the dormancy site. In many cases, characteristic behavioral patterns provide relatively accurate indices of diapause induction and termination. Because of the alteration in responses to food, pheromones, and baits that are used in traps for monitoring insect populations, care must be taken in interpreting data from traps for those species that enter diapause in the (monitored) adult stage.

Insects display considerable interspecific and intraspecific variability in their responses to food and mates during diapause. In most species that enter diapause in a free-living stage, responses are reduced to a very low level, so that no feeding or reproduction occurs even if food, mates, and oviposition sites are present. Others, such as certain species of mosquitoes, are not attracted to food during diapause, but they will feed if placed on a suitable host. By contrast, some diapausing insects continue to feed at a reduced rate that is ultimately regulated by token stimuli. Social insects sometimes make and store food for use during hibernation.

Although reduced reproductive behavior appears to be the rule among diapausing females, diapausing males vary in the degree of their reaction to reproductive stimuli. In some species, male mating behavior is completely suppressed during diapause; males of other species will mate and transfer sperm if a receptive female is present.

Reduced feeding and reproduction not only are characteristic of dormancy, but they are also essential components in the expression of migratory behavior. Thus, migration involves a simultaneous reduction in responsiveness to food and reproductive stimuli, and increased responsiveness to stimuli subserving directed movement. These behavioral changes (and thus migration) occur at a specific time during the course of diapause—usually during the prediapause phase or early during diapause induction.

Diapause-mediated migration varies from very short movements of a few meters or centimeters to long-distance, windborne flights of thousands of kilometers. It may occur in the stage that undergoes dormancy, or it may involve a stage, or even a generation, preceding the diapausing stage. In all cases, the movements are based on species-specific and highly evolved behavioral patterns that result in the transportation of the insect to and from a protective site in which they undergo dormancy.

4.3 Physiological Expression of Diapause

Insects inhabiting seasonal environments face a number of exigencies to which they would readily succumb if they remained in the active state. Among these exigencies are long periods during which an external source of energy is either greatly reduced or absent. In most cases, they overcome this deficiency by physiological alterations that characterize diapause: the prediapause accumulation of metabolic reserves, the suppression of developmental and reproductive functions, and reduced metabolic activity. These alterations seem to be highly efficient mechanisms for allocating energy because species that can survive periods of dormancy for several years are not uncommon among sawflies, midges, moths, and hymenopterous parasites (see Sections 7.1.1.2 and 9.2.2.1).

Another major problem confronting insects is how to tolerate the extreme physical conditions that they encounter during the unfavorable season(s). Insects have evolved several physiological adaptations that serve this purpose; included are cold-, heat-, and drought-hardiness. Usually, but not always, the physiological changes that subserve these functions are induced by environmental stimuli (token stimuli) acting prior to the onset of the unfavorable conditions. Thus, they occur in preparation for seasonal change, not as the result of it. Moreover, the physiological changes involved in the seasonal development of cold-, heat-, and drought-hardiness proceed in species-specific patterns as part of the diapause syndrome.

Below, we discuss the physiological adaptations that allow insects to withstand the long periods of dormancy without an external source of energy, and we explain how diapause serves in this function. Subsequently, we examine the physiological mechanisms that subserve the insect's ability to withstand the extremes of cold, heat, and drought during dormancy and we relate these adaptations to the concept of the diapause syndrome.

4.3.1 Adaptations to a seasonally restricted energy supply

Adaptation to a seasonally restricted source of energy involves the accumulation and conservation of reserves for use during and immediately after dormancy. Many factors influence these two functions, such as the life stage that undergoes diapause, the rate and duration of prediapause feeding, the degree of suppressed metabolism during diapause, the food requirements of postdiapause stages, and the availability of food before, during, and after dormancy. All of these factors show considerable variability.

4.3.1.1 Accumulation of metabolic reserves

Many insects undergo rapid or prolonged feeding during their prediapause period (Sections 3.3.1 and 4.2.1). However, the energy derived from prediapause feeding is not invested entirely in growth and development, but is channeled along alternative metabolic pathways to build up reserves. Reserves built up in this way may serve as a source of energy during diapause and during postdiapause development and oviposition.

During the prediapause period in adult insects, ovarial development and yolk formation are inhibited and the function of the fat body changes from active synthesis to storage of lipids, proteins, and other reserves (see Kono 1979, 1982). The buildup of metabolic reserves may also involve the breakdown of certain body tissues; wing muscles, ovaries, and testes degenerate and eggs are resorbed, while fat body and other storage organs undergo hypertrophy (e.g., Carson & Stalker 1948; Lees 1955; El-Hariri 1966, 1970; Tombes & Marganian 1967; Hodek 1973; Barnes 1976; Numata & Hidaka 1980, 1981; Pfaender *et al.* 1981; de Kort 1981). The haemolymph also serves to store large quantities of protein (Dortland & de Kort 1978), amino acids (Boctor 1981; Morgan & Chippendale 1983), lipids, and carbohydrates. In all cases studied thus far, the hormonal changes that regulate diapause also influence the synthesis and conversion of stored metabolites during the prediapause, diapause, and postdiapause periods (see Yamashita & Hasegawa 1976; Chippendale & Turunen 1981; de Kort 1981; Kono 1982).

Insects show considerable variation in the types of reserves and in the pathways and timing of their use. The silkworm, *Bombyx mori*, the rice stem borer, *Chilo suppressalis*, and the Colorado potato beetle, *Leptinotarsa decemlineata*, have been particularly well studied in this regard, and they present a paradigm for future studies with other species.

In the silkworm, diapause hormone stimulates the conversion of hemolymph trehalose to glycogen in the developing ovaries (Hasegawa & Yamashita 1967, 1970; Yamashita & Hasegawa 1970, 1976; Yaginuma & Yamashita 1980). Upon diapause initiation, glycogen is broken down by glycogen phosphorylase *a* to provide an initial substrate for sorbitol and glycerol formation (Yamashita *et al.* 1975). These polyols are maintained at high levels throughout diapause. At the end of diapause, NAP-dependent sorbitol dehydrogenase activity increases, and sorbitol and glycerol are utilized in the resynthesis of glycogen for resumed embryonic development (Chino 1957, 1958; Yaginuma & Yamashita 1978, 1979; Yamashita *et al.* 1981). Similar patterns of glycogen conversion to polyols and subsequent reconversion of polyols to glycogen have been reported for several species, including the rice stem borer, *Chilo suppressalis* (Tsumuki & Kanehisa 1980a, b). In all cases studied, the fat body appears to be the main tissue involved in the synthesis and conversion of diapause metabolites (see Tsumuki & Kanehisa 1979, 1980b).

In the Colorado potato beetle, flight and reproduction are the important energy consuming processes in adults, and the fat body plays an important role in these functions. Within two days of emergence, the rate of synthesis of yolk proteins (vitellogenesis) is high in nondiapause females, whereas it is low in prediapause beetles. However, in prediapause beetles three other proteins are synthesized in large amounts, and these proteins are stored in the fat body or hemolymph during diapause (Dortland 1978, 1979). The capacity to synthesize protein, as well as the nature of the synthesized proteins, is under hormonal control. When juvenile hormone titres are low, ovaries do not develop, the rate of vitellogenesis is low, and large amounts of storage proteins appear in the hemolymph or fat body. Flight muscles either do not mature or they degenerate, and the fat body loses its ability to synthesize proline, an amino acid that is essential for flight (Weeda *et al.* 1979, 1980; Weeda 1981). Juvenile hormone is very important to vitellogenesis; however, it is probably not the only hormone to have a role. Recent evidence suggests that ecdysteroid levels peak during diapause (Briers *et al.*

1982)—a situation that suggests some influence on metabolism during diapause.

Insects reveal considerable variation among species and also between the two sexes in their requirement of reserves for diapause and postdiapause development. In some cases, stored nutrients provide only enough energy for the successful completion of diapause and hatching or molting to the next feeding stage. When diapause occurs in mature larvae, pupae, or adults, sufficient nutrients must be stored to provide energy for diapause completion, development of the adult, and, in some cases, the initiation and maintenance of oviposition. Some species (e.g., some Lepidoptera that undergo diapause as mature larvae, pupae, or adults) do not feed after diapause; therefore, postdiapause oviposition depends entirely on the nutrients derived from reserves that have been carried over dormancy. In other species, such as *Chrysopa carnea*, the initiation of postdiapause oviposition is derived from nutrients carried over through dormancy, but sustained oviposition requires postdiapause feeding (Tauber & Tauber 1973c). Finally, certain ladybird beetles require postdiapause feeding to initiate and sustain oviposition.

Suppression of ovarian development in biting and other flies has been referred to as "gonotrophic dissociation" if it occurs in association with blood feeding and "gonotrophic concordance" if it occurs without blood feeding (see Section 4.2.1). Buildup of metabolic reserves that subserve survival of diapausing adult mosquitoes during dormancy does not appear to depend on prediapause blood feeding. However, it is not known whether or not prediapause blood feeding ultimately increases postdiapause fecundity. In general, a blood meal is required after diapause to initiate oviposition; autogenous females apparently do not enter diapause.

Recently, researchers have considered the function of sex-related differences in the carry-over of metabolic reserves during dormancy. In some species of mosquitoes, mites, and social Hymenoptera, only mated females undergo dormancy; the males, being unable to diapause, do not survive the dormancy period (see Clements 1963; Danilevsky 1965; Sections 3.1 and 6.2.1). Thus, any contribution to reproductive success by either sex of these species is dependent upon the ability of females to carry metabolites over dormancy. In contrast, both sexes of most species undergo diapause, and mating occurs after dormancy. In these cases, the carryover of reserves by both males and females may contribute to postdiapause reproductive success, especially if males have a large reproductive investment. In this regard, males vary considerably in the degree to which spermatogenesis is suppressed during diapause (see review by Pener & Orshan 1983). Testes or other male reproductive organs may regress considerably (Cloutier & Beck 1963; Ascerno *et al.* 1978; Numata & Hidaka 1980, 1981; Wellso & Hoxie 1981); however, in some species spermatogenesis continues, apparently at a temperature-dependent rate, throughout dormancy (see Hodek & Landa 1971; Bale 1979; Boiteau *et al.* 1979; Pfaender *et al.* 1981). In the spring-breeding carabid, *Pterostichus nigrita*, sexual development is inhibited and reproductive diapause is induced in both males and females by the long days of summer. Spermatogenesis begins after the beetles experience the short day lengths of winter; vitellogenesis takes place only under the subsequent long days of spring (Thiele 1966; Ferenz 1975a, b). In the boll weevil, post-diapause spermatogenesis occurs in the absence of feeding before males leave their winter quarters, but oögenesis does not begin until females have fed on seedling cotton (Brazzel & Newsom 1959). In *Sarcophaga crassipalpis*, although there is a considerable loss in fecundity during overwintering, there is no apparent reduction in the males' ability to fertilize females (Denlinger 1981b).

4.3.1.2 Reduced metabolism

After prediapause feeding has ended and diapause induction has begun, metabolic activity slowly declines until it levels off at a very low rate, where it remains until diapause ends. Thus, during the course of diapause, O_2 consumption in almost all insects that have been tested follows an approximately U-shaped curve (Wigglesworth 1972); this curve reflects diapause depth, as illustrated by independent tests. One exception, the rove beetle, *Atheta fungi,* appears to consume oxygen at equivalent rates during its diapause and nondiapause periods (Grigo & Topp 1980).

An important characteristic of respiration during diapause is that its rate is relatively independent of temperature. Oxygen uptake by diapausing bean leaf beetles, *Cerotoma trifurcata*, is equivalent at 25 and 30°C, whereas nondiapause beetles consume considerably more O_2 at 30 than at 25°C (Schumm *et al.* 1983). In addition, diapausing insects show reduced sensitivity to some metabolic inhibitors (de Wilde 1970), as well as reduced RNA synthesis (Wigglesworth 1972). The speed of response to diapause-terminating stimuli also can indicate changes in metabolic level (see Hodek & Hodková 1981).

The metabolic rates of diapausing insects as well as their patterns of change during dormancy are highly species-specific. In some hibernating insects, such as several species of coccinellid beetles, metabolic reserves (fat body) are utilized to a large degree in the fall, with less being consumed in the winter and early spring (Hodek & Čerkasov 1961, 1963; El-Hariri 1966, 1970; Stewart *et al.* 1967; Zar 1968; Hodek & Landa 1971; Barnard & Mulla 1978; Mills 1981; Pfaender *et al.* 1981; Ali & El-Saedy 1981). In some other species, e.g., *Semiadalia 11-notata*, the consumption of fat and glycogen reserves during hibernation is slow until one or two weeks before emergence from the overwintering quarters. At this time, consumption increases rapidly, especially in males, and the increase is correlated with heightened sexual activity by males (Hodek & Čerkasov 1958). Postdiapause oviposition by all the above ladybird species depends on an external food supply after dormancy. In the monarch butterfly, *Danaus plexippus*, fat body content declines from late November to late January, and subsequently levels off in late February when activity resumes (Chaplin & Wells 1982). It appears that the high thermal threshold for the postdiapause conversion of fat body reduces the rate at which metabolic reserves are consumed during the northward postdiapause flight (Barker & Herman 1976; see also James 1983). In the pink bollworm, *Pectinophora gossypiella*, total lipid content decreases during diapause, but the relative percentages of fatty acids do not change (Foster & Crowder 1980). However, in *Culex tarsalis* and *Anopheles freeborni* there are changes in relative levels of fatty acids (Schaefer & Washino 1969). In diapausing *Pieris brassicae* pupae, trehalose is the primary stored reserve, and there is a rapid turnover of this sugar during diapause. Low temperatures increase the rate of turnover (Moreau *et al.* 1981).

Once diapause is fully induced, metabolism remains at a relatively low level until diapause termination. Some species have been shown to have O_2-uptake cycles with periods of one hour to 14 days (Beck 1964; Denlinger *et al.* 1972; Hayes *et al.* 1972; Crozier 1979a, b)—none of which are related to feeding or other activity cycles. Thus, these metabolic rhythms are of considerable physiological interest, and several hypotheses have been proposed to explain their functional significance, including a possible role in water conservation and time measurement (see Denlinger *et al.* 1972; Crozier 1979a, b; Denlinger 1981a). Recent data imply that the O_2 cycles are regulated by juvenile hormone activity (Denlinger *et al.* 1984).

Recently, it was demonstrated that diapausing Japanese beetle larvae, *Popillia japonica*, are less susceptible to infection by *Bacillus thuringiensis* than are active larvae (Sharpe & Detroy 1979). Reduced susceptibility is associated with reduced metabolism; thus, fewer spores are ingested. It also is correlated with reduced alkalinity in the guts of diapausing larvae, which decreases the solubility of parasporal crystals. Similarly, a symbiotic relationship between the flagellate *Euglena* sp. and three species of damselflies that diapause as larvae has been shown to be seasonal. The hindguts of the larvae are inhabited by the *Euglena* only during the overwintering period, when metabolic activity and feeding are reduced (Willey *et al.* 1970). In the spring the *Euglena* either leave the winter host or they are discarded with the cast larval skin during the first postdiapause molt.

4.3.2 Cold-hardiness

Insects have evolved several behavioral, physiological, and biochemical mechanisms for coping with extremely low temperatures that could cause injury during hibernation. During the last 20 to 30 years, studies of this aspect of seasonal adaptations have advanced enormously. Sufficient results from a large number of species have fostered some generalizations, although further work is needed to elucidate the relationship between cold-hardiness and diapause (Chino 1957, 1958; Salt 1961, 1969; Takehara & Asahina 1961; Downes 1965; Asahina 1969; Danks 1978, 1981; Ring 1980, 1982; Zachariassen 1982; Baust 1982; Duman 1982; Duman *et al.* 1982; Block 1982; Sømme 1982; Adedokun & Denlinger 1984). Because of the recent and thorough reviews (above), our discussion deals with the generalizations produced thus far, and we refer only to selected studies illustrating major points. We emphasize the relationship between the development of cold-hardiness and diapause.

Among insects there are diverse physiological and biochemical mechanisms that underlie adaptation to cold. Some of these adaptations are stage-specific; others are not. Very often, closely related species and geographical populations of the same species exhibit considerable differences in their cryoprotective mechanisms (Asahina 1966; Baust & Lee 1981; Madrid & Stewart 1981). Moreover, in some cases, the cryoprotective mechanisms of a single population may differ from year to year (Duman 1984). In contrast, convergent evolution appears to be commonplace among the characteristics underlying cold-hardiness in insects. Thus, like diapause, cold-hardiness displays considerable evolutionary malleability, and generalizations derived from comparative studies of cold-hardiness may be fraught with pitfalls for those seeking evolutionary trends (see Chapter 8).

Intracellular ice formation is generally considered to be lethal for insects (see Ring 1980, 1982; Ring & Tesar 1980), and injury due to cold appears to be cumulative and unrepairable during subsequent exposure to favorable temperatures (Turnock *et al.* 1983). However, many insects are able to withstand freezing. Cold-hardy insects are thus generally divided into two categories with respect to their ability to withstand freezing: those that are freezing-susceptible (i.e., freezing of body fluids results in death) and those that are freezing-tolerant (i.e., freezing of extracellular fluids is tolerated). The majority of insects are freezing-susceptible, and although few species from arthropod groups other than the insects have been tested, no scorpions, araenids, acari, or myriapods have been shown to be freezing-tolerant (Block 1982). Similarly, no species from the Exopterygote insects have been found to survive freezing, even though species from the Collembola typically occur at both poles and in the alpine tundra of high mountains (Sømme 1979, 1982; Block 1982).

4.3.2.1 Freezing resistance

Avoidance of freezing by freezing-susceptible insects is essential, and it involves behavioral, physiological, and sometimes morphological adjustments. Previously, we discussed diapause-mediated migration and movement to protected sites for dormancy (Section 4.2.3). The selection of a suitable site is extremely important for avoiding freezing during dormancy; contact with water (even water condensed as dew or frost on the insect's surface) can prove fatal during hibernation because it may trigger inoculative freezing. Thus, specialized behavior patterns have evolved to serve in the selection of sites for dormancy. Some freezing-susceptible insects have also evolved morphological adaptations that aid in preventing direct body contact with water; these include thick cocoons and waxy secretions on the integument (see Sections 4.3.3 and 4.4.1). Although relatively poorly studied, these morphological adaptations may prevent inoculative freezing during dormancy.

Physiologically, avoidance of freezing is largely achieved through the ability to lower the supercooling point by accumulating antifreeze agents and removing ice nucleators from the body (Duman 1982). These can be extremely effective mechanisms for achieving cold-hardiness, and they are widespread among overwintering insects. Supercooling may occur in any stage of the insect life cycle; however, the degree to which insects can lower their supercooling point varies considerably among stages and among species (Sømme 1982). The supercooling points of the larvae of several gall-forming hymenopterans and dipterans from northern Canada and Alaska can be reduced to below −60°C (Ring & Tesar 1981), while that of the freezing-tolerant, hibernating eonymph of the poplar sawfly, *Trichiocampus populi*, only to −8.6°C (Tanno & Asahina 1964; Tanno 1967). Supercooling points of −30°C are not uncommon among hibernating insects of temperate regions (Sømme 1982).

The accumulation of a variety of different antifreeze agents is involved in supercooling. Among insects, glycerol is by far the most common, although other polyols (mannitol, sorbitol, and threitol) and sugars (trehalose, glucose, and fructose) are important (Ring & Tesar 1980; Sømme 1965, 1982). Recently, thermal-hysteresis proteins and amino acids have also been implicated as supercooling agents (Duman 1982; Duman *et al.* 1982; Hew *et al.* 1983; Morgan & Chippendale 1983). Proteins are common antifreeze agents in fish; among insects they appear to be most common in the Coleoptera, but they are not restricted to them (Duman 1982; Duman *et al.* 1982; Zachariassen & Husby 1982). Photoperiod and temperature, as well as diet and other factors that affect water balance in the insect body, can influence the accumulation of polyhydroxyl alcohols (polyols), sugars, and thermal-hysteresis proteins.

Avoidance of freezing also involves the elimination of ice nucleators from the body. Even with the accumulation of high levels of antifreeze agents, supercooling is substantially limited unless nucleating agents are either removed from the gut and body fluids or are masked in the cellular matrix (Sømme 1982; Zachariassen 1982). Some insects may fast or assume a nucleator-free diet prior to hibernation (Baust & Morrissey 1975; Sømme & Zachariassen 1981; Block & Sømme 1982; Sømme & Block 1982), or they eliminate the gut contents during diapause induction (Sømme 1982). Dehydration also may influence the insect's ability to depress the supercooling point, but few quantitative studies have considered the water content of various tissues during cold-hardening (Sømme 1982). There is a need to pursue experimentally the role of diapause both in the elimination of ice nucleators and in the partial dehydration of body tissues.

4.3.2.2 Freezing tolerance

Freezing tolerance occurs among the Diptera, Hymenoptera, Coleoptera, Neuroptera, and Lepidoptera (Asahina 1959, 1966; Block 1982; Miller

1982). It is a stage-specific characteristic that varies with the seasons (van der Laak 1982; Miller 1982). In one species of Arctic beetles freezing-tolerance occurs in two stages (larvae and adults); this variability is reflected in a similar variability in overwintering stage (Ring & Tesar 1980).

Closely related species may be quite different in their tolerance of freezing. Hibernating pupae of *Papilio xuthus* are susceptible to freezing, whereas those of *P. machaon* tolerate −196°C in a frozen state (Asahina 1966). In general, freezing-tolerant species do not have greatly reduced supercooling points; they seldom go below −10°C (Sømme & Zachariassen 1981; Duman 1982). However, this characteristic varies considerably among species (Block 1982; Ring 1982). Cryoprotectants (see above) accumulate in freezing-tolerant insects. Presumably, they serve several functions: (a) to regulate the temperature at which freezing occurs, (b) to prevent tissue damage during freezing, and (c) to prevent repeated freezing during periods of freezing and thawing or when freezing-tolerance may be reduced (Duman 1982; Block 1982).

In contrast to freezing-susceptible species, freezing-tolerant species retain ice nucleators (see van der Laak 1982). These chemicals serve to induce gradual ice formation in extracellular fluids, and by doing so, they reduce the likelihood of intracellular freezing (Duman 1982; Zacharaissen 1982). Internal nucleating agents may be proteinaceous, but further studies are needed to identify their structure-function relationship (Duman 1980, 1982). External ice nucleators can serve the same function as internal nucleators. Freezing-tolerant prepupae of *Sciara* sp. apparently take advantage of this by overwintering in moist surroundings, which presumably induce extracellular freezing at relatively high temperatures (Tanno 1977).

4.3.2.3 Anoxia tolerance

Freezing of groundwater and aquatic habitats often produces anoxic conditions that hibernating insects must withstand, and several insects have been investigated in this regard. Also, many species of soil-inhabiting alpine and Arctic insects tolerate anaerobic conditions when autumnal rains and hibernal or vernal thaws result in saturated soils. Collembola and mites overwinter in microscopic air spaces where O_2 deficiency is likely to occur. Some species have even been shown to survive several weeks without O_2 by using anaerobic metabolism with lactate as a major end product (Sømme & Conradi-Larsen 1977a, b). Overwintering adults of the carabid *Pelophila borealis* and the chrysomelid *Melasoma collaris* also have been shown to survive long periods in anaerobic conditions (Conradi-Larsen & Sømme 1973; Meidell 1983).

Aquatic habitats with little water flow during winter can also have anoxic conditions that are frequently associated with increased H_2S concentrations. In Sweden, the mayfly *Cloeon dipterum* occurs in shallow ponds that completely freeze over during the winter; as a result, diapausing larvae experience anoxic conditions for three to four months per year. As winter approaches, the larvae cease feeding and develop resistance to anoxia, both apparently in response to low temperature. In addition, diapausing larvae have altered phototaxic responses that allow them to take advantage of the periodic flow of oxygen-rich meltwater that accumulates under the ice of frozen ponds. In contrast, larvae of a very closely related species of *Cloeon* from the English Lake District are unable to develop resistance to anoxia; in this area ice cover is so brief that anoxia does not occur (Nagell 1980).

The mayfly *Leptophlebia verpertina* is also unable to survive long periods of anoxia at low temperatures. However, when the oxygen concentration is low, the larvae become both positively phototactic and negatively thermotactic and thus move to shore where oxygenated meltwater and food are available. Larvae feed and grow slowly during winter, suggesting that diapause may not be involved in this seasonal adaptation (Brittain & Nagell 1981).

4.3.2.4 Cold-hardiness and diapause

Many of the studies of insect cold-hardiness have been done with Arctic, Antarctic, and high-altitude species, some but not all of which retain a high degree of cold-hardiness all year round (see Baust 1982; Sugg *et al.* 1983). Thus, the relationship between diapause and cold-hardiness in these species may not reflect the general situation in temperate-zone species. Despite this bias, it has been demonstrated that insects undergo seasonal changes in their cold-hardiness, and these changes are related to seasonal changes in the amounts and types of cryoprotectants, the buildup or removal of ice nucleators, and the ability to withstand freezing (Salt 1961; Tanno 1965; Takehara 1966; Miller & Smith 1975; Rains & Dimock 1978; Zachariassen 1979; Baust 1982; Sømme 1982; Duman *et al.* 1982; James & Luff 1982; Shimada *et al.* 1984).

In general, diapause induction is associated with the production of cryosubstances, the elimination of gut contents, and the partial dehydration of body tissues or fluids, but the relationship between diapause and these functions is in some cases still obscure because low temperature may have a direct influence on cold-hardening (see Steele 1981; Izumiyama *et al.* 1983). Thus, we divide seasonal cold-hardiness into two categories: nondiapause cold-hardiness and diapause-associated cold-hardiness. This designation is in contrast to the three categories of Young and Block (1980). We found it difficult to distinguish between their second and third categories, both of which involve diapause, and we combined them. It is important to repeat here that diapause is a dynamic state that may terminate before midwinter without any overt sign. Thus, studies examining the relationship between cold-hardiness and diapause should include tests with insects that have been sampled at various times from throughout the entire overwintering period—not, as is often the case, tests with a single or a few samples from mid- or late winter.

Nondiapause cold-hardiness occurs in those insects in which cryosubstances and other cold-hardening mechanisms show a seasonal trend that is strictly temperature dependent. This is found in the calliphorid fly, *Lucilia sericata* (Ring 1972), bark beetle, *Scolytus ratzeburgi* (Ring 1977), and the Antarctic terrestrial mite, *Alaskozetes antarcticus.* All of these species can supercool under low temperature in any life stage (Young & Block 1980). It also occurs in the anholocyclic clones of aphids overwintering as nondiapause adults (Sutherland 1968; Griffiths & Wratten 1979; Tamaki *et al.* 1982a), the nondiapause Queensland fruit fly, *Dacus tryoni* (Meats 1976), and the two-spotted cricket, *Gryllus bimaculatus* (Izumiyama *et al.* 1983). In the Queensland fruit fly, acclimation to cold occurs only during two narrowly defined periods of morphogenesis—immediately prior to larval maturation and while the pharate adult is within the puparium (Meats 1983).

In other species, the development of cold-hardiness is, at least in part, influenced by token stimuli (photoperiod) and may also be hormonally influenced. Recognizing the enormous range of variation in reliance on token stimuli and hormonal changes in the development of cold-hardiness, we refer to this category as diapause-associated cold-hardiness. Examples include the eggs of the silkworm, *Bombyx mori* (Chino 1957, 1958), larvae of the rice stem borer, *Chilo suppressalis* (Tsumuki & Kanehisa 1981a), the banded woolly bear, *Pyrrharctia isabella* (Mansingh & Smallman 1972; Goettel & Philogène 1980), and the silkmoth, *Philosamia cynthia* (Hayakawa & Chino 1981), pharate pupae of the slug moth, *Monema flavescens* (Takehara & Asahina 1961; Takehara 1966; Asahina 1969; Takeda 1978a, b), and the flesh flies, *Sarcophaga crassipalpis* and *S. bullata* (Adedokun & Denlinger 1984), adults of the coccinellid, *Hippodamia convergens* (Lee 1980), and various stages of several other insects (Sømme 1964, 1965a, 1982; Patterson & Duman 1978; Horwarth & Duman 1982, 1983a, b, c; Duman *et al.* 1982; Shimada 1982). In most of these cases diapause-related processes and low temperatures interact in the full development of cold-hardiness. Low temperature stimulates or accelerates the production of polyols or other cryoprotectants. For example, the slug caterpillar does not form glycerol at a high temperature even if it is in diapause. However, in the silkworm, diapause hormone stimulates the accumulation of glycogen in the eggs, and during diapause induction glycogen is converted to sorbitol and glycerol, which serve as antifreeze agents. At the end of diapause, glycogen is resynthesized from the polyols and serves as an energy source during postdiapause development (Yaginuma & Yamashita 1979; Yamashita *et al.* 1981; Steele 1981). In *Ostrina nubilalis,* cold-induced glycerol accumulation is restricted to the stage that enters short-day induced diapause, the fifth instar, but it is not dependent on the induction of diapause or the presence of short-day conditions (Nordin *et al.* 1984). In the beetles *Dendroides canadensis* and *D. concolor*, exposure to either short day lengths or low temperatures elevates thermal hysteresis (Horwarth & Duman 1983a). Thus, although nondiapause beetles are able to achieve a high level of cold-hardiness, in nature the development of cold-hardiness and the induction of diapause are concurrent and perhaps related.

An especially close linkage between diapause and cold-hardiness is evident in the flesh flies *Sarcophaga crassipalpis* and *S. bullata* (Adedokun & Denlinger 1984). Diapause pupae that had been reared under warm conditions survived exposure to low temperature at much higher rates than did nondiapausing pupae. A period of cold acclimation was not necessary for the induction of cold-hardiness, and the attribute of cold-hardiness could not be separated from other features of the diapause syndrome.

Larval diapause in the rice stem borer, *Chilo suppressalis*, is controlled by the corpora allata and the prothoracic glands (see Section 4.1.3). In the absence of ecdysone, relatively high levels of juvenile hormone induce and maintain diapause; increase in ecdysone titres and decrease in juvenile hormone titres terminate diapause. Juvenile hormone and juvenile hormone analogues stimulate the conversion of glycogen to glycerol, whereas the reconversion of glycogen from glycerol occurs when diapause larvae are injected with β-ecdysone (Tsumuki & Kanehisa 1980c, 1981a, b). Hormonal control is not an absolute requirement for the conversion of glycogen to glycerol. Low temperatures also stimulate glycerol formation in nondiapause larvae, but the levels of glycerol are much higher in diapause than in nondiapause larvae (Tsumuki & Kanehisa 1980a). Thus, the neuroendocrine changes that induce diapause may also enhance cold-hardiness.

In many insects susceptibility to cold injury changes over time following the U-shaped curve similar to that for the development of other diapause symptoms (Behrendt 1963; Ilyinskaya 1968; Baust & Miller 1970; Bodnaryk 1978; Hansen *et al.* 1980; Suzuki *et al.* 1980; James & Luff 1982). This may signify a close relationship between diapause and cold-hardiness. In geographic populations of *Plodia interpunctella*, the degree of cold-hardiness (ability to survive under temperatures of 5-10°C) is to some extent correlated with the depth of diapause (Bell 1982). However, these correlations may not reflect a cause-effect relationship. The prediapause and diapause periods may merely serve as the times during which cold-hardiness develops, whereas the conditions that induce diapause may not necessarily induce cold-hardiness (Goettel & Philogène 1980; Young & Block 1980).

The relationship between diapause and cold-hardiness may be complex. Both involve dynamic processes that are subject to influence by external and internal factors (e.g., Baust 1982; Rojas *et al.* 1983). Thus, it is no surprise that the induction of diapause under artificial conditions may not necessarily result in the degree of cold-hardiness found in the field. Nor is it surprising that the development of diapause-associated cold-hardiness involves acclimation (see Fourche 1977) and that the cryoprotectants change as winter progresses. Just as responses to environmental stimuli change as diapause progresses, so is cold-hardiness induced at specific times during diapause and subject to change as diapause progresses. These changes depend on thermal and other conditions during winter and perhaps on the animal's changing state of diapause (see Baust & Lee 1981; Harper & Lilly 1982). Thus, the essential question in understanding the relationship between diapause and cold-hardiness is what factors induce and regulate cold-hardiness in nature? Are they the same as those regulating other facets of the diapause syndrome? How do they interact with the other environmental influences on the expression of diapause in nature? Are they essential to survival during diapause? Few field studies have analyzed these questions.

4.3.3 Drought- and heat-hardiness

Insects, like other poikilothermic animals, are subject to injury by extreme heat and loss of water, and they have evolved numerous mechanisms to cope with the broad range of humidity and temperature conditions they experience (Willmer 1982). Thermal balance is achieved through a number of controlling factors, including body size and shape, color, insulation, and behavior (posture, orientation, burrowing, basking, evaporation behavior, and flight); many of these characteristics are subject to seasonal variation. Similarly, water balance is affected by both size and shape, cuticular permeability, behavior, and physiological controls—again, factors that can vary seasonally. And thus, it is not unexpected to find that diapause is intimately involved in insect adaptation to the seasonal occurrence of high temperatures and drought conditions.

Unfortunately, experimental analyses of the association between drought-hardiness, heat-hardiness, and diapause are scanty. In some cases, adaptations to drought and heat are clearly unrelated to diapause; such is the case in the drought-resistant larvae of the African *Polypedilum* (see Section 2.1.1), the reduced number of larval instars in desert-inhabiting carabid beetles (Paarmann 1979b), and the ectothermic and endothermic regulation of body temperature by many insect species (see Heinrich 1974, 1975, 1981). Other instances, especially those in which one stage of the life cycle is particularly resistant, appear to be directly correlated with diapause or enhanced by diapause (Andrewartha & Birch 1954). In most cases, because heat and drought are usually associated with summer, an aestival diapause is involved.

Aestival diapause, like autumnal and hibernal diapause, exerts profound effects on the behavior, physiology, and sometimes morphology of insects, and thus can affect thermal and water balance in a number of ways (Masaki 1980). For example, burrowing or seeking shaded or protected sites may be essential to survival during periods of heat and water stress, but few studies have analyzed the behavioral adaptations associated with aestival diapause.

In addition to behavioral means, two physiological mechanisms enhance survival under drought conditions: resistance to desiccation and/or tolerance to water loss (see reviews by Schmidt-Nielson 1975; Edney 1977, 1980; Maddrell 1980), and diapause can be associated with both of these. The depressed metabolism, lowered water content, high fat content, and increased secretion of waxy coverings may confer diapausing insects with a high degree of resistance to desiccation. Furthermore, the intermittent opening of spiracles characteristic of diapausing insects may also be effective in decreasing water loss due to respiration (e.g., Schneiderman & Williams 1953; see Section 4.3.1.2).

Some diapausing insects show both resistance to desiccation and tolerance of water loss. Dormant eggs of *Austroicetes cruciata* lose water at an extremely low rate and are also highly tolerant of reduced water content during the hot, dry summer (see series of papers by Andrewartha, Birch, and Steele described in Andrewartha & Birch 1954). The cuticle of the egg and its associated wax remain highly resistant to water loss during the nine months of diapause. During this time, the suppressed growth rates reduce the embryo's needs for water.

Additional cases of presumed diapause-related drought- and heat-hardiness are found in eggs of the lucern flea, *Sminthurus viridis*, and the red-legged earth mite, *Halotydeus destructor*, both of which aestivate in semiarid areas of Australia (Andrewartha & Birch 1954; Wallace 1970a, b). In *Teleogryllus commodus*, diapausing eggs are somewhat more resistant to desiccation than eggs from which diapause has been eliminated. But the closely related nondiapause species, *Teleogryllus oceanicus*, is also resistant, and the resistance in *T. commodus* may be incidental to, rather than a function of, diapause (Hogan 1967).

Permeability of the cuticle is perhaps the single most critical factor in controlling water balance in insects (Willmer 1982, and references therein). Diapause-related production of wax apparently decreases the permeability of the cuticle to water during dormancy, and this can be a very conspicuous aspect of aestival and hibernal diapause. Aestivating nymphs of *Periphyllus* aphids (Hille Ris Lambers 1966), hibernating eggs of the mite *Petrobia latens* (Lees 1955), and hibernating pupae of the tobacco hornworm, *Manduca sexta* (Bell *et al.* 1975), and the bertha armyworm, *Mamestra configurata* (Hegdekar 1979), all produce copious amounts of wax. In some wax-producing species, construction of a hibernaculum or other protective covering may prevent abrasion of the waxy layer. A remarkable example of wax production is found in the "ground pearl," *Margarodes vitium*, which forms a hard, wax-coated cyst in which the nymph may survive more than ten years under dry conditions (Ferris 1919). Diapausing eggs of the acridid *Melanoplus differentalis* produce a secondary wax layer that covers the hydropyle and reduces water loss even if the egg is immersed in a hypertonic solution (Schipper 1938; Slifer 1949a, b). In the eggs of *Locustana pardalina*, a keratinlike membrane secreted by the serosal cells seems to be permeable to water when it is wet, but not when it is dry (Matthée 1951). The egg therefore is able to decrease water loss under dry conditions without impeding water absorption in moist conditions. Moreover, since the eggs tolerate a decrease in water content from 85 to 40%, very slight but well-distributed rainfall enables them to survive for several years.

Periods of water deficiency are not restricted to summer, and often hibernating insects must cope with drought, especially when temperatures fall below the level at which water freezes. In diapausing eggs of *Bombyx mori*, the chorion becomes almost airtight after oviposition (Okada 1971). Even when eggs are stored in a desiccator, water loss from the impermeable diapausing eggs is almost negligible. Females of the eastern tent caterpillar, *Malacosoma americanum*, cover their overwintering egg masses with spumaline—a frothy material that is produced by the accessory glands. This material is able to absorb water from the environment and presumably aids in preventing desiccation during dormancy (Carmona & Barbosa 1983). However, it cannot be assumed that all overwintering eggs in diapause are drought-hardy. In the false melon beetle, *Atrachya menetriesi*, diapausing eggs lose less water than do nondiapausing eggs, and yet most of them perish after exposure to a two-month dry period even at low temperatures (Ando 1978). In addition, dry conditions during hibernation may adversely influence subsequent survival and development. In the plum fruit moth, *Grapholitha funebrana*, postdiapause survival is greater when larvae have received precipitation during winter than when they have spent the winter in open-air but sheltered cages (Säringer & Szentkirályi 1980).

4.3.4 Summary

The physiological changes that characterize diapause enhance survival during periods of restricted energy supply, extremes in cold or heat, and water depletion. Among the metabolic adjustments are the accumulation and storage of metabolic reserves and a reduced rate of energy consumption. Insects display considerable variation in achieving these adjustments. Reserves generally accumulate during prediapause feeding and serve as a source of energy during both diapause and, in some species, postdiapause development and oviposition. Although the types of stored metabolites, their metabolic pathways, and the timing of their use differ among species, the fat body plays an essential role in all.

During diapause, metabolic activity declines to a very low rate, where it remains throughout diapause. Consequently, oxygen consumption, sensitivity to metabolic inhibitors, and RNA synthesis are substantially reduced. The level of suppressed metabolism during diapause varies among species and reflects the amount of activity and physical conditions during dormancy and the necessity for conserving metabolic reserves for postdiapause development and oviposition.

Metabolic reserves have important roles in the development of cold-hardiness in hibernating insects, but these are not the only means involved. Hibernating insects are either freezing-susceptible (i.e., freezing of body fluid results in death) or freezing-tolerant (freezing of extracellular fluids is tolerated).

To avoid freezing, freezing-susceptible insects undergo a variety of behavioral, physiological, and sometimes morphological adjustments. The most important physiological feature is the ability to supercool—that is, to lower the temperature at which body fluids freeze. Insects supercool by accumulating antifreeze agents (polyols, sugars, and thermal-hysteresis proteins) and by removing ice nucleators from the body.

Freezing tolerance among insects has been recorded, thus far, only from the Holometabola. It is a stage-specific character, but its occurrence varies greatly even among closely related species. Although freezing-tolerant species accumulate large amounts of cryoprotectants, they may not have very low supercooling points. They also tend to retain or produce ice nucleators that induce gradual ice formation in extracellular fluids.

The relationship between diapause and cold-hardiness varies considerably among insects. In some instances, diapause is clearly not involved and cold-hardiness develops strictly in a temperature-dependent manner. In others, the development of cold-hardiness depends on diapause because it is triggered by the same token stimuli and is mediated by the same neurohormonal changes that induce diapause. In most cases the full expression of cold-hardiness requires the induction of diapause as well as acclimation under falling temperatures prior to the coldest season.

Heat- and drought-hardiness are closely related adaptations. Thermal balance and water balance are achieved through several interrelated means, many of which vary seasonally and occur with diapause. As in the case of cold-hardiness, heat- and drought-hardiness can be either unrelated to diapause or, to a greater or lesser degree, dependent upon diapause. Few studies have used natural populations to analyze the relationship.

4.4 Morphological Expression of Diapause

Many species exhibit seasonal polyphenism in color and/or structure as an integral part of the diapause syndrome (Section 2.2.3.2). Diapause-mediated seasonal polyphenism is reported from all stages of the life cycle, but typically it occurs in a species-specific stage. In some cases two successive stages express seasonal polyphenism. The vapourer *Orgyia thyellina* offers such an example (Kimura & Masaki 1977). In northern Japan this moth has two generations per year and hibernates in the egg stage. Females emerging in the summer have functional wings and lay nondiapause eggs; those emerging in autumn are brachypterous and lay diapausing eggs that are larger and darker than nondiapause eggs.

In addition to stage specificity, diapause-mediated seasonal polyphenism is expressed at a species-specific period during the course of diapause. This period may occur several generations prior to diapause, as in the autumnal morphs of dioecious aphids (see Sections 2.2.3.2 and 3.3.2), or during the diapause stage itself, as in many other species. Moreover, diapause-mediated polyphenism may be expressed after dormancy. For example, the wing patterns of certain species of butterflies show postdiapause differences, some of which are related to thermoregulation. Several studies indicate a neuroendocrine basis for seasonal polyphenism (Fukuda & Endo 1966; Endo 1970, 1972; Pener 1970, 1974; Yin & Chippendale 1974).

The environmental stimuli that induce seasonal changes in phenotype are often the same as, or related to, the stimuli that induce diapause. An example occurs in the green lacewing, *Chrysopa carnea*, in which the depth of the photoperiodically induced diapause (as measured by its duration) is directly related to diapause color (Tauber *et al.* 1970b). In cases such as this, the close physiological relationship between diapause and seasonal polyphenism is evident. In other cases the relationship is not as clear as this because the stimuli that induce diapause may be different from those that induce seasonal polyphenism (see Section 3.3.2). However, when studying seasonal polyphenism, it is essential to consider the dynamic aspects of diapause. That is, the expression of diapause changes as diapause progresses through its various phases, and as this happens the insects' responses to stimuli change. Thus, no *one* factor can be said to regulate diapause. The full expression of diapause is controlled by a series of interacting factors—each of which has a specific role in developing particular aspects of diapause. Thus, the factors that influence diapause-mediated polyphenism need not be the same factors that serve to induce the other features of the diapause syndrome. They may be factors that intensify diapause or that act during the prediapause period, but in each of these cases the expression of the seasonal polyphenism is an essential part of diapause.

Although experimental evidence is lacking in most cases (however, see Wiklund 1975; Owen 1971), seasonal polyphenisms appear to serve three essential functions. First, color or structural changes can provide crypsis or physical protection from predators or from injury due to cold, heat, or desiccation during dormancy. Second, wing polyphenisms function in achieving efficient allocation of metabolic resources between dispersal and movement, on the one hand, and hibernation and reproduction on the other. Third, changes in color and size subserve pre- or postdiapause development during periods of seasonal transition. Finally, in some cases, such as the immaculate larvae of *Diatraea grandiosella* (Chippendale & Reddy 1972) and the dark larvae of *Protoparce quinquemaculata* (Hudson 1966), no function is known. We discussed many examples from each category in previous chapters (Chapters 2 and 3). Below we present additional examples.

4.4.1 Protective structures and coloration

Unlike many of their homothermic predators, poikilothermic insects are often much less mobile during dormancy than during their growing and reproductive phases, and thus they must rely less on movement and more on protective coloration and hiding places to avoid predation and parasitization during dormancy. Apparently to this end, many insects have evolved special prediapause patterns of behavior (discussed in Section 4.2.3) and special means for achieving crypsis during dormancy. Dormant morphs often differ substantially in color and shape from the equivalent nondormant morphs. Examples include the changes in adult coloration in *Chry-*

sopa carnea (Tauber *et al.* 1970b) and *Nexara viridula* (Michieli 1968; Michieli & Žener 1968), and *Tetranychus urticae* (= *T. telarius*) (Lees 1953; Parr & Hussey 1966), changes in larval and pupal color in *Lethe* species (Cardé *et al.* 1970), *Papilio polyxenes* (West et al. 1972; Hazel & West 1983), and *Papilio machaon* (Wiklund 1975), and changes in wing color in *Polygonia c-aureum* (Hidaka & Aida 1963; Hidaka & Takahashi 1967).

The adaptive significance of the seasonal color change is striking in the green lacewing, *Chrysopa carnea*, when it is compared with its sibling species, *Chrysopa downesi*, which does not undergo seasonal color change. During their reproductive period, which extends over most of the summer, adults of *C. carnea* are light green and they blend in well with the background foliage of their summer habitat, deciduous forest and meadows. In the fall, as the foliage assumes its autumnal coloration, *C. carnea* adults enter diapause and their color changes from bright green to reddish brown. They spend the winter at the edges of forests among the senescent foliage of deciduous trees. In contrast, *C. downesi*, which is associated with evergreen coniferous trees, does not undergo seasonal color change; it remains dark green all year round (Tauber & Tauber 1976c).

Although experimental studies on the function of cocoons and other "protective" structures are few, the adaptive importance of cocoons and modified pupal cuticle during dormancy can be deduced from their dimorphism. Insects that are destined to enter diapause as mature larvae and pupae often spin thicker cocoons than those that will not enter diapause. Examples are found in both parasitic and free-living insects (Schmieder 1939a; Schlinger & Hall 1960; Miyata 1974; Tauber *et al.* 1983).

Dimorphism in cocoon construction has reached an extreme state in the peach moth, *Carposina niponensis*. Prediapause larvae of this species spin very tough, compact cocoons that are ball-shaped and quite distinct from the elliptical, coarsely spun cocoons of the nondiapause larvae (Toshima *et al.* 1961). When diapause ends, the larvae bite through the winter cocoons and spin the nondiapause type for pupation. Thus, the construction of the ball-shaped winter cocoon represents a highly specialized behavior associated with diapause. A similar behavioral pattern may be found in certain tortricid moths that overwinter as mid-sized larvae. The immature larvae spin cocoons on the bark of trunks or branches for the sole purpose of hibernation (Oku 1966). Likewise, just after hatching, the prediapause larva of the spruce budworm spins a small cocoon in which it hibernates (Harvey 1957).

Cocoons are generally not highly effective insulators against temperature extremes, but because of their hydophobic nature, they may effectively protect insects from injury caused by desiccation, flooding, or freezing. They are probably especially important to freezing-susceptible insects in preventing inoculative freezing associated with external ice crystals. Modified cuticle structure and the waxy or other secretions produced by diapausing insects may serve a similar function (see Kono 1973; Section 4.3.3).

4.4.2 Seasonal allocation of energy

The physiological, behavioral, and morphological adaptations that enhance survival during stressful periods, as well as the morphological and other adaptations that subserve movement to and from dormancy sites, all require energy that must be allocated at the expense of other functions, such as development and reproduction (see Section 9.2.1.1). As a result, insects have evolved delicately balanced mechanisms for achieving efficient seasonal allocation of their energy resources.

The seasonal nature of the allocation of resources is most evident in seasonal wing polyphenisms. Among insect groups with polymorphic wing structure, monomorphic species or populations with short wings are typical of stable environments, whereas long-winged species are typical of unpredictable or unstable environments. Populations with seasonal wing polyphenism are characteristic of predictably ephemeral or seasonally variable habitats. In this case, energy resources are allocated, depending on seasonal cues, to wing formation and flight, or to reproduction *in situ.* Within this framework, wing polyphenisms involve alternation between short-winged, nondiapause, reproductive forms, and long-winged, diapausing, or nonreproductive, dispersing forms as in the psyllids (Oldfield 1970), gerrids (Vepsäläinen 1974a, 1978), and crickets (Masaki 1978a). Or they may involve the alternation of long-winged adults that produce nondiapause offspring with short-winged forms that produce diapausing offspring, as in many species of aphids (Hille Ris Lambers 1966; Blackman 1974a; Dixon & Glen 1971; Lees 1978) and in the moth *Orgyia thyellina* (Kimura & Masaki 1977). In these cases, energy resources, rather than being invariably expended on nonfunctional wings, are directly or indirectly allocated, on a seasonal basis, to reproduction.

4.4.3 Seasonal thermoregulation

Diapause-mediated polyphenisms may function to enhance growth or reproduction after dormancy. In these cases they mainly involve variations in body size and wing melanization. These characters are very important in thermoregulation—small size and dark wings being typical of cool, postdiapause periods, large size and light colored wings of warm, nondiapause periods. Diapause-mediated seasonal polyphenism is most common among the butterflies, primarily the Pieridae. Several examples are discussed in Sections 2.2.3.1 and 3.6.3 and by Shapiro (1976).

4.4.4 Summary

An integral part of the diapause syndrome for many species is a stage-specific change in color and/or structure. In instances in which diapause-inducing stimuli also induce phenotypic change, seasonal polyphenism is readily associated with diapause. In other species, such as those in which diapause-intensifying or other stimuli induce phenotypic change, the connection with diapause is less apparent, but perhaps just as close. It appears that the full expression of the diapause syndrome, including morphological changes, in most cases, is controlled by a series of interacting environmental factors—each of which has a specific role in developing particular aspects of diapause.

Although their adaptive value has been the subject of few experimental studies, seasonal polyphenisms appear to serve three functions. (1) They provide crypsis or physical protection during dormancy. Examples include seasonal color changes, dimorphic cocoons, and variation in waxy secretions. (2) They allow the efficient seasonal allocation of metabolic resources. Wing polyphenisms are notable examples. And (3) they may subserve prediapause or postdiapause development during periods of seasonal transition. Examples include seasonal variation in thermoregulatory features of the body—body size and wing melanization.

Chapter 5

Environmental Regulation of Seasonal Cycles

In the preceding chapters, we have demonstrated that the diapause syndrome comprises a repertoire of symptoms that are expressed at specific times during the course of diapause. These symptoms, such as the behavioral and physiological expressions of diapause, are both diverse and species-specific. In this chapter we will further develop the concept of diapause, and we stress that diapause involves intrinsic patterns of response to environmental stimuli. That is, while an insect's set of diapause responses is genetically based, environmental conditions usually determine whether or not, and to what degree, the symptoms of diapause will develop. Thus, the expression of diapause depends on both environmental and genetic factors. In this chapter we examine how seasonally changing environmental factors, acting over time, influence the dynamic expression of the species-specific symptoms of diapause.

The extent to which environmental conditions affect diapause varies considerably among and even within species. Examination of the interspecific variability in the genetic/environmental interactions that regulate diapause reveals a broad-ranging continuum in the types of response patterns. Variation ranges from almost complete reliance on environmental factors (multivoltinism with environmentally determined diapause) to genetic predominance (multivoltinism with diapause-free development, at one extreme, and univoltinism, with diapause in each generation at the other extreme). This variation has led researchers to categorize diapause as either "facultative" or "obligatory." The term "facultative diapause" denotes diapause that would be averted under certain environmental conditions; it describes all cases of diapause in which the controlling environmental factors are known and alterable. Thus, it is equally applicable to univoltine (one generation/year) and multivoltine (more than one generation/year) species. It is used in contrast to "obligatory diapause," a term that implies that diapause is expressed in every individual, in each generation *regardless* of environmental conditions. Some authors also use the term "obligatory diapause" in reference to seasonal cycles in which, under normal circumstances, diapause is induced in nature, in each generation, regardless of whether or not it can be averted under laboratory conditions.

We and others (House 1967; Tauber & Tauber 1976c) are dubious as to the usefulness of the terms "obligatory" and "facultative" diapause for two main reasons. First, the category of obligatory diapause has largely served as a catchall for cases of diapause in which the environmental influence on diapause induction has not been well studied. The number of species in this category is continuously decreasing as research on univoltine species progresses (e.g., Tauber & Tauber 1976c; Solbreck & Sillén-Tullberg 1981). Second, the terms tend to polarize thinking into rigid, contrasting categories, when, indeed, there is a continuum of interactions spanning a full spectrum of variability.

Arguing the "obligatory" versus "facultative" diapause problem is akin to arguing the "nature" versus "nurture" problem. Thus, we suggest avoiding the terms "obligatory" and "facultative" diapause and substituting them with the terms "univoltine" and "multivoltine" life cycles where appropriate. The productive approach is to elucidate the environmental and genetic mechanisms underlying seasonal adaptations and to determine how they interact to produce species-specific responses to seasonal changes.

In the following three chapters we attempt to do so, first by discussing the environmental control of seasonal cycles (Chapters 5 and 6) and then by considering, in detail, the genetic aspects (Chapter 7).

5.1 Photoperiod

Nearly all physical factors in the environment show seasonal changes, the patterns of which are marked by considerable annual variation. It is not uncommon for periods of unseasonally high or low temperature, or humidity, to occur in winter or summer and, commonly, autumn and spring have irregular temperature and moisture patterns. Of all the physical factors that change seasonally (e.g., light, temperature, humidity, rainfall), photoperiod is the most regular and therefore provides the most reliable long-term cue to future conditions. Except at the equator, each latitude shows a regular, seasonal pattern in day-length changes, and as one moves farther from the equator, these changes become more pronounced (see Beck 1980 for a table of day lengths and twilight periods for latitudes in the Northern Hemisphere).

Many, if not most, insects have evolved the ability to take advantage of the reliable and predictable seasonal pattern of photoperiodic progression, and thus photoperiod is a major factor influencing the seasonal cycles of insects. Not only are free-living insects such as external plant feeders and insect predators sensitive to day length, but internal parasitoids, insect parasites of vertebrates (Ternovoy 1978; Belozerov 1975, 1982), social insects, and soil-, wood-, fruit-, and water-inhabiting species are sensitive as well.

Photoperiod has an especially important influence on the seasonal cycles of species in the temperate zone, where seasonal changes in day length are large and highly correlated with seasonal changes in temperature, moisture, food supply, and other factors affecting development. Generally, for temperate species, temperature and other environmental factors modify the effects of photoperiod. The situation is often different among equatorial and tropical insects, in which the role of photoperiod may be superseded by the effects of temperature and moisture. But even here, some insects have evolved the ability to utilize the relatively small seasonal changes in day length near the equator (Norris 1965a). Thus, among the Insecta there is a broad and full spectrum in the degree of dependence on photoperiod as a seasonal cue.

Some species rely on photoperiod to regulate almost the entire life cycle. For example, development in the earwig, *Forficula tomis*, and in the antlion, *Myrmeleon formicarius* , is controlled by photoperiod throughout several stages of morphogenesis and throughout much of the year (Khaldey 1977; Furunishi & Masaki 1981). Other species depend on photoperiod, to varying degrees, to regulate a pivotal aspect of the life cycle, after which sensitivity to day length decreases and other environmental conditions assume the primary role. Because of its multifaceted nature, diapause itself has a dynamic interaction with photoperiod. For example, an insect may be responsive to specific photoperiodic cues that initiate prediapause changes in behavior and physiology, but these cues and, in fact, photoperiod itself, may have little effect on the subsequent induction of other symptoms of the diapause syndrome, such as cold- or heat-hardiness. Moreover, insects vary in their dependence on photoperiod as an indicator of seasonal conditions during and after diapause as well (see Geyspits & Simonenko 1970; Geyspits *et al.* 1974, 1978; Tauber & Tauber 1976a; Hodek 1983; Pener & Orshan 1983). Some species, like the ticks *Dermacentor marginalis* and *Argas arboreus* (Khalil 1976; Belozerov 1982) and the green lacewing *Chrysopa oculata* (Tauber & Tauber unpublished), lose their sensitivity to photoperiod relatively soon after diapause induction. Others, such as the European corn borer, *Ostrinia nubilalis* (McLeod & Beck 1963), and the lacewing *Chrysopa carnea* (Tauber & Tauber 1973d, e), gradually lose their sensitivity to photoperiod as diapause progresses. In the case of the damselflies *Lestes disjunctus* and *L. unguiculatus* (Sawchyn & Church 1973), photoperiodic sensitivity appears after diapause has been induced thermally, and photoperiodic cues influence only the second phase of diapause. In other insects, sensitivity to photoperiod persists throughout the whole course of diapause, and diapause is effectively terminated by photoperiodic cues. And finally, in the bugs *Aelia acuminata* and *Riptortus clavatus* and the grasshopper *Oedipoda miniata* (Hodek 1971c, 1979, 1981; Pener & Broza 1971; Numata & Hidaka 1982), sensitivity to photoperiod remains even after diapause has ended. Thus, clearly, the photoperiodic reactions of insects are both diverse and dynamic; each species has specific photoperiodic responses that change in a characteristic manner with time and with exposure to environmental factors. These responses serve to adapt insects to seasonal changes in external conditions, and in most cases they are intimately associated with the diapause syndrome.

5.1.1 Response curves and the critical photoperiod

Insects have evolved numerous ways to utilize photoperiod as a diapause-regulating cue. Various types of responses have been described (see Danilevsky 1965; Danilevsky *et al.* 1970; Lees 1968; Tauber & Tauber 1976a, 1978; Beck 1980; Saunders 1982); so has the variation in spectral sensitivity and perception of twilight (Norris *et al.* 1969; Lees 1971, 1981; Saunders 1982; Bradshaw 1972, 1974a; Takeda & Masaki 1979; Bradshaw & Phillips 1980; Claret 1982; Stark & Tan 1982).

Usually the photoperiodic responses of insects are represented by the percentage of individuals entering or terminating diapause as a function of stationary photoperiods. For example, when percentage diapause is plotted against photoperiod, a characteristic response curve results. The photoperiod that elicits 50% response is called the *critical photoperiod.*

Closely related species, even closely related species at the same locality, may have markedly different critical photoperiods for diapause induction. For example, the critical photoperiods for three species of green lacewings from Ithaca, New York, range from about 13 to 15.5 hours of light per day (Fig. 5.1). At 24°C, one species, *Meleoma signoretti*, has a critical photoperiod for diapause induction that falls between 15 and 16 hours of light per day; those for *Chrysopa oculata* and *C. carnea* are about 12.7 and 13.5 hours of light per day, respectively (Tauber & Tauber 1972a, 1975a, 1977c).

The response curve usually rises very steeply as it passes through the critical photoperiod, although there are several exceptions, such as the tortricid moth *Choristoneura fumiferana*, the mosquito *Anopheles maculipennis messeae*, the mite *Tetranychus telarius*, and the spider *Philodromus subaureolus* (Harvey 1957; Vinogradova 1960; Bengston 1965; Hamamura 1982). Researchers generally consider that this steepness reflects rigorous natural selection for the timing of diapause induction (Lees 1968; Saunders 1982). Indirect experimental evidence for this line of reasoning comes from studies with *Choristoneura fumiferana* in which there was increased variation in the photoperiodic responses of a laboratory-reared strain (Harvey 1957). The high level of variation can be attributed to the absence of natural selection for a specific diapause-inducing photoperiodic response.

Two major types of response curves have been shown: the long-day type and the short-day type. The long-day type of response (Type I of Beck 1980) is characteristic of insects that reproduce, grow, and develop under the long-day conditions of late spring and early summer and that go into diapause after experiencing the short days of late summer or autumn. The response curves of the three green lacewings (see Fig. 5.1) are typical of this type. Low temperatures tend to enhance the diapause-inducing effects of short day lengths in insects with this type of response curve (see section 5.2.2). The short-day type of response (Type II of Beck 1980) is less common and is usually characteristic of insects that grow and reproduce under short day lengths and that undergo aestival diapause (Masaki 1980; Ishii *et al.* 1983; Noda 1984). The effect of temperature on insects with a short-day type of

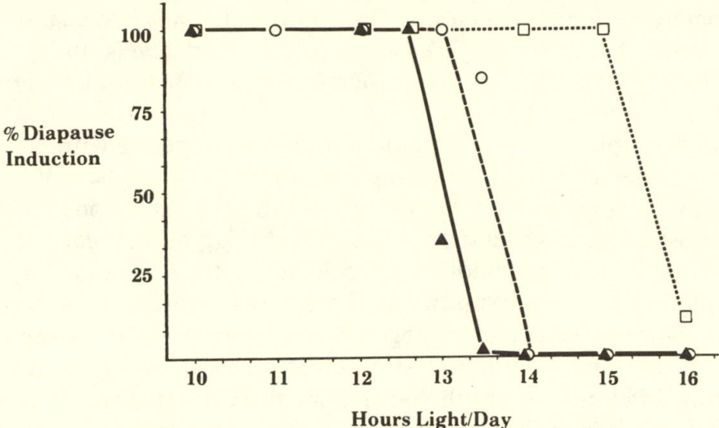

Fig. 5.1 Photoperiodic response curves for diapause induction in three green lacewing species (Chrysopidae) from Ithaca, New York. All are "long-day" species (Type 1 of Beck 1980). ▲ *Chrysopa oculata;* ○ *Chrysopa carnea;* □ *Meleoma signoretti.* Data from Propp *et al.* 1969; Tauber & Tauber 1972a, 1977c.

response to photoperiod is the opposite of that on insects with a long-day type of response. High temperatures tend to enhance diapause induction in short-day insects (see Section 5.2.2). Variations of these seasonal patterns are found in several species, such as the commercial silkworm, *Bombyx mori*, in which the sensitive stages occur far in advance of the diapausing stage. Thus, in this species, which has a short-day type of response to photoperiod, an autumnal-hibernal diapause is induced by exposure of the egg and early larval stages of the mother to long day lengths (Kogure 1933).

Some insects have both long-day and short-day responses to photoperiod—with summer diapause being induced by long days and winter diapause by short days. Development occurs under day lengths of intermediate duration during spring and/or autumn. Examples are found in some sawflies, leafhoppers, and moths (see Masaki 1980). In the noctuid *Mamestra brassicae*, summer and winter diapause are induced by long (> LD 15:9) and short (< LD 13:11) days, respectively; only within a narrow range of intermediate photoperiods (around LD 14:10) do a considerable proportion of pupae avert either type of diapause (Masaki 1956b; Masaki & Sakai 1965; Poitout & Bues 1977a, b; Furunishi *et al.* 1982). A similar type of response was found in laboratory experiments with the bivoltine ladybird beetle, *Coccinella novemnotata.* Day lengths of 10, 12, and 18 hours induce diapause in a high percentage of adults, whereas intermediate day lengths of 14 and 16 hours promote reproduction (McMullen 1967a, b). However, the pattern of photoperiodic response in *C. novemnotata* is in need of further study before it can be said to explain fully the bivoltine life cycle. Day lengths in northern California do not reach 18 hours; if the beetle's threshold sensitivity is not usually high or low, it perceives the longest summer day length as approximately 16 hours, which does not induce diapause. We suggest that either *C. novemnotata* is very sensitive to low light levels and perceives astronomical twilight as "day," or other factors interact with photoperiod to induce the summer diapause.

Beck (1980) depicts the photoperiodic response of species with both long-day and short-day responses on a single response curve; these are Type III and Type IV response curves. We do not use these designations for several reasons. Most of the cases designated as Types I, III, and IV cannot be distinguished under natural conditions; ecologically they are long-day types, and phenologically they are summer-active species. In addition, graded variations among the types occur in geographic populations of the same species and even in the same population at different temperatures (Masaki 1984), thus making differentiation of the types difficult. More important, however, in those cases in which there is a summer diapause that is induced by long days and an autumnal-hibernal diapause induced by short days, such as in *Mamestra brassicae* and *Abraxas miranda* (Masaki 1958, 1959; Masaki & Sakai 1965), the two types of diapause are distinct physiological states. The difference between the two states is not merely a quantitative one, as is implied by the Type III and Type IV designations. For example, in *M. brassicae* and *A. miranda*, responses to temperature differ greatly between individuals in summer diapause and those in winter diapause. Thus, we feel it is best to consider the two distinct responses separately from each other, with summer diapause induced by one set of responses and winter diapause by another.

Various species in the genus *Drosophila* also have both long-day and short-day responses. In *D. phalerata* from Leningrad, short day lengths induce an intense autumnal diapause. A less intense aestival diapause occurs in a proportion of the population in response to long days and reduced food availability. Uninterrupted development occurs only at intermediate day lengths (Geyspits & Simonenko 1970). By contrast, a laboratory population of *D. phalerata* from Leeds, England, does not enter diapause in response to long days either in the laboratory or under field conditions (Charlesworth & Shorrocks 1980). Unfortunately, the laboratory population used in these tests had been maintained under constant light for three generations, and it is possible that inadvertent selection had reduced the diapause response to long day lengths. It is also possible that food may influence diapause induction in this population.

Another species of *Drosophila, D. auraria*, enters diapause in response to both long and short day lengths under laboratory conditions. Development ensues only at intermediate day lengths. In this *Drosophila* it is not known how the long-day diapause differs from the short-day diapause. If adults experience a low temperature (15°C), the incidence of diapause under long days is much lower than if adults experience a high temperature (19°C). This response is not expected for a summer diapause. Moreover, it is unlikely that a long-day-induced aestival diapause is expressed in nature because the long-day critical photoperiod exceeds the day length of the summer solstice

(Minami *et al.* 1979). However, clearly, two responses—one to long-day and one to short-day—are present in this species of *Drosophila.*

It has generally been believed that within a population the critical photoperiods for diapause induction and termination are the same. This generalization is almost invariably based on early findings with *Antheraea pernyi* (Williams & Adkisson 1964). However, most studies, especially those that examined the photoperiodic responses of insects over the full course of diapause, have shown a difference in the critical photoperiods for diapause induction and diapause maintenance/termination. For example, the critical photoperiods for diapause maintenance/termination in *Wyeomyia smithii* and *Riptortus clavatus* are somewhat shorter than those for diapause induction (see Tauber & Tauber 1976a; Numata & Hidaka 1983), and in *Meleoma signoretti* the critical photoperiod for diapause induction is about two hours longer than that for diapause maintenance/termination (Fig. 5.2). Interestingly, in *Pyrrhocoris apterus,* the critical photoperiod for diapause maintenance in the laboratory is longer than that for diapause induction (Saunders 1983). These findings demonstrate the dynamic nature of diapause-controlling photoperiodic responses, and they underscore the importance of accurately describing the responses under study. A critical photoperiod applies to a *specific* response, and it is applicable only under specific conditions; for example, *at 24°C* the critical photoperiod *for diapause induction* in *Meleoma signoretti* falls between LD 15:9 and LD 16:8. It varies with temperature and other environmental conditions, and it is different from the critical photoperiod for diapause maintenance/termination.

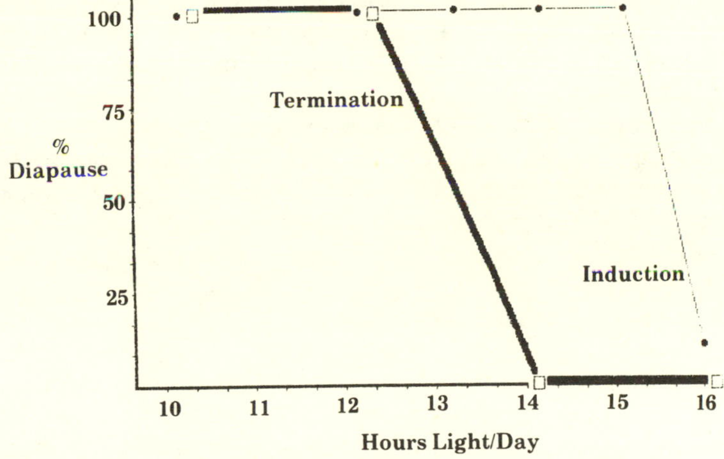

Fig. 5.2 Photoperiodic response curves for diapause induction and diapause maintenance in *Meleoma signoretti* from Ithaca, New York. The critical photoperiod for diapause induction falls between 15 and 16 hours of light per day (24°C). The critical photoperiod for diapause maintenance in individuals sampled from the field throughout autumn and winter is considerably shorter, between 12 and 14 hours of light per day (24°C). Data from Tauber & Tauber 1975a, 1977c.

5.1.2 Limitations of the photoperiodic response curve

As shown above, the response curve, including the critical photoperiod, is a useful concept in studying the photoperiodic control of diapause. However, it has some important limitations. First, despite the fact that it depicts quantitatively the relationship between percentage diapause and photoperiod, it does not indicate differences in the intensity of the diapause produced by the various day lengths. Some insects show an "all-or-none" response to day length, that is, days are measured as either "long" or "short," and all day lengths below the critical photoperiod induce diapause of equal depth (duration). In other insects, such as the green lacewing, *Chrysopa harrisii* (Tauber & Tauber 1974), the blowfly, *Calliphora vicina* (Vinogradova 1974), the grasshopper, *Oedipoda miniata* (Pener & Orshan 1980), and the two spotted lady beetle, *Adalia bipunctata* (Obrycki *et al.* 1983a), diapause-inducing day lengths near the critical photoperiod produce a weaker (i.e., shorter) diapause than day lengths well below the critical photoperiod. The reverse situation is true for the carnivorous midge *Aphidoletes aphidimyza* in which day lengths just below the critical photoperiod produce a more intense diapause than shorter day lengths (Havelka 1980). And, finally, still other insects (e.g., the green lacewing *Chrysopa carnea*, the arctiid *Spilarctia imparilis*, and the Mexican bean beetle, *Epilachna varivestis*), show a graded response to day length; namely, the depth of diapause varies as a function of day length (Tauber & Tauber 1972a, 1973e; Kimura *et al.* 1982; Taylor & Schrader 1984; also see below). In *S. imparilis* the effect of diapause-inducing photoperiods on the depth of pupal diapause is clearly separated from the effect of diapause-maintaining photoperiods on the pupae because all the pupae, which had been exposed as larvae to different photoperiods, were kept for the full duration of diapause at the same temperature under constant darkness.

The second limitation of the photoperiodic response curve is that it is based on unnatural, stationary light-dark cycles, and therefore does not provide information on the influence of changing day length on diapause (see Zaslavsky 1972, 1975). We discuss this problem below.

5.1.3 Responses to changing day length

Insect responses to photoperiodic changes fall into four categories:

A. The first category includes species for which it is not important whether day lengths are increasing, decreasing, or stationary; the only significant factor for these species is the duration of the day length in relation to the critical photoperiod (see Dickson 1949; Lees 1968). For example, the mite *Panonychus* (= *Metatetranychus*) *ulmi* enters diapause under constant short-day conditions. Neither decreasing day lengths above the critical photoperiod nor increasing day lengths below the critical photoperiod have any effect on diapause induction (Lees 1953). Similarly, in *Meleoma signor-*

etti and *Wyeomyia smithii*, day-length changes during diapause do not affect diapause maintenance, and diapause ends only after day lengths exceed the critical photoperiod for diapause termination (Tauber & Tauber 1975a; Smith & Brust 1971). For these species, photoperiodic response curves (both for diapause induction and for diapause termination) are very useful in predicting the beginning and end of diapause.

B. The second category includes insects that require a change in day length across a critical photoperiod to induce or maintain diapause. In these species, neither constant long days nor constant short days are sufficient to induce a high incidence of diapause, and if the insects are reared strictly under stationary photoperiods they show a very low rate of diapause induction. However, diapause occurs if the sensitive stage experiences a decrease in day length in which the short day length falls below a critical photoperiod. We refer to such species as having a "long-day/short-day requirement for diapause induction."

In the bollworm, *Heliothis zea*, constant photoperiods of LD 10:14, LD 12:12, and LD 14:10 (at 21°C) yield very few diapausing pupae (Wellso & Adkisson 1966). However, a high incidence of pupae enter diapause if the mother and egg stage are held under long day lengths (e.g., LD 14:10) and the larval stages under short days (LD 10:14) (Table 5.1). For diapause to be induced, the larval photoperiod must be below a critical level of 13 hours of light per day. This type of response ensures that diapause will be induced by relatively short day lengths in fall, but not by equally short day lengths in spring. The effect of the change in photoperiod is enhanced and the incidence of diapause reaches 94% when the insects experience a concurrent reduction in photoperiod and temperature (Roach & Adkisson 1970).

In *Heliothis punctigera*, only a very narrow range of constant day lengths of around 12 to 12.5 hours of light per day is effective in inducing diapause. Both longer and shorter days produce a very low incidence of diapause, but a decrease in day length from LD 14:10 to LD 10:14 during the early first

Table 5.1 Example of the "long-day/short-day requirement for diapause induction." In the bollworm, *Heliothis zea*, pupal diapause is induced in a substantial proportion of individuals only if they experienced long day lengths during the parental and egg stages and short day lengths as larvae and pupae. Data from Wellso & Adkisson 1966.

INCIDENCE OF DIAPAUSE (%)		
Photoperiod experienced by parents and eggs (27°C)	Photoperiod experienced by larvae and pupae (15–24°C)	
	LD 10:14	LD 14:10
LD 10:14	9.9 (N = 141)	0.0 (N = 142)
LD 14:10	67.5 (N = 240)	0.0 (N = 154)

instar results in a high level of diapause induction (Cullen & Browning 1978). Adults of *Dolycoris baccarum*, a univoltine species in Europe, enter diapause when immatures are reared under long day lengths and adults are transferred to short days soon after emergence. Neither constant short days nor constant long days induce diapause (Conradi-Larsen & Sømme 1978). Further, a large increase in the incidence of adult diapause occurs in the mosquito *Culiseta inornata* when young pupae experience a decrease in both day length and temperature (Hudson 1977). Field observations and laboratory studies suggest that the decreasing day length must reach a critical level for diapause to be induced. Similarly, diapause is induced in the migratory seed bug, *Neacoryphus bicrucis*, only if the nonmigratory larval instars II through V experience a relatively large decrease in day length. Constant short days by themselves do not induce diapause (Solbreck 1979).

Some species require changes in day length to develop the full intensity of diapause. A constant photoperiod of LD 12:12 produces a weak (short) imaginal diapause in the red locust, *Nomadacris septemfasciata*. A considerably longer diapause occurs at the same photoperiod if the immature stages had been reared under a day length of 13 hours or longer. Similarly, an LD 13:11 photoperiod produces a short diapause if it is preceded by an LD 13:11 or longer day length, whereas no diapause is induced if LD 13:11 is preceded by a shorter day length. This response pattern allows the locust to utilize photoperiod in synchronizing its life cycle with alternate dry and rainy periods even though day lengths in the natural environment range only between LD 13:11 and LD 12:12 (Norris 1965a).

An analogous response to change in day length is found in the temperate-zone lacewing *Chrysopa carnea*; a decrease in day length during the diapause induction or intensification periods results in an increase in diapause duration and in the intensity of diapause color (Tauber *et al.* 1970b). However, in this species the response to changing day lengths has its major function in maintaining autumnal diapause through the winter solstice (Tauber & Tauber 1973d, e).

Another interesting example showing the effect of decreasing photoperiods on diapause depth is found in the tobacco hornworm, *Manduca sexta*. This species, rather than responding to the duration of day length or to a change in day length, responds to the number of short days experienced during the relatively long sensitive stage (Denlinger & Bradfield 1981). Both the embryos and all the larval stages monitor day length; constant short days throughout the sensitive period result in a high incidence of diapause but the duration of this diapause is short. By contrast, hornworms transferred from long to short day lengths at late stages of larval development enter diapause at a lower rate, but the resulting diapause is of greater duration. The results are consistent with field rearings. Hornworms present in September receive more short days than those that occur in August, and thus they have a high incidence of diapause, but a weaker (shorter) diapause than hornworms from August. This mechanism prevents early-diapausing (August) individuals from terminating diapause prematurely in the autumn;

whether it also synchronizes emergence in the spring, as Denlinger and Bradfield propose, is open to study.

A different situation occurs in *Mamestra brassicae.* In this species, transfer from LD 12:12 to LD 16:8 (20°C) during the first four larval instars induces an aestival diapause of short duration. The reverse shift from LD 16:8 to LD 12:12 during the first or second instars results in a long, winter diapause and during later instars produces nondiapause pupae (Yagi 1975). In comparison with *M. sexta,* it is interesting that a stationary photoperiod of 14 hours (Furunishi *et al.* 1982), as well as a decrease in photoperiod during mid-and late instars, prevents both winter and summer diapause in *M. brassicae.* This response pattern apparently allows individuals that reach the midlarval stages before day lengths fall below a critical level to complete another generation before diapause intervenes. These findings illustrate the variability in the photoperiodic responses of insects as diapause progresses. They also underscore the need to study the photoperiodic control of insect diapause as a dynamic process characterized by regular, and sometimes large, changes.

C. The third category of response to photoperiodic change includes those insects that require a change in day length across a critical photoperiod to avert or terminate, as opposed to induce, diapause. If these species are reared under any stationary photoperiods, the response curve consists of a relatively straight line at approximately 100% diapause. However, if the sensitive stage undergoes an increase in day length from a short day (below the critical photoperiod) to a longer one (above the critical photoperiod), diapause is avoided or terminated. We refer to these animals as having a "short-day/long-day requirement for diapause prevention and/or termination."

The first case of this type of photoperiodic response explored experimentally was the imaginal diapause of the univoltine delphacid, *Stenocranus minutus* (Müller 1957, 1958b, 1960a). Ovarian development is arrested at an early stage if this insect experiences a constant long day (LD 16:8 or longer) throughout its life cycle. Short days after adult emergence allow some egg development to occur, but the full process of yolk formation and oviposition occurs only if the insect subsequently receives a period of long days. Interestingly, if the insect receives short days during nymphal development, a small morph is produced and diapause is averted.

Several other univoltine species, especially carabid and staphylinid beetles (Thiele 1966, 1968, 1971; Krehan 1970; Eghtedar 1970; Ferenz 1975a), the leafhopper, *Mocydia crocea* (Müller 1976b), the grasshopper, *Melanoplus sanguinipes* (Dean 1982), the spider, *Philodromus subaureolus* (Hamamura 1982), and the antlion, *Hagenomyia micans* (Furunishi & Masaki 1982), have been shown to have a short-day/long-day requirement for diapause prevention and/or termination. Typically, these species enter diapause in response to the long-day conditions of summer, and diapause persists throughout autumn and winter. Diapause development involves two phases, the first promoted by short autumnal days, the second by long vernal days. Breeding takes place the following spring. Some development may occur under short days in autumn, but it is usually of a limited extent and it is followed by the immediate resumption of diapause. Researchers in Germany commonly refer to this type of photoperiodic regulation of diapause as "photoperiodic parapause" (Müller 1970).

The univoltine green lacewing *Chrysopa downesi* also has a short-day/long-day requirement for prevention and termination of reproductive diapause. Diapause is avoided only if a sensitive stage (the last larval instar or pupa) experiences at least a four-hour increase in day length in which the long day length exceeds 12 hours (Table 5.2). Thus, both a critical photoperiod and a change in day length are involved in diapause prevention (Tauber & Tauber 1976b). Similarly, diapausing *C. downesi* adults must first be exposed to short days, and then long days, for diapause to terminate.

The ground cricket *Pteronemobius nitidus* from Japan, has similar photoperiodic requirements for avoiding or terminating nymphal diapause. In this case, the critical photoperiod lies between 14 hours and 14 hours and 45 minutes. Thus, a four-hour increase from LD 10:14 to LD 14:10 is not effective in preventing diapause, but the same magnitude of increase to LD 14¾ : 9¼ averts diapause. When nymphs that had been reared under LD 14:10 are transferred to LD 14¾ : 9¼ they fail to avert diapause; however, transfer to LD 16:8 is highly effective (Tanaka 1978a). Thus, it appears that both a critical photoperiod and changing day lengths are involved in diapause prevention. In parallel with the growth-promoting effect, increasing photoperiod enhances development of long hindwings (Tanaka 1978b).

The water strider *Gerris odontogaster* from southern Finland also responds to changes in day length as well as to a critical photoperiod (Vep-

Table 5.2 Example of the "short-day/long-day requirement for diapause prevention and/or termination." Adults of the green lacewing *Chrysopa downesi* enter diapause if they have been reared under any constant day length. However, if the early stages are reared under short day lengths and then the third instar larvae or the pupae are transferred to long day lengths, diapause is averted. A similar increase in day length is required to terminate diapause. Data from Tauber & Tauber 1976b.

INCIDENCE OF DIAPAUSE (%)				
Photoperiod experienced by eggs and larvae until cocoon spinning (24°C)	Photoperiod experienced by third instar larvae in cocoons, pupae, and adults (24°C)			
	LD 10:14	LD 12:12	LD 14:10	LD 16:8
LD 8:16	—	100[a]	20	14
LD 10:14	100	100	42	0
LD 12:12	—	100	100	22
LD 14:10	—	—	100	100
LD 16: 8	—	—	—	100

[a]In all conditions N = 4–19 pairs.

säläinen 1971, 1974b). Short-winged, nondiapause individuals are produced only if the early instars experience incremental changes in day length and if the fourth instar experiences a day length greater than LD 18:6. Otherwise long-winged, diapausing individuals are produced. Such a mechanism ensures that nondiapause, short-winged individuals occur only in early summer.

The semivoltine antlion, *Hagenomyia micans*, has an interesting set of photoperiodic responses that includes a short-day/long-day requirement for diapause termination. Young larvae (first or second instars) enter diapause at the end of summer in response to short day lengths; growth resumes the following spring and summer irrespective of photoperiod. However, third instars do not pupate until they have experienced a series of short days followed by long days. The short days must be less than 14 hours of light, and there is no conclusive evidence that the larvae respond to a relative change in day length. The differential responses of the larval instars ensure maintenance of a semivoltine life cycle (Furunishi & Masaki 1983).

The above examples largely involve spring-breeding insects. Insects that enter aestival diapause and that reproduce in autumn or winter may also require reverse changes in day length that cross a critical photoperiod in order to avert or terminate diapause. These species have a "long-day/short-day requirement for diapause prevention and/or termination." For example, decreasing day lengths that cross a critical photoperiod appear to play a role in terminating aestival diapause in the chrysomelid *Galeruca tanaceti* (Siew 1966). Similarly, the European, autumn-breeding carabids, *Patrobus atrorufus* and *Nebria brevicollis* (Thiele 1969, 1971), and the North African winter-breeding carabids *Broscus laevigatus* and *Orthomus barbarus atlanticus* (Paarmann 1974, 1976a) may require long days, and then short days, for full ovarian maturation. The latter species also requires a decrease in temperature simultaneous with the decrease in day length.

D. The final category contains insects that respond to changes in day length without reference to a critical photoperiod, i.e., they respond to the actual day length *changes*, not to particular day lengths. From the above studies it is clear that insects react not only to the actual length of day or night (i.e., as expressed by the response curve and critical photoperiod), but that they also perceive and respond to changes in day length. However, all the above examples involve at least one critical photoperiod. This leaves the question whether the *change* in day length *alone* can induce or terminate diapause independently of a critical photoperiod.

It has been shown that *Chrysopa carnea*, which has a critical photoperiod for diapause induction, can enter diapause in response to an abrupt decrease in day length that does not encroach on the critical photoperiod. Both LD 18:6 and LD 14:10 fall above the critical photoperiod, and when larvae and adults are maintained under either of these photoperiods, diapause is averted. However, if larvae that had been reared under LD 18:6 are transferred to LD 14:10, diapause is induced in 29% of the animals; conversely a change from LD 8:16 to LD 12:12, both of which are below the critical photoperiod, completely averts diapause (Tauber & Tauber 1970a). Diapause is induced in *C. carnea* at the end of August, before the critical photoperiod is reached, and it is therefore possible that response to decreasing day length plays a role in diapause induction. Moreover, diapause in *C. carnea* is maintained by the actual duration of short autumnal and hibernal day lengths and it terminates without a photoperiodic stimulus (Tauber & Tauber 1973d, e).

A gradual increase in day length below the critical photoperiod reduces the incidence of diapause in the onion fly, *Hylemyia antiqua*, but a decrease above the critical photoperiod does not increase it (Vinogradova 1978; see also Vinogradova 1976a). Studies with the bean bug, *Riptortus clavatus*, show that at least some individuals (23-25%) are capable of responding to an actual change in day length. In this species, as in *C. carnea*, both a critical photoperiod and response to changing day lengths are involved in measuring seasonal time, and also as in *C. carnea*, the ecological significance of the responses is not clear (Numata & Hidaka 1983). It therefore remains to be shown that insects normally enter or terminate diapause in response to day-length changes alone, without the influence of a critical photoperiod.

Although the data are not entirely clear, it is possible that in the semivoltine carpet beetle, *Anthrenus verbasci*, which apparently has a circannual rhythm of development, day-length changes *per se* are important in regulating the seasonal cycle (Blake 1958, 1959, 1960, 1963). Here, however, they may act as a "Zeitgeber" for the circannual rhythm, rather than as a token stimulus for diapause induction or termination (see Saunders 1982). In view of recent claims, especially by Soviet scientists, regarding annual cycles of photoperiodic sensitivity (see Section 3.4.3), this type of response deserves further study. Attempts to find a circannual rhythm of development in Japanese populations of *A. verbasci* have not been successful (Masaki unpublished).

5.1.4 Photoperiod and the prediapause period

An essential feature of each insect's life cycle is its ability to reach the diapausing stage at the appropriate time of year (see Sections 3.3.1 and 4.1 to 4.4). Two features are involved: (1) the cessation of growth and development in time to avoid the unfavorable effects of seasonal exigencies and (2) the attainment of the diapausing stage with sufficient time to make the metabolic, physiological, and other preparations for adverse seasonal conditions. In some cases, insects use photoperiod to regulate their prediapause developmental rates and to keep growth and development in time with the progression of the seasons. For example, photoperiod controls the rate of nymphal development and the number of nymphal molts in several species of crickets that overwinter as eggs. Because morphogenesis to the adult stage is accelerated by short days and decelerated by long days (Masaki 1966,

1967a, 1972, 1978a), the primary function of this response pattern apparently is to ensure that adults emerge and deposit diapausing eggs during autumn, before conditions become too cold. A similar developmental reaction has also been found in the univoltine species, *Pteronemobius nitidus*, that overwinters as a nymph. Nymphs grow faster, undergo fewer molts, and reach the diapausing stage earlier under short than under long days (Tanaka 1978a, b, 1979). Photoperiodic control of prediapause growth rates is also known for *Gryllus campestris*, another species of crickets that undergoes diapause in the nymphal stage (Fuzeau-Braesch & Ismail 1976; Ismail & Fuzeau-Braesch 1976a). Additional examples occur in various Lepidoptera, such as the sod webworm *Crambus tutillus* (Kamm 1972) and the noctuid *Mamestra genistae* (Akhmedov & Abdinbekova 1977), in the linden bug, *Pyrrhocoris apterus* (Saunders 1983), and perhaps in the apple leaf miner, *Phyllonorycter ringoniella* (Ujiye 1983).

In contrast to the above, short day lengths cause some species to decelerate their rate of growth and extend their feeding period prior to diapause. This occurs in the pecan scorch mite, *Eotetranychus hicoriae*, prior to adult diapause (Micinski *et al.* 1979). Similar photoperiodic retardation of prediapause growth and ecdysis can also precede larval diapause; such is the case in the sarcophagid fly *Sarcophaga argyrostoma* (Denlinger 1972b; Saunders 1972) and among certain mosquitoes, such as *Toxorhynchites rutilus*, that undergo diapause as fourth-instar larvae (Bradshaw & Holzapfel 1975). Similarly, under diapause-averting, long-day conditions, the noctuid *Amathes c-nigrum* completes its entire larval development of five to seven instars within 21 to 33 days. Diapause-inducing short day lengths result in the extension of the prediapause developmental period of the first four larval instars to about 27 days, followed by diapause of about 26 days in the fifth instar. Postdiapause development involves the sixth to eighth instars and requires another 14 days (Honek 1979b).

A parallel situation may occur in certain moths in which short day lengths experienced during early larval stages reduce the rate of oviposition prior to diapause induction in larvae of the next generation (Deseö & Sáringer 1975b). In these species, there appear to be two series of sensitive stages. First, the embryo and first instar larvae are sensitive to day lengths that are below a critical level and that affect oviposition rates (Deseö & Sáringer 1975a). Later, second- and third-instar larvae of the following generation presumably have a shorter critical photoperiod that functions in inducing prepupal diapause (Jermy 1967).

In addition to the regulation of prediapause growth rates, photoperiod sometimes exerts significant influence on prediapause morphological changes—notably the development of wings. The cricket *Pteronemobius taprobanensis* provides an interesting example in which short day lengths prior to diapause favor the production of macropterous forms (Tanaka *et al.* 1976; Tanaka 1978a, b). Perhaps the most striking photoperiodically induced prediapause morphological changes occur in the aphids (see Section 2.2.3.2).

5.1.5 Photoperiodic induction and intensification of diapause

As shown above (Sections 5.1.1 to 5.1.3), numerous studies have examined the photoperiodic induction and intensification of diapause in the laboratory. From these studies it has become very clear that photoperiod has a profound effect on the timing of diapause induction and the ultimate depth of diapause in many, if not most, insect species. It has also become clear that seasonal changes in day length affect diapause induction and intensification in diverse ways. Despite these significant advances, few studies have attempted to apply knowledge derived from laboratory studies to the prediction of diapause induction in natural populations in the field. In part, this situation exists because conditions are much more complex in the field than in the laboratory. Numerous variables come into play under natural conditions; among them are the relative photoperiodic sensitivity of the various stages (see Section 3.2), the insects' differential responses to various photoperiods above and below the critical photoperiod, the effects caused by changing day lengths and by the previous photoperiodic experience of the insects, spectral sensitivity as it pertains to the perception of twilight, and the interaction between photoperiod and other environmental variables. Thus, it is not surprising that very few studies have attempted to isolate and define experimentally the effects of photoperiod on diapause induction in natural populations.

To start with, such an effort requires the sampling of individuals from field populations during the prediapause and diapause induction periods and the testing of their responses to numerous photoperiodic conditions, including natural photoperiod. During these tests temperature and other variables are held constant, so that the effects of photoperiod on diapause induction can be defined and quantified. Conversely, alteration of temperature or other variables, while photoperiods are held constant, would allow the assessment of the effects of these other factors on diapause induction and intensification. We are aware of no such studies, and as a result the ability to predict diapause induction and to characterize diapause intensification in field populations lags far behind our knowledge of photoperiodic responses in the laboratory.

5.1.6 Photoperiod and diapause maintenance

Photoperiod is one of two major factors that act to maintain diapause, and some species rely almost exclusively on photoperiod for this function. As in the case of diapause induction, numerous laboratory studies have demonstrated the importance of day length in maintaining diapause, but in contrast to the situation with diapause induction, researchers have sampled and tested the photoperiodic responses of insects undergoing diapause in the field. As a result, the role of photoperiod as a diapause-maintaining factor is becoming better known (Tauber & Tauber 1976a).

Perhaps the first report describing the comprehensive sampling of overwintering insects to examine the photoperiodic maintenance of diapause was that of McLeod and Beck (1963). This study, which tested the photoperiodic responses of the European corn borer, *Ostrinia nubilalis* (long days versus constant darkness), throughout autumn, winter, and early spring, illustrated that photoperiod plays the major role in preventing diapause termination during autumn. As diapause proceeds, the larvae lose their responsiveness to day length, and by mid-December, when temperatures are well below developmental thresholds, all response to photoperiod ends.

In a somewhat similar study involving the testing of diapausing individuals under long and short day lengths throughout late summer, autumn, and winter, Lutz and Jenner (1964) demonstrated that long day lengths maintain the early phases of aestival diapause in larvae of the dragonfly *Tetragoneuria cynosura.* After the autumnal equinox, there is a reversal in response, and short day lengths take over the diapause-maintaining function. As autumn and winter proceed, response to photoperiod diminishes.

The above technique of transferring field-collected, naturally diapausing individuals into stationary day lengths throughout hibernation has demonstrated in a number of species photoperiodic maintenance of diapause during autumn and the gradual loss of sensitivity to photoperiod as diapause progresses. Among them are the odonate, *Lestes eurinus* (Lutz 1968, 1974), the green lacewings, *Chrysopa oculata* and *C. harrisii* (Propp *et al.* 1969; Tauber & Tauber 1974), the linden bug, *Pyrrhocoris apterus* (Hodek 1971a, 1974), the flesh fly, *Sarcophaga bullata* (Denlinger 1972a), and the acridoid, *Tetrix undulata* (Poras 1973, 1976). However, this test does not provide information on how the insects perceive day length, and it does not indicate how the insects respond to naturally changing day lengths.

To solve these problems, the technique was broadened to include tests of the responses of diapausing insects to naturally changing photoperiods, as well as to a full range of constant day lengths (Tauber & Tauber 1973e). When the expanded test was used on the green lacewing, *Chrysopa carnea*, it indicated that autumnal diapause in this species is maintained by the actual duration of the decreasing day lengths. There was an inverse relationship between day lengths below the critical photoperiod and diapause duration; natural photoperiods gave results equivalent to those of the shortest day of the year (Fig. 5.3). Thus, although diapause is not maintained by a change in day length *per se*, it is clearly affected by the natural progression of decreasing day lengths throughout autumn. We refer to this type of response as a quantitative or graded response to day length (see examples in Table 5.3).

Other species whose responses to a range of photoperiods have been tested over the natural course of diapause show a different type of reaction to day length—one that involves a distinct critical photoperiod. Species in this category show discrete responses to day length. For example, during autumnal-hibernal diapause, all day lengths shorter than the critical photoperiod maintain diapause equally well, whereas those longer than the critical

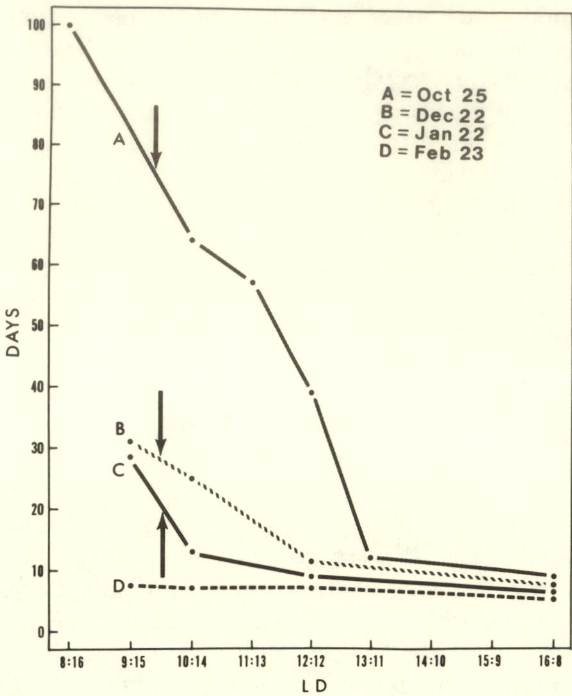

Fig. 5.3 Quantitative response to day length by overwintering *Chrysopa carnea.* Periodical samples (A through D) were transferred from the field to various photoperiodic conditions in the laboratory (points on curves) and a heated greenhouse (arrows on curves) (~24°C). The response curves for diapause termination illustrate that diapause depth decreases as autumn and winter progress and that the natural progression of day lengths maintains diapause until sometime in February. Data from Tauber & Tauber 1973e.

photoperiod terminate diapause. Such is the case in the grass spider, *Agelena limbata,* which overwinters as second-instar spiderlings in the egg sack. When diapause terminates, the spiderlings emerge from the egg sack. The critical photoperiod for diapause maintenance gradually decreases during winter, but all day lengths below the critical photoperiod remain equal to each other in their ability to maintain diapause. As the critical photoperiod decreases, the duration of the developmental time required for emergence also gradually decreases (Kurihara 1979). In addition to the grass spider, several other species that have been sampled from the field possess this type of response (Table 5.3, categories 1A and 1B). In some of these, the critical photoperiod for diapause maintenance does not change or changes only somewhat throughout autumn and winter, and ultimately spring day lengths that exceed the critical photoperiod are responsible for diapause termination (see Section 5.1.7), and in others, such as the European corn borer, the green lacewing, *Chrysopa carnea,* and the grass spider, *A. limbata,* the crit-

Table 5.3 Types of photoperiodic responses involved in maintenance and termination of autumnal and hibernal diapause in the field. For aestival diapause, reciprocal responses prevail.[a]

Stimulus	Response under field conditions	Studies that tested field populations
1. Critical photoperiod (long vs. short days)	All-or-none[b]	
A. Sensitivity to photoperiod persists throughout diapause	A. Short days maintain diapause and long days terminate diapause	*Wyeomyia smithii* (Smith & Brust 1971) *Meleoma signoretti* (Tauber & Tauber 1975a) (prob) *Coleomegilla maculata* (Obrycki & Tauber 1979) *Pteronemobius nitidus* (Tanaka 1983)
B. Sensitivity to photoperiod diminishes as diapause progresses	B. Short days maintain diapause but long days do not time its termination in the spring	*Chrysopa harrisii* (Tauber & Tauber 1974) *Tipula subnodicornis* (Butterfield 1976) *Perilitus coccinellae* (Obrycki & Tauber 1979) *Agelena limbata* (Kurihara 1979) *Tetrastichus julis* (Nechols *et al.* 1980)
2. Absolute duration of day length	Graded[c]	
A. Sensitivity to photoperiod persists throughout diapause	A. Short or decreasing day lengths maintain diapause; increasing day lengths accelerate the rate of diapause development and terminate diapause in spring	*Chrysopa downesi* (Tauber & Tauber 1976c) *Tipula pagana* (Butterfield 1976) *Adalia bipunctata* (Obrycki *et al.* 1983a)
B. Sensitivity to photoperiod diminishes as diapause progresses	B. Decreasing day lengths decelerate diapause development and therefore maintain diapause; photoperiod has no active role in terminating diapause in spring	*Chrysopa carnea* (Tauber & Tauber 1973d, e) (prob) *Pyrrhocoris apterus* (Hodek 1974) *Tetragoneuria cynosura* (Lutz & Jenner 1964; Lutz 1974)
3. Direction of change in day length	?	None

[a]For example, long days maintain aestival diapause, whereas short or shortening days terminate it.

[b]The rate of diapause development is either fast or slow depending on which side of the critical photoperiod the day length occurs.

[c]The rate of diapause development is quantitatively related to daylength.

ical photoperiod for diapause maintenance diminishes as diapause progresses and diapause ends spontaneously. Diapause in the midge *Chironomus tentans* (Englemann & Shappirio 1965) is photoperiodically maintained and it may also be photoperiodically terminated, but samples were not taken throughout the overwintering period to confirm this claim.

The importance of evaluating the responses of field-collected insects to a full range of photoperiods is illustrated by recent findings with the linden bug, *Pyrrhocoris apterus*. This species was found to exhibit an unusual response to photoperiod in the laboratory, but previous studies of field samples were not sufficiently comprehensive to demonstrate that the response occurs in nature. In the laboratory, the critical photoperiod for diapause maintenance in *P. apterus* increases rather than decreases during the early phases of diapause; that is, the critical photoperiod for diapause induction for a population from Bohemia falls around 15.75 hours of light per day, whereas after three to four weeks of short days, diapausing adults exhibit a critical photoperiod for diapause maintenance (= diapause termination of Saunders 1983) of about one hour longer (Saunders 1983). Whether such a phenomenon occurs in naturally diapausing linden bugs is not known because Hodek (1971a, 1974) used only long and short day lengths in his tests of field populations.

5.1.7 Photoperiodic termination of diapause

In examining the role of photoperiod in diapause regulation, it is important to keep in mind the different functions of diapause maintenance and diapause termination. It can be argued that if short day lengths maintain diapause, then the absence of short days, which is synonymous with the presence of long days, terminates diapause. This reasoning may be applicable to laboratory experiments that involve tests of diapause responses at a particular time during the course of diapause. But, it does not apply when the photoperiodic responses are viewed *in toto* and in their dynamic context—that is, over time as diapause progresses. Thus, responsiveness to photoperiod may diminish gradually as diapause progresses, as described above for several species, with photoperiod playing no role in diapause termination. In these cases, some other factor, such as temperature, may act as a second diapause-maintaining factor or diapause development may just run its course to completion. In other cases, photoperiodic sensitivity may persist throughout diapause or it may become important as a diapause-maintaining factor during a second phase of diapause, as was shown for the damselflies *Lestes disjunctus* and *L. unguiculatus* (Sawchyn & Church 1973). In these cases, photoperiod not only maintains diapause, but a specific photoperiodic condition is required before diapause terminates. Without the appropriate photoperiodic cue, diapause would persist long beyond the time when development should begin. It is these cases, therefore, that we are concerned with when we discuss the photoperiodic *termination* of diapause. An example is provided by the univoltine saturniid *Antheraea yamamai*, which hibernates in the egg stage and aestivates at the pupal stage. Aestival diapause in the pupae is induced if the larvae experience long days (LD 16:8); transfer of diapausing pupae to short day lengths terminates diapause within a few days (Kato & Sakate 1981). However, when aestivating pupae are maintained under long days, diapause persists more than three months. Under natural day lengths, adults emerge mainly during late August or September, one to two months after pupation. Therefore, it appears that, in the field, photoperiod is responsible for both maintaining and terminating the aestival diapause (see also Kato *et al.* 1979).

The distinction of diapause maintenance from diapause termination, at first glance, may seem quite subtle and perhaps unimportant, but this is not the case. For phenological studies to predict the posthibernation or postaestivation activity of insects, or for predictive phenological models to be useful in insect pest management, it is essential to know what factors regulate diapause termination, when diapause terminates in nature, and when post-diapause heat accumulations begin.

Most studies examining the photoperiodic termination of diapause provide only indirect evidence for the phenomenon. Studies of laboratory populations have led to the logical conclusion that photoperiod may play a role in diapause termination in some species. For example, diapausing individuals of *Antheraea pernyi* (Williams & Adkisson 1964), *Aedes solicitans* (Anderson 1970), and the damselflies *Enallagma hageni* and *E. aspersum* (Ingram 1975) remain sensitive to day length after long periods of exposure to low temperature, and from this it is concluded that long day lengths, after the winter cold, terminate diapause.

Studies that provide direct evidence for photoperiodic termination of diapause in the field are scarce, but they exist. Photoperiodic termination of winter diapause has been demonstrated or logically concluded through tests of the photoperiodic responses of naturally diapausing individuals of the cricket, *Pteronemobius nitidis* (= *Nemobius yezoensis*) (Masaki & Oyama 1963; Tanaka 1978a, 1983), the culicid, *Wyeomyia smithii* (Smith & Brust 1971), the lacewings, *Meleoma signoretti* and *Chrysopa downesi* (Tauber & Tauber 1975a), and the tortricid, *Platynota idaeusalis* (Rock & Shaffer 1983).

As illustrated in Table 5.3, the photoperiodic termination of diapause involves two main mechanisms. The differences between the mechanisms are shown by comparing the two lacewing species, *Meleoma signoretti* and *Chrysopa downesi*, both of which are univoltine and both of which use photoperiod to time the end of diapause in late winter-early spring. Diapause in *M. signoretti* ends when day lengths exceed a critical length (see Fig. 5.2); in the field, this occurs at the end of March. By contrast, the rate of diapause development in *C. downesi* during winter is quantitatively related to day length (Fig. 5.4). Thus, as winter day lengths increase, diapause development accelerates; this results in diapause terminating at about the time of

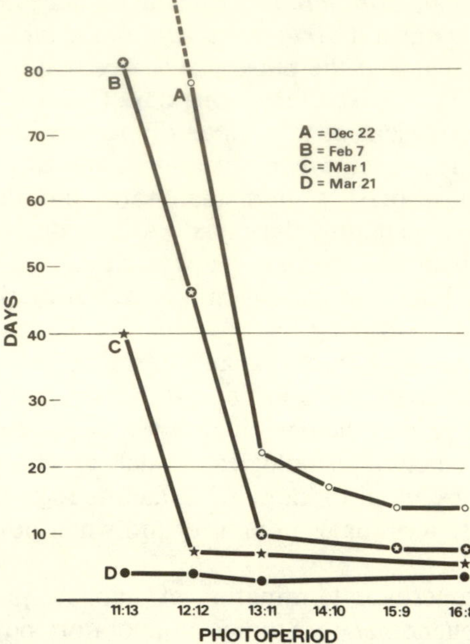

Fig. 5.4 Diapause termination by photoperiod in *Chrysopa downesi.* Periodical samples (A through D) were transferred from the field to various photoperiodic conditions in the laboratory and a heated greenhouse (~24°C). As day lengths increase during late winter, the rate of diapause development increases and diapause terminates in the field at about the time of the vernal equinox. Data from Tauber & Tauber 1976c.

the vernal equinox, when day lengths are increasing at the maximum rate (Tauber & Tauber 1975a).

There is good evidence that photoperiod ends diapause in several other univoltine or semivoltine species that require a sequence of short days followed by long days (spring breeders) or long days followed by short days (autumn breeders) (Müller 1960a; Thiele 1968, 1973, 1977; Krehan 1970; Hoffmann 1970; Neudecker & Thiele 1974; Ferenz 1975a, 1977; Bale 1979; Furunishi & Masaki 1983). Whether photoperiod acts in these cases to terminate diapause through a critical photoperiod or through changing day lengths is not known. It is also of interest to note that at least in some of the spring-breeding species (e.g., the beech leaf mining weevil, *Rhynchaenus fagi*, and the carabid, *Pterostichus nigrita*), only the females require a long day stimulus to terminate diapause. Sexual maturation occurs in males either spontaneously or after they have experienced short day lengths (Ferenz 1975b; Thiele 1977; Bale 1979).

Summer diapause is much more likely than hibernal diapause to require a specific stimulus for termination (Tauber & Tauber 1976a; Masaki 1980). This generalization is illustrated by the craneflies *Tipula subnodicornis* and *T. pagana*, which undergo diapause during winter and summer, respectively, and which vary accordingly in their diapause-terminating responses to photoperiod (Butterfield 1976). Short day lengths induce and maintain autumnal-hibernal diapause in *T. subnodicornis*; response to day length diminishes as winter progresses, and photoperiod apparently plays no role in diapause termination (Butterfield 1976; Coulson *et al.* 1976). By contrast, autumnal emergence by *T. pagana* is synchronized by the photoperiodic termination of aestival diapause. A quantitative response to the shortening day lengths of autumn accelerates diapause development and results in a decrease in variation around the mean dates of pupation and emergence (Butterfield 1976).

Photoperiodic (short-day) termination of aestival diapause also occurs in the sawfly, *Neodiprion sertifer* (Minder 1980). Most of the diapausing eonymphs complete development and give rise to adults during late summer or early autumn. Eonymphs that do not terminate diapause in the late summer undergo prolonged diapause and regain their sensitivity to day length over winter; subsequently, long days of late spring and early summer delay emergence until day lengths begin to decrease in late summer. Other species in which photoperiod apparently serves to terminate aestival diapause are the lady beetle, *Coccinella novemnotata* (McMullen 1967a, b), the fungus fly, *Drosophila phalerata* (Geyspits & Simonenko 1970), the grasshopper, *Oedipoda miniata* (Pener & Orshan 1980), and the papilionid, *Luehdorfia japonica* (Ishii & Hidaka 1982, 1983).

Danilevsky *et al.* (1970) and Masaki (1980) give several examples of summer diapause for which photoperiod is the terminating cue. In none of these examples has it been established which aspect of late summer, or autumnal photoperiod (i.e., either day lengths below a critical photoperiod or decreasing day lengths) is the important factor. However, in the laboratory, decreasing day lengths that cross the critical photoperiod appear to be more effective in ending summer diapause in the chrysomelid *Galeruca tanaceti* than any short stationary photoperiod (Siew 1966).

5.1.8 Photoperiod and postdiapause development

Although day length can influence development, fecundity, and male sexual behavior during the prediapause and nondiapause periods, it is generally accepted that photoperiod does not affect the rate of morphogenesis of postdiapause insects as they develop to the next stage of metamorphosis. For example, after hibernation the development of postdiapause larvae leading to pupation occurs at equal rates under both long and short day lengths in the European corn borer, *Ostrinia nubilalis* (McLeod & Beck 1963), and the green lacewings, *Chrysopa oculata* and *C. nigricornis* (Tauber & Tauber unpublished).

Similarly, in most cases, the postdiapause oviposition rates of insects with imaginal diapause are similar under long and short day lengths; examples are found in the Colorado potato beetle, *Leptinotarsa decemlineata* (de Wilde *et al.* 1959), the green lacewing, *Chrysopa carnea* (Tauber *et al.* 1970a), the milkweed bug, *Oncopeltus fasciatus* (Dingle 1974b), and the ladybird beetle, *Semiadalia undecimnotata* (Hodek & Růžička 1977). However, in certain other species that undergo diapause as adults, such as the pentatomids, *Aelia acuminata* (Hodek 1971c, 1979, 1981) and *Dolycoris baccarum* (Hodek 1977), the bean bug, *Riptortus clavatus* (Numata & Hidaka 1982), the grasshopper, *Oedipoda miniata* (Pener & Broza 1971; Orshan & Pener 1979a, b; Pener & Orshan 1980), and the ladybird beetle, *Coccinella septempunctata* (Hodek *et al.* 1977), day length influences postdiapause oviposition and it can reinduce diapause after a period of oviposition. With the exception of *O. miniata*, these species all have long-lived adults and presumably can undergo a second winter in diapause (Hodek 1976).

5.1.9 Summary

Diapause is a genetically determined pattern of response to environmental stimuli, and thus the expression of diapause is subject to both environmental and genetic factors. Categorization of diapause into either "facultative" or "obligatory" diapause tends to obscure our vision of the full range of environmental-innate interactions that characterize the diapause syndrome.

Among the environmental stimuli that influence diapause, photoperiod is by far the most prominent. Insects vary greatly in the ways in which they respond to photoperiod and also in the degree to which they depend on photoperiod to regulate their life cycles. Because of its multifaceted nature, diapause has a highly dynamic relationship with photoperiod—that is, responses to photoperiod may undergo major changes as the prediapause period progresses and as diapause is induced, maintained, and finally terminated.

The response curve and the critical photoperiod are useful in describing the responses of insects to static photoperiods, and they indicate a variety of responses to day length. However, they have severe limitations in that they do not convey information regarding the depth or intensity of diapause induced by various photoperiodic conditions, nor do they convey the diverse influences of changing day lengths.

Although numerous studies have examined the photoperiodic induction of diapause in the laboratory, few have followed the changes in photoperiodic response during the full course of diapause under natural conditions. The variables involved in such a study are numerous, but with careful planning some of the components can be isolated and tested.

Photoperiodic maintenance and termination of diapause have been studied under natural conditions for a few species. In all cases, the results indicate a dynamic role for photoperiod over the course of diapause and a wide variety in the types of photoperiodic responses involved. In most cases, diapause termination is characterized by the loss of all response to photoperiod; however, there are some cases of imaginal diapause in which photoperiod influences the subsequent rate of oviposition.

The dynamic action of photoperiod on the full expression of the diapause syndrome makes it apparent that diapause has evolved to take advantage of the seasonal progression of photoperiods. It is not centered on just one "critical photoperiod" to which all expressions of the diapause syndrome are linked; rather it involves responses to a series of photoperiods (and/or other factors) that exert their influence as the course of diapause proceeds. In many cases, prediapause developmental responses, which ensure the proper timing of diapause induction, are regulated by photoperiodic conditions that are separate from those that actually induce diapause. Diapause induction itself may require a series of photoperiods, and the photoperiodic conditions that maintain and, in some cases, terminate diapause are different from the photoperiodic conditions that act earlier during diapause. Thus, the interaction between diapause and photoperiod must be viewed as complex and dynamic. That is, the numerous expressions of the diapause syndrome are induced over time in response to series of photoperiodic or other stimuli, and the responses to these stimuli may change drastically as diapause proceeds through its full course.

5.2 Temperature

Next to photoperiod, the most important factor affecting diapause is temperature. Although unpredictable for any given day, temperature conditions have a regular seasonal pattern, and insects have evolved various response patterns that use seasonal changes in temperature as cues for the regulation of their life cycles.

Temperature influences diapause in a variety of ways throughout the course of diapause. In this section, we discuss the types of influences, and we distinguish between the role temperature has as a token stimulus and that as a modulator of the rates of diapause induction, diapause development, and postdiapause development. We illustrate the diversity of ways in which temperature influences diapause and we stress that, like photoperiodic responses, responses to temperature change drastically over the course of the seasonal cycle.

5.2.1 Temperature as a diapause-inducing stimulus

Although photoperiod is the major diapause-inducing environmental stimulus for most arthropods, for others it has little or no role, and temperature may be the primary diapause-inducing cue. In such cases, temperature not only has an immediate effect on the rate of growth or development, but it also serves to induce the diapause syndrome and prepare the insect for approaching seasonal changes. In this role, temperature acts as a token stimulus that is perceived by a sensitive stage(s), and whose effect is not expressed until the diapausing stage is reached.

The function of temperature as the *primary* diapause-inducing stimulus has been shown for relatively few temperate-zone species. Photoperiod appears to have little effect on diapause induction in the muscoid fly *Chortophila brassicae*; however, the temperature conditions that the subterranean larvae experience are important in inducing pupal diapause. When larvae experience constant temperatures below 15°C, the incidence of pupal diapause is high. Delay in pupal development is not a direct effect of low temperature on development because it persists in pupae after they are transferred to 20°C (Missonnier 1963).

Low temperature plays the reverse role and acts as a diapause-averting stimulus in another soil-inhabiting species. Under mild temperatures (22°C), mature larvae of the scarab *Anomala cuprea* enter diapause without respect to photoperiod. However, if late second-instar or third-instar larvae experience cold conditions (5°C), diapause is averted. Low temperature also serves to hasten diapause termination in this species (Fujiyama & Takahashi 1973a, b, 1977). Larvae of the ground beetle *Pterostichus vulgaris* appear to have a similar response to temperature. At temperatures of 15°C or higher, metamorphosis is halted at the third instar; dormancy is terminated and development resumes after the larvae experience temperatures below 7°C (Thiele & Krehan 1969; Krehan 1970). Diapause in other species of ground beetles is regulated by a similar mechanism (see Thiele 1977); German researchers refer to the phenomenon as thermal "parapause."

Low temperatures appear to play an important role in diapause regulation of wood and bark-inhabiting beetles (e.g., Dyer & Hall 1977; see also Ring 1977). However, often the design of experiments does not allow photoperiod to be ruled out as a cue. Similarly, temperature often dominates the seasonal cycles of insect pests found in houses and warehouses. In some species, such as the Indian meal moth, *Plodia interpunctella* (Bell 1976a, c, 1982; Bell *et al.* 1979), and the tropical warehouse moth, *Ephestia cautella* (Bell & Bowley 1980), temperature reactions are of major importance because photoperiod acts only within a very narrow temperature range, but in other species temperature alone induces and regulates diapause (Howe 1962; Baker 1982). By contrast, some pests of stored products use photoperiod (Bell 1976b, 1983) or population density (see Section 5.3.2) as cues for diapause.

Analogous types of responses to temperature occur in insects near the equator, where seasonal changes in day length are small. If larvae and pupae from equatorial populations of the muscoid fly *Poecilometopa punctipennis* are reared at 20°C, development proceeds without diapause. At 18°C a few pupae enter diapause, and as the temperature decreases the incidence of diapause increases. Diapause is not the result of low-temperature action on the pupal stage because young pupae transferred to cold conditions do not enter diapause (Denlinger 1974). Similarly, low temperatures induce and maintain diapause in *Leptopiliana boulardi*, a tropical cynipid parasite of *Drosophila* (Claret & Carton 1980). High temperatures may function as an inducing stimulus for aestival diapause in tropical and subtropical insects, such as in the North African carabid *Pogonus chalceus* (Paarmann 1976b, 1979a).

Still other cases of diapause induced by high temperatures are reported for the harvestman, *Mitopus morio* (Tischler 1967), and for the acridid, *Locustana pardalina* (Matthée 1978), and the carrot fly, *Psila rosae* (Brunel 1968). In *L. pardalina* the incidence of egg diapause is directly related to the duration of exposure to high temperatures. Constant temperature of 32.5°C results in no diapause; two hours of 45°C produces a low (~ 5%) incidence of diapause; and 30 hours of 45°C induces diapause in over 75% of the eggs. In the striped ground cricket, *Allonemobius fasciatus,* high temperatures induce aestival diapause in an early stage of embryonic development; the higher the temperature, the earlier the stage at which diapause occurs. Aestival diapause is terminated by low temperatures and is followed by a second, hibernal diapause (Tanaka 1984).

In some insects in which larval development spans a considerable period, temperature exerts its influence by determining the stage at which diapause intervenes. If, for example, Japanese beetles, *Popillia japonica*, are reared at 25°C, diapause occurs during the third instar. At 20°C, growth is interrupted during the second or third instar, and at 15°C during the first instar (Ludwig 1932). Similar responses to temperature are found in the lymantriid *Orgyia gonostigma* (Kozhantshikov 1948), the bug *Reduvius personatus* (Readio 1931), and the cricket *Pteronemobius nitidus* (Tanaka 1979). In nature, such a response pattern probably increases the survival rate of late-developing individuals by allowing them to enter diapause early in their metamorphosis. It is of interest to note the similarity in this thermal response and an analogous photoperiodic response. Several species that enter diapause in any of several middle instars tend to do so at an earlier instar under short day length than under long days. Examples include the ground cricket, *Pteronemobius nitidus* (Tanaka 1979), the arctiid, *Spilarctia imparilis* (Sugiki & Masaki 1972), and the butterfly, *Hestina japonica* (Shiotsu 1977). If the diapause stage were fixed at a particular instar in any of these species, individuals that failed to reach the diapausing instar would perish. If they are able to enter diapause at an earlier instar in response to low temperatures or shorter day lengths, they can survive.

Temperature influences which stage overwinters in the katydid *Leptophyes punctatissima.* This species has two periods of embryonic diapause, the initial period of which is induced by temperature. If newly laid eggs experience warm conditions (30°C), diapause in the early stages of embryogenesis is averted, and only the fully formed embryo enters diapause. In this case, the egg hatches the following spring, and a full life cycle is completed in one year. However, if the newly laid eggs are kept at 25°C, there is a delay in the start of embryogenesis. This delay, which constitutes the initial diapause, is greatly extended under lower temperatures of 16 to 20°C. Further reduction in temperature to a level between 8 and 12°C greatly facilitates subsequent embryonic development. Thus, eggs that are laid during late summer are not subjected to the high temperatures that avert the initial diapause, and their embryonic development does not begin until spring; the second diapause, which intervenes in the fully formed embryo, extends the life cycle to two years (Deura & Hartley 1982).

Temperature affects the overwintering stage in a somewhat similar manner in the cricket, *Gryllus campestris.* Growth is slow and uninterrupted by diapause at 20°C; at 30°C growth is rapid, but diapause intervenes. As a result, the total duration of the nymphal stage is almost the same at the two temperatures. Early-developing nymphs encounter high temperatures in their middle instars; these individuals enter diapause and hibernate. Late-developing nymphs are exposed to low autumnal temperatures before they reach the diapause stage; these individuals hibernate in a nondiapause condition (Fuzeau-Braesch & Ismail 1976; Ismail & Fuzeau-Braesch 1976a).

5.2.2 Thermal modification of photoperiodic responses

When temperature is not the major diapause-inducing stimulus, it may nevertheless interact with photoperiod in several ways to induce diapause. From an ecological viewpoint, the significance of this interaction is evident in the effects temperature has on the seasonal timing and duration of the sensitive stages and in its modification of the photoperiodic requirements for diapause induction.

Temperature influences growth rates and therefore exerts a strong effect on the proportion of individuals completing their photosensitive stages before or after the occurrence of diapause-inducing day lengths. Such an effect determines not only the numbers of individuals entering photoperiodically induced diapause, but, as shown earlier, the characteristics of the diapause thus induced (see Section 5.1.3).

In most species examined, the interaction between photoperiod and temperature is expressed in thermal alterations of the critical photoperiod (see reviews by Danilevsky 1965; Lees 1968; Beck 1980; Saunders 1982). In long-day insects (i.e., insects that develop without interruption under long days and that enter diapause in response to short days), low temperatures tend to promote diapause, whereas high temperatures tend to prevent it. In the noctuid moth *Acronycta rumicis*, a long-day insect, the critical photoperiod for diapause induction lengthens by 1½ hours with each five-degree Celsius drop in the temperature under which the insects are reared; that is, longer days are necessary to avert diapause at lower temperatures (Danilevsky 1965). Numerous other insect species have similar shifts in critical photoperiod in response to temperature (see Danilevsky 1965; Beck 1980; Saunders 1982), but there is considerable interspecific variation in the extent of the thermal effect on the critical photoperiod. Although *A. rumicis* (the Belgorod population) shows a shift of three hours in the critical photoperiod with an increase in temperature from 15 to 25°C, *Pieris brassicae* (the Leningrad population) has a shift of only one hour over the same range of temperatures (Danilevsky 1965).

In some long-day insects, temperature apparently does not alter the critical photoperiod to any great degree, but it has an important effect on whether or not the insects respond to photoperiod at all. In the muscoid fly *Sarcophaga argyrostoma*, which enters pupal diapause when larvae are reared under short days, all photoperiods at 25°C produce almost no diapause (Saunders 1971). However, under larval rearing conditions of both 15 and 20°C, the critical day length is approximately 13.5 to 14 hours of light per day. *Laspeyresia* (= *Grapholitha*) *molesta* enters diapause in response to short days only in the temperature range between 21 and 26°C. Short days at either 12 or 30°C do not induce diapause (Dickson 1949). *Diatraea grandiosella*, the southwestern corn borer, has a similarly narrow thermal range for photoperiodic induction of diapause. However, the response is somewhat different. In the Missouri population at 25°C a photoperiodic response is clearly evident; at 27°C all larvae develop without diapause, and at 23°C all larvae enter diapause regardless of photoperiod (Chippendale & Reddy 1973).

A little-studied phenomenon with regard to thermal influence on diapause induction is the effect of temperature changes. In *Sarcophaga argyrostoma*, an increase in temperature from 16 to 28°C during the early pupal stage reduces the incidence of diapause, and a reduction of six degrees Celsius increases the incidence of diapause (Gibbs 1975). Similar reactions are found in the blowfly *Calliphora vicina* and other sarcophagids (Vinogradova & Zinovyeva 1972a, b), and in *Heliothis punctigera* (Cullen & Browning 1978). Similarly, seasonal changes in temperature modify the diapause-inducing effect of changing day lengths. A decrease in temperature, in conjunction with a decrease in day length, greatly increases the incidence of diapause in the bollworm, *Heliothis zea* (Roach & Adkisson 1970), and in the mosquito, *Culiseta inornata* (Hudson 1977). By comparison, hibernal maturation of the gonads of the carabid, *Orthomus barbarus atlanticus*, in North Africa requires a simultaneous decrease in day length and a lowering of temperature (Paarmann 1976a).

Temperature also affects photoperiodic reactions during diapause termination, and again, high temperatures tend to enhance the growth-promoting effects of long day length in insects with hibernal diapause. *Drosophila transversa* can respond to diapause-terminating long day lengths at temperatures of 5, 10, or 15°C, but not at 0°C (Kuznetsova & Tyshchenko 1979). An interesting exception to the generalization regarding the enhanced effect of high temperature on diapause termination in hibernating insects is found in the dragonfly, *Plathemis lydia* (Shepard & Lutz 1976). Diapausing final-instar larvae are very sensitive to diapause-terminating (growth-promoting) long day lengths at temperatures between 15 and 20°C; above 20°C the effect of long day length declines. This type of response is in keeping with the species' life cycle in that larval maturation begins in spring while water temperatures are still relatively low, and it is completed by the time air temperatures are favorable for the adult.

In short-day insects or insects that aestivate, temperature's role is the reverse of that in autumnal and hibernal diapause; in these cases high temperatures enhance diapause induction and low temperatures promote diapause termination and growth (Masaki 1980). Summer diapause occurs in pupae of the noctuids *Mamestra brassicae* and *M. oleracea* when larvae are reared under long days; high temperatures increase both the incidence and duration of this diapause (Masaki & Sakai 1965; Poitout & Bues 1977a, b).

Another example is found in the aestival diapause of the grasshopper *Oedipoda miniata.* In this species, low temperatures in northern latitudes overcome the effect of long day lengths, and they result in early diapause termination, whereas high temperatures that prevail in the southern latitudes tend to prolong diapause. Such a mechanism allows diapause to terminate in synchrony with the seasonal occurrence of favorable local conditions throughout the whole distribution of *O. miniata,* without the evolution of a geographic cline in the critical day length for diapause termination (Pener & Orshan 1980).

5.2.3 Responses to thermoperiod

Most studies dealing with the effects of temperature on diapause have only considered constant temperatures. But temperature, like daylight, has diurnal and seasonal patterns of change, and these patterns can play an important role in influencing diapause. Beck (1983a) recently provided a comprehensive review of the effects of thermoperiod (24-hour cycles in temperature) on insect behavior and development, including its influence on diapause. From the data currently available, which Beck stresses are meagre, it appears that although thermoperiods may not induce diapause *per se,* they strongly influence the timing of diapause induction and the development of many symptoms of the diapause syndrome, such as cold-hardiness.

Thermoperiods usually enhance or diminish the effect of light cycles. Generally, cold nights tend to increase the incidence of diapause in long-day insects; this effect is especially noticeable at photoperiods near the critical photoperiod. Under LD 15:9, the critical photoperiod for its diapause induction, the European corn borer, *Ostrinia nubilalis,* shows a high incidence of diapause when it experiences constant 21°C or a thermoperiod of 31°C during the day and 21°C during the night. However, a day temperature of 21°C combined with a night temperature of 31°C results in a much lower incidence of diapause (Beck 1962). The same thermoperiod in constant darkness gives a low incidence of diapause. Numerous other species show a similar effect (see Beck 1983a; also Veerman 1977; Vinogradova 1975a; Zinovyeva 1976). Such responses, although they do not indicate a thermoperiodic effect, do show that night temperatures influence diapause induction (Beck 1983a, b).

Thermoperiodic induction of diapause in the absence of light has been demonstrated in very few species, for example, the European corn borer, *O. nubilalis* (Beck 1962, 1982, 1983b, 1984), the parasitic wasp, *Nasonia vitripennis* (Saunders 1973b), the pink bollworm, *Pectinophora gossypiella* (Menaker & Gross 1965), the southwestern cornborer, *Diatraea grandiosella* (Chippendale *et al.* 1976), *Pieris brassicae* (Dumortier & Brunnarius 1977), *Plodia interpunctella* (Masaki & Kikukawa 1981), and the tufted apple budworm, *Platynota idaeusalis* (Rock 1983). In these cases, thermoperiod clearly acts as a periodic stimulus triggering the internal regulation of development, rather than through a direct influence on metabolic rate. Its effects simulate the photoperiodic regulation of diapause.

5.2.4 Temperature and diapause intensification

For some species, temperature can modify the effect of photoperiod after the photoperiodic stimulus has been received by the sensitive stage, but before the insect achieves the full expression of diapause. Two types of response to temperature are known to occur during the prediapause period and very early phases of diapause. First, temperature extremes at these times can cancel the photoperiodic programming of diapause. In several species of *Heliothis*, pupal diapause occurs when late-stage larvae experience short day lengths and low temperatures. However, diapause is averted, regardless of larval conditions, if the prediapause insects are subjected to high temperature for a short period between larval-pupal apolysis and pupal ecdysis (Browning 1979; Hackett & Gatehouse 1982). Exposure to low temperature prior to full diapause induction can have a similar effect (Rakshpal 1962). And, in the Australian plague locust, *Chortoicetes terminifera,* exposure to either high or low temperature during the prediapause period may reverse the photoperiodic induction of egg diapause (Wardhaugh 1980b; Hunter & Gregg 1984).

Eldridge (1966) reported that prediapause *Culex pipiens* mosquitoes that had been reared under diapause-inducing short day length and low temperature failed to suppress ovarian development (did not undergo gonotrophic dissociation) after a blood meal if they were kept under high temperatures (25°C) after feeding. Approximately half of those kept under low temperature (15°C) suppressed ovarian development. By contrast, Sanburg and Larsen (1973) found that high temperature could negate the diapause-inducing influence of short day lengths only if the day lengths used to induce diapause were longer than LD 10:14. If the mosquitoes were reared at LD 10:14, diapause was induced regardless of temperature. The differences in the results obtained by Eldridge (1966) and Sanburg and Larsen (1973) could be related to the differences in genetic stock; their laboratory colonies originated from different geographical areas and were of unspecified age in the laboratory. Nevertheless, in both cases, temperature conditions during the very early phases of diapause had profound effects on diapause induction.

Second, temperature may influence diapause during its early phases by altering diapause intensification. Temperatures prevailing during diapause induction appear to influence the final diapause intensity in *Teleogryllus (Acheta) commodus* (Hogan 1960a) and the parasite *Hexacola* sp. (Eskafi & Legner 1974). And temperature, as well as photoperiod, affects the percentage of individuals switching from the short summer diapause to prolonged diapause in the sawflies *Neodiprion sertifer* and *N. abietis* (Sullivan & Wallace 1967; Wallace & Sullivan 1974).

The process of diapause intensification (apart from diapause induction) occurs within species-specific temperature limits. If the bertha armyworm, *Mamestra configurata*, experiences very low temperatures during early diapause induction, diapause does not achieve its normal depth (Bodnaryk 1978). Similarly, pupae of the fall webworm, *Hyphantria cunea*, undergo a more intense (longer) diapause at 15°C than if they are held under either 1°C or room temperature during early diapause (Masaki 1977). Larval diapause in the codling moth, *Cydia* (= *Laspeyresia*) *pomonella*, is more intense if larvae are reared and maintained under 26°C than either 19 or 21°C (Sieber & Benz 1980a). And, finally, the intensity of egg diapause in *Teleogryllus emma* is increased by exposing eggs to 30°C for a few days just before or after the onset of the diapause stage (Masaki 1962).

Adults of *C. carnea* that have been reared under diapause-inducing short day lengths, develop a deep diapause color and an intense diapause if the newly emerged adults experience a gradual decrease in temperature (from 24 to 21 to 18°C) during the diapause intensification period (first two to three weeks after adult emergence) (Tauber & Tauber unpublished; see also Honek 1973). From these data we conclude that the diapause syndrome develops fully if short day lengths are combined with low temperatures in a specific temporal pattern.

5.2.5 Temperature and diapause maintenance

To varying degrees, insects rely on temperature to maintain and terminate diapause. Among most insects, thermal reaction curves for diapause development are quite different from those for nondiapause development and growth (Andrewartha 1952; Lees 1955). Usually, during autumnal diapause both the upper and lower thermal thresholds, as well as the optimum temperature for the progression of diapause development, are shifted downward. In the parasitoid *Nasonia* (= *Mormoniella*) *vitripennis*, diapause terminates fastest (about 10 weeks) at 2°C; temperatures above 15°C and below −6°C are ineffective in promoting diapause development (Schneiderman & Horwitz 1958). In the fruit fly *Rhagoletis indifferens,* 3°C is the optimum temperature for diapause development; 0, −3, and temperatures above 6°C are less effective (Van Kirk & AliNiazee 1982). The threshold for postdiapause development is approximately 8°C (Van Kirk & AliNiazee 1981). Thus, in these species the entire thermal reaction curve for diapause development is lower than that for nondiapause or postdiapause development.

Although the temperature range for rapid diapause development is sometimes considerably below that for nondiapause development, there is wide interspecific variability in this characteristic. In addition to the examples above, the grasshopper *Austoicetes cruciata* shows an optimum range of 6 to 13°C for diapause development, whereas the lower thermal threshold for nondiapause development is higher than 13°C (Andrewartha 1943). In

Melanoplus bivittatus (Church & Salt 1952), the differences in thermal responses during diapause and nondiapause development are not as extreme, and although the optimum temperature for diapause development is low, the ranges for diapause and nondiapause development overlap considerably. In still other species, diapause development proceeds at temperatures similar to those for nondiapause growth; examples include *Ostrinia nubilalis* (McLeod & Beck 1963), *Leptinotarsa decemlineata* (de Wilde 1969), *Teleogryllus* spp. (Masaki 1963, 1965; Masaki *et al.* 1979), *Teleogryllus* (= *Acheta*) *commodus* (Hogan 1960a), *Sarcophaga argyrostoma* (Fraenkel & Hsiao 1968), *S. bullata* (Denlinger 1972b), *Chrysopa carnea* (Tauber & Tauber 1973c), *Diatraea grandiosella* (Chippendale & Reddy 1973), *Aedes atropalpus* (Kalpage & Brust 1974), and *Agromyza frontella* (Tauber *et al.* 1982).

Despite the above variation in thermal optima for diapause development, many textbooks and research reports continue to stress or imply that exposure to low temperature (= "chilling") is generally required for termination of hibernal diapause and that it is diapause termination by chilling that synchronizes postdiapause development or activity (e.g., Gillott 1980; Morden & Waldbauer 1980; Borror *et al.* 1981). Such generalizations can impede an understanding of the physiological and ecological adaptations accompanying diapause. Evidence indicates that the adaptive value of low thermal thresholds during diapause lies primarily in their function of maintaining diapause and therefore preventing premature development under warm conditions before winter (see Tauber & Tauber 1976a). High temperatures at this time either retard diapause development in insects with low thermal thresholds or, in some cases, they may even reverse it (Church & Salt 1952; Schneiderman & Horwitz 1958; Masaki 1962; Selander & Weddle 1972; Ando 1978), but in all cases they prevent premature development. The importance of lowered thermal responses to diapause maintenance is illustrated particularly well by the finding that the thermal responses of many overwintering insects return to nondiapause levels sometime during late fall or early winter (e.g., Nechols *et al.* 1980; Tauber *et al.* 1982; Collier & Finch 1983a). Low temperatures prevailing at this time have no further influence on diapause; rather they (a) prevent premature development, (b) prevent the use of metabolites that are essential for maintaining cold-hardiness or for promoting postdiapause development (see McLeod & Beck 1963; Goettel & Philogène 1980), and (c) synchronize the initiation of postdiapause development among individuals (e.g., Bartelt *et al.* 1981; Collier & Finch 1983a, b).

In those cases in which thermal reactions during autumnal diapause are not substantially different from those during the nondiapause period (e.g., *Leptinotarsa decemlineata, Chrysopa carnea*), photoperiod or some factor other than temperature may serve to maintain diapause (see Tauber & Tauber 1976a). By contrast, an interesting situation occurs in the alfalfa blotch leafminer, *Agromyza frontella*, in which diapause development proceeds at high temperatures, and yet temperature is the primary diapause-maintaining factor. In this case, low temperatures, rather than hastening the termination of autumnal diapause, act to maintain it. As a result of this mechanism, *A. frontella*, which has a relatively short generation time and a very malleable diapause, can foreshorten diapause if conditions are warm during late August or September and produce an additional late-season generation (Tauber *et al.* 1982; Nechols *et al.* 1983).

Malleability in a thermally controlled diapause also occurs in the carrot fly, *Psila rosae*, but this species has added flexibility in that two developmental stages—nondiapause larvae and diapausing pupae—can overwinter successfully. If larvae developing in autumn experience temperatures between 10 and 16°, they enter diapause on pupation. Higher temperatures promote pupal development. Thus, a third generation of flies emerges in autumn if temperature conditions are favorable. Nondiapausing, overwintering larvae appear to lose sensitivity to diapause-inducing temperatures gradually as autumn and winter progress; as a result, overwintered larvae that pupate under cool conditions in spring do not enter diapause but undergo imaginal development as soil temperatures rise (Burn & Coaker 1981).

Similarly, some tropical insects rely on low temperature for diapause maintenance. The tropical flesh fly, *Sarcophaga spilogaster*, enters and remains in diapause in response to low temperatures (Denlinger 1974). Return of warm conditions allows diapause development to proceed within three days. A similar situation occurs in the tropical cynipid *Leptopiliana boulardi*, which parasitizes *Drosophila* species (Claret & Carton 1980).

In addition to maintaining autumnal-hibernal diapause, temperature may also function to maintain aestival diapause. And, in this regard, it may act through two different mechanisms. First, during aestival diapause, lower thermal thresholds and optima for diapause development may be quite high, thus ensuring that diapause does not end before summer temperatures reach their height. The Australian predaceous mite *Bdellodes lapidaria* and its host, the phytophagous mite *Halotydeus destructor*, both use this mechanism for maintaining aestival diapause. Both species have very high thermal thresholds (30 and 52°C, respectively) for diapause development during summer (Wallace 1970a, 1971). Apparently the high thermal thresholds prevent the premature termination of diapause in late spring or early summer, and they ensure that diapause development takes place during summer. After summer, the eggs of both species require exposure to relatively low temperatures before they hatch; we discuss this phenomenon further in the next section.

In contrast to the species above, the lepidopterans *Mamestra brassicae* and *Heliothis armigera* use a second thermal mechanism for maintaining aestival diapause. These species have thermal thresholds and optima for diapause development that are lower than the prevailing summer temperatures. Thus, diapause is maintained until the low temperatures of autumn arrive (Masaki & Sakai 1965; Hackett & Gatehouse 1982).

5.2.6 Changing responses to temperature during diapause

The responses of diapausing insects to temperature are by no means static. At the beginning of diapause the thermal response curve (thermal optimum and the upper and lower thresholds) for diapause development may be shifted downward as diapause intensifies and proceeds. This is the case for diapausing eggs of *Atrachya manestriesi*. If the eggs are kept at 25°C and then chilled, the optimum temperature for diapause development is about 7.5°C; both 0 and 15°C are much less effective than 7.5°C in promoting diapause development. However, after the eggs are held at 7.5°C for 10 or more days, 0°C is highly favorable for completing diapause development (Ando 1978). Thus, it appears that the lower threshold temperature for diapause development decreases as diapause proceeds. A similar situation appears to occur during pupal diapause in the papilionid *Battus philenor* (Sims & Shapiro 1983a). In other species, such as the parasitoid *Tetrastichus julis*, diapause is fully induced and the lower thermal threshold is at its lowest by the beginning of autumn; subsequently, the thermal response curve gradually shifts upward as diapause proceeds (Nechols *et al.* 1980).

In insects with a low thermal threshold for diapause development, the return to nondiapause thermal responses may take place either gradually or relatively abruptly, and this happens at various times during the course of diapause. In diapausing eggs of the odonates *Lestes sponsa* and *Aeshna mixta*, thermal thresholds change slowly over 15 weeks in autumn (Corbet 1956; Schaller 1968). However, in the thermally maintained diapause of the grasshopper *Austroicetes cruciata*, there is a sudden change in the ability to hatch at high temperatures immediately before winter (Birch 1942). In the sarcophagid *Wohlfahrtia magnifica*, thermal maintenance of diapause is completed by the middle of winter (Ternovoy 1978). But, in the gypsy moth, *Porthetria* (= *Lymantria*) *dispar*, the rate of diapause development at a constant low temperature decreases with time, and there is a gradual transition from the thermally maintained diapause to the postdiapause period (Masaki 1956a). At the beginning of diapause, the European red mite, *Panonychus ulmi*, has an optimum temperature for diapause development that is between 0 and 5°C, well below the lower threshold for postdiapause morphogenesis, whereas during the final part of diapause the optimum temperature is near 9°C, which is above the lower threshold for morphogenesis (Cranham 1972). Similarly, the geometrid *Chesias legatella* has a relatively low optimum temperature for diapause development at the beginning and a higher optimum during the latter part of diapause (Wall 1974).

Studies by Hussey (1955) with the seed chalcidoid *Megastigmus spermotrophus* show interesting seasonal changes in response to temperature. When diapausing larvae are stored at a low temperature, their thermal responses shift downward for the first 15 weeks of diapause. During this time, both the percentage of larvae remaining in diapause after transfer to warm conditions and the number of days required for emergence decreases. However, after 15 weeks of cold storage, the diapausing larvae begin to show an upward shift in thermal responses, that is, both the number of larvae remaining in diapause and the number of days for emergence increase. Such responses suggest an endogenous rhythm, and they provide interesting problems for further study.

Morden and Waldbauer (1980) claim that thermally controlled diapause termination synchronizes embryogenesis and hatching by postdiapause eggs of the psychid moth *Thyridopteryx ephemeraeformis*. However, we suggest that *T. ephemeraeformis'* lowered thermal thresholds during autumn serve to maintain diapause and that the termination of diapause occurs gradually over a long period of time—beginning with some individuals at the end of December and ending in others at the end of April. After diapause termination, low temperatures prevent postdiapause development. Thus, as Morden and Waldbauer state, temperature serves to synchronize hatching in the spring, but it does this not only through its influence on diapause development but primarily through its influence on the initiation and rate of postdiapause development.

5.2.7 Temperature as a diapause-terminating stimulus

As mentioned above, the downward shift in thermal thresholds that characterizes autumnal-hibernal diapause in many species has led to the common assumption that "chilling" functions as a diapause-terminating cue. While it is true that for many species exposure to low temperatures can be used in the laboratory to hasten diapause termination (as in *Hyalophora cecropia*, Williams 1956), it cannot be assumed that "chilling" acts to terminate diapause in nature. In general, temperature acts primarily to maintain diapause by regulating the rate of diapause development over relatively long periods (several months). During autumn the thermal response curve for diapause development gradually rises; this occurs as environmental temperatures fall. Thus, temperature (= "chilling") does not generally act as a diapause-terminating *signal*, but as a factor regulating the *rate* of diapause development, and finally, when diapause is complete, as a factor determining when and how fast postdiapause development will proceed.

The above statement is not intended to imply that temperature, even low temperature, never serves as a diapausing-terminating cue. On the contrary, there are well-documented cases in which temperature does act as a specific cue to end diapause. Although most of the existing studies were done under laboratory conditions, many appear to relate well to field conditions. Nevertheless, future studies should aim to determine the responses of naturally diapausing insects by collecting diapausing insects at regular intervals and testing their thermal responses as diapause progresses.

An interesting case in which temperature has been shown to act as a diapause-terminating cue occurs in the cricket *Teleogryllus commodus*, which undergoes hibernal diapause in the egg stage. In this species diapause persists for long periods (60-80 days) at 20°C, but its termination is remarkably accelerated by brief exposure to high temperatures (Figs. 5.5 and 5.6). The duration of high temperature needed to terminate diapause varies as a function of temperature. Only three days at 30°C are sufficient; 30°C is more effective than 25°C in terminating diapause. Therefore, the process of diapause termination is evoked by transfer to a high temperature, and its rate is temperature dependent. Diapause is completed in a very brief period, and high temperatures after winter apparently act as a diapause-terminating cue. High temperature certainly is not a requirement for postdiapause development because the lower threshold for postdiapause development is lower than 20°C (Masaki *et al.* 1979).

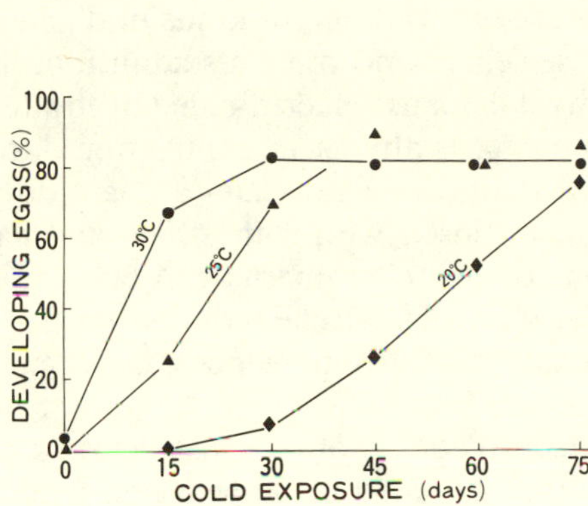

Fig. 5.5 Termination of diapause in *Teleogryllus commodus* eggs by high temperatures (20, 25, and 30°C) after exposure to low temperature (10°C) for various periods of time. The results indicate that diapause is not terminated by cold. Rather, it is terminated after transfer to high temperature, and the rate of diapause termination is positively related to temperature. Each point represents the results from 250 eggs. Data from Masaki *et al.* 1979.

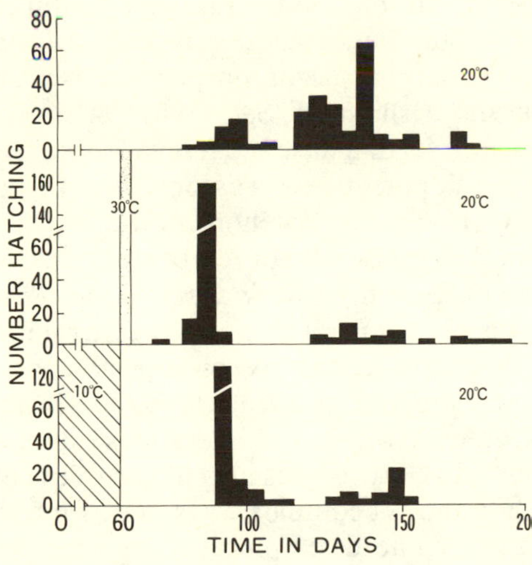

Fig. 5.6 Diapause termination in *Teleogryllus commodus* by exposure to 30°C for only three days. Note that more eggs hatched and that eggs hatched sooner after brief exposure to high temperature (30°C) than did so after chilling for 60 days at 10°C. Data from Masaki *et al.* 1979.

The requirement of such a thermal stimulus for the termination of the final phases of hibernal diapause may be more common than previously thought. Examples are now known to occur in the lepidopteran *Heliothis armigera* and the katydid *Leptophyes punctatissima*. In pupae of *H. armigera* from the Namoi Valley of New South Wales, the first phase of hibernal diapause proceeds at relatively low temperatures, but postdiapause development does not occur unless temperatures exceed the threshold for nondiapause pupal development (Wilson *et al.* 1979). Similarly, in *L. punctatissima* there is a lag in postdiapause embryonic development unless the temperature rises well above the lower threshold for postdiapause development (Deura & Hartley 1982).

Temperature also may function to terminate diapause in insects that undergo aestival diapause. The papilionid *Luehdorfia puziloi inexpecta* is on the wing in early spring, and egg and larval development is completed in June. Subsequently, the pupa lies dormant during summer, autumn, and winter. This lengthy pupal stage comprises two different kinds of diapause. Soon after pupation, the first diapause sets in and prevents adult morphogenesis during the summer. This aestival diapause is terminated by low temperatures, and adult morphogenesis proceeds almost to completion during autumn, during which time all pigments in the wing scales are laid down. The pharate adult, however, remains enclosed within the thick and hard pupal cuticle where it enters the second diapause to hibernate. A brief period of exposure to low temperatures (near 0°C) seems to be required for partial digestion of the pupal cuticle; subsequent adult emergence may occur at temperatures lower than 10°C (Kimura 1975).

The turnip maggot *Hylemyia floralis* provides another example of the termination of aestival diapause by temperature. Adults appear and oviposit during September in the northern Honsyu region of Japan. Prediapause larvae mature and form puparia in the soil during October. Diapausing pupae do not resume development in response to high temperatures after hibernation but remain in diapause during summer. If these pupae are kept at high temperatures, they fail to develop, but they begin morphogenesis when exposed to low temperatures in autumn (Ishitani & Sato 1981). A similar type of thermal stimulus terminates aestival diapause in *Heliothis armigera* from Sudan Gezira. Postdiapause development proceeds only if temperatures drop from 34°C to around 26°C (Hackett & Gatehouse 1982).

Very few studies have determined the precise role of temperature in diapause termination. For example, the mites *Halotydeus destructor* and *Bdellodes lapidaria* both appear to require low temperatures to end aestival diapause (Wallace 1970a, 1971), but this is not certain because postdiapause development also requires fairly low temperatures (Wallace 1970b). In the clover root curculio, *Sitona hispidulus*, relatively low temperatures subserve termination of aestival diapause, but short day lengths also play a role (Leibee *et al.* 1980). Considering the subtle and yet complex ways in which temperature can function, this area of research needs emphasis.

5.2.8 Temperature and postdiapause development

After diapause ends (i.e., after all response to token stimuli ends and the thermal responses of insects are at a level characteristic of nondiapause growth and development) temperature is usually the primary environmental factor governing the rate at which diapause characteristics (e.g., cold-hardiness, accumulated metabolic reserves) are lost and the rate at which postdiapause growth and development occur (Tauber & Tauber 1976a). The thermal thresholds and temperature-dependent growth rates specific to the postdiapause period have been studied for many species, for example, the two-striped grasshopper, *Melanoplus bivittatus* (Church & Salt 1952), several dragonfly species (Corbet 1963; Sawchyn & Gillott 1974; Trottier 1971), the sawfly, *Neodiprion sertifer* (Wallace & Sullivan 1963), the ichneumonid parasitoid, *Pleolophus basizonus* (Griffiths 1969a), the winter moth, *Operophtera brumata* (Embree 1970), the fall webworm, *Hyphantria cunea* (Morris & Fulton 1970a), the mite, *Halotydeus destructor* (Wallace 1970b), the arctiid, *Spilosoma lubricipeda* (Baker 1971), the leafhopper, *Nephotettix cincticeps* (Hokyo 1971), the mayfly, *Ephemerella ignita* (Bohle 1972), the larch sawfly, *Pristiphora erichsonii* (Heron 1972), the moth, *Exoteleia nepheos* (Hain & Wallner 1973), the forest tent caterpillar, *Malacosoma disstria* (Ives 1973), the green lacewing, *Chrysopa carnea* (Tauber & Tauber 1973c), the eulophid parasitoid, *Tetrastichus julis* (Nechols *et al.* 1980), the tobacco budworm, *Heliothis virescens* (Potter *et al.* 1981), the Japanese beetle, *Popillia japonica* (Régnière *et al.* 1981), the western cherry fruit fly, *Rhagoletis indifferens* (Stark & AliNiazee 1982), the European corn borer, *Ostrinia nubilalis* (Anderson *et al.* 1982a), the European red mite, *Panonychus ulmi* (Herbert & McRae 1982), the red turnip beetle, *Entomoscelis americana* (Gerber & Lamb 1982), tortricid moths in the genus *Choristoneura* (e.g., Volney *et al.* 1983), and the pecan nut casebearer, *Acrobasis nuxvorella* (Ring & Harris 1983; Ring *et al.* 1983a, b).

Many studies use laboratory-derived thermal thresholds and growth rates to predict the timing of postdiapause events in the field. This procedure often requires precise information on diapause termination, especially for species whose thermal thresholds for diapause development overlap those for postdiapause development (see Section 10.1.1). To be accurate, prediction should be based on knowledge of thermal reactions during diapause and the postdiapause transitional period, an understanding of when diapause ends in nature, as well as information on the thermal requirements for postdiapause development.

An added complication in calculating heat accumulations for postdiapause development is found in the wheat stem sawfly, *Cephus cinctus*, in which high temperatures immediately after diapause termination can reinduce diapause (Church 1955; Salt 1947). Moreover, temperature conditions during diapause, as well as the duration of diapause, may influence postdiapause thermal reactions. Such responses are found in the ichneumonid parasitoid *Pleolophus basizonus* (Griffiths 1969a), several species of meloiids (Selander & Weddle 1972), and the katydid *Ephippiger cruciger* (Dean & Hartley 1977a, b); response patterns such as these and their effects must be considered when predictions of postdiapause development are made.

5.2.9 Summary

Insects display an enormous variety in their seasonal responses to temperature, and like photoperiodic responses, thermal responses change drastically during the course of the seasonal cycle.

For some insects, especially equatorial species and species that inhabit subterranean environments, temperature is the main diapause-inducing cue. In these cases, temperature acts as a token stimulus rather than as a regulator of developmental rates, and it serves to induce diapause in advance of seasonal change. More commonly, especially among temperate-zone insects, temperature acts as a modifier of photoperiodic responses. Usually, there are upper and lower thermal limits within which insects can respond to photoperiodic stimuli; these limits may determine when photoperiod will act as a diapause-inducing or -averting factor. Temperature may also shift an insect's critical photoperiod. Thus, the temperatures during the photoperiodically sensitive stages greatly influence the timing, induction, and depth of diapause.

Temperature conditions between the time the diapause-inducing stimulus is perceived and the time diapause begins to be expressed can modify or nullify the diapause-inducing effects of photoperiod. Subsequently, during the early phases of diapause induction, temperature can intensify or reverse various diapause symptoms, such as the depth or duration of diapause, the level of cold-hardiness, and the intensity of diapause color.

Temperature frequently has a very important role in maintaining hibernal and aestival diapause; however, even in the temperate zone, insects vary considerably in the degree to which they rely on temperature to maintain diapause. Some have evolved a high degree of dependence on temperature and show complicated thermal reactions during diapause. Others rely on the interaction of temperature, photoperiod, or other factors, either simultaneously or serially, to maintain diapause; and still others do not utilize temperature for diapause maintenance.

During thermally maintained hibernal diapause, thermal reaction curves for diapause development are usually lowered so that warm, growth-promoting conditions do not result in the resumption of development. As autumnal conditions become colder, diapause development proceeds and, usually by early or mid winter when temperatures are generally below those necessary for growth and development, thermal responses return to the non-diapause level. Subsequently, the prevailing low temperatures prevent development and inhibit the loss of the various diapause symptoms, such as reduced metabolism, cold-hardiness, and diapause color. They also decelerate the utilization of metabolites that are essential to postdiapause development, and they serve to synchronize the initiation of postdiapause development.

From the above, we conclude that although "chilling" can result in a high percentage of insects successfully terminating diapause under laboratory conditions, it is not correct to assume that diapause is terminated in nature by cold. In general, lowered thermal responses are most pronounced before temperatures fall below the thermal threshold for postdiapause development. Warm conditions in autumn thus maintain diapause. As autumn proceeds, the thermal responses return to nondiapause levels. The timing of the return to nondiapause thermal reactions varies among species. In some there is an abrupt change in thermal responses at a particular time during diapause, and in other species there is a gradual transition from diapause responses to the nondiapause level.

In very few cases has it been shown that temperature acts as a diapause-terminating cue at the end of hibernal or aestival diapause in nature. However, this dearth of examples may not accurately reflect the actual frequency of insects having thermal requirements for diapause termination. It is more likely a function of the difficulty in distinguishing the role of temperature as a regulator of the rates of metabolic and other physiological processes from its role as an environmental cue that triggers some internal regulation of development. In addition, diapause may terminate in response to relatively subtle changes in temperature. Therefore, well-controlled, precisely defined experiments are necessary to elucidate thermal requirements for diapause termination.

After diapause is finished, temperature is the primary factor regulating the rate at which diapause symptoms are lost and the speed at which postdiapause development proceeds. Thermal requirements for postdiapause development have been determined for many species, but in order to develop predictive models, it is necessary to establish when diapause ends in nature and to determine the thermal requirements for diapause and postdiapause development.

5.3 Other Environmental Factors

In the past, studies of the environmental influences on seasonal cycles have concentrated on photoperiod and temperature. Such studies have made it clear that these two factors serve as the primary stimuli regulating the seasonal cycles of most temperate-zone insects, and they may also exert major influences on the seasonal cycles of many tropical insects. However, three other external environmental factors have also been shown to have important influences on insect seasonal cycles; they are food, population density, and moisture. All three, like temperature, can influence growth, development, and reproduction directly, and they are involved in nondiapause seasonal migration, polyphenism, and quiescence (see Chapter 2). However, they also may have more or less regular seasonal cycles and thus may provide token stimuli that signal seasonal changes. As a result, insects have evolved mechanisms that utilize these cues to regulate, to various degrees, diapause mediated changes in the life cycle.

Some important examples of seasonal responses to these factors are found among insect parasitoids. The intimate relationship that parasitoids have with their hosts deserves separate attention, and we deal with them in a separate section (Section 6.1).

5.3.1 Food

In some instances, food can be a reliable purveyor of seasonal cues, and insects have evolved different ways of utilizing alterations in the quantity and quality of their diet as factors regulating dormancy (Masaki 1980). Food may be especially important as a seasonal cue in regions near the equator where temperature and photoperiod show little annual change. It may act as a primary diapause-inducing factor, but more often it modifies the diapause-inducing effects of another stimulus such as photoperiod and/or temperature. Sometimes food also influences postdiapause development.

5.3.1.1 Food as a major diapause-regulating factor

Food has been shown to be a major diapause-regulating factor for only a relatively small number of insects, most of which undergo aestival diapause. In the *mohave* strain of *Chrysopa* carnea, which is native to California, both long day lengths and the presence of prey promote continuous reproduction. Short day lengths induce reproductive diapause in 100% of adults tested; however, prey must be present for reproduction to occur after termination of the photoperiodically controlled phase of diapause (Tauber & Tauber 1973b, 1982). Furthermore, under long-day conditions, lack of prey results in about 60 to 80% diapause; introduction of prey terminates this food-induced diapause (Table 5.4). Therefore, during summers when prey populations are low, a high proportion of the lacewing adults enters a food-mediated, aestival diapause. Subsequently, the whole population enters the photoperiodically controlled autumnal diapause. Reproduction occurs in the spring when days are long and prey are available. With the introduction of intensive agricultural practices into California—procedures, such as irrigation, that promote high populations of prey during summer—the aestivating *mohave* strain has apparently been swamped out in agricultural areas by the introduced, nonaestivating *carnea* strain (Tauber & Tauber 1973b, 1982).

In what may be a similar situation, Hagen (1962) stated that food during the summer controls aestivation in the ladybird beetle, *Hippodamia convergens*. In this species, as in the *mohave* strain of *C. carnea*, aestival diapause occurs if prey populations are low (Stewart *et al.* 1967). However, the occur-

Table 5.4 Influence of food and photoperiod on diapause induction in the *mohave* strain of *Chrysopa carnea* from Strawberry Canyon, Alameda County, California. Note that both short day lengths and the absence of prey result in a high incidence of reproductive diapause, whereas prey in the presence of long days averts diapause. Data from Tauber & Tauber 1973c.

	INCIDENCE OF DIAPAUSE UNDER VARIOUS DIETS (%)		
Photoperiod	Protein[a]	Wheast[b]	Aphids[c] and wheast[b]
LD 10:14	100[d]	100	100
LD 16:8	100	88	0

[a]Protein hydrolysate of yeast (Type A), sugar, water.

[b]Wheast, sugar, water.

[c]*Myzus persicae.*

[d]In all conditions N = 25 to 45 pairs.

rence of large populations of aphids on irrigated crops during the summer allows continued development throughout the summer. Reduced access to prey appears to enhance diapause induction in another ladybeetle *Semiadalia undecimnotata* (Iperti & Hodek 1974; Rolley *et al.* 1974). It is also claimed for this species that the age of the host plant may influence diapause through its effect on the host; however, the test on which this claim is based did not quantify the prey taken from the host plants of various ages. It is possible that the aphids on senescent plants were more mobile or otherwise more adept at avoiding predation than those on younger hosts, and thus the higher incidence of diapause in the predators on senescent plants may be associated with a reduced amount of feeding, not with a reduction in prey quality *per se.*

Food also affects aestival diapause in two Australian species, the collembolan *Sminthurus viridis* and the mite *Halotydeus destructor.* In these cases increased maturity of the host plant apparently induces diapause (Wallace 1968, 1970a), but the chemical basis for this response is unknown. Such diapause-inducing effects of senescent vegetation may be related to plant growth hormones (e.g., Ellis *et al.* 1965; Visscher *et al.* 1979; Visscher 1980) or to the altered chemical composition of the host plant (Haglund 1980). Further research in this area would be very fruitful.

Host plant condition (maturity) may play a dominant role in diapause induction in certain tropical stem borers, such as the white rice borer, *Rupela albinella* (van Dinther 1962; see also Hummelen 1974), the maize stemborer, *Busseola fusca* (Usua 1973), and two pyralid stalk borers of maize, *Chilo zonellus* and *C. argyrolepia* in Kenya (Scheltes 1976). In two other maize stalk borers, *Chilo partellus* and *C. orichalcociliella*, there is a positive correlation between low levels of precipitation and diapause incidence; however, it appears that moisture acts through the host plant to influence diapause (Scheltes 1978a, b). In the Indian meal moth, *Plodia interpunctella*, hibernal diapause terminates in response to rice bran extract (Tsuji 1966). As in the other examples, further research on the nutrient and chemical bases for these responses is needed.

Although food most commonly influences aestival diapause, its effects are not restricted to diapause during this season. Hibernal diapause in the chaoborid, *Chaoborus americanus*, ends in response to the simultaneous occurrence of prey and long day length (Bradshaw 1969, 1970). The presence of prey also acts as an important regulating factor in the hibernal diapause of larvae of the univoltine antlion, *Myrmeleon formicarius* (Furunishi & Masaki 1981). However, unlike the situation with *C. americanus*, short day length, low temperature, and scarcity of food act additively to retard development in *M. formicarius*.

In the pitcher plant mosquito, *Wyeomia smithii*, larval food levels and conditioning of the water in the larval habitat influence diapause induction. Under photoperiodic conditions that would be expected to cause 56% of individuals to enter diapause, the incidence of diapause varies from 15 to 100% dependent on the amount of food available. Food levels that are low enough to cause a reduction in fitness result in a high incidence of larval diapause (Istock *et al.* 1975). These responses seem to be related to the autogenous habit of *W. smithii* adults (females oviposit without having had a proteinaceous meal), and they have an important influence on the evolution of the species' life history (Istock *et al.* 1975; Istock 1981; see Section 9.2.2.1).

Host plant condition has an important role in determining seasonal responses in aphids. It is one of several factors, including photoperiod, temperature, and crowding, that affect prediapause morph and wing-length changes and that ultimately lead to the deposition of diapausing eggs (see Dixon 1971b; Lees 1966; Mittler & Dadd 1966; Mittler & Sutherland 1969; Mittler 1973; Harrewijn 1978).

5.3.1.2 Food as a modifier of photoperiodic responses

Food modifies the primary response to photoperiod in several species of insects, thereby altering the incidence and duration of diapause (see reviews by Andrewartha 1952; Lees 1955; Danilevsky 1965; Laudien 1973; Tauber & Tauber 1973a; Saunders 1982). When females of the mite *Panonychus* (= *Metatetranychus*) *ulmi* are reared on senescent or "bronzed" foliage, they lay a high proportion of diapausing eggs even when they experienced diapause-averting temperatures and photoperiods (Lees 1953). Similarly, photoperiodic induction of diapause in the Colorado potato beetle, *Leptinotarsa decemlineata*, is enhanced when the insects feed on physiologically aged rather than young or mature potato leaves (de Wilde *et al.* 1969; Hare 1983).

Larval diet alone has little effect on diapause induction in the pink bollworm, *Pectinophora gossypiella.* However, when larvae are reared under short, diapause-inducing day lengths, high lipid content in the diet substantially increases the incidence and duration of diapause (Adkisson *et al.* 1963; Foster & Crowder 1980). High lipid content in the experimental diet may simulate the increased oil content during maturation of the cotton bolls.

A similar effect occurs in the beetle *Epilachna vigintioctopunctata*, when it is reared on sliced potatoes—a high starch diet that is low in lipids, proteins, and carotenoids (Kono 1979). Adults fed sliced potatoes respond to short day lengths by stopping ovarian development, but their fat body does not enlarge and they do not enter diapause. Short-day adults fed a high-protein diet of host leaves exhibit enlarged fat body and they do enter diapause.

Food constitutes one of at least four interacting determinants of diapause induction in *Hyphantria cunea* from New Brunswick and Nova Scotia (Morris 1967); however, a population of *H. cunea* recently introduced into Japan appears to enter diapause primarily in response to photoperiod (Masaki *et al.* 1968; Masaki 1977). Diet may also be important to diapause induction in the boll weevil, *Anthonomus grandis*, both in the laboratory and in the field. When diet is held constant, short day lengths and low temperatures induce diapause (Earle & Newsom 1964; Sterling 1972). However, the short days necessary to induce diapause in the laboratory do not correspond well with natural day lengths prevailing when diapause is induced in the field, and it is possible that food may alter the timing of diapause induction. Laboratory studies have shown that the type and maturity of adult food plant affects diapause induction in this species (Earle & Newsom 1964; Tingle & Lloyd 1969; Tingle *et al.* 1971; Sterling 1972). Similarly, host plant quality is one of four factors important in diapause induction in the codling moth, *Cydia* (*Laspeyresia*) *pomonella* and one of three factors affecting diapause induction in the phytophagous alfalfa ladybeetle, *Subcoccinella 24-punctata.* Both species have a higher incidence of diapause when they are fed a nonpreferred host than when they receive a preferred host (Gambaro-Ivancich 1958; Ali & Sáringer 1975).

5.3.1.3 Food and postdiapause development

The degree of dependency on an external source of food for the completion of postdiapause development is a species-specific characteristic. Some species that overwinter in the larval stage must feed prior to postdiapause molting. Others, such as the European corn borer, *Ostrinia nubilalis*, and the woolly bear caterpillar, *Pyrrharctia isabella*, do not have this food dependency, and postdiapause pupal and adult development as well as postdiapause oviposition occur without postdiapause feeding.

Many species that overwinter in the adult stage require food after diapause, before they oviposit; in others, food may not be required to initiate postdiapause oviposition, but it is required for sustained oviposition (see Section 4.3.1.1).

5.3.2 Density

Very often, the effects of crowding, or density, are difficult to separate from the effects of reduced food quality and/or quantity. Despite this problem, there are well-documented cases in which density has been shown to influence diapause. Most of these cases indicate that density is one of several interacting factors that affect diapause, but for some gregarious insects and also insects that occur in confined habitats, crowding may play a major role in diapause induction.

The influence of population density on diapause is particularly noticeable in pests of stored products (Takeda & Masaki 1976). An example is found in *Ephestia cautella*, a lepidopteran pest of citrus pulp stored in warehouses (Hagstrum & Silhacek 1980). Diapause is induced in larvae of this species by a residue produced during larval feeding. Diapause occurs when larvae are reared in crowded cultures, when they are reared singly on a small amount of fresh diet, or when they are reared on a large amount of fresh diet containing residual diet from crowded cultures. The effect of crowding does not require interaction with other individuals; the stimulus remains in the larval medium when no larvae are present. These results suggest that the token stimulus for diapause induction may be high concentrations of a pheromone, such as has been reported for the nematode *Caenorhabditis elegans* (Golden & Riddle 1982, 1984).

E. cautella's density-dependent diapause forms the basis for a seasonal cycle that is well synchronized with seasonal fluctuations in the availability of food in warehouses. Only diapausing individuals survive when the warehouses are empty and food is not present. When the warehouses are refilled in March, diapause terminates and adults emerge. As the season progresses, the number of individuals entering diapause increases in response to the increased production of dietary residues from larval feeding. By the end of the storage period (December), the incidence of diapause is high, and thus the infestation carries over in the empty warehouses until the next storage season.

Both short day lengths and low temperatures (20°C) induce diapause in the Indian meal moth, *Plodia interpunctella* (Tsuji 1963; Bell 1976a). Under these conditions, larval density does not influence diapause induction; this "density-independent" diapause terminates when the animals experience 30°C. However, when larvae are reared at 30°C, crowding of early larval stages induces a "density-dependent" diapause in some individuals (Tsuji 1963). Thus, this pest of grain has a dual protective mechanism: a "density-independent" diapause for avoiding mortality resulting from seasonal changes in the environment and a "density-dependent" diapause for avoiding mortality associated with increased population growth under warm, crowded conditions. It is interesting that termination of both types of diapause is influenced by the addition of rice bran extract. The codling moth, *Cydia* (= *Laspeyresia*) *pomonella*, has a similar dual protective mechanism. Photoperiod and temperature are the primary inducers of diapause in late-season larvae, but under diapause-averting day lengths and temperatures of midseason, nutrition and larval density strongly influence diapause induction (Brown *et al.* 1979).

The density-dependent stimulation of diapause has a reverse effect in the dermestid *Trogoderma variabile.* Larvae of this species pupate without diapause if they are reared in groups, whereas larvae reared individually enter diapause. High larval density terminates diapause through the action of mutual stimulation of larvae (Elbert 1979).

Crowding of nondiapause aphids strongly influences the production of alate forms and it may interact with other factors to influence the production of prediapause alates (see Section 2.1.1), but in one species crowding also induces and maintains diapause. In the sycamore aphid, *Drepanosiphum platanoides*, as in most aphids, short day lengths induce the development of autumnal sexual forms that produce the diapausing, overwintering eggs. However, in this species the second posthibernal generation enters a reproductive diapause during summer. This diapause is timed to occur when sycamore leaves are mature and of poor quality, but before they have become senescent and regain their high nutritive value. Dixon (1975) showed that second-generation adults are morphologically and physiologically prone to enter aestival diapause. They have smaller abdomens, longer guts, greater fat reserves, less developed ovaries, and longer preoviposition periods than first-generation females, even if both generations are reared under identical conditions. However, Dixon also demonstrated that crowding is the primary environmental factor that induces and maintains the aestival diapause. Crowded conditions during nymphal development result in diapause induction, and diapause is prolonged by crowded conditions during adult life. The length of the diapause is also increased by maternal crowding conditions (Chambers 1982).

Diapause in *Daphnia* spp. is induced through the action of two token stimuli—photoperiod and an unknown stimulus associated with population density (Stross & Hill 1965; Stross 1969b). In an Arctic strain of *D. middendorffiana*, an increase in density intensifies the photoperiodic response by shifting the critical photoperiod upward (Stross 1969a). In other *Daphnia* species, density effects may override photoperiodic effects (Stross 1969b).

Larvae of the predaceous ladybeetle *Chilocorus bipustulatus* live in groups, and population density affects larval growth rates. Larvae in large groups grow more slowly than those reared individually. Although photoperiod is the primary regulator of imaginal diapause, it is possible that group size may have a role in diapause induction. Both the group effect on larval development and the photoperiodic effect on diapause induction are modified similarly by temperature (Zaslavsky & Fomenko 1973).

Pupal diapause in the rice green caterpillar, *Naranga senescens*, is induced by short day lengths, low temperatures, and mature host plants. However, even when these factors signal diapause avoidance, a relatively high proportion of individuals enters diapause if the larvae are reared in groups (Iwao 1968). Two larvae can constitute a "crowd" in that a "group effect" is noticed even when only two larvae are reared together; however, 10 larvae per container had the strongest effect on diapause induction. At higher densities diapause induction is reduced.

5.3.3 Moisture

The influence of moisture on diapause regulation remains obscure and its role has been studied primarily in relation to egg diapause. Because of the complexities involved, there have been few cause-effect relationships demonstrated for moisture and diapause induction and/or diapause termination. One of the main problems is that in many species water absorption or intake is a normal requirement for development, including egg development. Thus the designation of water intake (or the lack of it) as a diapause-regulating factor (token stimulus) or as a requirement for postdiapause development has been difficult.

Although insects have evolved numerous mechanisms for maintaining water balance (e.g., Edney 1977, 1980; Maddrell 1980; Willmer 1982), few have been directly linked to diapause. Often, quiescence caused by dry conditions and diapause caused by other stimuli occur in distinctly different stages of development, even different stages of embryonic development, and thus they appear to be unrelated (Lees 1955). For example, postdiapause eggs of the leaf beetle, *Atrachya menetriesi*, and the western corn rootworm, *Diabrotica virgifera*, require moisture to complete embryogenesis. When deprived of contact moisture after diapause, the eggs undergo some development and then enter a state of quiescence that is separate from the diapause stage (Ando 1972; Krysan 1978). The same is true for *Locustana pardalina* eggs (Matthée 1951). Moreover, return of the quiescent insects to moist conditions results in rapid water uptake (Ewer 1979) and causes the immediate resumption of development, whereas the termination of egg diapause is usually dependent on other internal and external factors. In the Australian plague locust, *Chortoicetes terminifera*, eggs laid in dry soil become quiescent after completing about one-fourth of their development. If some water is absorbed at this point, development proceeds to a second stage of development when another delay may occur if conditions are dry. Presumably, these temporally separated moisture thresholds ensure that eggs will not hatch unless there has been sufficient rainfall to promote germination of host plants (Wardhaugh 1970).

In general, water absorption is reduced during egg diapause; insects, nevertheless, have evolved a variety of methods for dealing with water needs before, during, and after diapause. Ando (1972) divided diapausing eggs into three categories: (1) eggs in which water is not absorbed after oviposition and therefore is not necessary for embryonic development or diapause termination, such as the commercial silkworm, *Bombyx mori*, and the gypsy moth, *Lymantria dispar*; (2) eggs in which almost all the water necessary for postdiapause development is taken up before the egg enters diapause, such as *Teleogryllus commodus* (Browning 1965), *Chorthippus brunneus* (Moriarty 1969), and *Zonocerus variegatus* (Chapman & Page 1978; Modder 1978); (3) eggs in which water is required for initiation of postdiapause development, such as *Diabrotica virgifera* (Krysan 1978), *Locustana pardalina* (Matthée 1951), and *Atrachya menetriesi* (Ando 1972). In addition to the above, we add a fourth category: (4) eggs in which water uptake is essential for diapause induction. This category apparently includes the Australian plague locust, *Chortoicetes terminifera.* Although the thermal and photoperiodic conditions under which the parents are reared largely determine whether or not diapause will occur in *C. terminifera* eggs, significant reductions in diapause induction are caused by low moisture conditions. Warm, moist conditions are essential for diapause induction (Wardhaugh 1980b).

Further studies are needed with mosquitoes, such as *Aedes vexans*, and other insects, such as the cutworm *Euxoa sibirica* and univoltine sciomyzids like *Knutsonia lineata*, that need water for initiating postdiapause embryogenesis (Horsfall *et al.* 1973; Berg & Knutson 1978; Oku 1982) to determine if a second phase of diapause is ended by contact with moisture or if moisture is a requirement for postdiapause development. Presumably, it is a postdiapause requirement because nondiapause eggs also need water for embryogenesis. An additional consideration for mosquitoes is the association of hatching with reduction of dissolved oxygen and the factors resulting in oxygen reduction (Judson *et al.* 1966; Trpis & Horsfall 1967; Wilson & Horsfall 1970; Gillett *et al.* 1977).

Diapause in stages other than the egg can be affected by moisture. Larval diapause in the European corn borer, *Ostrinia nubilalis*, is controlled initially by photoperiod. Subsequently, water intake is required to initiate postdiapause morphogenesis (Beck 1967). The requirement of water can be classified as a requirement for diapause development or for postdiapause development depending on the definition of diapause termination (Beck 1980). We would classify it as a requirement for postdiapause development if studies were to show that nondiapause larvae have a similar requirement for morphogenesis, even if the requirement must be met before postdiapause endocrine function has begun. Similar criteria can be used to determine if moisture requirements for larval development and pupation in other species, such as the pink bollworm, *Pectinophora gossypiella* (e.g., see Wellso & Adkisson 1964; Henneberry & Clayton 1983), are related to diapause termination or postdiapause development. In some larvae, such as the southwestern corn borer, *Diatraea grandiosella*, water has been shown to promote postdiapause development (Reddy & Chippendale 1973; see Tauber & Tauber 1976a).

High soil moisture content may play a role in the termination of larval diapause in the wheat blossom midges *Contarinia tritici* and *Sitodiplosis mosellana* because the larvae are sensitive to moisture only during a six-week period at the end of winter. If soil moisture remains low during this time diapause persists for an additional year (Basedow 1977). Claims that moisture terminates diapause in the sorghum midge, *Contarmia sorghicola* (Baxendale & Teetes 1983a) are unsubstantiated. Tests were conducted with hibernating insects that had been stored under cold conditions for six to eight months, and it is likely that diapause had terminated before the tests were started.

Despite the problems in distinguishing the direct effect of moisture on diapause from that on postdiapause development, the ecological consequences are similar in both cases; development is delayed until moisture conditions become favorable for development.

5.3.4 Summary

Food and moisture vary with the seasons, and insects time their activities to take advantage of the seasonal availability of these resources. In most cases, especially among temperate zone species, photoperiod and temperature provide reliable indicators of the seasonal occurrence of food and moisture resources. But in some instances these factors themselves act as the primary "token stimuli" that herald seasonal change. Population density, too, can vary in a regular, seasonal pattern, and in some cases insects have evolved mechanisms that rely on density-related interactions with conspecific individuals to indicate seasonal change.

Food and moisture act as major diapause-regulating factors primarily among insects with aestival diapause and insects in tropical areas. However, there are examples in which prey availability is very important in regulating aspects of hibernal diapause in temperate-zone species. Moreover, food quality commonly modifies the diapause-inducing effects of photoperiod and temperature, and it strongly affects postdiapause development in many species that undergo diapause as adults or larvae.

Very few studies have separated the diapause-regulating effects of population density from those of food. However, there are good examples of density-dependent diapause among stored products pests and other gregarious species. In some cases, density is the overriding factor; in others it is secondary to photoperiod and/or temperature.

Insects vary greatly in their developmental requirements for moisture, and few investigations have differentiated the effects of moisture on diapause from those on nondiapause development. The few studies that have been made deal primarily with egg diapause, and they illustrate a wide range of variability in responses to moisture. Eggs of some species do not absorb water after oviposition and therefore do not require water for either diapause or postdiapause development. A second group contains insects that absorb water before diapause and do not require it for postdiapause development, and a third group requires it for the initiation of postdiapause development. A fourth group may require moisture to induce diapause; further research is needed to make this clear.

Chapter 6

Seasonal Adaptations—Special Cases

We have distinguished five groups of insects whose seasonal adaptations are special in one way or another, and they therefore require separate attention. Although these five groups are very diverse, the study of their seasonal cycles shares several essential features. First, the seasonal cycles of each group have unique aspects, both in the environmental and genetic factors that underlie their characteristics and in the selective pressures that act to shape their responses. Thus, each of the groups is of special interest to our understanding of the evolution of seasonality.

Second, study of the seasonal cycles of each of these five groups requires special techniques and/or approaches. For the parasitoids, it is the intricate relationship between host and parasitoid that is paramount in the analysis. For the social insects, communication among colony members is a major consideration. For analysis of seasonal cycles of tropical insects, environmental factors that are often difficult to manipulate must be taken into account. Arctic and desert-inhabiting species present logistical, among other, problems.

Third, and perhaps foremost, is the fact that all aspects of the seasonal cycles in each of the five groups of insects is in general need of investigation. The studies we examine here involve only a minute percentage of the species in each of the groups, and any generalizations drawn from these studies are tentative at best. Our discussion therefore has as its goal the stimulation of future work on the seasonal cycles of these important groups of insects, especially since they pose intriguing evolutionary problems and, in some cases, important practical considerations.

We begin with parasitoids and social insects. As in most other insects, environmental conditions influence the seasonal cycles of parasitic and social insects. But, in addition to abiotic factors, parasitic and social insects are affected by other individuals with which they have special, intimate relationships: in the case of parasitoids, an interspecific relationship with the host, and in the case of social insects, intraspecific interactions with other members of the colony. These intimate associations provide special means whereby information concerning the seasonal progression can be transmitted to individuals. Moreover, the mechanisms involved in the parasitoid's or social insect's perception of seasonal changes in the environment may differ qualitatively from the mechanisms found in nonparasitic and nonsocial insects, and they may also differ quantitatively. Thus, they provide an important dimension to the analysis of insect seasonal cycles.

The three other groups of insects whose seasonal cycles we consider in this chapter are the tropical, Arctic, and desert-inhabiting species. All three of these groups have unique, seasonally related problems that are overcome in a variety of ways. In many instances, diapause is involved in adapting these insects to their unique seasonal environments; for these cases, the environmental and genetic influences on diapause, as well as their importance to the evolution of diapause, are discussed elsewhere (Chapters 5, 7, and 8). But for many other species in these three groups, diapause may not be the primary seasonal adaptation, and we therefore emphasize the comparison between diapause and nondiapause seasonal adaptations to these special environments.

6.1 Parasitoids

The life cycles of parasitoids (see definition, Doutt 1959) are closely adapted to those of their hosts (e.g., Flanders 1944; Askew 1971; Fisher 1971; Doutt *et al.* 1976; Price 1980; Beckage 1985). Among the various host/parasitoid interactions, a primary feature is the seasonal synchrony between host susceptibility and availability and parasitoid activity (see, e.g., Griffiths 1969b; Hassell 1969; Münster-Swendsen & Nachman 1978; Bryan 1983). However, the interaction becomes more complicated when it is considered that interactions within and among three trophic levels are involved in the ultimate outcome: a food source for the host, an insect host or hosts, and the parasitoid.

A parasitoid's host requirements may be highly specific, in terms of both species specificity and stage specificity; in addition, the availability of suitable hosts is often highly ephemeral (e.g., see Strand *et al.* 1980). Moreover, free-living adult parasitoids themselves may be short-lived and active for only brief periods. As a result, parasitoids generally require precise timing of their development in relation to the host.

To overcome stressful periods and to keep in synchrony with the seasonal occurrence of their biotic requisites, parasitoids have evolved three major adaptations: (1) regulation of development and reproduction, particularly through diapause, (2) use of alternative hosts (Doutt & Nakata 1973; Chantarasa-ard 1984), and (3) the ability to switch to a new trophic level (facultative hyperparasitism, autoparasitism) (Askew 1971; Fisher 1971; Starý 1972; Doutt *et al.* 1976). Relatively little work has been done to elucidate the environmental factors that cause parasitoids to switch to an alternative host or that result in hyperparasitism or autoparasitism. Generally, these changes appear to occur on an aseasonal basis, in response to crowding or lack of available primary hosts. Thus, they are virtually equivalent to the responses of nonparasitic insects to aseasonal exigencies in the environment. However, in other cases, they may happen as a part of the regular seasonal cycle, and although it has not been demonstrated, it is possible that they are regulated by token stimuli. The factors controlling parasitoid switches to alternative hosts and to hyperparasitism and autoparasitism are in great need of study. Therefore, in this chapter we concentrate on the first adaptation—the seasonal regulation of development and reproduction.

6.1.1 Synchronization of host and parasitoid development

Doutt *et al.* (1976) point out that many parasitoids keep in step with their hosts, generation for generation, throughout the seasons. Such synchrony may be based on a variety of mechanisms. One example is found in the tachinid flies that lay eggs on the food plants of their host. Hatching by these eggs does not occur until after they have been ingested into the digestive tract of the host. This prehatching period can be fairly extensive. In *Sturmia sericariae*, a parasitoid of the silkworm, eggs can survive up to 50 days under favorable conditions. Once ingested, eggs hatch within 10 to 30 minutes of reaching the midgut (Ishikawa 1936).

More commonly, however, synchrony between host and parasitoid life cycles is based on adaptations that allow parasitoids to alter their developmental rates in response to the physiological condition of their hosts. For example, complete larval development by the tachinid *Sturmia sericariae* requires about one month if the egg is ingested by a second-instar caterpillar, but development requires only two weeks in a fifth-instar host (Machida 1935). Similarly, the braconid *Chelonus curvimaculatus*, which oviposits only in lepidopteran eggs, maintains its development in close step with development of its primary host, the potato tuber moth caterpillar. Furthermore, it can also synchronize its development with other host species that grow at slower rates than its preferred host (Broodryk 1969). These examples of flexibility in larval growth rates demonstrate the very close developmental relationship that can exist between a parasitoid and its host (Doutt *et al.* 1976).

After an endoparasitic egg hatches, the first instar often delays growth and development until the host reaches a specific stage in its metamorphosis (see Schoonhoven 1962; Doutt *et al.* 1976; Nechols & Tauber 1977). This type of arrested growth has a number of advantages for the parasitoid. First, it ensures a host of adequate size and physiological condition in which the parasitoid can complete its development. Second, it synchronizes host and parasitoid generations. If the host enters diapause at an early stage, the parasitoid will also delay development until after the host terminates diapause and undergoes some postdiapause development.

There have been different interpretations of the pause in growth and development during the first instar. Andrewartha (1952) and Lees (1955) equate it roughly with a nondiapause quiescence, whereas Schoonhoven (1962), Danilevsky (1965), and Mellini (1972) call it diapause. The pause in development by first-instar parasitoid larvae appears to have many characteristics of diapause. Especially significant is the fact that development is curtailed before environmental conditions deteriorate (i.e., before parasitoid development causes depletion of host resources). Thus, there is an anticipatory aspect to the phenomenon, and host condition (size, developmental stage, hormonal state) becomes a token stimulus for the parasitoid (see Beckage 1985). The main difference between this form of diapause and that associated with hibernation or aestivation lies in the depth or duration of the suppressed development—one is frequently measured in days, the other in months (Mellini 1972). It is interesting to note that the ichneumonid parasitoid, *Trogus mactator,* has both a host-mediated developmental delay in the first instar and a photoperiodically controlled diapause in the second or third instar (Omata 1984).

Just as temperature (high or low) acts both as a development-regulating factor and as a token stimulus (see Section 5.2), the host's hormones, or other internal factors (even substances that are generally considered nutrients), might function both to regulate the immediate course of parasitoid development and as token stimuli signaling future conditions. The analogy is carried further when we consider that plant hormones, as well as the nutritional value of plants, may influence the diapause and development of phytophagous species (see Section 5.3.1).

6.1.2 Characteristics and regulation of diapause in parasitoids

Diapause in parasitoids shares many essential features with that observed in nonparasitoids. In both, diapause occurs in all stages of development (Doutt 1959), but for any one species the diapausing stage is fixed and genetically determined. Similarly, in parasitoids, diapause-inducing stimuli are perceived before the diapausing stage is reached (see Doutt 1959; Saunders 1982). As in nonparasitic insects, the sensitive stages may be widely separated from the diapausing stage, or they may overlap. In *Nasonia vitripennis* the environmental conditions that the female parasitoid experiences determine whether or not her offspring will enter larval diapause (Schneiderman & Horwitz 1958; Saunders 1965). A maternal effect on diapause induction also occurs in *Trichogramma evanescens* (Zaslavsky & Umarova 1981). By contrast, diapause in the first-instar larvae of *Microctonus vittatae* is largely determined by the temperature and photoperiodic conditions that the eggs and young first-instar larvae experience (Wylie 1980). For other examples of sensitive stages in parasitoids see Simmonds (1948), Jackson (1963), Ryan (1965); Rabb and Thurston (1969), Claret (1973), McNeil and Rabb (1973), Anderson and Kaya (1974), Eichhorn (1976), Schopf (1980), and Omata (1984).

Another feature of diapause that parasitoids and nonparasitoids share is the considerable interspecific and intraspecific variation in the expression of the diapause syndrome. In egg parasitoids of the genus *Trichogramma*, diapause ranges from a relatively intense, photoperiodically controlled diapause in all or a large part of the population, to a thermally controlled, less intense arrest of development in a small proportion of individuals (Maslennikova 1959; Burbutis *et al.* 1976; Curl & Burbutis 1977; Lopez & Morrison 1980; Zaslavsky & Umarova 1981). Finally, some species or strains of *Trichogramma* appear to lack diapause altogether (Oatman & Platner 1972). Intraspecific variation also involves differences in responses to diapause-inducing and -terminating stimuli and in the ability to enter prolonged diapause (see Hoy 1975b; Weseloh 1982; Chapters 9 and 10).

Parasitoids, like nonparasitic insects, frequently express seasonal polyphenism in association with diapause. Examples are found in the polymorphic forms of *Melittobia* spp., chalcid parasitoids of wasp and hornet larvae (Schmieder 1933; Freeman & Ittyeipe 1982), the polymorphic larvae and cocoons of *Sphecophaga burra*, an ichneumonid parasitoid of *Vespula* (Schmieder 1939a, b), the tough, brown and thin, white cocoons of diapausing and nondiapausing aphid parasitoids *Praon palitans* and *Trioxys utilis* (Schlinger & Hall 1959, 1960, 1961), the dark and light cocoons of the aphidiid *Lipolexis* (Shuja-Uddin 1977), and the dark and light brown cocoons of *Bathyplectes curculionis*, an ichneumonid parasitoid of the alfalfa weevil (Parrish & Davis 1978). In addition, there are often size differences between diapausing and nondiapausing individuals (e.g., Weseloh 1973).

Finally, the mechanisms involved in controlling diapause in parasitoids are analogous to those that influence diapause in other insects. Parasitoids can respond to a wide variety of diapause-regulating stimuli; in some cases they have evolved specialized mechanisms for perceiving environmental stimuli (e.g., Claret 1982), including the physiological status of the host.

Three types of interactions characterize the relationship between hosts and their parasitoids during diapause (categories adapted from Maslennikova 1968; see also Lees 1955; Askew 1971; Doutt *et al.* 1976; Tauber *et al.* 1983): (1) parasitoid diapause primarily regulated by abiotic environmental cues and therefore independent of host's diapause, (2) simultaneous dependence of the parasitoid's diapause on the host's physiological state (diapause vs. nondiapause), as well as on external cues, and (3) dependence of parasitoid diapause on the host's physiological state. These three conditions are not rigidly delineated, and there is a continuum in the degree to which parasitoids rely on their hosts for diapause-regulating cues.

6.1.2.1 Regulation of parasitoid diapause by abiotic cues

Most insect parasitoids that have been studied rely primarily on abiotic diapause-inducing signals. For example, diapause induction is relatively independent of host condition in various chalcidoid parasitoids of the pine sawfly (Eichhorn & Pschorn-Walcher 1976), the aquatic mymarid parasitoid, *Caraphractus cinctus* (Jackson 1963), and the ichneumonid parasitoid, *Agrothereutus minousubae* (Shiotsu & Arakawa 1982).

Photoperiod is the most common diapause-inducing stimulus among parasitoids, as it is among nonparasitoids; numerous examples are reported (see

Askew 1971; Fisher 1971; Doutt *et al.* 1976; Saunders 1982; Beck 1980; Tauber *et al.* 1983). The ichneumonid parasitoid, *Pimpla instigator*, is extremely sensitive to red light—a situation that is rare among insects but that allows perception of light transmitted through the dense host cuticle (Claret 1982). Two types of photoperiodic responses have been shown in parasitoids: the long-day type (Type I of Beck 1980) (see Danilevsky 1965; Saunders 1966; Beck 1980) and the long-day/short-day type (Type IV of Beck 1980) (see Anderson & Kaya 1974). And, the ichneumonid wasp, *Pimpla turionellae*, responds to a change in day length; a decrease in day length induces a deeper (longer) diapause than that produced by constant short days (Schopf 1980).

Photoperiod and temperature may interact in diapause induction in parasitoids (see Maslennikova 1958; Zeleny 1961; Griffiths 1969a; Ryan 1965; Messenger 1969; McNeil & Rabb 1973; Anderson & Kaya 1974; Eskafi & Legner 1974; Claret 1978; McPherson & Hensley 1978; Parrish & Davis 1978; Wallner 1979; Wylie 1980). The types of interactions appear to be similar to those found in nonparasitoids. The importance of thermoperiod has also been demonstrated for parasitoids (Saunders 1973; Kalmes 1975). In a few cases, temperature alone apparently acts to induce diapause (see Simmonds 1948; Claret & Carton 1980; Flint 1980).

Food (Schmieder 1933) and parasitoid density (Freeman & Ittyeipe 1982) function in morph determination and delayed development in two species of *Melittobia* (Eulophidae) that attack solitary bees and wasps. However, it is not clear if the delay in development is associated with diapause. Food or parasitoid density, or both, may influence diapause in the ichneumonid *Sphecophaga burra*, which parasitizes hornet and yellow jacket larvae. This parasitoid undergoes larval diapause within tough, brown cocoons in the host cell. Nondiapausing parasitoid larvae also spin cocoons in the host cell, but these cocoons are delicate and white (Schmieder 1939a). If there are one to three parasitoid larvae per cell, most of the larvae spin brown cocoons and enter diapause, but when more than four parasitoids occur in one host, the number of diapausing larvae usually does not exceed three, and the number of nondiapausing larvae increases. It is difficult to conclude from these data whether crowding or the nutritional quality of the host influences diapause induction (Schmieder 1939b), but clearly one or the other or both factors are involved.

There is some indication that humidity has a role in diapause induction and termination in *Bathyplectes curculionis*, a parasitoid of the alfalfa weevil, in *Hexacola* sp., a parasitoid of *Hippelates* eye gnats, and in *Pimpla turionellae*, an ichneumonid parasitoid of lepidopteran pupae (Parrish & Davis 1978; Eskafi & Legner 1974; Schopf 1980). However, as in nonparasitoids, further experimental evidence is needed to elucidate the importance of moisture in diapause regulation in parasitoids.

Once diapause is initiated, its maintenance and termination in some parasitoids is regulated by external environmental factors independent of the host. Low temperatures apparently hasten the rate of diapause development in several species (Birch 1945; Henderson 1955; Schneiderman & Horwitz 1958; Jackson 1963; Griffiths 1969a; Weseloh 1973; Tyler & Jones 1974; Anderson & Kaya 1975; Renfer 1975; Wylie 1977; Claret & Carton 1980; Nechols *et al.* 1980). In other species it does not (Claret 1973; Obrycki & Tauber 1979). However, with the exception of the examples below, very few parasitoid species have been studied under natural conditions to determine the actual role of temperature or photoperiod in diapause maintenance and termination in the field.

Photoperiod acts to maintain diapause in the braconid *Perilitus coccinellae* and its ladybird beetle host, *Coleomegilla maculata*, but the photoperiodic regulation of diapause ends three months earlier in the parasitoid than in the host (Obrycki & Tauber 1979). Diapause in this parasitoid ends without the intervention of a special stimulus from the environment or from the host. A similar situation occurs in the ichneumonid *Pimpla instigator* (Claret 1973). By contrast, in the cereal leaf beetle parasitoid, *Tetrastichus julis*, both photoperiod and temperature regulate the rate of diapause development (Nechols *et al.* 1980). Low temperatures act to maintain diapause during early autumn, whereas short day lengths continue to exert their influence until the beginning of winter. Postdiapause development in both *P. coccinellae* and *T. julis* begins when temperature exceeds 9°C, and both species require similar heat accumulations for postdiapause emergence (Tauber *et al.* 1983).

6.1.2.2 Dependence of parasitoid diapause on host and on abiotic cues

Both abiotic environmental cues and cues emanating from the host influence diapause in some parasitoids. The host's influence on parasitoid diapause varies considerably with the species and the strain involved. The photoperiod and temperature that the parasitoid's mother experiences are the primary factors inducing larval diapause in the pteromalid, *Nasonia vitripennis* (Schneiderman & Horwitz 1958; Saunders 1965). Long-day females produce nondiapause progeny; however, if females are transferred to short day lengths, they begin to produce diapausing offspring after several days. The period of time required for the female parasitoid to make the switch is influenced by the species of blowfly host it receives. Apparently, the host exerts an effect on diapause induction in *Nasonia* larvae through a dietary influence on the parasitoid mother (Saunders *et al.* 1970).

A somewhat similar interaction occurs in the braconids *Aphaereta minuta* and *Alysia manducator* that also attack various species of blowflies (Vinogradova & Zinovyeva 1972a, b). Although both species may respond to the physiological status of some species of hosts, photoperiod and temperature constitute the primary diapause-inducing factors.

In comparing responses of two hyperparasitoids of the tobacco hornworm, McNeil and Rabb (1973) showed that for one species (*Cataloaccus aeneoviridus*), four factors—photoperiod, temperature, maternal age, and diapause status of the host—play significant roles in diapause induction. A higher proportion of *C. aeneoviridis* individuals enter diapause when the primary parasitoid, *Apanteles congretatus*, is in diapause rather than in the nondiapause condition. The physiological status of the host does not play a significant role in diapause induction of the second hyperparasitoid, *Hypopteramalus tabacum* (McNeil & Rabb 1973).

Induction of larval diapause in the pteromalid parasitoid of butterflies, *Pteromalus puparum*, depends on temperature, photoperiod, and the physiological status of the host. If the parasitoids are exposed to diapause-inducing photoperiod and temperature, diapausing hosts frequently give a higher incidence of parasitoid diapause than nondiapause hosts. However, in some instances the photoperiodic and thermal reactions of the parasitoid are entirely independent of the host (Maslennikova 1958).

In contrast, the tachinid *Meigenia bisignata*, which attacks the sawfly *Altalia rosae*, shows a strong dependence on its host's physiological status. When this species is reared on nondiapause hosts it does not enter diapause even under diapause-inducing photoperiods; however, it has a critical photoperiod and thermal requirements for diapause induction that are different from those of its host (Zinovyeva 1972). In some instances the type of response shown by the parasitoid depends on the species of host it attacks. When the braconid *Apanteles glomeratus* parasitizes *Pieris brassicae*, it enters diapause as a mature larva in response to short day lengths. But, when *Aporia crataegi* is the host, the first larval instar of the parasitoid enters diapause in response to the physiological status of the host (Maslennikova 1958).

In the eupelmid *Mesocomys pulchriceps*, which parasitizes saturniid eggs, diapause induction appears to depend on photoperiod whereas diapause maintenance may involve interaction with the host (van den Berg 1971). The host's physiological state can also influence postdiapause development. If the ladybird host, *Coleomegilla maculata*, receives a suboptimal diet at the end of dormancy, postdiapause development by the parasitoid, *Perilitus coccinellae*, is delayed and survival decreases (Tauber *et al.* 1983).

6.1.2.3 Primary dependence of parasitoid diapause on host's physiological status

Some parasitoid species depend almost entirely on their host's physiology to regulate their own diapause induction and/or termination. This type of interaction is particularly common among parasitoids that exhibit delayed development in the first instar (see Schoonhoven 1962; Fisher 1971; Mellini 1972; Jacquemard 1976b; Baronio & Sehnal 1980; Weseloh 1984; Beckage 1985).

Various types of data may demonstrate the host's influence on the phenology of the parasitoid (Schoonhoven 1962; Mellini 1972; Beckage 1985). Indirect evidence comes from the following types of observations: (1) The rate of parasitoid development is correlated with the rate of host development. (2) Although the age of the host at parasitization varies, emergence of the parasitoid is synchronized with a particular stage of the host. (3) The rate of parasitoid development varies depending upon the species of host. (4) Parasitoid development varies with host strain. (5) Host and parasitoid emergence times coincide in nature. Examples of these categories are given by Schoonhoven (1962), Harbo and Kraft (1969), Fisher (1971), and Mellini (1972); see also Hummelen (1974), Torchio (1975), Eichhorn (1977a, 1983), Gurjanova (1979), Legner (1979, 1983), Jervis (1980), and Beckage (1985).

The above examples suggest that parasitoids may utilize their host's physiological status as a token stimulus to regulate their own phenology, but they do not provide conclusive evidence, nor do they reveal mechanisms, that is, hormonal, nutritional, or other influences of the host on the parasitoid. In contrast, direct, experimental methods can demonstrate that the host does influence parasitoid development, and they can elucidate the mechanisms involved. Experimental methods aimed at showing hormonal or nutritional interactions between hosts and parasitoids include extirpation, implantation, and ligature experiments, application of hormones, and transplantation of parasitoids into artificial hosts (see Schmieder 1933; Schneider 1950, 1951; Schoonhoven 1962; Maslennikova & Chernysh 1973; Salt 1975; de Loof *et al.* 1979; Ascerno *et al.* 1980; Freeman & Ittyeipe 1982). Other methods involve experimentally manipulating host food, density, and physical conditions and subsequently observing the effect on the parasitoid.

Through a series of experiments in which he transplanted parasitoids into new hosts, Schmieder (1933) was able to deduce a nutritional role for the host in the host-regulated diapause of *Melittobia chalybii*, a chalcid parasitoid of wasp and hornet larvae. In contrast, Schneider (1950, 1951) attributed diapause in the ichneumonid *Diplazon* to an inhibitory substance in the host's blood. Schoonhoven (1962), working along similar lines, demonstrated a hormonal interaction between the tachinid parasitoid *Eucarcelia rutilla* and its geometrid host *Bupalus piniarius* during diapause. Diapause in *Eucarcelia* was maintained when there was a physical connection between the host and the parasitoid, but not if the parasitoid was isolated from the source of hormones in the host.

Reduced levels of ecdysone may play a role in diapause regulation in some endoparasitoids (see Claret *et al.* 1978). Maternally determined larval diapause in *Nasonia* is associated with a deficient ecdysiotropin-ecdysone system (de Loof *et al.* 1979). Juvenile hormone does not seem to play an essential role in diapause regulation in this species, but the maternally produced factor(s) causing the inhibition of the larval ecdysiotropin-ecdysone system is unknown. Levels of ecdysteroids and juvenile hormone in the host appear to be involved in the regulation of diapause in the tachinid *Gonia cinerascens* (Baronio & Sehnal 1980). Evidence for the role of an inhibitory endocrine factor is also found in the pteromalid *Pteromalus puparum* (Maslennikova & Chernysh 1973) and the tachinid *Compsilura concinnata* (Weseloh 1984).

6.1.3 Influence of parasitization on host development

Parasitization can sometimes cause premature termination of diapause in the host (see Fisher 1971; Vinson & Iwantsch 1980). A high incidence of premature diapause termination occurs in blowflies when they are parasitized by the braconid *Aphaereta minuta* (Zinovyeva 1974). This may result from an effect of the parasitoid's hormonal system on the host tissues, or it may result from the influence of the parasitoid on the host's hormonal system.

It may be argued that premature diapause termination and molting by the host provides some sort of protection for the parasitoid. However, experimental evidence for this claim is lacking. Early diapause termination by parasitized overwintering adult insects may also play a role in increasing host utilization by parasitoids. When adults of the flea beetles *Phyllotreta cruciferae* and *P. striolata* are parasitized by the braconid *Microtonus vittatae*, they emerge from their overwintering sites earlier than when they are unparasitized (Wylie 1982). Emigration from overwintering sites is associated with diapause termination in the parasitoid, and early emergence by the parasitoid may improve the synchrony between the parasitoid's seasonal cycle and that of its host. Obrycki and Tauber (1979) reported a similar situation involving the braconid parasitoid *Perilitus coccinellae* overwintering in *Coleomegilla maculata* ladybird beetles, but in this case the overwintering beetle population is subject to parasitization at two times—once before hibernation (in the late summer or autumn) and again after hibernation in the spring. Such host/parasitoid interactions could have profound effects on the population levels of both pests and beneficial species. However, experimental data from field populations are not available.

6.1.4 Summary

The seasonal cycles of insect parasitoids are influenced by the same range of environmental factors that influence nonparasitic insects. However, they are also strongly influenced by interactions with the host, which, in turn, have a strong role in the evolution of parasitism. First, seasonal synchrony between parasitoid and host, in some cases, is achieved through a physiological, for example, hormonal, connection between host and parasitoid diapause. In these cases, selection pressure on the host's seasonal cycle results in simultaneous pressure on the parasitoid's cycle. Second, for all parasitoids, synchrony with the host is achieved through the interactions of numerous life-history traits (see Chapter 9); as a result, the evolution of parasitoid seasonality is inseparable from the evolution of the host/parasitoid interaction.

Diapause in parasitoids occurs in all stages of metamorphosis, and it shares all the features of diapause in nonparasitic insects. In this regard, three types of interactions characterize the relationship between parasitoids and their hosts. In many parasitoid species, abiotic environmental cues are the primary inducers of diapause, and the parasitoid's diapause is independent of the host's physiological state. In other species, diapause in the parasitoid is highly dependent on the host's physiological state (diapause vs. nondiapause), and in others there is simultaneous dependence on both external cues and the host's physiological state. Parasitization also may influence diapause in the host; in such cases the host/parasitoid interaction can have especially important effects on the population dynamics of both hosts and parasitoids.

6.2 Social Insects

Seasonality is of prime importance in the evolution of social behavior in insects. The seasons set limits to social interaction and to the production of reproductives. Seasonal timing of reproductive activity (production of sexuals, flight and mating by sexuals, swarming, foundation of new colonies) must occur during a favorable season, but it cannot be delayed too long. Similarly, growth and development of the colony must be synchronized with favorable seasonal conditions and curtailed or reduced during unfavorable periods. Successful fulfillment of these requirements requires coordinated seasonal responses among individual colony members. Such coordination is achieved through the insects' perception of and response to external seasonal cues, through communication between members of the colony, and perhaps through endogenous rhythms.

Because of the interaction between the seasonal and social components of the life cycle, seasonality must be considered an essential factor in the evolution of social behavior (Brian 1965a, 1977, 1979; West-Eberhard 1978; Eickwort 1981; Seger 1983; Brockmann 1984). Trends in the evolution of sociality tend to run in parallel with trends in seasonal cycles. The colonies of primitively social wasps and bees (species in which castes intergrade and in which caste differentiation is largely behavioral rather than morphological) have seasonal cycles that are clearly different from those of highly social forms (species in which there are discrete caste differences in size and behavior). In general, temperate zone species of primitively social wasps and bees have annual colonies that are reestablished each year. Because these nests exist for only one year, the degree of sociality that can be expressed within them is clearly limited. In contrast, highly developed social behavior is based on interactions in nests that persist over many years. Moreover, in some cases, survival of the colony during harsh periods depends on highly evolved patterns of social behavior. It is clear that the evolution of sociality cannot be separated from the evolution of seasonal adaptation.

Although some tropical species of social insects (e.g., the polybiine wasps and certain bees) do not have obvious seasonal cycles (e.g., Wilson 1971; Michener & Amir 1977; Michener *et al.* 1979; Brian 1979), all social insects in the Temperate Zone and many in the tropical zone have clearly marked periods of seasonal activity (Brian 1965a; Watson 1974; Baker 1976; West-Eberhard 1982; Ackerman 1983; Lepage 1983). The seasonal cycles of numerous social insects are summarized by Wilson (1971), Brian (1965a, 1977, 1979), Sakagami and Hayashida (1968), and Sakagami (1976).

Social insects characteristically overwinter in one of two ways: either as hibernating queens in or away from the nests or as adults in perennial colonies with or without brood. Recent work dealing with seasonal variation in sex ratios has predicted that eusociality will arise more frequently in haplodiploid species that overwinter as inseminated females than in those that spend the winter as larvae or unmated adults (Seger 1983). The taxonomic distribution of sociality in the Hymenoptera is largely in agreement with this prediction (Brockmann 1984). Among the temperate-zone social wasps (e.g., *Polistes* and Vespinae) only fertilized females overwinter, and these do so in diapause. In the spring the foundresses start nests; in the case of *Polistes*, but not in the Vespinae, auxiliaries may aid in the initiation of nests.

Social bees exhibit a range of seasonal cycles and overwintering habits, with the primitively social forms (e.g., most Bombicinae and Halictinae) hibernating as mated gynes in diapause usually, but not always, away from the nest (e.g., Sakagami 1976; Brockmann 1984). The degree of sociality varies greatly among the temperate-zone Halictinae; however, all overwinter only as mated females. Some species have two distinct broods per year, with gynes being produced only in the last brood (e.g., Sakagami & Hayashida 1968). In other species there is continuous production of brood after the first workers emerge and there is no distinct temporal separation in worker and gyne production (Brian 1979). The highly social bees, such as the metaponine bees and *Apis mellifera*, have perennial nests in which the mated females and workers overwinter with or without brood.

Ants typically overwinter in their perennial colonies, some with brood, some without. In species of *Myrmica*, *Leptothorax*, and *Tetramorium*, the third-instar larvae hibernate along with the adults. In the other groups, such as *Palagiolepis*, several larval stages comprise the overwintering brood and in still others, such as *Formica*, the perennial colonies do not contain brood during winter. Development of cold-hardiness is clearly evident in some species, and castes may differ in the degree of resistance to cold (Erpenbeck & Kirchner 1983). The fire ants, *Solenopsis invicta* and *S. richteri* may produce brood all year round, but these species have distinct seasonal cycles in the extent of brood production, in brood-rearing behavior, and in the production of reproductives (Lofgren *et al.* 1975).

Like the ants, termites also overwinter in their nests. In most species, the reproductives, workers, soldiers, and certain nymphal stages overwinter. Such is the case in the Australian harvester termite *Drepanotermes perniger*, in which the only nymphs to hibernate are in the fourth stage (Watson 1974). After diapause, these nymphs develop into winged sexuals that fly in early summer.

Because of the difficulties in rearing colonies of social insects under controlled conditions, there are relatively few experimental studies showing the cause-effect relationship between environmental cues and seasonal activity in social insects. Generalizations are often based on observation and deduction rather than experimental studies, and they are generally derived from few species. Thus, what is presented here and elsewhere (e.g., Brian 1977, 1979; Wilson 1971) about seasonal cycles of social insects should be regarded as starting points for further investigations.

Many aspects of diapause in social insects might be the same or similar to those of nonsocial insects. Thus, it is likely that both genetic and environmental factors underlie diapause characteristics. Available data, although scanty, indicate that photoperiod, temperature, and food provide environmental cues that synchronize the seasonal cycle with the external environment. However, what appears almost unique to social insects (see also locusts; Nolte 1974; Pener 1974) is their ability to transmit photoperiodic and other information among individuals in the colony. Some species also have evolved the ability to store food in their nests for use during diapause. In addition, many social insects, especially the highly social forms, have the ability to regulate to some degree the temperature of their nests (Seeley & Heinrich 1981). This ability is developed only slightly in species with small colonies, but it is very conspicuous among species with large, complex societies. These specific aspects of seasonality in social insects, as well as a possible hormonal connection between caste determination and diapause regulation, and the interaction between seasonal life-history traits and the evolution of social structure are untapped but potentially very fruitful areas for future research dealing with the evolution of sociality in insects.

6.2.1 Hibernating queens

Among species in which the queen overwinters, there is an evolutionary trend away from control of the seasonal cycle by external, abiotic factors toward control by factors endogenous to the nest (Breed 1975; Brian 1977). For species with primitive social structure (e.g., species of *Polistes*), photoperiod, temperature, and food probably are the dominant factors that directly influence the production and survival of diapausing autumnal gynes (Bohm 1972; Grinfeld 1972; Grechka & Kipyatkov 1983; Ishay *et al.* 1984; see Brian 1979). In general, the sensitive stages and the mechanisms involved in diapause induction in the primitively social species are unknown, although Bohm (1972) demonstrated that both immature and adult *Polistes metricus* perceive diapause-inducing stimuli.

Among the highly social insects, factors within the nest, specifically the queen's age and physiological status, become influential in the production of diapause-destined gynes. As the queen ages over summer, the worker:larva ratio increases and workers shift from worker cell to queen cell construction; workers also alter their nursing behavior. These changes can be mediated through signals from the queen, including queen pheromones, and they ultimately result in production of new gynes that enter diapause and overwinter (Free 1955; Ikan *et al.* 1969; Ishay 1975; Röseler 1970). The factors controlling the seasonal timing of pheromone production by the queen and the factors controlling diapause in the new gynes are largely unknown, but photoperiod is involved in two species of *Bombus* (Grinfeld & Zakharova 1971).

Species that overwinter as hibernating queens also produce males on a seasonal basis, usually from mid- to late summer. The environmental and physiological factors influencing oviposition of the haploid, male-producing eggs are unknown for most species (Spradbery 1971). However, in a singular series of studies on the effects of photoperiod on male-egg and female-egg production in the polistine wasp, *Polistes chinensis antennalis*, Suzuki (1981, 1982) demonstrated that male-egg production by foundresses is clearly affected by photoperiod and enhanced under short-day conditions. Foundresses can resume female-egg production when the photoperiod is shifted from LD 12:12 to 18:6. Therefore, the production of male eggs is not due to exhaustion or annihilation of the sperm storage. This example illustrates that the change in oviposition from fertilized (female) to unfertilized (male) eggs requires accumulation of photoperiodic information, just as does diapause induction. Since only fertilized females overwinter to found a new nest in the spring, male production in late summer and autumn might represent a unique part of the diapause syndrome in this social wasp.

As in species that overwinter in perennial nests, storage of food has also been reported for species that overwinter as diapausing females away from the nest. *Polistes annularis* females store honey in their nests in autumn. They return to their nests on warm winter days, eat the honey, and defend it from nonsisters. Honey deprivation decreases survival during the winter, and those females that do survive without honey build small nests in spring (Strassman 1979).

6.2.2 Perennial colonies

Among species that overwinter in perennial colonies, there are seasonal cycles in brood production, in the production of the various castes, and in the execution of various behaviors, such as swarming, drone eviction, and foraging. Typically, there are two types of perennial colonies—those that overwinter without brood and those that overwinter with brood (Brian 1977, 1979).

Honeybees, *Apis mellifera*, overwinter in perennial colonies, and they illustrate the high degree to which seasonal adaptations have evolved in social insects. Honeybee hibernation apparently involves two phases—an initial autumnal phase during which no brood are present and an hibernal phase during which brood production is begun and continued into spring. Presumably the first phase involves a diapause-mediated arrest in reproduction, whereas the second represents postdiapause reproduction.

In experiments in which day length was artificially altered, Cherednikov

(1967) and Kefuss (1978) showed that photoperiod plays a dominant role in initiating autumnal diapause in workers. Specifically, short or decreasing day lengths result in reductions in feeding and metabolism, buildup of fatty tissue, and changes in worker behavior that halts the development of the bee colony. In addition, the oviposition rate of short-day queens is reduced substantially. Later work by Kefuss (1978) suggests that the direction (increase or decrease) of day-length change, not the day length itself, is important in determining the brood-rearing cycle of the honeybee. The findings are consistent with the observation that brood-rearing decreases in autumn and then recommences in early spring before temperatures rise greatly.

In addition to entering autumnal diapause, honeybees are able to regulate the temperature of their colony during all seasons (Seeley & Heinrich 1981). During autumn and early winter, low temperatures (below 18°C) cause colony members to cluster in a very characteristic manner. The size and tightness of the cluster varies with ambient temperature, and using stored honey as fuel, the bees generate heat to maintain the temperature of the periphery of the cluster above approximately 9°C. At hive temperatures of around 14°C the temperature of the cluster fluctuates with ambient temperature (Johansson & Johansson 1979; Seeley & Heinrich 1981; Southwick 1983). After midwinter a second period of hibernation begins, and a dramatic change occurs in the thermoregulation of the colony. Instead of merely maintaining the temperature of the periphery of the cluster at a conservative 9°C, the bees now regulate temperatures throughout the brood area at between 30 and 35°C (Seeley & Heinrich 1981). These temperatures are maintained even in the face of severe environmental conditions, and they represent an enormous cost in energy (honey) consumption.

Honeybees are tropical in origin, and in the temperate zone wild honeybees (as opposed to managed bees in man-made hives) suffer severe mortality during winter (Seeley 1978, 1983). Survival of the colony is dependent upon the initial swarm occupying a suitable site early in the season and storing sufficient honey for the winter (see Seeley 1983). Successful preparation for overwintering involves the cessation of reproduction, the storage and utilization of food, and the behavioral and perhaps physiological changes necessary for thermoregulation. The relationship between these parameters and diapause is an area in need of study.

Swarming, drone production, and drone eviction are other seasonal events in honeybee hives. Drone production is related to worker number and the presence of larvae in cells. In the spring, drones are not produced while worker populations are low. During summer, as the number of workers increases, drone brood increases, and at the end of summer, when brood rearing decreases, drone production again decreases (Free 1977). Apparently, food, queen age, adult density, and social interaction provide important cues for the seasonal timing of drone production (see Brian 1979). The seasonal rejection of drones from colonies with reproducing queens is directly related to the amount of food collected from outside the colony

(Free & Williams 1975). Experimental reduction in the amount of forage results in early eviction of drones, whereas artificial prolongation of forage availability results in delayed drone eviction.

Several groups of ants overwinter in perennial colonies without brood; among them is the genus *Formica.* Species in this genus rear sexuals from the first eggs laid in the season. Apparently, low hibernal temperatures (below 15°C), along with other factors, have an effect on oögenesis and fertilization, resulting in large, gyne-biased eggs; at temperatures of 15°C and above, largely male-biased eggs are laid (Gösswald & Bier 1957). Such is the situation in *Formica polyctena,* which has an interesting seasonal cycle that probably involves both exogenous control by environmental conditions, such as temperature, and endogenous control by factors within the nest (Schmidt 1974). This species' life cycle is characterized by two vertical migrations and by oviposition of two types of eggs. The first migration occurs in spring and is followed by oviposition of large eggs directly on the surface of the soil. These eggs were formed in the queen's ovaries during the previous autumn, and they result in gynes that produce new nests. Oviposition of large eggs terminates after several days, apparently when the oögonia are depleted of large eggs, although other factors may also be involved. After oviposition of the first batch of eggs is complete, the queen returns to the nest interior and begins laying the second batch of eggs. These eggs are smaller and they produce workers. This second period of egg laying ends sometime in July. The termination of oviposition could be due to a number of factors: conditions external to the nest, the lack of response by the second brood of workers to the queen's solicitation for food, or factors within the queen herself.

Among the ants that overwinter in perennial colonies with brood, *Myrmica rubra* is by far the best studied, and yet there exists considerable controversy as to what controls the seasonal cycle. Summer brood invariably gives rise to workers, whereas winter brood (diapausing third-instar larvae) produces both sexes, including both forms of females (workers and gynes). Large third-instar larvae are gyne-biased and predisposed to diapause from the egg stage; however, both low temperature (below 23°C) and reduced food quantity strongly enhance diapause induction (Brian 1963, 1965b, 1975; Brian & Kelly 1967). For the large, overwintered larvae to produce gynes, rather than large workers, diapause development must be completed under low temperatures before development is induced; larvae that are forced to develop prematurely in autumn produce workers (Brian 1955). Some small, worker-biased larvae also overwinter, but they do not enter diapause, apparently because they were underfed by the new autumnal workers. Brian and Kelly (1967) and Hand (1983) tested the effect of long and short day lengths (LD 18:6 and LD 6:18) on the incidence of diapause, but found none. Thus, from their studies, Brian and co-workers (Brian 1963, 1965a, 1977, 1979; Hand 1983) concluded that temperature and the influence of the queen on worker behavior toward larvae are the two dominant factors influencing diapause induction and gyne production in the overwintering brood.

Kipyatkov (1973, 1974, 1976) differs with Brian's conclusions regarding the influence of photoperiod on diapause induction. His experiments indicate that *M. rubra*'s seasonal cycle is largely controlled by photoperiod, and they illustrate that photoperiodic information is transmitted among colony members. Short day lengths (LD 10:14) act directly to induce diapause in the workers. In turn, the induction of diapause in workers causes third-instar larvae to enter diapause. Diapause in the queen requires both the direct perception of short day length by the queen herself and the induction of diapause in workers. If the queen perceives a long day (LD 18:6) or if the workers are not in diapause, she will not enter diapause. The discrepancies between the results of Brian and those of Kipyatkov concerning the influence of day length on diapause is not easily resolved on the basis of the information available (see Hand 1983). But the finding by Kipyatkov (1979) that photoperiodic sensitivity in *M. rubra* is itself subject to seasonal variation may shed some light on the problem. Photoperiodic responses are virtually absent from posthibernation colonies in the spring; as the season progresses responsiveness is gradually regained, and by mid-July it is fully expressed. If the experiments by Brian and co-workers regarding photoperiodic induction of diapause were run with posthibernation colonies, then the results would be consistent with those of Kipyatkov. Clearly, more work with this species is needed.

In tropical species of ants, brood may be produced all year round, but seasonal cycles in the production of reproductives are largely related to the wet and dry seasons. Tropical dorylines, army ants, are typified by ~ 40-day cycles of nomadic and stationary phases that involve foraging and reproduction. These cycles continue throughout the year; thus, the ants appear to be aseasonal. However, tropical dorylines do have seasonal cycles, because although oviposition and foraging continue throughout the year, sexual broods are produced only seasonally—usually at the end of the dry season (Schneirla & Brown 1950, 1952; Rettenmeyer 1963; Raignier *et al.* 1974). Larval diapause and the influence of humidity on the queen may be important factors governing this seasonal cycle, but experimental evidence is lacking. Schneirla (1963) proposed that the annual temperature cycle regulates activity of a doryline ant, *Neivamyrmex nigrescens*, in southeastern Arizona. Springtime emergence from dormancy is associated with a rise in soil temperature, whereas low temperatures reduce nighttime foraging and thereby curtail oviposition. Again experimental evidence for these claims is lacking.

Seasonal regulation of soldier production by the ant *Pheidole pallidula* (a species that overwinters with brood) is similar to the regulation of drone production by the honeybee, *Apis mellifera* (a species that overwinters without brood). That is, soldier production is dependent on the abundance of nursing workers in the colony. The presence of more than 4 or 5% adult soldiers in posthibernation colonies inhibits the rearing of new soldiers and favors the production of workers. During summer, as the spring-reared workers emerge and join the workforce, the inhibition of soldier production is removed. In addition, the threshold (percentage of soldiers) above which there is an inhibition of soldier production rises throughout the year, so that in autumn a higher percentage of soldiers is tolerated without inhibiting soldier production than in spring (Passera 1977).

6.2.3 Summary

The seasonal cycles of social insects are influenced by the same range of environmental factors that influence nonsocial insects. However, they are also strongly influenced by interactions between colony members, and it appears that seasonal cycles have a major role in the evolution of sociality. Social insects have the apparently unique ability to transmit photoperiodic and other information among individuals in the colony. Some species also have unique abilities to store food for use during diapause and to thermoregulate their colonies.

Social insects characteristically overwinter either as fertilized solitary queens or in perennial colonies, with or without brood. Experimental evidence on the factors that regulate the seasonal cycles of social insects is difficult to obtain. From the few studies that have been made, photoperiod, temperature, food, and humidity appear to be important external factors; however, factors within the nest (e.g., the degree of queen dominance, queen pheromones, behavioral responses of workers, physiological state of workers), and endogenous rhythms may also be important.

6.3 Tropical Insects

In the tropics, biological events follow seasonal patterns that are quite different from those in the temperate zone. Temperate regions are characterized by substantial, rather clearly defined, and relatively predictable seasons during which conditions are generally either favorable or unfavorable for growth and reproduction. The main physical factors that, either directly or indirectly, determine seasonal cycles in the temperate zone are the incremental changes in day length and in temperature. Most growth and reproduction are reduced during winter when days are short and temperatures low. As a result, the life cycles of temperate-zone insects are closely adjusted to seasonal change; growth and reproduction either occur or do not occur at all during a particular season.

The situation is different in tropical regions. Here, especially near the equator, day length remains nearly constant year round and seasonal changes in temperature are relatively small. This does not imply that the tropics are without seasons. Quite the contrary! The tropics are marked by dramatic seasonal alternation in dry and wet seasons, and in some areas the contrast between the wet and dry seasons is spectacular (see Phipps 1968; Owen 1971; Frankie *et al.* 1974; Leigh *et al.* 1982; also Wolda & Galindo

1981). Most plant growth occurs during the wet season, but flowers are relatively rare at the height of the rains. During the dry season, many trees and shrubs shed their leaves, and many weeds and small plants die back completely. In addition, the amount of available sunlight varies considerably from the dry to wet seasons, although tropical forests may be less affected by this factor than open areas. Seasonal variations in physical factors such as these not only have a direct effect on the rate and type of plant growth and on the interactions between phytophagous insects and their hosts, but they also exert secondary effects on the interactions between predators and prey and between parasitoids and hosts (Brakefield & Larsen 1984; Janzen 1984).

Unlike in the temperate zone, where insect growth and reproduction are usually completely precluded during specific seasons, in the tropics physical conditions may be amenable to continuation of these functions year round. Moreover, some host plants may remain available all year, and thus it is not uncommon for insects to grow and reproduce throughout the year. However, most tropical insect species show distinct seasonal cycles in the level of these activities (Phipps 1968; Derr 1980; Young 1982; Wolda 1983; Wolda & Denlinger 1984). Most of these cases can be related to the fact that very often host plants are more abundant or more suitable during specific seasons (e.g., see Janzen & Waterman 1984). Many insects also exhibit seasonal cycles in coloration and/or body form (Shapiro 1976; Brakefield & Larsen 1984) that provide crypsis or other protection from predators in a seasonally variable environment. In many, but not all, tropical species, diapause serves an important function in synchronizing the life cycle with seasonally variable conditions. Other species have evolved specific, nondiapause mechanisms that adapt their life cycles to the seasonal fluctuations in their tropical environment. Still other species appear to have evolved both diapause-mediated and nondiapause adaptations to unpredictable environmental fluctuations. We discuss these various adaptive pathways below.

6.3.1 Diapause in tropical environments

Diapause in tropical insects is generally more difficult to study than diapause in temperate zone species. Thus, before delving into questions of the functions and adaptive value of diapause in the tropics, it is advisable to discuss certain characteristics of tropical diapause. First, the responses of tropical insects to seasonal stimuli are often difficult to quantify (e.g., see Wolda & Denlinger 1984). Insects in the temperate zone rely primarily on seasonal changes in day length to provide diapause-inducing stimuli. These changes are pronounced and strongly correlated with relatively drastic changes in temperature, food, and other important requisites, and the responses of temperate zone species are generally clear-cut and readily observable in relatively simple laboratory experiments (see Chapter 5). In contrast, insects in tropical or equatorial areas rely on subtle changes in photoperiod (Norris 1965a) or on changes in temperature, moisture, population density, and composition of food as diapause-inducing stimuli (e.g., van Dinther 1962; Usua 1973; Hummelen 1974; Ankersmit & Adkisson 1967; Denlinger 1974, 1978; Jacquemard 1976a, b; Scheltes 1976, 1978a, b; Paarmann 1977, 1979; Derr 1980). Compared with temperate-zone photoperiods, these factors are more difficult to work with in the laboratory and to measure in the field.

Second, diapause in tropical species may go undetected because it may be relatively short or involve quite subtle changes in physiology or behavior. For example, tropical army ants typically continue their foraging and nest-moving activities year round. To observers the colonies appear to be fully active. However, most species have clear seasonal cycles in reproduction—that is, the oviposition and rearing of worker- and/or reproductive-destined offspring (Schneirla & Brown 1950, 1952; Rettenmeyer 1963; Raignier *et al.* 1974). The seasonal cycle of reproduction in these ants involves larval diapause. Similarly, tropical butterflies, mosquitoes, and other insects are commonly found during all months, but there are clear seasonal fluctuations in their abundance or activity (Owen 1971; Janzen 1973; Wolda 1980, 1982; Akinlosotu 1982; Lowman 1982; Smythe 1982; Young 1982; Levings & Windsor 1982; Sempala 1983; Guerra *et al.* 1984). Imaginal diapause may occur in some of the species, but until specific tests or dissections are performed, it will not be known if these seasonal changes in abundance are the result of diapause or some other factors (e.g., see Young & Thomason 1974).

Third, although temperature, moisture, and food composition and abundance typically have seasonal cycles in the tropics, their cycles do not always produce extreme conditions that entirely prevent growth or reproduction. Recurring, but unpredictable, environmental changes are conducive to the evolution of polymorphic seasonal cycles (e.g., Bradshaw 1973; see Chapters 7 and 9). Because of this, diapause in tropical species is often quite variable and may occur only in a relatively small proportion of the population. Numerous studies have demonstrated the genetic variability in diapause characteristics of tropical insects (Ankersmit & Adkisson 1967; Rabb 1969; Blackman 1972, 1974b; Cantelo 1974; Denlinger 1974; Krysan *et al.* 1977; Kurahashi & Ohtaki 1977; Ae 1978; Masaki 1978a; Dingle 1978a; Rankin 1978; Branson *et al.* 1978; Ando 1979; Bell *et al.* 1979; Paarmann 1979; Dingle *et al.* 1980a, b; Dingle & Baldwin 1983; Kikukawa & Chippendale 1983). The nondiapause portion of the population remains active or moves to areas that are favorable for development and reproduction.

Such variation in diapause responses brings up the question of the function and adaptive value of diapause in the tropics. Is it a relict of adaptation to temperate-zone conditions, or does it have an adaptive function in the tropics? In some species there are readily apparent advantages to diapause-regulated seasonal cycles. Examples include insect species with seasonally restricted host plants, such as *Diabrotica virgifera* (Branson *et al.* 1982), *Diatraea grandiosella* (Kikukawa & Chippendale 1983), and *Anthonomus grandis* (Guerra *et al.* 1984). However, the adaptive advantage, if any, for diapause in other species often is not as apparent, and in only a few cases has it come under experimental study.

From comparative morphological, physiological, and genetic studies of the cricket *Pteronemobius fascipes* from throughout the Japanese archipelago, it was concluded that diapause characters are maintained, though at a very low level, in subtropical populations without the intrusion of morphological characteristics from more northern populations (Masaki 1978a). Thus, it appears that maintenance of the genetic basis for diapause in subtropical populations of this species does not require gene flow from temperate regions.

Artificial selection and field studies with the univoltine corn rootworm *Diabrotica virgifera* indicate that tropical populations from southern Mexico have the genetic variability to evolve a multivoltine or nondiapause life cycle (Branson 1976; Branson *et al.* 1982). However, they retain a long diapause in the tropics even though diapause duration is shorter among populations in temperate regions. Apparently, the long diapause is very important in maintaining synchrony between the beetle's life cycle and the phenology of its host plants in the tropics (Branson *et al.* 1982). Similarly, selection for nondiapause development in tropical flesh flies with thermally and thermoperiodically controlled pupal diapause results in a substantial reduction in the incidence of diapause within relatively few generations (Denlinger 1979). Thus, the genetic potential for nondiapause development resides in tropical flesh fly populations, but natural selection apparently prevents it from evolving. From these studies, we conclude that diapause is adaptive as a seasonal trait in the tropics and that its occurrence is maintained by natural selection—not by the fortuitous intrusion of genes from temperate zone populations.

A possible explanation for the variable incidence and intensity of diapause in some tropical species may be extrapolated from several studies of subtropical populations (Masaki 1978a; Ichinosé & Negishi 1979; Ichinosé & Iwasaki 1979). These insects experience well-defined seasonal cycles of temperature, but temperatures do not persistently fall below the activity threshold and the food supply is not completely depleted even in the coldest month. Under such circumstances, it is possible for individuals to overwinter either in diapause or in a nondiapause state. The genetic variability allowing such diverse responses is maintained because the relative fitnesses of the two types of overwintering individuals vary from year to year. Similar situations may be produced in the tropics by the recurrent, but less regular, occurrence of droughts that deplete the food supply. Such an hypothesis is open to long-range study.

6.3.2 Nondiapause adaptations

Many tropical insects may tolerate seasonally reduced abundance and/or quality of their food without entering diapause. Rather, they are forced to reduce their level of activity (growth and/or reproduction) and "wait out" the less favorable period. In some species this type of life cycle is associated with the ability of a particular stage to withstand a relatively long period of starvation or retarded development. Alternatively, a nondiapause life-style may be maintained by short-distance migration that facilitates the utilization of alternative hosts.

Nondiapause life histories have been well studied in certain groups of bugs that occur in tropical Africa and America. Although some *Dysdercus* species are reported to enter diapause (see Dingle & Arora 1973; Derr 1980), many apparently colonize a succession of Malvales host plants throughout the year without entering diapause. When fruits of the preferred host are not available, the adults move to another host that has fruit. When none of the preferred hosts is in fruit, the bugs migrate to a reservoir plant, one of the less-preferred hosts that flowers and fruits all year (Fuseini & Kumar 1975). The preferred hosts provide abundant food at specific times of the year, facilitating quick increase in population size. The less-preferred hosts are sufficient for reproduction and development, but act only as a reservoir between the fruiting seasons of the preferred hosts (Fuseini & Kumar 1975; Edmunds 1978). Thus, *Dysdercus* growth, development, and reproduction may occur year round, but the rate and circumstances are seasonal.

Migration in *Dysdercus* spp. appears to be facultative—occurring when food (fruits) becomes scarce (Dingle & Arora 1973); feeding, oögenesis, and oviposition take place when food and moisture are available. Depending on the range of acceptable hosts and on the seasonal availability and abundance of food, species with the *Dysdercus* pattern of migration vary in body size and other life-history traits (Dingle & Arora 1973; Derr *et al.* 1981).

Tropical milkweed bugs in the genus *Oncopeltus* also have nondiapause adaptations to seasonal conditions. The species *O. fasciatus* is a well-studied example (see Dingle 1978, 1981; Blakley 1980; Miller & Dingle 1982; Dingle & Baldwin 1983). Unlike temperate ones, populations of this species from the tropical islands of Puerto Rico, Jamaica, and Guadeloupe do not enter diapause in response to short day lengths, nor do they undergo long-distance migration. However, nondiapause adults of *O. fasciatus* are resistant to relatively long periods of starvation (Rankin 1978). Therefore, reproduction may occur year around, in conjunction with considerable movement among the locally ephemeral and sparsely distributed sources of food—fruiting milkweed plants. Other tropical species of *Oncopeltus*, such as *O. cingulifer, O. sandarachatus*, and *O. unifasciatellus*, vary in a number of nondiapause life-history traits (e.g., propensity for migration, host specificity) that adapt them to transient food resources (Root & Chaplin 1976; Dingle 1978a, 1981; Dingle & Baldwin 1983).

An important factor impinging on the life histories of these nondiapausing bugs is the relationship between body size and migration. In general there is an association between large size and flight (Dingle *et al.* 1980b). Large bugs survive longer than smaller ones, suggesting that large size may be an advantage during periods of food deprivation and/or migration.

Smaller nymphs can eclose sooner, but large adults survive stress better and migrate farther. In *Oncopeltus,* variation in growth rate occurs when nymphs develop on host plants that differ in nutritional quality, thus leading to variation in adult size and reproductive success (Blakley 1981; Blakley & Goodner 1978). There are considerable trade-offs associated with large size—that is, extended nymphal developmental time and lower survival to adult eclosion are balanced against increased survival under starvation and increased ability for prolonged dispersal. Each of these factors plays an especially important role in the evolution of nondiapause seasonal cycles.

Urania fulgens, a day-flying moth in tropical Central and South America, undergoes nondiapause seasonal flights that apparently adapt the species to a seasonally variable host plant. The larvae of *U. fulgens* have restricted food preferences; feeding is limited to *Omphalea* spp. (Euphorbiaceae). Data suggest that after three generations of larval feeding the host plants react chemically, so that larval growth is delayed. Female moths may be able to distinguish hosts that are of low toxicity or high nutritive value, and the periodic flights by moths may be an adaptation to the seasonal timing of biochemical changes in the host plants (Smith 1982).

In addition to the species discussed above, other tropical insects appear to have evolved nondiapause seasonal cycles. The African tabanid fauna contains examples of both diapausing and apparently nondiapausing species that provide the basis for interesting comparisons of life histories and their role in the epidemiology of disease (Bowden 1976). Tropical flesh flies show similar variation. Some equatorial species enter pupal diapause that is primarily controlled by temperature or photoperiod (Denlinger 1974, 1978; Kurahashi & Ohtaki 1977); others, such as *Euboettcheria trejosi* and *Pattonella intermutans*, from Panama, fail to enter diapause in response to the photoperiodic, thermal, and moisture cues that induce diapause in other Sarcophagidae. These nondiapausing species have a markedly variable and long period of wandering during the final larval instar, and they are very long lived as adults. Such attributes, which promote a relatively uniform seasonal distribution, may offer an alternative to diapause (Denlinger & Shukla 1984).

Among species that inhabit rain pools and other temporary aquatic environments in the tropics, the threat of desiccation is extreme, and the insects that utilize these habitats have evolved a variety of mechanisms subserving drought resistance. In some cases the adaptation may be associated with diapause; in others it is not. The aquatic larvae of the chironomids *Polypedilum vanderplanki* and *Chironomus imicola* provide spectacular examples of nondiapause drought resistance (Hinton 1951; McLachlan 1983). Larvae of *P. vanderplanki* are able to survive under extremely dry, hot conditions for several years, but upon contact with water, they resume growth and development immediately (Hinton 1960). This species has a relatively long larval stage, females produce few eggs, and adults do not disperse widely before oviposition. By contrast, *C. imicola*, which does not have the extreme drought resistance of *P. vanderplanki*, has a greatly foreshortened larval (aquatic) period, and the adults have high fecundity and high dispersal ability (McLachlan 1983). Both species are well adapted to utilizing ephemeral pools, although through very different means.

Extended periods of oviposition by long-lived adults and large variability in egg hatch are typical of several tropical corixids that breed and develop all year round. And, thus, all stages co-occur throughout the year. Temperate-zone species usually overwinter only in the adult stage (presumably in diapause); these populations tend to show structured age-class distributions. Populations from intermediate areas (e.g., the margins of warm salt marshes and shallow lagoons) may share characteristics reported for both tropical and temperate corixids—a polymorphic population in which one cohort reproduces year round and another cohort overwinters as nonreproductive adults (Balling & Resh 1984).

Aedine mosquito eggs also can be very resistant to drought; both nondiapause and postdiapause eggs can survive for long periods under very dry conditions. Once they experience the proper conditions of moisture and oxygenation, hatching occurs quickly (Clements 1963). Thus, the population fluctuations of some tropical mosquitoes closely follow the seasonal pattern of rainfall (e.g., Sempala 1983).

Our final example of a nondiapause adaptation to seasonal changes in the tropics is the seasonal polyphenism of lepidopteran adults. The wings of both butterflies and moths may vary greatly in their background coloration as well as in the patterns of their markings (see Shapiro 1976; Brakefield & Larsen 1984; Janzen 1984). In one case, the saturniid moth *Rothschildia lebeau*, adult coloration is related to the temperature conditions that the pupae experience during development. Dark-winged moths are produced under low temperatures, whereas pupae that develop under high temperatures produce morphs with rust- or light-colored wings. Temperature conditions are associated with the occurrence of the dry (hot) and wet (cool) seasons and with the seasonal change in background color in the forests where the moths live. Studies indicate that the thermally induced seasonal changes in wing coloration provide an excellent means for the moths to avoid predation by vertebrates (Janzen 1984).

6.3.3 Summary

Tropical insects, unlike temperate-zone insects, are not often faced with seasonal extremes in physical conditions that completely prevent growth and reproduction. However, they do experience marked, perhaps less regular, seasonal alternations in dry and wet seasons, with associated changes in the availability of suitable food. Insects have evolved a variety of means for coping with the seasonal fluctuations of a tropical environment. Many have a seasonally regulated diapause; others do not.

Diapause is sometimes expressed in somewhat subtle ways in tropical species, and is more difficult to analyze than in temperate species. In those tropical species that enter diapause, there often is considerable within-population variation in the propensity to enter diapause. For many species, only a small proportion of individuals in a population may enter diapause, while the remainder remain active. Artificial selection and field studies have demonstrated that diapause is adaptive in tropical populations, and thus its occurrence is maintained by natural selection—not by the fortuitous intrusion of genes from temperate-zone populations.

Nondiapause life cycles in tropical insects may be associated with specific life-history adaptations that subserve survival and synchrony with requisites under relatively mild seasonal conditions. The ability to withstand long periods of starvation and drought, short-distance migration to alternative hosts, and the prolongation of particular metamorphic stages appear to be especially important seasonal adaptations for nondiapausing tropical insects.

6.4 Arctic and Desert Insects

Like the tropics, Arctic and desert regions provide some unique phenological problems for insects. There can be no doubt that insects have met these problems with considerable success, as witnessed by the predominance of insects in both environments. However, very few studies have examined in detail the mechanisms and evolutionary history of seasonal adaptations in Arctic and desert-inhabiting insects. As a result, the discussions that follow here are brief.

6.4.1 Arctic insects

Constant daylight in summer, constant darkness in winter, low temperatures with relatively small diurnal fluctuations, and relatively low incoming radiation per unit time are the physical conditions of the Arctic region. In addition, Arctic habitats are characterized by a minimum of shelter and the presence of permafrost (Remmert 1980). These characteristics reflect immense seasonal variation in the physical conditions that are conducive to insect growth and reproduction. The Arctic growing season is short and cool; the effects of the extreme cold of the long winter are difficult to escape; cues (especially photoperiodic changes) that accurately herald seasonal change in the temperate zone may not function in the Arctic as they do further south (e.g., see Wielgolaski 1974). Thus, insects inhabiting the Arctic region appear to have made unique adjustments in their life cycles that permit them to survive and, in fact, thrive under the extreme conditions of the far north. These adaptations are described in detail by Downes (1965), Danks (1971a, b, 1981), and Remmert (1980), and we examine some of them here.

Among the significant factors that Arctic species must contend with are the shortness and coolness of the Arctic growing season—a situation that results in a small heat budget (see Downes 1965). Some species, such as bivoltine aphids and univoltine mosquitoes and chironomids, have relatively short life cycles that are completed within one summer (Downes 1965). Other species require two or more seasons to complete development. These species must overwinter two or more times in two or more different developmental stages. Examples of prolonged life cycles are found in Arctic Collembola and Lepidoptera, and Arctic and Antarctic chironomids, some of which can take up to seven years to complete development (Downes 1965; Butler 1982; Sugg *et al.* 1983). In one case, the bumble bee, *Bombus* sp., adults can accelerate development of larvae by raising the temperature of the nest with metabolic heat from their own bodies.

Because of the short growing season, it is a distinct advantage for Arctic species to be able to resume development in spring as soon as conditions permit and to be able to prolong development as long as possible at the end of summer. Downes (1965) presents the argument that temperature acts as the primary cue for both the termination and initiation of dormancy in Arctic species, primarily because of the lack of photoperiodic cues above the Arctic circle. Remmert (1980) disputes this contention and emphasizes the role of daily changes in light levels as *Zeitgeber* for photoperiodic rhythms. He cites studies by Stross (1969a, b) on *Daphnia* (see also Stross & Kangas 1969; Ferrari & Hebert 1982), by Vaartaja (1959) on North American trees (*Pinus* and *Betula*), and by Krüll (1976a, b, c) on birds, all of which demonstrate the importance of photoperiod as a seasonal cue in the Arctic. Boreal species of drosophilid flies also respond to photoperiod (Lakovaara *et al.* 1972; Lumme 1978). Clearly, the responses of other Arctic insects should be examined in this regard.

The role of diapause in regulating the seasonality of Arctic species is another area that should be examined. Among the bivoltine and univoltine species mentioned above, overwintering occurs during a specific metamorphic stage, and diapause apparently plays an important role in regulating the life cycle. However, many Arctic species appear to be able to overwinter in a number of metamorphic stages (at least several larval stages). In these species it is not certain whether overwintering is accomplished in the state of diapause or in a state of quiescence. In some cases, such as "spring species" of Chironomidae, there is good evidence for the occurrence of diapause in at least one of the several overwintering stages (Danks & Oliver 1972). But for other species, data are lacking. These problems need experimental study; mere observations of temperature-related postdormancy emergence patterns are not sufficient to explain what regulates the initiation, maintenance, and termination of dormancy, nor are they sufficient to illustrate the presence or absence of diapause.

The second factor that Arctic insects, with the exception of avian and mammalian ectoparasites, must overcome is the extreme winter cold (see Danks 1971a; Remmert 1980). Conditions during Arctic winters are such that overwintering insects must be frost tolerant. Although some Arctic species have a remarkable ability to supercool (Chapter 4), it is highly unlikely that many of them overwinter in an unfrozen state (see Downes 1965). Supercooling and its associated accumulation of glycerol and other cryoprotectants probably function to protect tissues from the effects of ice crystal formation and to reduce the temperature at which freezing occurs.

Frost formation rarely occurs during Arctic summers, and although snow may fall during summer, it rarely persists. Thus, the requirement for cold-hardiness varies seasonally. Some high-Arctic species express a relatively high degree of frost resistance year round but many that tolerate freezing during the winter readily succumb to frost during the summer (Remmert & Wisniewski 1970; Danks 1971b; Baust 1982; also Sugg *et al.* 1983 for an Antarctic species).

The final factor that may play a role in shaping the seasonal cycles of Arctic insects is interspecific interaction, for example, with parasitoids, predators, and competitors. Remmert (1980) points out that the Arctic summer is not undifferentiated as is sometimes suggested, and high-Arctic regions with a growing season of little more than a month have clearly differentiated "spring," "midsummer," and "late-summer" species. Typically, sciarids and other very small Diptera emerge early, and other flies and parasitic Hymenoptera follow later in the season. An interesting study of two sibling species of *Chironomus* that breed in Arctic Alaskan tundra ponds shows that the extremely long life cycles of the two species (seven years) result from slow growth during an annual open water season of about 90 days. However, neither food nor temperature appear to limit the rate of growth. It was suggested that the prolonged life cycle results in stabilized larval populations, a situation that may have evolved in response to benthic-feeding waterfowl in the ponds (Butler 1982).

6.4.2 Desert-inhabiting insects

Deserts present insects with particular phenological problems because the occurrence and persistence of the biotic and abiotic requisites that they provide for insect growth and reproduction are variable, unpredictable, and often short term. Rainfall is not only scarce; it is extremely irregular in timing, locality, and amount. The seasonal occurrence of desert plant growth and reproduction is often the direct result of the addition of water to the soils at a particular time (e.g., Koller 1969; Ackerman & Bamberg 1974; Beatley 1974). As a result, insect growth and reproduction, too, are intimately timed to the occurrence of rainfall. Deserts may also experience extreme daily and seasonal fluctuations in temperature. Thus, the physical conditions of desert regions have both strong direct and strong indirect effects on the seasonal cycles of the insects that inhabit them (e.g., see Falkovich 1979).

Very few studies have examined experimentally the ecophysiological adaptations of desert-inhabiting insects to the seasons, and the relative importance of diapause versus nondiapause seasonal adaptations is largely unknown. However, studies with a few species offer an idea of the range of seasonal adaptations desert-inhabiting insects have evolved. Many desert inhabitants seek protected habitats during the hottest periods and are active only at dusk or at night (see Willmer 1982). The desert-inhabiting burrowing cockroaches, *Arenivaga* sp., insects that live in desert sand dunes in California, exhibit seasonal cycles in their digging behavior. During spring, summer, and fall, adults burrow during the day at a depth of 20 to 60 cm, while at night they occur within 1 to 3 cm of the soil surface. In the winter, they are rarely found near the surface at any time of the day (Hawke & Farley 1973).

Some desert-inhabiting species are able to regulate their body temperature and rate of oxygen consumption (e.g., Cohen & Cohen 1981; Chappell 1983), whereas other species have specialized life cycles such as those found in certain carabid beetles. These beetles are particularly susceptible to desiccation during their developmental stages. The dry season is passed in a state of reproductive diapause and reproduction is timed to the rainy season (Paarmann 1979a). Additional specializations occur in the carabid *Thermophilum sexmaculatum.* This species produces big eggs that are able to absorb large amounts of water. Larval development, which encompasses a reduced number of larval instars, occurs within a short period. The first-instar larva may feed for only five days and then molt to a nonfeeding second instar that pupates without additional feeding (Paarmann 1979b). These adaptations apparently allow the insects to take advantage of moisture in the upper layer of soil which is available only for short periods.

Prolonged (or carry-over) diapause (see Section 3.4.1) is another important adaptation of desert-inhabiting insects because of the typically irregular pattern of seasonal conditions. The situation is similar to that found in plants in which whole populations (seedbanks) remain dormant for a year or more; resumed development is tied to the occurrence of favorable seasonal conditions. Powell (1974) provided examples of arid-region insects, notably yucca and ethmiid moths, that exhibit such a diapause.

Two other lepidopteran species provide additional interesting examples of prolonged diapause—the butterflies *Pieris napi microstriata* in California and *Papilio alexanor* in southern Europe and the Middle East. Through field studies with *P. n. microstriata* during and after the California drought of 1975-77, Shapiro (1979) concluded that prolonged pupal diapause accounted for survival of populations during the unfavorable period. At one locality, the first year of the drought saw a large, reproductively successful flight of *P. n. microstriata*; during the second year, the adult population was almost nonexistent despite the large burst of reproduction the previous year. Host plant biomass was greatly reduced, and no successful reproduction was observed that year. However, rather than becoming extinct, as would be expected for an annual species, the population was back in predrought numbers during the third (a rainy) season, suggesting that most of the population had undergone prolonged diapause. The advantages of remaining dormant and forgoing reproduction during the unfavorable year are obvious, but quantitative data on the mechanisms involved in the adaptation are needed.

In the other arid-zone lepidopteran, *P. alexanor*, the larvae feed exclusively on the flowers of the perennial herb, *Ferula*. Plants in this genus produce shoots each year, but they flower only in years of sufficient rainfall. Years in which the plants do not flower occur frequently and sometimes in succession. Pupae of *P. alexanor* apparently remain in diapause during those years in which flowering does not occur (Nakamura & Ae 1977).

Anticipation of seasonal change is a significant aspect of the life cycles of some desert-inhabiting species that have been studied, and we presume that most, if not all, of these cases involve diapause. These studies have been summarized by Crawford (1981), and our discussion below follows his analysis relatively closely.

The reliability of photoperiod and temperature as predictors of rainfall in the deserts is not known, but both factors do appear to have roles as seasonal cues for desert species, especially in synchronizing the susceptibility of dormant insects to the stimulating effects of rainfall or food availability. Such is possibly the case in certain desert-inhabiting *Orthoporus* millipeds, elaterids, and scarabs, but there is no experimental evidence to substantiate this claim (cf. Nilsen & Muller 1982, in relation to plant dormancy). Analysis of this type of response in desert-inhabiting insects would benefit from studies similar to those done with the *mohave* strain of *Chrysopa carnea*, an insect that does not inhabit deserts, but is adapted to unpredictable, arid conditions (Tauber & Tauber 1973c).

Some insects appear to rely on moisture as a cue to environmental changes in the desert. Some scorpions, mites, wasps, lepidopterans, carabid beetles, and tephritid flies become active almost immediately after the onset of rainfall. The butterfly *Callophrys macfarlandi* provides an interesting example. This usually univoltine species was reported to produce an unusual second generation during an exceptionally wet summer when its host plant flowered a second time. Such a coincidence strongly suggests that the butterfly, as well as its host plant, relies on rainfall as a cue to favorable conditions (Crawford 1981). Other arid zone Lepidoptera in which rainfall may be implicated in diapause termination are ethmiid and yucca moths, *Papilio alexanor*, and *Pieris napi microstriata* (see above), as well as numerous other species occurring in southern California (Emmel & Emmel 1973).

Photoperiod, moisture, and temperature appear to have roles in regulating the seasonal cycle of the Australian plague locust, *Chortoicetes terminifera*, an arid-zone species with a multivoltine life cycle (Wardhaugh 1977, 1980a, b). Females insert nondiapausing eggs more deeply into the soil in the summer than they do in the winter, and thus nondiapausing summer eggs avoid the stimulating effect of moisture that might occur with light summer rains. Rains of similar magnitude are effective in promoting hatching of nondiapausing winter eggs. When diapause eggs are produced, the locust egg batches contain both diapausing and nondiapausing eggs—a situation that results in hatching over a long period of time. Both photoperiod and temperature are involved in inducing diapause. Thus, the locust, through its sensitivity to photoperiod, temperature, and moisture, increases the chances of survival for its offspring by spreading the risk of hatching over several periods.

The extreme physical conditions of the desert and the seasonal occurrence of food are not the only factors that act to shape the seasonal cycles of desert-inhabiting species. The desert scorpion, *Paruroctonus mesaensis*, presents an interesting case in which seasonal activity and stage-specific population abundance are fashioned in important ways by intraspecific interactions. The seasonal and nightly patterns of surface activity (foraging and mating behavior) of adults are highly correlated with temperature and prey abundance, but this relationship does not hold true for young scorpions. The age-specific differences in seasonal and nightly patterns of activity tend to minimize the interaction between adults and young and may serve as important means whereby young scorpions avoid cannibalism by adults (Polis 1980).

6.4.3 Summary

Arctic and desert-inhabiting species face especially extreme seasonal conditions that insects from other areas usually do not experience. Despite the fact that the seasonal adaptations of these insects offer interesting material for ecophysiological and evolutionary analysis, few experimental studies have examined them. Thus, in most cases the role of diapause and the function of environmental cues in regulating these special seasonal cycles have not been delineated.

Arctic insects must contend with a very short, cool growing season, with very cold winter conditions, and with the absence of photoperiodic cues as they occur in temperature latitudes. Insects have adapted to these constraints in a number of ways. Some species are able to complete a full life cycle within the short, cool Arctic season; others have longer life cycles that include two or more periods of overwintering in two or more different metamorphic stages. Whether or not all of the several overwintering stages enter diapause is unknown, but at least in some species, one stage does so. The cues that induce and maintain diapause in Arctic insects are not known; both temperature and daily changes in light intensity could be important.

The presence of permafrost throughout most of the Arctic virtually excludes the possibility of Arctic insects escaping subfreezing temperatures during winter (avian and mammalian ectoparasites being the exceptions). Thus, the development of cold-hardiness (i.e., both supercooling and freezing tolerance) is an important seasonal requirement for most Arctic insects. However, the extreme cold is not the only factor limiting the life histories of Arctic insects; interspecific interactions also are important.

Few studies have examined the environmental cues that synchronize desert species with the often variable and short-term seasonal occurrence of their requisites for growth and development. Photoperiod, temperature, and moisture are important in a few cases, but further experimental studies are needed.

Chapter 7

Variability and Genetics of Seasonal Adaptations

Previous chapters have stressed the dynamic aspects of diapause by illustrating how insects constantly undergo physiological and behavioral changes that keep them in tune with seasonal changes in their environment. By analogy, insect seasonal cycles are also dynamic over evolutionary time. Although many insect species have broad distributions over large areas having diverse environmental conditions, at each locality their seasonal cycles are kept in synchrony with local conditions. Moreover, even within one area, seasonal conditions are subject to variation among years and gradual change over time. This chapter therefore introduces the second part of our book, which analyzes evolutionary questions dealing with seasonal cycles by discussing the variability and genetics of insect seasonal adaptations.

7.1 Variability in Seasonal Cycles

Widespread species generally encounter great diversity in climatic conditions among localities, and in most localities insect populations experience considerable variation in seasonal conditions from year to year. These types of environmental variations are often reflected in correspondingly large variations in seasonal cycles among geographical populations and within local populations. Such variation in seasonal cycles is based on a number of contributing factors, including environmental influences, phenotypic plasticity, genetic differences, and interaction between genetic and environmental factors.

The environmental control of geographic differences in seasonal cycles is illustrated by the craneflies *Ripula subnordicornis* and *Molophilus ater*. Populations of both of these species from various localities along an altitudinal gradient have clearly different postdiapause emergence times—those at higher altitudes emerging considerably later than those at lower altitudes. However, individuals transferred from one altitude to another emerge in correspondence to the temperature conditions of the new locality, not according to date of emergence at their home site. Temperature differences among the localities are sufficient to account for differences in emergence dates; there do not appear to be any innate differences among populations in their prepupal or pupal developmental rates (Coulson *et al.* 1976).

Nongenetically based variability is also found in widely distributed species having flexible seasonal cycles controlled by diapause-regulating mechanisms that do not vary substantially among diverse geographical populations. Two notable examples are the grasshopper *Oedipoda miniata* (Pener & Orshan 1980, 1983) and the cynipid parasitoid of *Drosophila, Leptopiliana boudardi* (Carton & Claret 1982). In both species diapause is strongly influenced by temperature, and geographical gradients in temperature may result in phenotypic gradients in diapause regulation, apparently without the need for genetically controlled seasonal variation among populations. A similar, nongenetic mechanism subserving adaptation to new localities has been proposed for *Agromyza frontella*, an alfalfa pest recently introduced into North America from Europe (Tauber *et al.* 1982; Nechols *et al.* 1983). Diapause in this species, as in the others above, is strongly affected by temperature.

In contrast to the few examples of environmentally controlled variation in seasonal cycles are the numerous cases of genetic variability. Workers in the USSR (Danilevsky and co-workers, see Danilevsky 1965) were the first to analyze in a comprehensive manner the nature of the variation in the ecophysiological responses that underlie geographical differences in seasonal cycles. One of their important findings was that the ecophysiological mechanisms that regulate the synchronization of the life cycle with the seasonal rhythm of climate are subject to genetic variation. Their studies quantified the variation in seasonal responses, examined the genetic bases for the variation, and related the findings to seasonal conditions in the field. They also provided a solid basis for future analyses of intrapopulation and geographical variation in the genetic control of seasonal cycles.

Such genetic variation can be viewed from two different standpoints. First, as the basis for diversity it provides a means for analyzing the genetics of seasonal traits. Second, as the material for the "building blocks" of evolution, it provides a means of reconstructing evolutionary pathways in the formation of seasonal cycles. Therefore, the first part of this chapter discusses the various types of genetic variation in seasonal cycles. The second part builds on this information to analyze the genetic control of diapause. Subsequent chapters consider specific evolutionary questions such as the evolution of diapause and seasonal cycles, the coevolution of seasonal and life-history traits, and the role of seasonal cycles in speciation.

7.1.1 Intrapopulation variation

Intrapopulation variation in insect seasonal adaptations is expressed in a wide variety of ways. At one end of the spectrum, it involves quantitative aspects of dormancy with continuous, within-season variation—that is, variation in response around a single mode. This type of variation is typical of populations with monomorphic seasonal cycles. At the other extreme, intrapopulation variation includes bi- or trimodal patterns of response that characterize polymorphic seasonal cycles. Often, the latter type of variation involves sets of traits that are both directly and indirectly related to seasonal cycles.

7.1.1.1 Continuous variation

Continuous intrapopulation variation in seasonal cycles characterizes a number of seasonal responses, a well-studied example being the photoperiodic response curve for diapause induction. For most species this curve rises or falls very sharply as it passes through the critical photoperiod, thus indicating a small degree of variability in the photoperiodic responses that determine diapause. In the arctiid moth, *Spilosoma menthastri*, the curve drops steeply from 100% diapause at LD 15:11 to 0% at LD 16:8—resulting in a maximum possible variation of only ½ hour on either side of the critical photoperiod. Similar, small degrees of variation around the critical photoperiod are reported for numerous other species (see Danilevsky 1965; Helle 1968). By contrast, the number of diapausing females of the mosquito *Anopheles maculipennis messeae* gradually falls as day lengths increase from 15 to 20 hours per day, and thus the variation around the critical photoperiod may reach approximately two hours (Vinogradova 1958, in Danilevsky 1965).

The biological significance of variation around the critical photoperiod has not been analyzed. Apparently, it is maintained within family lines. Analysis of photoperiodic and thermal effects on diapause induction in the butterfly *Colias alexandra* showed high variability both within and between sibling groups (Hayes 1982b). Occasionally, such variability results in the occurrence of a few nondiapausing larvae, or larvae with very short periods of diapause. These individuals appear in nature and in laboratory stocks and have been useful in selecting diapause-averting laboratory populations of *Choristoneura fumiferana* (Harvey 1957), *Porthetria dispar* (Hoy 1977), *Atrachya menetriesi* (Ando 1979), and *Aglais urticae* (Niehaus 1982). But these individuals generally do not survive in nature; perhaps their production is a cost related to maintaining genetic variation (see Section 9.2.2). When they do survive, they may provide the source for the evolution of multivoltine lines from univoltine populations, as appears to be the case in *Colias alexandra* (Hayes 1982b; see also Hoy 1977).

Sex may exert considerable influence on diapause. In species that undergo diapause as adults, the diapause-regulating responses of males and females are often similar, and as a result males and females are synchronized in their readiness to mate before and/or after diapause (Pener & Orshan 1983). However, in some cases, there may be considerable differences in the diapause-controlling responses in the two sexes. In the carabid beetle, *Pterostichus nigrita*, the females require a sequence of short days followed by long days to terminate diapause, but males end diapause after exposure only to short days (Ferenz 1975b; Thiele 1977a; Bale 1979). In the papilionid, *Bat-*

tus philenor, males are less likely to enter diapause, diapause is less intense in males, and males emerge sooner than females (Sims & Shapiro 1983b). And in the milkweed bug, *Oncopeltus fasciatus*, diapause may end sooner in males than in females (Hayes & Dingle 1983). The significance of these sex-related differences is open to study.

In some species, such as the blow fly *Lucilia caesar* (Ring 1971) and the flesh fly *Sarcophaga bullata* (Denlinger 1981b), the difference between males and females in the responses that control diapause induction is reflected in male-biased sex ratios in overwintering populations. Denlinger examined the reasons for these sex differences and presented evidence that they are related to the relative reproductive investment by the two sexes after diapause (see Section 9.1.3).

Differences between the sexes in the responses that control diapause may occur among species that undergo diapause as immatures. Such a situation occurs in the southwestern corn borer, *Diatraea grandiosella*, which has a larval diapause (Takeda & Chippendale 1982a). Under photoperiods near the critical photoperiod for diapause induction, a higher incidence of diapause occurs among females than males. However, females undergo one more instar than males, and thus they enter diapause later. After diapausing, females resume development sooner than males.

There are also sex-related differences in diapause in the mosquito *Aedes geniculatus*, which has both an egg and a larval diapause. Males are less likely to enter diapause, and they have a less intense diapause than females (Sims & Munstermann 1983). Another mosquito, *Aedes triseriatus*, also shows small, sex-related differences in the photoperiodic responses that induce egg diapause. In addition, the eggs of this species have very large sex-related differences in their response to postdiapause hatching stimuli (Shroyer & Craig 1981), which ultimately result in seasonal variation in the adult sex ratio. Males respond more readily to the suboptimal hatching stimuli that prevail in spring; thus, a large cohort of female eggs persists until late in the season.

Characteristics of dormancy other than the critical photoperiod express continuous intrapopulation variation. Among them is diapause duration in both hibernal and aestival diapause. In studies with winter eggs of the mite *Panonychus ulmi* from six orchards in Kent, England, Cranham (1973) found a wide range of variation in median dates of hatching at field temperatures. Laboratory studies indicated that variation in hatching time results from differences in diapause intensity and response to low temperatures during diapause. Quantification of the variation enhanced the accuracy with which postdiapause hatch may be predicted.

The green lacewing, *Nineta flava*, from Europe also has considerable continuous intrapopulation variation in the duration of diapause, but this variation involves aestival diapause. This univoltine species overwinters as diapausing third-instar larvae within cocoons. When posthibernal pupae and adults are subjected to short day conditions, oviposition proceeds in all individuals within 20 days. Long day lengths produce a reproductive diapause of very variable duration; thus, the preoviposition period in the summer ranges from 11 to 111 days. Such variability partially explains the occurrence of all stages of this univoltine species throughout the summer (Canard 1982), but the adaptive significance of the variability remains unknown. Perhaps it is a form of "bet-hedging" in an environment in which competition for food may be intense.

Species that end diapause in midwinter and subsequently enter a thermally controlled, postdiapause quiescence may also show considerable variation in the duration of diapause (Masaki 1965; Morris & Fulton 1970b; Dingle 1974a; Tauber & Tauber 1976a; Morden & Waldbauer 1980; see our Section 5.2.6). For example, diapause in field populations of the green lacewing *Chrysopa carnea* ends any time between mid-January and mid-March; subsequently, low temperatures prevent the initiation of postdiapause processes (Tauber & Tauber 1973d). Only when temperatures exceed the morphogenetic threshold does the transition from diapause to the postdiapause state begin. Therefore, variation in the timing of the end of the physiological state of diapause will not cause similar variation in the loss of the overt expressions of diapause (i.e., loss of cryobiosis) or in the timing of postdiapause reproduction. Thus, under field conditions there seems to be little reason for natural selection to act strongly to restrict in any precise manner the variation in diapause duration, and this trait may not be an important direct component of fitness.

Numerous species end diapause over a similarly broad period of time, and in laboratory experiments they show considerable variation in the age at first reproduction after diapause (e.g., Dingle *et al.* 1977; Dingle 1978a). In some cases, the duration of diapause is highly correlated with the insects' responses to diapause-inducing photoperiods (Tauber & Tauber 1972a; Masaki 1958, 1959; Dingle 1974a; Denlinger & Bradfield 1981; Furunishi *et al.* 1982; Kimura *et al.* 1982), and the variation in diapause depth may be the result of variation in responses around the photoperiodic response curve. Thus, care must be exercised in attributing biological significance to results dealing with this form of variability.

From the above, it appears that there may be two different categories of continuous variation in diapause characters (see also Istock 1983). One is manifest (though it sometimes escapes our notice) in the field and is directly significant to the ecology or the survival of the species. The other is concealed and revealed only under artificial conditions. This type of variation may not directly affect fitness, but it may be associated with other characteristics that do. In either case, traits with continuous intrapopulation variability are very responsive to artificial selection (Dingle 1974a; Hoy 1977; Section 9.2.1), and they exhibit high additive genetic variance (Dingle 1974a).

7.1.1.2 Disjunct expressions of intrapopulation variation

In addition to continuous variation, seasonal cycles also vary in a disjunct manner. This type of intrapopulation variation is characterized by polymodal patterns of response and it underlies polymorphic seasonal cycles. Such variation appears especially important in spreading over time the risks involved with terminating dormancy and initiating growth, development, and reproduction in an unpredictable environment (see Chapter 9). It avoids the situation in which the individual has "put all its eggs into one basket" (e.g., see Waldbauer 1978; Wise 1980) and subjected them all simultaneously to the possibility of encountering unfavorable environmental conditions. These unpredictable environmental risks stem from variations in both physical factors, such as temperature and moisture (Bradshaw 1973; Shapiro 1979), and biotic factors, such as parasitoids, predators, competitors, and suitable hosts (Maeda *et al.* 1982; Shiotsu & Arakawa 1982; Leinaas & Bleken 1983). There are also risks and costs involved in delaying the emergence from dormancy, and these factors must be balanced by the benefits. Thus, the proportion of the population that delays growth or reproduction is influenced by the severity of the environmental hazards and by the frequency of their occurrence (Takahashi 1977).

Perhaps the most notable example of this kind of variation involves emergence times. Waldbauer (1978) categorized the various types of polymodal diapause termination strategies exhibited by insects; two types involve within-season variation and one involves between-year variation in emergence. Waldbauer's first category includes bi- or trimodal emergence by individuals of a single age class, all of which entered diapause in the same season. For example, pupae of the moth *Hyalophora cecropia* enter diapause at the end of summer. Adults emerge the following spring and summer in a bi- or trimodal pattern, depending on the timing of diapause termination (Waldbauer 1978; Waldbauer & Sternburg 1978; Waldbauer *et al.* 1978; Nechols & Tauber 1982).

Several populations of the cabbage root fly, *Delia radicum*, also show disjunct variation in emergence from hibernation. Some geographical populations require only 14 days at 20°C after storage at 4°C for most postdiapause flies to emerge. Other populations have large numbers of late emerging flies (requiring more than 100 days at 20°C) (Finch & Collier 1983). Apparently diapause in the late-emerging flies is not completed under the low temperature of 4°C. Such variation is genetically controlled and transmitted between generations (Finch & Collier 1983). It results in prolonged posthibernation emergence and causes difficulties in predicting emergence of field populations. Similar patterns of bimodal emergence are found in the sawfly *Diprion pini* and the papilionid *Papilio glaucus* (Eichhorn 1979; Hagen & Lederhouse 1984). However, in the sawfly, the early emerging cohort overwinters in a more advanced stage of development than the late emergers. Thus, emergence patterns can be predicted by monitoring the developmental stage of overwintering eonymphs (Eichhorn 1979).

The first kind of disjunct variation in the termination of diapause also occurs in the chaoborid fly, *Chaoborus americanus*, but in this case phenological variation is associated with morphological differences. Two forms of larvae overwinter in shallow ponds and kettle holes (Bradshaw 1973). In the spring, after the winter thaw, the larval habitat is subject to unpredictable periods of freezing. Early developing morphs, which are large and yellow, do well when the spring is mild and ponds do not refreeze; however, they suffer heavy mortality if a cold period occurs. Under these conditions, the late-developing morph, which is small and pale, does well. The two morphs appear to have different responses to the diapause-terminating stimuli of day length and food (Bradshaw 1970).

Waldbauer's second type of within-season variation in emergence involves insects that overwinter in more than one age class. Spring emergence by the dragonfly *Anax imperator* represents metamorphosis to the adult stage by nymphs belonging to an age cohort that had passed two winters as larvae, the second winter being in the final instar. Approximately a month later, a second cohort emerges. These larvae had experienced only one winter and they did so as penultimate instars (Corbet 1963). This form of biomodal emergence involves variation in aspects of the seasonal cycle other than in the timing of diapause termination.

Mesocomys pulchriceps, an eupelmid parasitoid of emperor moth eggs, provides an example of variation in emergence pattern in which two age classes overwinter. This parasitoid can have seven generations per year. Apparently, diapause is induced by short day lengths, but some individuals that develop from eggs laid under summer conditions have life cycles that extend slightly more than one year. Emergence by these individuals is asynchronous with the emergence of individuals with a short-day induced diapause (van den Berg 1971).

Eggs of the variegated grasshopper, *Zonocerus variegatus*, in Nigeria hatch from September to April; however, the population consists of two distinct cohorts. The first contains the dry-season eggs that do not enter diapause and that hatch from September to November; the second cohort comprises wet-season eggs that hatch from November to April. Adults of the two cohorts overlap from February to July, and they interbreed freely (Iheagwam 1983).

Waldbauer's third category of variation in emergence involves prolonged diapause (diapause lasting more than one year). Such variation is common among insects, and it may spread emergence over as many as 12 or more years (Barnes 1952; see Powell 1974; Eichhorn 1977a, 1979, 1983; Sunose 1978; Surgeoner & Wallner 1978; Havelka 1980; Wise 1980; Muona & Lumme 1981; Lopez 1982; Hedlin *et al.* 1982; Maeda *et al.* 1982; Torchio & Tepedino 1982; Baxendale & Teetes 1983b; Sims 1983b). Usually only a small percentage of individuals remain in diapause while the major proportion of the population becomes active (Rivnay & Sobrio 1967; Sullivan & Wallace 1967; Powell 1974; Wallace & Sullivan 1974; Torchio 1975; Sunose 1978; Waldbauer 1978; Hedlin *et al.* 1982); however, under some conditions it may involve a large percentage or all of the population (see Powell 1974; Shapiro 1977, 1980b; Annila 1982; Torchio & Tepedino 1982).

An interesting example of prolonged diapause occurs in the tettigoniid, *Ephippiger cruciger*, which undergoes a two-stage egg diapause. A portion of the population in the egg stage responds to low temperatures during the first winter by terminating the initial diapause, whereas the remaining eggs do not (Dean & Hartley 1977a). Once the first diapause has ended, embryonic development proceeds until a second diapause intervenes. All individuals in the second diapause respond to cold and terminate diapause within one season. Thus, the first diapause ensures that hatching will be spread over several years, whereas the second diapause ensures that hatching will be timed to the appropriate seasonal conditions of the year in which hatching occurs (Dean & Hartley 1977b).

Broods of the megachilid bees *Osmia montana, O. californica*, and *O. iridis* exhibit variable life cycles with emergence occurring over one or two years and with diapause involving two different life stages. Univoltine individuals overwinter as diapausing adults and emerge the following spring; semivoltine individuals spend the first winter as mature larvae in diapause and the second winter as diapausing adults in the nest. The proportion of one- and two-year individuals is very variable and not easily explained on the basis of spatiotemporal heterogeneity (Torchio & Tepedino 1982).

The expression of prolonged diapause in insects appears to be influenced by both environmental as well as genetic factors—neither of which are well understood. It is clear that the offspring of a single female may include individuals that enter prolonged diapause and those that do not (Maeda *et al.* 1982; Torchio & Tepedino 1982; Eichhorn 1983). Elucidation of the genetic component will be particularly difficult because of the time span required for experiments and because of our lack of knowledge concerning the relevant environmental parameters. However, research on this problem is needed for both theoretical and practical reasons (see Chapters 9 and 10).

The disjunct expression of dormancy is not restricted to emergence times; it is also expressed in the induction of diapause and therefore in voltinism. In temperate zones, all or a very high percentage of individuals usually enter hibernation. Those that do not, probably suffer injury or die. However, exceptions are known or suspected, in which both diapausing and nondiapausing individuals overwinter (Ismail & Fuzeau-Braesch 1976a; Khoo 1968). In Japan, the citrus red mite, *Panonychus citri,* contains two strains. One occurs on deciduous fruit trees and undergoes hibernal diapause in the egg stage, and the other is found on citrus and does not enter diapause (Uchida & Shinkaji 1980; Takafuji & Kamezaki 1984). Clones of the green peach aphid, *Myzus persicae*, even from the same locality vary considerably in their overwintering characteristics. Some clones produce sexual forms that mate and lay overwintering eggs. Others do not have sexual forms (or produce only a few males); these clones overwinter as parthenogenetic females. In the temperate region, overwintering by parthenogenetic females is successful in mild years, but under severe winter conditions the females perish (Sutherland 1968; Blackman 1971, 1974b; Kolesova *et al.* 1980; Tamaki *et al.* 1982a). Similarly, quiescent larvae of the leafhopper *Euscelis incisus* are able to overwinter under mild conditions, but only diapausing eggs survive during harsh winters (Müller 1981).

The Oriental chinch bug, *Cavelerius saccharivorus*, in the subtropical area of Okinawa, Japan, exhibits an interesting type of variation in its life cycle that could be confused with disjunct variation. At this locale the bugs overwinter as nondormant adults that lay eggs. Eggs deposited in late autumn can undergo diapause at both high and low temperatures, but eggs laid after this time enter diapause only in response to low temperatures. Hatching occurs synchronously in the spring, which is probably the main function to be accomplished by the egg diapause. This life cycle involves disjunct overwintering states—nondiapausing adults and diapausing eggs, but the final result is synchronous hatching in the spring. This happens because eggs laid later in the fall have a less intense diapause than those laid earlier (Hokyo *et al.* 1983). Therefore, we cannot say that this is a case of disjunct polymorphism even though two different life stages coexist throughout winter.

The lepidopteran *Dendrolimus spectabilis* in central Japan offers another example of a life cycle that appears to be polymorphic, but upon examination is shown to be homogeneous. This species has both univoltine and bivoltine life cycles. The progeny of univoltine parents hibernate at a more advanced larval stage than the progeny of bivoltine parents, so that after hibernation the latter take a univoltine course and the former take a bivoltine course. Therefore, the univoltine and bivoltine patterns of development appear alternately in individual lines. Although univoltine and bivoltine individuals coexist spatially and temporally, the situation cannot be regarded as a case of polymorphism because the population is homogeneous. All lines produce three generations every two years, and there is no seasonal isolation between the lines (Habu 1969).

Perhaps the most common expressions of polymorphism in diapause induction are associated with aestival diapause, density-dependent diapause, and the occurrence of a second diapause. When aestival diapause occurs, or when a second diapause intervenes, it often appears only in a part of the population, while the other part continues development or reproduction (Tsuji 1963; Harvey 1967; Hodek & Honek 1970; Tauber & Tauber 1973c, 1982; Jorgensen 1976; Hodek & Růžička 1979; Masaki 1980; Hodek *et al.* 1981). For example, in the *mohave* strain of *C. carnea*, 40 to 60% of the individuals enter a food-mediated aestival diapause in response to lack of prey, whereas the others continue to reproduce even when prey densities are low (Tauber & Tauber 1973c, 1982).

Similarly, in *Gerris odontogaster*, a water strider that inhabits temporary waters, a portion of the population, the long-winged form, enters premigratory reproductive diapause during summer, whereas the short-winged part of the population reproduces. The long-winged form survives drought conditions by migrating to more favorable sites; the short-winged form reproduces successfully in the ponds that do not dry up. Both environmental and genetic factors influence the induction of diapause and the formation of long wings (Vepsäläinen 1974a, 1978).

The density-dependent diapause of the Indian meal moth, *Plodia interpunctella*, presents another example of disjunct variation in diapause induction. Some members of the population enter diapause in response to overcrowding and lack of food, whereas others remain in the active stage, even though nondiapause individuals may suffer heavy mortality (Tsuji 1963).

It is important to point out that, in many cases, it is difficult to distinguish disjunct and continuous variation. In some cases disjunct (all-or-none) variation may be based on continuously varying genetic and physiological mechanisms. Thus, the physiological responses that control the propensity to enter diapause may vary continuously, but the expression of the variability may be dependent upon a threshold. Those individuals above a certain threshold value may enter diapause; those below it may avert diapause. Similar variation may also occur in diapause termination, especially when a specific stimulus is required for completion of the process. The expression of discontinuity in nature may depend on the relationship between the distribution of the response threshold in the population and the range of natural environmental conditions.

The degree to which insect species express polymorphism in their seasonal cycles varies greatly—often as a measure of the unpredictability of their environment (see Stearns 1976; Southwood 1977; Istock 1978, 1981; Vepsäläinen 1978; Tsuda 1982). Examples of "bet-hedging" seasonal cycles are found among the milkweed bugs (Dingle *et al.* 1977, 1982), the mosquitoes (Istock 1978, 1981; Kingsolver 1979), the green lacewings (Tauber & Tauber 1982), crickets (Walker 1980), and copepods (Hairston & Munns 1984). In all these examples, the polymorphic populations occupy habitats with unpredictable environmental extremes, and seasonal polymorphism allows the population to exploit favorable periods while maintaining a fail-safe structure in the event of a catastrophe. Many of the examples of intraspecific variation described above involve coordinated sets of both seasonal and other life-history characteristics, and they illustrate that genetically based polymorphisms allow populations to spread their chances of success along more than one line of adaptation. Such polymorphisms therefore serve as an excellent means of analyzing the evolutionary interaction between seasonal cycles and life-history traits (see Chapter 9).

7.1.2 Geographical variation

Adaptation of geographical populations is notably expressed in the number of generations per year (voltinism) that characterize species with wide latitudinal distributions. This variability (univoltine, bivoltine, multivoltine, and nondiapause life cycles) may result from geographical differences in environmental conditions (see Section 7.1 above), but its primary cause lies in genetic differences among populations. Similarly, genetic variability also underlies most cases of geographical differences in the seasonal timing of phenological events such as emergence from the overwintering stage.

In general, southern populations of most species have short critical photoperiods and an associated tendency toward multivoltinism and in extreme cases nondiapause life cycles (e.g., McMurtry *et al.* 1976; Canard 1982), whereas northern populations tend to have longer critical photoperiods and an associated high percentage of univoltinism. Some studies have related the degree of change in latitude to corresponding changes in photoperiodic responses. Populations of the noctuid moth, *Acronycta rumicis*, taken from four localities in the Soviet Union along a gradient of approximately 20° latitude have critical photoperiods for diapause induction that range from 14.5 hours for the southernmost population to 19 hours for the northernmost. This amounts to an increase in the critical photoperiod of approximately 1.5 hours for each 5° increase in latitude (Danilevsky 1965). The critical photoperiod of another species of noctuid, *Mamestra* (= *Barathra*) *brassicae*, increases by about one hour for every 5° increase in latitude (Danilevsky 1965), and for *Drosophila littoralis* it is about 1.3 hours for every 5-degree increase in latitude (Lankinen & Lumme 1984). Those for several species of mosquitoes show geographical variation in the same range of magnitude: *Aedes sierrensis* larval diapause—1 hour/5-degrees latitude (Jordan & Bradshaw 1978); *Aedes triseriatus* (northern U.S. populations) egg diapause—1.2 hours/5-degrees latitude (Shroyer & Craig 1983); *Wyeomyia smithii* larval diapause—1 hour/5-degrees latitude (Bradshaw & Lounibos 1977).

Despite the above similarities in geographical variation, it is important to stress that there are considerable differences among species in the degree to which the variation in their photoperiodic response curves can be correlated with latitude. Indeed, some species, like the cabbage white butterfly, *Pieris brassicae*, from European USSR (Danilevsky 1965) and the mosquito *Toxorhynchites rutilis septentionalis* from the United States (Trimble & Smith 1979) show very little variation over large geographical areas. Other species show geographical variation of intermediate magnitude. The critical photoperiod for imaginal diapause in the green lacewing, *Chrysopa carnea*, increases by only 0.62 hours per 5-degree increase in latitude (Tauber & Tauber 1982). And, in the mosquitoes *Wyeomyia smithii* and *Aedes triseriatus*, the psocid *Peripsocus quadrifasciatus*, and the sawfly *Diprion pini*, the correlation between critical photoperiod for diapause induction and latitude is clearly influenced by the effects of altitude (Bradshaw 1976; Bradshaw & Lounibos 1977; Eertmoed 1978; Eichhorn 1979; Holzapfel & Bradshaw 1981). Studies of the malaria mosquito, *Anopheles maculipennis messeae*, in the USSR (Vinogradova 1960) and the green lacewing, *Chrysopa carnea*, in North America (Tauber & Tauber 1982) illustrate significant influence of local conditions on variation in photoperiodic and other responses. Thus, it is clear that the patterns of geographical variation are not simple, but are based on complex interactions between latitude, altitude, local conditions, and the genetic flexibility of the species (Tauber & Tauber 1982).

Seasonal characteristics that show intraspecific clinal variation fall into two categories that correspond to the two types of intrapopulation variation discussed above. First are quantitative aspects of dormancy, such as (a) latitudinal variation in the *length* of the critical photoperiod (see Masaki 1961;

Croft 1971; Tauber & Tauber 1972a, 1978; Maslennikova & Mustafayeva 1971; Mustafayeva 1974; Vinogradova 1975b; Hoy 1975b; Bradshaw 1976; Eichhorn 1977b, 1978; Eertmoed 1978; Tauber *et al.* 1983; numerous others); (b) depth or *duration* of diapause (Goldschmidt 1934; Masaki 1956b, 1963, 1965, 1967a, 1973; Rabb 1969; Tauber & Tauber 1972a, 1978, 1982; Poitout & Bues 1977c; Sternburg & Waldbauer 1978; Minami & Kimura 1980); (c) *duration* of the sensitive stage (Beach 1978); (d) thermal regulation of the *rate* of diapause development (see Andrewartha 1944, 1952; Masaki 1961; Beck & Apple 1961; Danilevsky 1965; Watson *et al.* 1974; Krysan *et al.* 1977; Ando 1979; Finch & Collier 1983); (e) body *size* (Masaki 1967a, 1973, 1978a, b); (f) *rate* of prediapause development (Masaki 1967a, 1978a; Eichhorn 1979); and (g) *rate* of postdiapause development (Kishino 1970a, b; Heron 1972; Baker & Miller 1978; Boller & Bush 1974; Nechols *et al.* 1980; Tauber & Tauber 1976a, 1978). Thermally controlled nondiapause developmental rates also can vary geographically (Campbell *et al.* 1974; Trimble & Smith 1978; Obrycki & Tauber 1982; Tauber & Tauber 1982). Because of its quantitative nature most of this variability may be based on gradients in polygene frequencies (see Section 7.2.2.1), although in some cases the variation may involve a cline in a genetic switch mechanism (see Section 7.2.2.2).

Second, in contrast to clinal variation in quantitative characteristics, geographical variation can also involve disjunct or qualitative differences in seasonal characteristics. In some species there is geographical variation in the occurrence of nondiapausing individuals at the end of the season (Azaryan 1966; Blackman 1974b; Dingle 1978a; Ichinosé & Negishi 1979; Dingle *et al.* 1980a; Morimoto & Takafuji 1983; Takafuji & Morimoto 1983; Takafuji & Kamezaki 1984). In others, the overwintering stage varies geographically (Bradshaw & Lounibos 1977; Ferrari & Hebert 1982). The sensitive stage and the timing of diapause termination may also vary geographically (Bradshaw & Lounibos 1977; Heron 1972; Vinogradova 1975b; Sternberg & Waldbauer 1978; Ichinosé & Negishi 1979). Or, there may be geographical differences in the occurrence of aestival diapause. Adults of the ladybeetle *Coccinella septempunctata* enter a short-day-induced diapause in Europe and a long-day-induced aestival diapause in central Honshu, Japan (Okuda & Hodek 1983). Populations of the green lacewing, *Chrysopa carnea*, in western North America may show a prey-mediated and/or photoperiodically controlled aestival diapause, the incidence of which varies among geographical populations. The populations in drought-prone areas with unpredictable prey populations tend to have a higher incidence of aestival diapause than those in irrigated agricultural areas with relatively reliable, high levels of prey (Tauber & Tauber 1978, 1982). Such variation constitutes geographical variation in polymorphism for seasonal cycles that are adapted to specific geographical conditions.

In addition to diapause-mediated dormancy, both diapause-mediated migration and polyphenism may vary geographically. Certain southern populations of the migratory milkweed bug, *Oncopeltus fasciatus*, do not migrate but overwinter in the area in which they reproduce (Dingle 1978a; Dingle *et al.* 1980b). There are also geographical differences in the environmental control of changes in pupal color that are associated with diapause in the papilionid butterfly *Battus philenor* (Hazel & West 1983; Sims & Shapiro 1983) and in the wing length of water striders (*Gerris*) (Vepsäläinen 1974a, 1978) and planthoppers (*Prokelisia*) (Denno & Grissell 1979).

The ecophysiological variation cited above underlies the expression of locally adaptive differences in seasonal cycles (e.g., differences in voltinism or in the expression of diapause-mediated seasonal traits). However, clinal variation can also result in a uniform life cycle throughout a large north-south distribution such as that found in species like the gypsy moth *Porthetria* (= *Lymantria*) *dispar* (Goldschmidt 1934), the field crickets *Teleogryllus emma* (Masaki 1967a, 1978a), and the corn rootworms *Diabrotica virgifera zea* and *D. v. virgifera* (Krysan 1982) that have stabilized univoltine life cycles over broad geographical ranges. For example, throughout the emma field cricket's (*T. emma*) north-south distribution covering the length of the Japanese archipelago, univoltinism is an unvarying, dominant part of the life cycle. This life cycle character is maintained, despite large variation in major environmental factors such as temperature and photoperiod, because of the clinal variation in the genetic and ecophysiological mechanisms that control univoltinism (Masaki 1978a). Contrary to what is the case for many insects (Danilevsky 1965; Tauber & Tauber 1972a), diapause in southern populations of *T. emma* is intrinsically longer than in northern populations. In the north, a long diapause is not necessary, because winter and spring conditions are cool and prevent egg development until summer, but in the south a long diapause delays development until late in the season, even though conditions may be favorable for development before then. Moreover, in the south, nymphal development is slower and more sensitive to photoperiod; as a result, development is delayed and univoltinism persists in the south, but the adult crickets are larger (Masaki 1978a). Thus, life-cycle uniformity in the face of large environmental diversity is achieved by genetic variation in diapause traits, especially diapause intensity and prediapause growth rates. A similar situation occurs in the corn rootworm, *Diabrotica virgifera*. Populations from southern Mexico have a much more intense diapause than those from South Dakota (Krysan *et al.* 1977), and there is a smooth cline in the variation of diapause duration among geographical populations from 36° to 46°N latitude (Krysan 1982). From the above examples and several others in which there is a less intense diapause where the cold season is longer than where it is shorter (Masaki 1961; Holtzer *et al.* 1976; Watson *et al.* 1974; Keeley *et al.* 1977), it is clear that populations of univoltine species from regions with a long growing season may have a relatively long diapause that prevents production of more than one generation per year, whereas those from areas with short growing seasons may have a short diapause because environmental conditions (e.g., low temperatures) directly prevent production of a second generation. In cases of such variation, it appears that the clines in diapause intensity are relatively smooth.

It is usual among insects for only one stage in the life cycle to enter diapause and survive the rigors of the dormancy period. As a result, it is clearly disadvantageous for multivoltine species to produce incomplete generations at the end of the growing season because individuals that do not reach the diapause stage almost invariably succumb. Thus, insect life cycles are closely synchronized with the growing conditions of local areas. Such adaptations prevent the wasteful production of partial generations, and they often result in geographical clines in insect seasonal responses with stepped or saw-toothed gradients even along relatively smooth environmental gradients (see Endler 1977; Roff 1980). Characteristics such as the critical photoperiod for diapause induction, the influence of photoperiod on prediapause growth, diapause depth, stage of diapause, rate of postdiapause development, and other factors that can hasten or prevent the production of a full generation often vary abruptly among populations from relatively close localities, especially where they are associated with a change in voltinism (Kishino 1970a, b; Masaki 1973, 1978a, 1979; Bradshaw & Lounibos 1977; Kidokoro & Masaki 1978; Jordan 1980b). Stepped or saw-toothed clinal variation reflects the number of complete generations that can be produced in the various localities throughout the distribution of the species.

Comparative studies of the geographical variation in several species of crickets from Japan illustrate clearly the influence of voltinism on the patterns of variation. Species such as *Teleogryllus emma* and *T. yezoemma* that are univoltine throughout the full length of Japan decrease in size from south to north (Masaki 1965, 1967a, 1978a), presumably because faster growth rates and less intense diapause are selected under cool northern conditions (see above). Despite the smooth latitudinal gradient in temperature that occurs throughout their entire range, the species *Pteronemobius nigrofasciatus* (formerly *fascipes*) and *P. mikado* (formerly *taprobanensis*) have variable voltinism and saw-toothed patterns of size variation that reflects variation in voltinism (Masaki 1978a, b, 1979; Kidokoro & Masaki 1978). In the *P. taprobanensis* group, the lawn ground crickets that occur along the full length of the Japanese islands, there is an abrupt increase in adult size and more conspicuously in ovipositor length at about 28°N latitude. Corresponding changes in life-history traits also occur at this latitude. These changes coincide with the change from a multivoltine subtropical form in which the nymphs overwinter to a bivoltine temperate form that overwinters as diapausing eggs. Northward, body size decreases slightly and then conspicuously increases between 33° and 39°N—the range of latitudes at which the life cycle switches from bivoltine to univoltine. Body size of the univoltine crickets decreases north of 39°N latitude.

Geographical populations of the rice stem borer, *Chilo suppressalis*, in Japan, express stepped variation in certain diapause-related characters that result in univoltinism in the north and bivoltinism in the south. In this species, the critical photoperiod for diapause induction varies along a smooth north-south cline. By contrast, the rates of postdiapause and nondiapause development vary in a stepped manner—the univoltine populations having slower developmental rates and the bivoltine ones having faster rates (Kishino 1970a, b). The intermediate area has both univoltine and bivoltine individuals.

In some species more than one stage can undergo dormancy, and in these instances there may be geographical variation in the characteristics and the incidence of the two diapauses. Such is the case in the mosquito *Aedes triseriatus*, which ranges throughout central and eastern North America. Overwintering is accomplished over most of its range by diapausing eggs (Clay & Venard 1972; Holzapfel & Bradshaw 1981; Sims 1982). Populations from Georgia northward have a high incidence of response to photoperiods that induce embryonic diapause, whereas populations from Florida have a much lower incidence of embryonic diapause even under very short day lengths. In northern populations the critical photoperiod for embryonic diapause increases approximately one hour for each 4.2-degree increase in latitude (Shroyer & Craig 1983).

In addition to an embryonic diapause, *A. triseriatus* has a fourth-instar larval diapause throughout its range, and the critical photoperiod for this diapause also varies geographically, but to a lesser degree than that for the egg (Holzapfel & Bradshaw 1981; Sims 1982). The southern populations that have a low incidence of embryonic diapause commonly overwinter as diapausing larvae. In northern populations, overwintering by larvae is rare or sporadic, and it is presumed that larvae cannot survive severe winter conditions. However, the fourth instar larval diapause has an adaptive advantage in the far north. If overwintered eggs hatch during the winter when day lengths are still short, the resulting larvae may enter diapause when they reach the fourth instar. This larval diapause thus serves to prevent premature pupal and adult development during a time when their habitats are subject to unpredictable freezing and thawing.

A similar trend in geographical variation—larval diapause in the south and egg diapause in the north—occurs in another mosquito species, *Aedes togoi*, that occurs throughout Japan (Mogi 1981). Natural populations at the intermediate locality of Nagasaki overwinter in both embryonic and larval diapause. Diapause in each stage is controlled by different photoperiods, but geographical variation in the controlling mechanisms has not been shown.

7.1.3 Summary

Insect seasonal cycles exhibit considerable intraspecific variability both between years and among localities. Variation in seasonal cycles involves environmental influences and phenotypic plasticity, but the most important factor underlying the variation is genetic variability. It occurs within populations in which it is expressed as continuous variation in quantitative characteristics or disjunct variation in qualitative traits, and it also occurs among geographical populations in which it is expressed as smooth or saw-toothed clines in both quantitative and qualitative traits.

Continuous intrapopulation differences in seasonal cycles are most frequently expressed as variation around the critical photoperiod for diapause induction. Species vary in the degree to which their critical photoperiods vary—some show very large differences among individuals, others show very little. Characteristics other than the photoperiodic response curve for diapause induction, such as diapause duration and response to temperature during diapause, also vary continuously within populations. Some of the variation is of direct importance to fitness; other aspects, although related to components of fitness, may not be directly subject to natural selection and may not be of direct importance to fitness. Thus, care should be exercised in ascribing biological significance to such variation.

Disjunct intrapopulation variation is characterized by polymodal patterns of response, and it forms the basis for polymorphic seasonal cycles. Such variation allows individuals to spread, over time, the risks involved in terminating dormancy and initiating growth, development, and reproduction, in an unpredictable environment. In the event of harsh physical or biological conditions, individuals that delayed reproduction or growth by remaining in dormancy may have a very strong selective advantage. Such variation is most commonly illustrated by polymodal emergence times, both within years and between years (prolonged diapause), but it also occurs in other seasonal traits such as the occurrence of aestival diapause, hibernal diapause, and density-dependent diapause.

Geographical variation involves both continuously varying, quantitative characteristics of the seasonal cycle and disjunctly varying, qualitative aspects, including diapause-mediated dormancy, migration, and polyphenism. Photoperiodic response curves vary greatly among geographical populations, with northern populations usually having longer critical photoperiods for diapause induction than southern populations. Also, species differ in the degree to which their critical photoperiods vary with latitude. Some have an increase in critical photoperiod of as much as 1.5 hours for each 5-degree increase in latitude; other species have critical photoperiods that show considerably less or no geographical variation with latitude.

In addition to geographical clines in seasonal cycles, geographical variation in seasonal traits can underlie uniform seasonal cycles over large distances and in the face of large differences in major environmental factors. This feature is especially true of insects that maintain univoltine life cycles over large geographical areas.

Finally, because it is usual in insect life cycles for only one developmental stage to enter diapause, geographical variation is often expressed in saw-toothed clines with abrupt changes in life-history traits occurring in areas where voltinism changes. These saw-toothed clines often involve a number of interrelated characteristics that vary simultaneously.

7.2 Genetics of Seasonal Cycles

Each species has a characteristic set of genetically determined ecophysiological responses that underlies its seasonal cycle, and as shown in the chapters above, the diversity in seasonal responses is enormous. And, as might be expected from this diversity, the genetic basis for insect seasonality is complex (Masaki 1961, 1978a; Dingle 1974a; Hoy 1978a; Tauber & Tauber 1979, 1982).

Because of its central role in seasonality, diapause has been a major focus of genetic studies. With a few notable exceptions (e.g., Goldschmidt 1932, 1934), it was scientists in Japan and the USSR who pioneered the genetic study of insect diapause, and most of their work is reviewed by Lees (1955), Tazima (1964), and Danilevsky (1965). Additional reviews were provided by Masaki (1961), Dingle (1974), Hoy (1978a), and Tauber and Tauber (1979).

7.2.1 Prerequisites for genetic analysis

The variety and the complexity of insect seasonal cycles contribute to the difficulty in analyzing the genetics of seasonal adaptations. Seasonal traits are subject to considerable environmental influence and care must be exercised to control conditions to minimize this source of variability during genetic studies. Thus, a substantial body of experimental data must be assembled before genetic analysis can be initiated. These data should include such items as the following:

A. The sensitive stage(s). Information on sensitive stages is especially important because diapause in some insects has strong maternal influence (Saunders 1965, 1982; Ando & Miya 1968), which if undetected can be confused with sex-linkage. Vinogradova and Tsutskova (1978) and Zinovyeva (1980) have thoroughly discussed this problem. In addition, genetic and nongenetic paternal effects have been identified (see Ando & Miya 1968; King 1974; Ando 1978).

B. Environmental influence on the expressivity and penetrance of seasonal responses. Conditions under which early stages are reared may influence the expression of photoperiodic response patterns in subsequent stages (Lumme 1978; Lynch & Hoy 1978; Zinovyeva 1980; Kono 1982; Kimura 1983; Tsuchida & Yoshitake 1983a, b). Similarly, the penetrance and/or expressivity of other genetically determined, seasonally expressed life-history traits, such as autogeny (egg maturation without a blood or highly proteinaceous meal) and anautogeny in adult mosquitoes, are influenced by larval food and temperature conditions (e.g., Spielman 1957; Lea 1963; O'Meara & Krasnick 1970; Trpis 1978; Nayar *et al.* 1979).

C. Specific responses that underlie seasonal cycles. Genetic studies of the seasonal response patterns in insects have generally emphasized only one or two traits, that is, the geographically variable critical photoperiod for diapause induction and the duration of diapause (see Hoy 1978a and section below). These characters, although of considerable significance to fitness, may however, not represent the major responses controlling the seasonal cycle. They may represent refinements, for example, of some primary, diapause-controlling photoperiodic response pattern, and as such the genes controlling these traits may comprise sets of modifiers, and not the major gene(s) underlying the seasonal traits (see Shapiro 1976).

To achieve a comprehensive analysis of the genetics of seasonal cycles, it is essential to identify additional ecophysiological response patterns that vary among hybridizable species and strains of insects (e.g., see Danilevsky 1965 for mosquitoes; Tauber *et al.* 1977 and Tauber & Tauber 1976b for *Chrysopa*; Masaki 1978a, 1983, 1984 for crickets). For example, specific responses such as the short-day/long-day requirement for diapause avoidance in *Chrysopa* (Tauber *et al.* 1977; Tauber & Tauber 1976b), the photoperiodic responses and behavioral characteristics of *Culex* mosquitoes (Vinogradova 1961; Danilevsky 1965), and the maternal influence on diapause induction in *Calliphora* (Vinogradova & Tsutskova 1978; Zinovyeva 1980) have been delineated and analyzed genetically. It is also important that the diapause-controlling responses of the various phases of diapause be examined separately. It is likely that the responses that initiate diapause may have different genetic bases from those regulating diapause maintenance and termination (Tauber *et al.* 1977; Dingle *et al.* 1982). Therefore, preliminary to genetic analysis, studies must identify specific components of diapause and they must separate diapause-inducing mechanisms from diapause-maintaining and -terminating mechanisms.

Another word of caution in dealing with the genetics of insect seasonal cycles is appropriate here. It is well known that after only a few generations, even under conditions of presumably neutral selection, the diapause characteristics of laboratory stocks may change drastically from those found in wild populations. These changes are most frequently the result of inadvertent selection or the relaxation of natural selection under laboratory conditions (see Section 8.2.1), although the possibility of circaannual rhythms has not been excluded (see Section 3.6.4). It is essential to be aware of such variability in diapause responses and to design controlled experiments accordingly.

7.2.2 Genetic analysis of seasonal cycles

In general, the genetic analysis of seasonal adaptations has relied on two methods: (a) hybridization and backcross tests between strains or closely related species that exhibit differences in diapause characteristics and (b) artificial selection for specific characteristics of diapause and subsequent hybridization and backcross of the selected lines. Both methods have advantages as well as disadvantages. It is often difficult to hybridize naturally diversified strains or species, and there are only a few studies in which hybridization between species was successful and the interspecific variation in seasonal cycles was subject to genetic analysis. All of these cases involved very closely related species, usually in the same subgenus or species-group (Fraser & Smith 1963; Ohmachi & Masaki 1964; Danilevsky 1965; Harvey 1967; Masaki & Ohmachi 1967; Tauber *et al.* 1977; Dingle 1978a; Masaki 1978a; Kimura 1983; Takafuji & Fujimoto unpublished). The species-pairs of these groups have very different seasonal adaptations whose genetic analysis, although difficult, is very interesting. By contrast, hybridization and analysis of artificially selected lines is relatively easy, but study is limited to those traits alterable over short periods of time by selection in the laboratory.

As expected for an adaptation with wide-ranging effects, the genetic basis for diapause is not simple, and several types of genetic mechanisms play a role in determining diapause characteristics; basic to each are polygenic inheritance and Mendelian inheritance. Each of these modes appears to influence particular types of diapause characteristics and examples are discussed below.

7.2.2.1 Polygenic inheritance

Evidence for the polygenic inheritance of diapause characteristics is relatively substantial and comes from genetic analysis of both naturally diversified populations and artificially selected lines. In general, the characteristics subject to polygenic control are quantitative aspects of diapause: primarily, the *length* of the critical photoperiod, the *rate* of diapause development, and the *duration* of diapause (Tauber & Tauber 1979).

In his classic work with the moth *Acronycta rumicis*, Danilevsky (1965) showed that Leningrad and Sukhumi populations have basically similar types of responses to photoperiods, that is, short days induce diapause whereas long days avert it. However, the critical photoperiod for diapause induction differs by about four hours for the two populations. Both F1 and F2 hybrids, as well as backcross offspring, retain the basic photoperiodic responses of the parental populations, and their critical photoperiods are always intermediate to that of their parents. Although F1 males gave some indication of sex-linkage, there was no filial segregation in F2 or backcross progeny. These results are generally consistent with those for polygenic inheritance (see Lees 1968, Beck 1980).

Similar hybridization tests show that differences in critical photoperiods exhibited by geographic strains of other species in the Lepidoptera (Danilevsky 1965; Showers *et al.* 1972; Rabb 1969; Hong & Platt 1975; Raina *et al.* 1981), Coleoptera (Hsiao 1981), Diptera (Kurahashi & Ohtaki 1977), Orthoptera (Masaki 1984), and parasitic Hymenoptera (Hoy 1975b; Weseloh 1982) are also attributable to differences in polygenes.

Hybridization tests between geographic strains have shown that the duration (or depth) of diapause is also under the control of polygenes. For example, reciprocal crosses between strains of the cricket *Teleogryllus emma* that have long- and short-duration diapause, yielded hybrids with diapause of intermediate duration (Masaki 1965, 1978a). A similar situation is found in the western corn rootworm, *Diabrotica virgifera* (Krysan & Branson 1977). Prolonged diapause in the common pine sawfly, *Diprion pini*, has four levels of expression that are under genetic control (Eichhorn 1983); the mode of inheritance is unknown, but presumably polygenes are involved. In *Choristoneura* spp., the propensity for a second larval diapause and thus a two-year life cycle appears to be under polygenic control (Harvey 1967).

As in *A. rumicis*, the diapause characteristics of several species appear to be, at least in part, sex-linked. Sex-linkage is apparent or implied in the diapause characteristics of the lymantriid moth, *Leucoma salicis* (Danilevsky 1965), the tortricid moth, *Choristoneura orae* (Harvey 1967), the stonefly, *Diura bicaudata* (Khoo 1968), the tobacco hornworm, *Manduca sexta* (Rabb 1969), the codling moth, *Cydia* (= *Laspeyresia*) *pomonella* (Wildbolz & Riggenbach 1969), the fall webworm, *Hyphantria cunea* (Morris & Fulton 1970b), the tobacco budworm, *Heliothis virescens* (Proshold & LaChance 1974; Stadelbacher & Martin 1981), the European corn borer, *Ostrinia nubilalis* (Arbuthnot 1944; Sparks *et al.* 1966; McLeod 1978; Reed *et al.* 1981), the pyralid *Pionea forficalis* (King 1974), the pitcher-plant mosquito, *Wyeomyia smithii* (Istock *et al.* 1976), the gypsy moth, *Porthetria dispar* (Lynch & Hoy 1978), the gypsy moth parasitoid, *Apanteles melanoscelus* (Hoy 1975b), the pink bollworm, *Pectinophora gossypiella* (Langston & Watson 1975; Raina *et al.* 1981), the alfalfa leafcutter bee, *Megachile rotundata* (Parker & Tepedino 1982), the milkweed bug, *Oncopeltus fasciatus* (Hayes & Dingle 1983), and the southwestern corn borer, *Diatraea grandiosella* (Takeda & Chippendale 1982a).

The silkworm presents an interesting case in which voltinism is apparently under the control of three alleles at one sex-linked locus and paired alleles at each of three unlinked autosomal loci. The sex-linked alleles exhibit epistasis, whereas the effect of the autosomal alleles is additive (Tazima 1964; also see Lees 1955). In hybrids, various combinations of the genes are expressed as a continuous gradient in voltinism, from univoltine to multivoltine.

Analyses of the results of artificial selection for quantitative characteristics of diapause have given very high heritability estimates (= high additive genetic variance) (Morris & Fulton 1970b; Dingle 1974a; Holtzer *et al.* 1976; Istock *et al.* 1976). In the migrant milkweed bug, *Oncopeltus fasciatus*, diapause duration and the critical photoperiod for diapause induction show very high heritability (Dingle 1974a; Dingle *et al.* 1977). However, despite the responsiveness of these characters to directional selection, high additive genetic variance is maintained and apparently provides the basis for the "genetic rheostat" that adjusts the timing of diapause induction to the variable seasonal conditions that the migrant bugs encounter (Dingle *et al.* 1977). Both seasonally reversing selection pressures and genetic interchange among populations seem to play a role in maintaining genetic variance in this species (Section 9.2.2.1).

In the pitcher-plant mosquito, *Wyeomyia smithii*, variation in both developmental rate and diapause is under strong genetic influence. Selection for fast development produces a profound decrease in the propensity to enter diapause; selection for slow development results in a slight decrease in the tendency to diapause; selection for diapause causes a dramatic increase in the incidence of diapause. Data suggest that polygenic, perhaps additive, variation underlies the difference between fast and slow development and that the genes in this polygenic system are somehow linked with the genes controlling responses to diapause-inducing photoperiods (Istock *et al.* 1976). Natural selection also appears to maintain genetic variance for diapause and developmental time in this species (Istock *et al.* 1976; see also Section 9.2.2.1).

7.2.2.2 Mendelian inheritance

In contrast to the characteristics above, in which hybrids show intermediate responses in certain quantitative aspects of diapause, some seasonal characteristics are inherited in typical Mendelian fashion. Mendelian inheritance of diapause characteristics has been demonstrated in relatively few studies, but the rarity of its demonstration may not accurately reflect its relative abundance among insects. Although it is possible that Mendelian inheritance may not be a common form of inheritance for seasonal traits, there have been few studies that have done the necessary preliminary work for analyzing the inheritance of specific seasonal traits. In all the instances in which Mendelian inheritance has been demonstrated, clearly defined, discretely varying responses have been identified before their inheritance was analyzed.

In the case of the two sibling species of *Chrysopa*, studies showed that the multivoltine *C. carnea* reproduces without the intervention of diapause when it is reared and maintained under constant long-day conditions; the univoltine *C. downesi* requires a sequence of short days followed by a sequence of long days to avert diapause (Tauber & Tauber 1976b). The differences in diapause between these species and their hybrids are discrete—without intermediates— and thus there is no evidence for polygenic inheritance. F1 hybrids between the two species show that *C. carnea*'s long-day response is dominant over *C. downesi*'s short-day/long-day requirement for reproduction. F2 hybrids and backcross progeny segregate into phenotypic classes in ratios expected for the segregation of two unlinked autosomal loci. Dominant alleles at either of the two loci result in a *C. carnea*-like response to photoperiod, whereas recessive alleles at both loci result in the response typical of the univoltine *C. downesi* (Tauber *et al.* 1977). Thus, relatively small genetic differences underlie gross differences—multivoltine versus univoltine life cycles—in the seasonal characteristics of the two species.

Mendelian inheritance of diapause-controlling responses has also been demonstrated in a comprehensive series of experiments with diapausing and nondiapausing strains of the dipteran *Calliphora vicina* (Vinogradova

& Tsutskova 1978; Zinovyeva 1980). Diapause is expressed in this species only if both the mother and her offspring as larvae experience appropriate photoperiod and thermal conditions. Thus, larval diapause is influenced by the larva's own genotype and by the maternal genotype. Monofactorial segregation of hybrid progeny was observed for both the maternal influence on diapause of the subsequent generation and for the larval effect on diapause. For both characters the ability to enter diapause is dominant, and the dominant, diapause-producing factors must be present in both the female and her progeny for diapause to be expressed. Based on measurements of diapause depth, and on other data, it was also concluded that the maternal and larval influences on diapause in *C. vicina* are controlled by different, but linked, loci.

In addition to diapause-controlled dormancy, Mendelian inheritance may also regulate seasonal polyphenism and other seasonal traits associated with diapause. This is illustrated by the green peach aphid, *Myzus persicae*. Most *M. persicae* females are holocyclic—that is, they respond to short day lengths by producing sexual males and females that mate and lay diapausing eggs. In contrast, when females from androcyclic clones experience short days, they continue to produce mainly parthenogenetic females, but they do produce a few males. Under mild winter conditions the parthenogenetic females may overwinter successfully; under harsh conditions they perish. Results from F1 and F2 backcross tests between androcyclic males and holocyclic females are consistent with monofactorial inheritance of a single recessive switchgene that determines nondiapause, parthenogenetic reproduction as opposed to sexual reproduction followed by diapause (Blackman 1972). The switchgene operates under environmental conditions that would otherwise produce only sexual reproductives, and it completely suppresses the production of sexual females and partially suppresses the production of males.

A somewhat similar type of switchgene may underlie the expression of diapause in autogenous and anautogenous culicine mosquitoes. In general, autogenous mosquitoes (those that oviposit a batch of eggs without a blood or high protein meal) do not have the ability to enter adult diapause, whereas their anautogenous relatives do. For example, *Culex pipiens pipiens*, which has a northern distribution, reproduces anautogenously, and females clearly react to short day lengths by entering diapause. In contrast, *C. p. molestus*, which has a more southern distribution, is autogenous but cannot enter diapause. By performing one-way crosses between *C. p. molestus* females and *C. p. pipiens* males, Vinogradova (1961) showed that the ability to enter diapause is determined by monofactorial inheritance, with the diapause characteristic being recessive. Similarly, autogeny appeared to be inherited by a monofactorial, recessive unit. Other studies with culicine mosquitoes have also demonstrated that relatively few genes, perhaps in association with modifier genes, control the tendency for autogeny (Spielman 1957, 1979; Trpis 1978).

Autogeny in the aedine mosquito *Aedes atropalpus* also is inherited monofactorially (O'Meara & Craig 1969), whereas the fecundity of autogenous females is under polygenic control (O'Meara 1972). However, this species differs from the culicine mosquitoes above in that diapause occurs in the egg stage and apparently is not related to the female's ability to oviposit autogenously.

Incomplete evidence exists for Mendelian inheritance of diapause and diapause-mediated traits in other species. Two closely related species of green blowflies differ in their response when larvae experience dry conditions. *Lucilia caesar* reacts with a mean rate of diapause of 55 ± 20%; whereas in *L. illustris* the rate is only 17 ± 22%. The response of the F1 hybrids is similar to that of *L. caesar*—77 ± 22%—a situation that suggests dominance (Fraser & Smith 1963). Similarly, geographical strains of *Apanteles melanoscelus*, a braconid parasitoid of the gypsy moth, differ in their ability to enter diapause. European populations have clear photoperiodic responses, and under day lengths of 16 hours or less they enter diapause in the cocoon stage (Hoy 1975b; Weseloh 1982). In contrast, a strain from India is almost free of diapause at all photoperiods tested. F1 European-Indian hybrids are similar to the Indian strain in that they do not enter diapause in response to photoperiod (Weseloh 1982). Differences in the critical photoperiod among the European and American populations are consistent with a polygenic mode of inheritance, but the absence of photoperiodic response in the Indian strain suggests the possibility of Mendelian inheritance. Backcross tests are needed. Substantial evidence suggests that the inability of the eggs of certain *Aedes triseriatus* strains to enter diapause in response to photoperiod is monogenetically controlled (Shroyer & Craig 1983). Hybridization and backcross tests with this strain show discrete inheritance of the nondiapause trait and they demonstrate that the nondiapause condition is recessive to the ability to enter photoperiodically induced diapause.

It has also been proposed that the switch mechanism controlling the seasonal occurrence of diapause and associated wing-length changes in *Gerris* water striders involves Mendelian inheritance (Vepsäläinen 1974a, 1978). However, the interaction between environmental and genetic factors in *Gerris* seasonal cycles is very complex, and evidence for this claim is not conclusive (see Harrison 1980; Dingle 1982).

Apparently, in some cases, several genes controlling diapause have become closely linked, and the resulting "supergene" is inherited in Mendelian fashion (Lumme 1978, 1982). In *Drosophila littoralis* originating between 42° and 69°N latitude there is a continuous gradient in photoperiodic reactions from photoperiodic "neutrality" (no diapause) to multivoltine and univoltine strains with critical day lengths of various durations (Lumme *et al.* 1975). Hybrids between multivoltine and univoltine strains show multivoltinism to be dominant (Lakovaara *et al.* 1972), and the factors controlling the critical day length segregate as single, autosomal Mendelian units. Allelic variation within the unit apparently is sufficient to form a continuous cline in photoperiodic reactions over large geographic areas

(Lumme *et al.* 1975; Lumme & Oikarinen 1977). Recent studies (Lumme 1981) localized the "supergene" on the right arm of the fused third and fourth chromosomes and identified the closest flanking markers, as well as a modifying factor on the first chromosome of some stocks.

Lumme and Keränen (1978) also report preliminary data showing monofactorial inheritance of photoperiodic diapause in another species, *Drosophila lummei*; but in this species the diapause-controlling unit is located on the X chromosome rather than an autosome as in *D. littoralis.*

Drosophila species contain an untapped wealth of opportunities for unraveling the genetic control of seasonal cycles. Unfortunately, few studies have made the essential ecophysiological studies that must precede genetic analysis (see Dobzhansky & Epling 1944; Carson & Stalker 1948; Kambysellis & Heed 1974; Begon 1976; Allemand 1977; Kimura *et al.* 1978; Toda & Kimura 1978; Minami *et al.* 1979; Charlesworth & Shorrocks 1980; Ichijo *et al.* 1980; Iwao *et al.* 1980; Minami & Kimura 1980; Kimura & Minami 1980; Ochando 1980; Kimura 1982, 1983, 1984; Lumme 1982; Lumme & Lakovaara 1983; Fukatami 1984). However, the above cited work has set the stage for what can readily become a very fruitful area of study for students of *Drosophila* genetics and seasonality.

In addition to *Drosophila* diapause, the genetically controlled "unitary system" of *Gryllus campestris* also shows some characteristics of supergenic inheritance. Hybrids between the fast-growing, diapausing strain and the slow-growing, nondiapausing strain exhibit dominance and recessiveness; the traits are amenable to artificial selection (Ismail & Fuzeau-Braesch 1972, 1976a).

7.2.2.3 Genetics of seasonal polymorphism

Several genetic mechanisms can account for the numerous examples of disjunct variation that underlies polymorphic seasonal cycles, and the study of these mechanisms constitutes a relatively new, active area of research (Dingle 1978a; Istock 1981). As described above, in some cases the disjunct variation is based on certain crucial aspects of the insects' seasonal cycle that are controlled by relatively few genes. In such cases, the disjunct phenotypic polymorphism involves discretely segregating Mendelian units. For example, this type of inheritance is found in the photoperiodically controlled univoltine and multivoltine life cycles of the *Chrysopa carnea* species complex, in the nondiapause and diapause strains of *Calliphora vicina,* and in some species, such as *Myzus persicae* and others, that overwinter as both diapausing and nondiapause individuals (see Section 7.2.2.2 above).

In other cases, the phenotypic polymorphism may be structured on a threshold response (see Dingle *et al.* 1977; Masaki 1978a) and thus result from continuous genetic variation. Threshold mechanisms can serve as the basis for polymorphic life cycles by acting in a discrete, all-or-none manner, and they may vary geographically through differences in the threshold for response (Masaki 1978a; Denno & Grissell 1979). However, the expression of a polygenically controlled mechanism as a polymorphic trait requires a high degree of variability in the threshold responses among individuals within a population, and indeed this is the case. Offspring from individual *Gilpinia polytoma*, *Tetranychus urticae*, and *Colias alexandra* females have demonstrated considerable among-family variation in the tendency to avoid diapause and produce a second generation (Prebble 1941; Helle 1968; Hayes 1982b). Populations of *Choristoneura* spp. from western North America have significant within-family variation in their thermal requirements for postdiapause emergence. In areas where white fir is the only host, within-family variation is greater than among-family variation, but the situation is reversed in a mixed stand of Douglas fir and white fir (Volney *et al.* 1983).

Istock (1981), working with the mosquito *Wyeomyia smithii*, also showed considerable among-family variation in the tendency to enter diapause, but in his study, which extended over a full summer, the variation decreased during the summer because those individuals with high genetic propensity for diapause enter diapause early and do not contribute offspring to late summer generations. Late season populations therefore are composed mainly of individuals from parents that were biased toward nondiapause (or a longer critical photoperiod). In the laboratory, artificial directional selection diminishes genetic variation for development time and diapause (Istock 1978, 1981). However, in the field, the population is subject to the fluctuating-stabilizing selection that is typical of an unpredictable environment. This type of selection serves to maintain genetic variability in polygenic traits, and it provides the basis for *W. smithii*'s polymorphic seasonal cycle.

7.2.3 Summary

Diapause has been the main focus of studies dealing with the genetics of seasonal cycles, but because of its diversity and dynamism, it presents special problems for genetic analysis. Therefore, a substantial body of data regarding environmental influences on the expression of diapause must be amassed before meaningful genetic analysis can be carried out. Studies that have identified specific, seasonal traits that differ among populations or hybridizable species have been particularly rewarding.

Most quantitative aspects of diapause are under polygenic control; examples include the length of the critical photoperiod, the rate of diapause development, and the duration of diapause. In some cases sex-linkage is involved. Such polygenically controlled characters have high additive genetic variance and therefore are amenable to rapid alteration by directional selection. In nature, seasonally variable selection pressures and genetic interchange among populations apparently serve to maintain high genetic variance in these characters.

Mendelian inheritance of seasonal characteristics has been demonstrated in a few studies that were able to identify and focus on seasonal traits that vary discretely among populations or closely related species. This mode of inheritance may or may not be common for seasonal traits, but it is probably much more common than currently assumed simply because very few studies have completed the necessary analysis of specific seasonal traits that must precede significant genetic investigation.

In some cases, the polygenes controlling diapause apparently have become closely linked and are inherited in Mendelian fashion as a "supergene." Allelic variation within the "supergene" apparently is sufficient to form a continuous cline in photoperiodic reactions over large geographical areas.

Polymorphic seasonal cycles are based on variety of genetic systems. In some cases, the underlying genetic variation is restricted primarily to one or a few Mendelian units that control discrete aspects of the seasonal cycle. Other polymorphic seasonal cycles involve genetic variation in polygenically controlled traits. These cases require some type of threshold mechanism that allows continuous genetic variation to be expressed in a discrete, all-or-none manner. They also require the maintenance of high genetic variance among individuals within local populations, a function that in nature is served by the fluctuating-stabilizing selection that is typical of unpredictable environments.

Chapter 8

Evolution of Seasonal Cycles

Given the great interspecific and intraspecific variation in insect seasonal cycles, and the difficulties in making genetic analyses of seasonal traits, can we elucidate the origin of diapause and unravel the evolutionary development of insect seasonal cycles? This is a crucial question because diapause is the primary seasonal adaptation for most insects, and it is difficult to conceive that insects would have achieved their present widespread distribution and notable abundance without it (e.g., Waterhouse & Norris 1980). Moreover, seasonal cycles are critical to such important evolutionary functions as adaptation, diversification, and speciation. Because of this, knowledge of the evolutionary history of seasonal cycles has special significance in improving our abilities to predict and manage the geographic distributions, abundance, and seasonal cycles of important insect pests and beneficial species.

Despite its importance, analysis of such evolutionary questions presents a very difficult challenge not only because of the complexity and diversity of seasonal cycles but also because many details of the physiology and genetics of diapause remain unknown. In examining its evolution, we consider diapause to be a developmental process, and we separate its evolutionary origin from its subsequent evolutionary modification. The origin of diapause involves the evolution of complex genetic and physiological functions that control major developmental patterns; its evolutionary modification encompasses relatively discrete genetic and physiological variation in specific aspects of larger developmental processes (see Wallace 1985, pp. 34–35, for a fine discussion stressing the importance of distinguishing the two approaches to evolutionary and genetic questions).

Differences between the two phenomena become clear if we draw an analogy between the evolution of dispause and the evolution of a specific structure, e.g., legs. The genetic interactions, developmental processes, and evolutionary steps leading to the expression of legs are complex, numerous, and largely unknown; those involved in changes in leg length (or other defined characteristics of legs) are simpler and tractable. The situation with diapause is identical. It leads us to separate our evolutionary considerations into two parts—one dealing with the origin of diapause *per se*, and the second dealing with evolutionary alterations of diapause and other seasonal adaptions.

Not surprisingly, the origin of diapause has been considered in very few reviews, and the evolution of seasonal cycles has only recently been discussed in a comprehensive manner (e.g., Tyshchenko 1977; Dingle 1978b; Tauber & Tauber 1981a; Brown & Hodek 1983). Thus, in this chapter we delve into aspects of these evolutionary questions, first by reviewing previously proposed schemes for the evolution of diapause and by presenting a new, comprehensive model that takes into account the expanded view of diapause as the mediator of diverse seasonal responses, including dormancy, seasonal migration, and seasonal polyphenism. In the second part of the chapter we consider the evolution of seasonal cycles from two general viewpoints: studies of recent changes in seasonal cycles in response to artificial and natural selection and comparative analyses of insect seasonal cycles.

8.1 Origin of Insect Diapause

Two important considerations arise when the origin of diapause is sought. First, diapause occurs in many animals that are considered more primitive than the insects [e.g., nematodes (Schad *et al.* 1973; Evans & Perry 1976) and marine copepods (Marcus 1982a, b, 1984)], and second, many tropical species have diapause characteristics of one type or another (Section 6.3). Thus, it is highly probable that various groups of insects already had some form of diapause when they first appeared in the Devonian period. As a result, to find the true origin of diapause will probably require studies with animals that are much lower on the evolutionary tree than even the most primitive insects. Verification of this suggestion will have to wait until the diapausing abilities and the seasonal characteristics of primitive insects, especially primitive insects in the tropics, are elucidated. But these considerations do indicate that when we speak of the origin of diapause in most groups of insects, we refer to some sort of modification of a diapause response that was already present in an ancestral population, not the *de novo* evolution of diapause in the derived group.

8.1.1 Previous schemes for the evolution of diapause

Because an enduring dormancy is the most common expression of diapause, it is not unexpected that the simple form of dormancy—quiescence—has frequently been proposed as the starting point for the evolution of diapause. Among the early proponents of this view was Emme (1953). He stated that the ability to undergo quiescence in response to the tropical dry season enables insects to expand into the temperate region because dehydration of the body often enhances cold-hardiness; such a quiescence presumably could serve as a preadaptation for the evolution of winter diapause. Emme (1953) also proposed that independent of the tropically derived diapause, the ability to tolerate a long period of cold during quiescence might give rise to firm "obligatory" diapause in northern univoltine species.

Müller (1970, 1976a), in classifying various forms of dormancy, also takes the view that some form of quiescence serves as the evolutionary basis for diapause. He recognizes four categories of dormancy: quiescence, oligopause, parapause, and eudiapause. Among these four categories, the dependence on external conditions for the expression of dormancy apparently becomes more intricate, and from this, Müller concludes that the four categories represent a series of states with increasing evolutionary complexity. Mansingh (1971) and Ushatinskaya (1976a) proposed somewhat different schemes for classifying dormancy, but they expressed a similar view that "sleep" or quiescence, oligopause, and diapause form an evolutionary sequence of increasing complexity.

None of these hypotheses is supported by physiological, ecological, or genetic evidence (Tauber & Tauber 1978, 1981; Beck 1980). First, diapause has as its primary function the timing of several different types of physiological, biochemical, behavioral, and morphological changes; clearly its function is not restricted to merely protecting the insect from adverse conditions (Section 2.2). Within this context, diapause-controlled dormancy is the result of response to anticipatory cues occurring during the active phase of the life cycle (see Section 3.2). Even the most "primitive" forms of thermally controlled diapause in tropical insects are based on anticipatory cues (e.g., Denlinger 1974, 1979, 1981a; Paarmann 1976a; see Section 6.2.1). Therefore, the argument that the mechanisms controlling diapause originated from physiological events after the onset of quiescence is not convincing. Second, diapause may occur even in warm areas where there is no persistent period of harsh weather that induces quiescence. For example, tropical and subtropical populations of the corn rootworm, *Diabrotica virgifera*, have univoltine life cycles with long embryonic diapause even though the physical conditions of the environment are suitable for nondiapause, multivoltine life cycles (Krysan *et al.* 1977; Branson *et al.* 1982). And, third, there are many groups of insects in which diapause has evolved without the intervening steps of either quiescence, "sleep," parapause, or oligopause (e.g., see later discussion of work by Bradshaw & Lounibos 1977; Dingle 1978a; Krysan *et al.* 1977; Masaki 1978a, 1984). Thus, we conclude that the origin of diapause resides in responses other than those involved in quiescence.

Because the anticipatory aspect of diapause is its universal and unique property, the evolution of diapause seems to be profitably approached by tracing the evolution of response to seasonally recurring token stimuli. Such an approach apparently was initiated by Levins (1968), who proposed a highly hypothetical scenario for the evolution of diapause through selection for the ability to use a signal to anticipate an environmental exigency. His explanation was relatively simple (see Shapiro 1976), especially since it did not recognize the elaborate and presumably evolutionarily constrained steps involved in translating an anticipatory signal into an adaptive response.

Tyshchenko (1977) indirectly considered these problems and proposed an interesting scheme for the evolution of diapause based on the evolution of seasonal photoperiodism. His model goes back to the very beginning of response to light, and thus many of the evolutionary steps undoubtedly could have occurred long before insects evolved. Tyshchenko's argument (Fig. 8.1) originated in the premise that the photoperiodic time-measuring system involved in seasonality is derived from the circadian system that governs the daily rhythm of activities. Through evolution, these endogenous circadian rhythms become integrated (Fig. 8.1, Step IV) and in doing so they assume control over neuroendocrinologically mediated physiological processes of a quantitative nature, such as developmental and growth rates. The intensity of these processes is controlled by the photoperiodically determined phase relationship of component oscillations in the circadian system. Thus, the relationship forms a basis for utilizing information gained from external (token) stimuli that vary quantitatively over the season to influence internal, physiological processes in a quantitative manner.

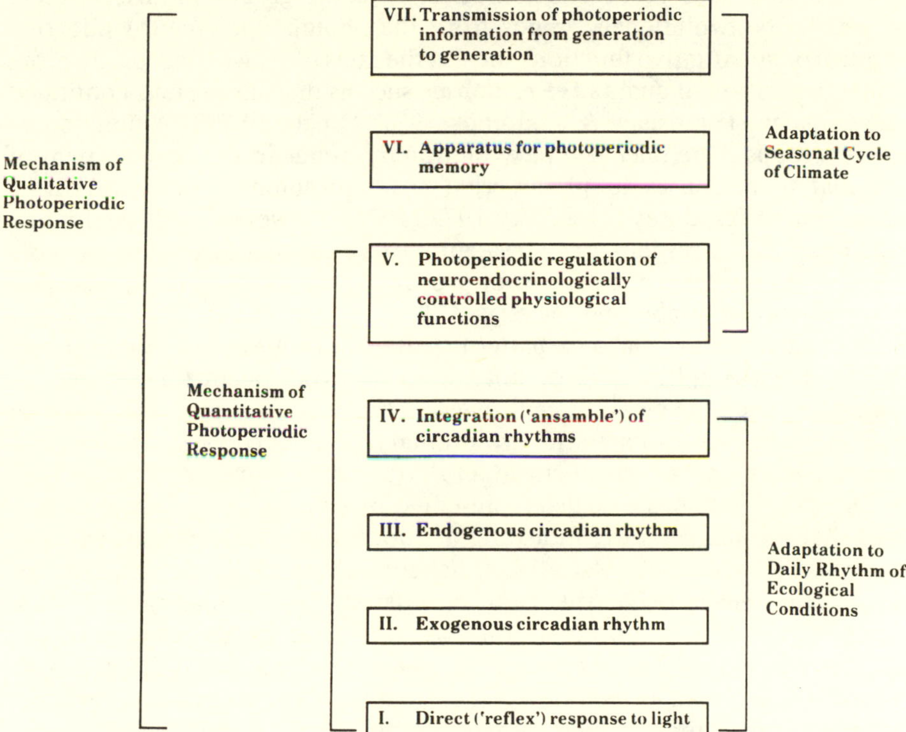

Fig. 8.1 Tyshchenko's model for the evolution of photoperiodic adaptation in insects (Tyshchenko 1977). It contains two essential features: (1) the evolution of seasonal adaptations based on daily rhythms and (2) the evolution of qualitative photoperiodic responses from the presumably more primitive quantitative responses.

According to the next step in Tyshchenko's model, the quantitative response is elaborated to perform as a "qualitative" response, directing the life cycle along one of two different pathways—continuous development or diapause (Fig. 8.1, Step V). This step is the one most directly involved in the evolution of diapause, and it requires a functioning "memory" system in which environmental information is stored and later translated into neuroendocrine function. Unfortunately, this is the least-known aspect of diapause; presumably polymerized biological molecules such as polypeptides or nucleotides are utilized for this function (see Section 4.1.5). The qualitative response does not necessarily exclude the ancestral quantitative response that may persist as an integral part of the insect's overall adaptation to seasonal changes. Still higher evolutionary complexity is achieved with the transmission of stored photoperiodic information from one generation to the next (Fig. 8.1, Step VI).

Given the present state of our knowledge, it is impossible to verify or invalidate Tyshchenko's model. There is strong evidence from numerous species for the participation of the circadian system in photoperiodic time measurement (see Section 4.1.5). Also, in some groups of insects there appear to be evolutionary trends from the photoperiodic-neuroendocrine control of quantitative functions such as the rates of growth or reproduction, to the regulation of qualitative responses such as diapause versus continued development (Bradshaw & Lounibos 1977; Dingle 1978a). Among some insect groups there may also be evolutionary trends in the development of qualitative responses to photoperiod from presumably more primitive quantitative responses (Zaslavsky 1972, 1975). However, both qualitative and quantitative responses can be adaptive (see Chapter 5), and on the basis of current knowledge the derived or primitive state of specific response patterns cannot be established for certain.

Some insects have diapause patterns that do not conform to Tyshchenko's model. For example, aphids do not appear to involve a circadian system in photoperiodic time measurement (Lees 1970, 1973). However, it is possible that these responses represent evolutionary divergence from an ancestral mechanism with its basis in circadian rhythmicity (Saunders 1978b).

Diversity in diapause is also represented by tropical and subtropical species that use temperature, moisture, or food as anticipatory seasonal cues (Denlinger 1974, 1979; Usua 1973; Scheltes 1976, 1978a, b; van Dinther 1962; Hummelen 1974; Ankersmit & Adkisson 1967; Paarmann 1979a). These factors also retain important diapause-regulating roles in numerous species that extend into the temperate zone as well (Sections 5.2 and 5.3). In a few cases, such as the European corn borer, *Ostrinia nubilalis*, and the cabbage white butterfly, *Pieris brassicae*, the thermal responses have a circadian component (Beck 1974a, b, 1975, 1977, 1982, 1983b; Dumortier & Brunnarius 1981), and thus their action could be adapted to Tyshchenko's model. In addition, in the lady beetle *Chilocorus bipustulatus*, it is possible that photoperiod and population density influence diapause through the same mechanism (Zaslavsky & Fomenko 1973). In other species, the mode of action of the nonphotoperiodic diapause-influencing factors is not known, and thus they may or may not be applicable to Tyshchenko's model. Clearly, diapause has evolved numerous times in insects, and Tyshchenko's model may illustrate one of many routes that lead to the evolution of diapause in insects. Because of the diversity of diapause controlling mechanisms there is no reason to assume that diapause has only one evolutionary pathway.

Work by Saunders (1981a, b) describing a detailed model for the photoperiodic regulation of diapause can also be applied to the evolution of diapause. According to Saunders, minimal requirements for seasonal photoperiodism are (1) a pigment and photoreceptor system to distinguish light from dark, (2) a clock to measure the duration of the day or night period, and (3) an effector system to regulate the appropriate humoral control mechanism. Diversity appears within any and all of these systems, thus implying multiple origins and evolutionary pathways for insect diapause. Although Saunders presents strong argument for the involvement of the external coincidence model (= a single oscillation interacting with a cyclic environmental cue) of the circadian photoperiodic clock in seasonal time measurement, he recognizes the diversity provided by the internal coincidence model (= two or more oscillations whose relationship changes with the seasons) (see, e.g., Pittendrigh 1981) and the hourglass mechanism (see Lees 1971, 1973). He proposes that no matter which mechanism is used to measure day or night length, an essential component of the photoperiodic control of diapause is a counter—a mechanism that records and sums the number of, for example, long nights needed to induce diapause. Such a photoperiodic counter is temperature compensated, subject to interactions with nonphotoperiodic factors, and quite variable among species and strains.

In addition to the counter, the model has another requirement. Once recorded, the photoperiodic information must be stored and translated into neuroendocrine function; these activities are the function of a memory system. Saunders proposes some neuroendocrine components of such a system for *Sarcophaga argyrostoma*, but basic questions concerning the memory system still remain. Moreover, the system Saunders described for *Sarcophaga* is not applicable to other species whose prediapause (sensitive) stages are not lengthened or shortened by photoperiodic conditions as they are in *Sarcophaga*.

8.1.2 New, comprehensive model

Given the diversity of diapause-mediated functions and the evolutionary lability of insect diapause, we propose a comprehensive model for the evolution of diapause in insects (Fig. 8.2). Our proposal expands Tyshchenko's model described above, and it incorporates ideas presented by Lees (1971), Beck (1983b), Saunders (1981a, b), and Pittendrigh (1981) regarding the origin and control of circadian and other seasonal time-measuring systems. It

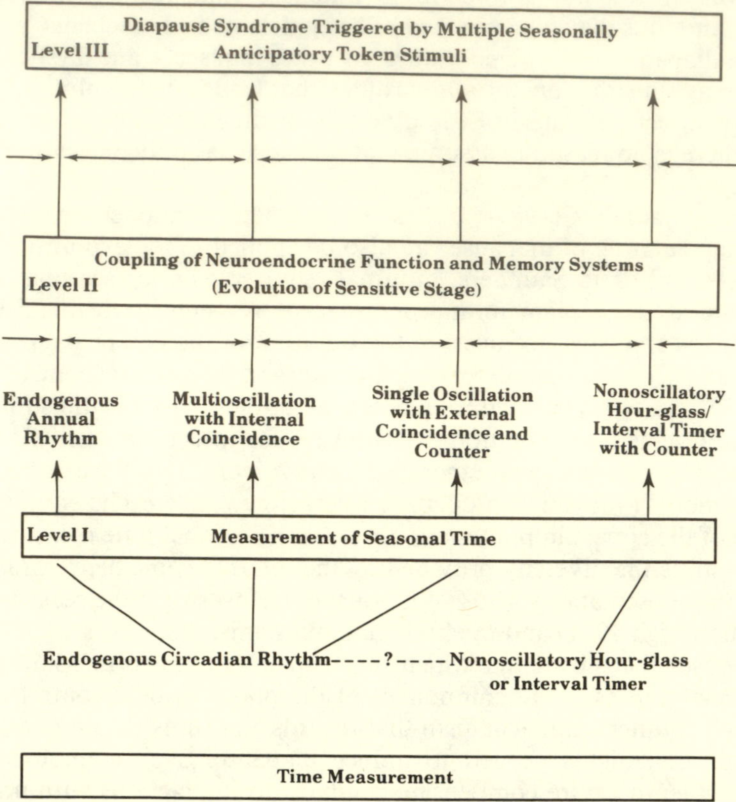

Fig. 8.2 Model of multiple pathways in the evolution of insect diapause. Level I is expressed in the neuroendocrine control of simple functions. Level II involves the coordination of neuroendocrine functions controlling the rates of development, growth, and reproduction, behavioral responses, color and body form, and development of resistance to environmental extremes. These responses, which are adaptive under some conditions (e.g., certain seasons) but not others, are environmentally controlled. Level III involves the coupling of these responses to reliable seasonal indicators—the result being the evolution of an integrated repertoire of species-specific symptoms that constitute the diapause syndrome. *Zeitgeber* for the endogenous rhythms and stimuli for the hour-glass/interval timer are diverse; they include photoperiod, thermoperiod, food, moisture, and population density. The horizontal arrows indicate that they may readily change and combine before and after coupling with the memory system. Pathways are reversible. The scheme is based on ideas from Tyshchenko (1977), Saunders (1981a, b), Shapiro (1978), Pittendrigh (1981), and Beck (1980).

also includes aspects of Shapiro's (1978) scheme for the evolution of seasonal polyphenism. By integrating ideas from these various sources, we can formulate a comprehensive scheme allowing several possible pathways for linking *diverse* seasonal adaptations to control by *diverse*, seasonally recurring token stimuli. Some of the token stimuli function through a circadian system (similar to that proposed by Tyshchenko), but others do not. For the latter, nonoscillatory systems ("hourglass" or "interval timer") and perhaps circaannual rhythms provide the basis for the seasonal time-measuring function inherent in diapause.

In essence, our model comprises three steps. First is the evolution of a time-measuring system that is sensitive to seasonal changes. It appears that at least some insects rely on an oscillatory or circadian clock system for this function (see Pittendrigh 1981; Saunders 1978a, b, 1981a, b). Circadian systems that can measure seasonal changes fall into two types—those that include a single oscillation interacting with an environmental cycle (= external coincidence model) and those that include two or more (e.g., dawn and dusk) oscillations whose relationship changes with the seasons (= internal coincidence model) (Saunders 1981a, b; Tyshchenko & Tyshchenko 1982). Both the external coincidence and the internal coincidence models have been used to explain photoperiodic induction of diapause in insects (Pittendrigh 1981; Saunders 1981a, b). Combined with the counter system of Saunders (described above), they provide rich sources of variation in the time-measuring system, and thus they are compatible with the diversity in seasonal time-measuring systems observed among insects (Saunders 1981a, b).

Thermoperiodism can also function through these circadian systems. Some aspects of the response to thermoperiod imply homology between thermoperiodically and photoperiodically responsive biological clocks (Masaki & Kikukawa 1981; Beck 1983a), but such a relationship is not universal and analogous systems are suggested (Dumortier & Brunnarius 1981; Beck 1983a). Thus, our model proposes that, although there may be a single intracellular biological clock system, the mechanisms that insects use to measure *seasonal* time may have multiple origins. The multiple origins may be most notably reflected in the diversity of receptor mechanisms involved in seasonal time measurement (see Masaki & Kikukawa 1981).

As stated above, some insects seem to rely on a nonoscillatory time-measuring system (e.g., hourglass or interval timer) that is sensitive to night length (Lees 1971, 1973; Hillman 1973; Skopik & Bowen 1976; Thiele 1979b; Vaz Nunes & Veerman 1982; Kikukawa & Masaki unpublished). Saunders (1981b) pointed out the functional similarities between the "hourglass" and the external coincidence model, and it is possible that the difference between the two may be due to differences in the "dumping" rates of the oscillators involved. If so, it is possible that the hourglass evolved from a circadian based system (see also Thiele 1979b; Masaki 1984), but evidence for this is not conclusive. Therefore, our model depicts the separate origin of the hourglass system, although we recognize that it could be a derived condition.

Finally, some insects appear to use endogenous circaannual rhythms in their measurement of seasonal time (see Section 3.6.4). "Zeitgeber" for such rhythms may be very diverse, including photoperiod and thermoperiod, temperature, food, moisture, and population density. We depict an independent origin for the role of these rhythms in diapause, while recognizing their possible derivation from a circadian system.

The second step in our model involves the coordination of neuroendocrine functions with the time-measuring mechanism in simple quantitative functions. This step comprises the evolution of a neuroendocrinologically controlled response and coordinated sets of responses to environmental factors. These responses include a variety of developmental and behavioral patterns; numerous examples are found in the photoperiodic regulation of growth and reproduction rates in insects (see Beck 1980).

The third step involves the coupling of the neuroendocrinologically controlled functions to a reliable seasonal cue(s). It includes linkage of neuroendocrine function to a memory system. As stated earlier, the linkage of neuroendocrinological responses to a memory system is perhaps the step most crucial to the evolution of diapause, and yet it is the step which is least known. It is the step that allows the anticipation of seasonal change and it allows the evolution of coordinated sets of adaptive traits. This is an area of insect physiology that should prove very fruitful for research in the years to come.

The seasonal cycles of the various insect species, even closely related species, place different degrees of emphasis on dormancy versus migration and the other ecological expressions of diapause (Chapter 2). Thus, our model proposes that token stimuli can come to control not just rates of development, but numerous, diverse seasonal adaptations, such as behavioral responses, changes in color and body form, and the development of resistance to environmental extremes. Any of these functions can develop as quantitative responses that become altered to act in a qualitative fashion and then become linked to a memory system. Through natural selection, some of these functions come to be controlled in a quantitative fashion because intermediates are not adaptive. Moreover, they also become linked to seasonally anticipatory cues. Depending on the ecological requirements and life history patterns of the species, such coupling may involve attributes that enhance dormancy, such as the seasonal development of cold-, heat-, or drought-hardiness, or it may involve specific behavioral patterns, such as migration, or specific morphological or color changes, any of which may take the leading role in the insect's seasonal cycle.

In the case of diapause-mediated migration, which is characterized by several physiological and behavioral alterations, including reduced responsiveness to vegetative and reproductive stimuli and reduced metabolism (Section 2.2.2.2), control by token stimuli is known to be exerted through the neuroendocrine system in a manner similar to the way in which token stimuli influence diapause-mediated dormancy (see Rankin 1978; Section 4.1.1). Thus, we propose that the diverse responses associated with seasonal migration and dormancy come under the control of seasonal token stimuli through the same evolutionary pathways—that is, through the evolution of neuroendocrinologically controlled responses and through the coupling of neuroendocrine function to a memory system.

Conceivably, such a mechanism could result in the evolution of seasonal migration or polyphenism that is independent of dormancy. Thus, our model differs from the scheme presented by Shapiro (1978) for the evolution of seasonal polyphenism, which implies that diapause-controlled dormancy evolves first and that seasonal polyphenism becomes linked secondarily with diapause. In fact, according to our model, either seasonal migration or seasonal polyphenism may be the single or primary function under the control of seasonal token stimuli. Such appears to be the case in the pierid butterfly *Nathalis iole*, in which seasonal polyphenism for thermoregulatory wing coloration is under photoperiodic control, and yet the species has no diapause-mediated dormancy (Douglas & Grula 1978). We propose that the photoperiodic control of this seasonal polyphenism evolved in a manner very similar to the evolution of diapause-mediated dormancy.

Our model does not imply that for each species only one pathway will lead to the fully developed diapause syndrome. On the contrary, several coordinated neuroendocrine functions can become coupled to the same, or to different controlling mechanisms and ultimately become integrated under the control of a single or serial seasonal cues. Thus, the multifaceted diapause syndrome may be the result of the simultaneous evolution of various components along a variety of pathways.

The environment clearly exerts its influence on diapause through several different means, including seasonal changes in photoperiod, temperature, food, moisture, and population density (Chapter 5). These factors provide a wide variety of cues to which insects can respond, and insects have evolved a large variety of appropriate responses to the cues. Among insects, diversity is the hallmark of diapause-controlling responses. However, neither the seasonal cues provided by the environmental factors nor the responses of insects to these cues are without constraints. Light and temperature, which have both daily and annual cycles of fluctuation, can provide a relatively large number of different cues to which insects can evolve responses, but this number is limited. Factors such as food, moisture, and population density, which do not have daily fluctuations, are probably even more limited in the number and types of reliable seasonal cues they can provide.

Similarly, all of the diapause-mediated seasonal adaptations of insects are based on three interrelated physiological systems—the time-measuring system, the neuroendocrine system, and the memory system. Very little is known about the origins of the time-measuring system and the memory system; they remain virtual "black boxes," and thus we cannot at this time discuss the variety or constraints of the mechanisms associated with them.

However, they too would be limited in their variability. Similarly, the insect neuroendocrine system, even with its interspecific diversity, does not have unlimited variation in its parts and functions, and the basic neuroendocrine system common to all insects does seem to provide some constraints on the evolutionary development of diapause and diapause-mediated responses. As a result, convergent and parallel evolution are commonplace in the evolution of insect seasonal adaptations (Masaki 1978a, 1983, 1984; Tauber & Tauber 1978, 1981). Our model, which is compatible with the diversity of diapause-controlling responses among insects, also incorporates the factors that place environmental constraints on these phenomena.

The final point we make concerning the model is the reversibility and interchangeability of the environmental controls. We envision that the processes depicted in the model (see Fig. 8.2) are reversible under certain circumstances, so that a function, for example, that was once under photoperiodic control, could be freed of photoperiodic regulation. Coupling and uncoupling of seasonal polyphenism and diapause-mediated dormancy is known from pierid butterflies (Hoffmann 1974, 1978; Shapiro 1976, 1978, 1980a, c) and crickets (Masaki 1984). Shapiro (1976, 1978) postulates that polyphenism in the pierids is subject to selection for genetic modifiers that affect its coupling to a reliable seasonal indicator.

8.1.3 Summary

Despite several proposals to the contrary, numerous studies have shown that the origins of diapause lie in responses other than those controlling quiescence, the most simple form of dormancy. Because the anticipatory aspect of diapause is its typical and usual property, it follows that the origins of diapause lie in the evolution of the ability to perceive and respond to environmental cues that herald an approaching seasonal change. Such is the shared view of Levins (1968) in deriving his hypothetical scenario for the evolution of diapause and of Tyshchenko (1977), Saunders (1981a), and Shapiro (1976) in presenting their ideas on the evolution of photoperiodically controlled seasonal traits.

We propose a comprehensive scheme for the evolution of insect diapause that integrates previously proposed views and allows for the linkage of *diverse seasonal adaptations* to control by *diverse, seasonally recurring token stimuli*. This general model comprises three steps. *First* is the evolution of a time-measuring system that is sensitive to seasonal change. At present, oscillatory (circadian and circaannual) and hourglass mechanisms cannot be arranged properly in an evolutionary sequence. Thus, our model treats them independently. *Second* is the evolution of environmentally controlled neuroendocrinological responses that are adaptive under some, but not other, conditions. The responses include physiological and behavioral functions that could ultimately subserve dormancy, seasonal migration, seasonal polyphenism, or other seasonal adaption.

The model's *third* step involves the coupling of neuroendocrinologically controlled responses to a seasonally reliable indicator(s). This is the step most closely associated with the evolution of diapause, and yet it is very poorly known.

Our scheme depicts multiple pathways for the evolution of diapause responses—leading to the multifaceted diapause syndrome characteristic of insects. This may explain the variability and the diversity of responses within and among insect species, and yet it is consistent with the evolutionary constraints set by the limits of variation in physiological systems involved in diapause—the time-measuring system, the neuroendocrine system, and the memory system. Finally, the steps in the model are reversible, at least at some levels, thus allowing for the decoupling of diapause-mediated functions from control by specific cues. Recoupling to another or the same seasonal cue is also possible.

8.2 Evolutionary Changes in Seasonal Cycles

At this stage in the development of our knowledge, it is impossible to verify the validity of our scheme for the evolution of diapause. Keeping in mind that diapause is a physiological phenomenon, information on its component parts, especially the time-measuring mechanisms and the memory systems, is too scanty (see Section 4.1.5); the neuroendocrine control of diapause has been examined for too few species; and sufficient comparative data from diverse taxa are not available. Despite these problems, can we, nevertheless, identify the origins of insect diapause and derive evolutionary trends for its change? Is there sufficient knowledge to designate areas of potentially fruitful research in the future? Our answers to these questions are "no" and "yes"; we illustrate below.

Three major types of studies provide insight into the evolution of diapause: artificial selection experiments, analyses of adaptive changes among colonizing species, and comparative studies of seasonal adaptations among taxa. Investigations within these categories thus span from the genetic control of seasonal adaptations to the ecological manifestations of seasonal adaptations, and they provide a relatively broad base of knowledge upon which to construct evolutionary hypotheses. Unfortunately, studies dealing with these three categories almost universally omit the physiological and neurohormonal connections between the genetic and ecological manifestations of seasonal cycles. Thus, their results do not apply directly to verification of models for the evolution of diapause, and they cannot, at this time, identify steps in the evolutionary development of the physiological state of diapause.

However, at the genetic and ecological levels, considerable information is available. Some systems and taxa have been sufficiently well examined so that evolutionary trends in seasonal adaptations are inferred with reasonable certainty. From the point of view of evolutionary models for diapause, the main value to be obtained from discussion of these systems is identification of the areas of study as well as taxa that could contribute to future comparative physiological research. But from the point of view of evolutionary ecology, they also provide valuable insight into the types of selective pressures to which seasonal cycles respond and the changes that they can undergo. Models are of particular importance in elucidating the coevolutionary interaction between the various seasonal adaptations that constitute seasonal cycles. Changes in one component often elicit coordinated changes in others, and thus they strongly influence the directions in which life histories evolve (see Chapter 9).

Below we examine the evolutionary changes in seasonal cycles as they can be inferred from the three main types of studies. First, we consider the results of experiments dealing with artificial selection. We ask, what types of responses are amenable to alteration by selection and what constraints and problems are associated with selection under laboratory conditions? Second, we analyze the changes in seasonal adaptations resulting from natural selection acting on colonizing species—that is, native and introduced species that are undergoing evolutionary changes associated with range extension into areas with diverse seasonal characteristics. Finally, we ask at which taxonomic levels are comparative studies most meaningful. Our answers are illustrated with some particularly well-studied examples.

8.2.1 Response to artificial selection

Insect seasonal cycles provide excellent examples of characteristics that are highly susceptible to natural selection. The previous chapter (Chapter 7) gave ample evidence for this in the numerous cases of geographical variability involving various seasonal and life-history parameters. Almost all workers who have examined the seasonal cycles of insects have found variation in the incidence and/or duration of diapause if they collected their samples from different climatic areas. Such variations are frequently clinal and well correlated with climatic gradients, and they undoubtedly represent response to climatic selection.

The genetic variability underlying seasonal adaptations is also amenable to artificial selection. Table 8.1 lists more than 40 cases in which artificial selection acting over relatively few generations produced significant changes in laboratory populations (see also Hoy 1978a). In most of these studies, selection has been aimed at establishing nondiapause strains from populations inhabiting regions where diapause is essential for survival. For example, multivoltine or nondiapause strains have been produced from univoltine stock by selection for an increasingly shorter duration or lowered incidence of diapause (Table 8.1).

In most cases it is not known what diapause-controlling mechanism (e.g., critical photoperiod or other photoperiodic responses) was altered by artificial selection. Wherever this question has been examined, it was found that selection influenced both the critical photoperiod and the duration of diapause. In the milkweed bug, *Oncopeltus fasciatus*, selection for reduced age at first reproduction under diapause-inducing and -maintaining photoperiods resulted in simultaneously reducing both the duration of diapause and the critical photoperiod for diapause induction (Dingle 1974a; Dingle *et al.* 1977). Similar correlations between diapause incidence and duration occur in the green lacewing, *Chrysopa carnea* (Tauber & Tauber unpublished), the flesh fly *Sarcophaga bullata* (Henrich & Denlinger 1982b), and the anise swallowtail, *Papilio zelicaon* (Sims 1983).

Plodia interpunctella, the Indian meal moth, presents an interesting case in which selection may act simultaneously on three different diapause-controlling mechanisms: response to photoperiod, temperature, and population density. After 38 generations of rearing, under diapause-averting conditions of LD 16:8 and 25°C, the incidence of diapause in the laboratory colony generally decreased. Tests of the photoperiodic responses of individuals in the colony showed that the response curve had changed from that of the earlier generations (Kikukawa unpublished). In this species, diapause also occurs in response to high temperature and high population density (see Section 5.3.2). Three years of continuous rearing at 30°C decreased the incidence of the density-dependent diapause (Tsuji 1960). Selection of early-emerging individuals at 20°C likewise established a nondiapause strain that had also lost its sensitivity to population density. However, both of the selected strains entered diapause when midinstar larvae experienced a temperature shift from 20 to 30°C. The above studies with *P. interpunctella* suggest that natural selection may act on the sensitivity to multiple environmental cues and that responsiveness to the various cues may be selected independently of each other.

The ease with which artificial selection alters diapause characters indicates that inadvertent selection against diapause occurs even when stock cultures of insects are reared continuously under nondiapause conditions in the laboratory. A colony of *Teleogryllus commodus*, derived from a temperate area of Australia, almost completely lost its ability to enter egg diapause after 15 years of rearing under constant temperature (Cousin 1961). Similarly, a once predominantly diapausing strain of *Pseudosarcophaga affinis* turned into a predominantly diapause-averting strain after 200 generations of rearing in the laboratory (House 1967). And, a diapausing colony of *Las peyresia* (*Grapholitha*) *molesta*, Oriental fruit moth, was reduced to ~ 0% diapause after five years and ~ 60 generations under nondiapause conditions (Glass 1970). Since the culture conditions for these colonies almost completely prevented diapause, selection against diapause would be expected to be minimal. Therefore, it is possible that the inadvertent selection resulted from the pleiotropic effect of the diapause genes on the innate capacity for increase (*r*), either through its effect on fecundity or developmental time.

From the above and other studies (see Table 8.1), it appears that it may be rather difficult to maintain laboratory stocks of insects for long periods

Table 8.1 Examples of diapause altered by artificial selection.

Diapause stage	Species	Character selected	Reference
	ORTHOPTERA		
L	*Gryllus campestris*	nondiapause & survival	Ismail & Fuzeau-Braesch 1972, 1976a
E	*Locusta migratoria gallica*	diapause incidence and duration	LeBerre 1953
E	*Melanoplus differentalis*	high & low incidence of diapause	Slifer & King 1961
E	*Melanoplus sanguinipes*	nondiapause	Pickford & Randell 1969
E	*Pteronemobius fascipes*	increased incidence of diapause	Masaki 1978a
E	*Teleogryllus commodus*	nondiapause (inadvertent)	Cousin 1961
	HEMIPTERA		
A	*Aelia acuminata*	low incidence of diapause	Honek 1972 Hodek & Honek 1970
A	*Aelia rostrata*	low incidence of diapause	Honek 1972
A	*Oncopeltus fasciatus*	diapause duration	Dingle 1974a; Dingle *et al.* 1977
A	*Pyrrhocoris apterus*	macroptery (independent of diapause)	Honek 1979a
	HOMOPTERA		
A	*Aleyrodes asari*	nondiapause	Bährmann 1977
E	*Dysaphis anthrisci*	reduced incidence of prediapause sexuals	Azaryan 1966
	NEUROPTERA		
A	*Chrysopa carnea*	reduced duration of diapause and increased response to photoperiod	Tauber & Tauber 1982
	COLEOPTERA		
A	*Anthonomus grandis*	nondiapause (inadvertent)	Lloyd *et al.* 1967
E	*Atrachya menetriesi*	low incidence of diapause low & high incidence	Ando & Miya 1968 Ando 1978

A	*Conotrachelus nenuphar*	low incidence of diapause	Featherston & Hays 1971
E	*Diabrotica virgifera*	nondiapause	Branson 1976
		diapause duration	Krysan & Branson 1977
A	*Hypera postica*	nondiapause (inadvertent)	Huggans & Blickenstaff 1964
A	*Leptinotarsa decemlineata*	incidence of diapause; incidence of prediapause oviposition (inadvertent)	de Kort *et al.* 1980
L	*Trogoderma granarium*	presence and absence of density independent diapause	Nair & Desai 1973
	LEPIDOPTERA		
A	*Aglais urticae*	nondiapause	Niehaus 1982
P	*Antheraea pernyi*	univoltinism (high incidence of diapause)	Chetverikov 1940; Tanaka 1951
E	*Bombyx mori*	voltinism	Muroga 1951
L	*Choristoneura fumiferana*	low incidence of diapause	Harvey 1957; Lyon *et al.* 1972
L	*Cydia* (= *Laspeyresia*) *pomonella*	low incidence of diapause	Wildbolz & Riggenbach 1969
L	*Diatraea grandiosella*	nondiapause and increased incidence of diapause	Takeda & Chippendale 1982a Kikukawa & Chippendale 1983
L	*Ephestia elutella*	multivoltinism	Waloff 1949
P	*Heliothis virescens*	low incidence of diapause	Benschoter 1970
P	*Heliothis zea*	reduced incidence and duration of diapause	Benschoter 1970 Herzog & Phillips 1974 Holtzer *et al.* 1976
P	*Hyalophora cecropia*	duration of diapause	Waldbauer & Sternburg 1973
L	*Laspeyresia* (*Grapholitha*) *molesta*	low incidence of diapause	Glass 1970
P	*Mamestra brassicae*	depth of summer diapause	Poitout & Bues 1977c
P	*Mamestra oleracea*	high and low incidence of summer diapause	Poitout & Bues 1977c
L	*Ostrinia* (= *Pyrausta*) *nubilalis*	univoltinism	Arbuthnot 1944
P	*Papilio zelicaon*	incidence and duration of diapause, critical photoperiod, and diapause intensity	Sims 1980, 1983a, b

Table 8.1 Examples of diapause altered by artificial selection (Continued).

Diapause stage	Species	Character selected	Reference
L	*Pectinophora gossypiella*	critical photoperiod; reduced diapause duration; reduced incidence of diapause	Pittendrigh & Minis 1971 Langston & Watson 1975 Barry & Adkisson 1966
L	*Pionea forficalis*	high & low incidence of diapause	King 1974
L	*Plodia interpunctella*	reduced incidence of diapause	Tsuji 1960; S. Kikukawa unpublished
E	*Porthetria* (= *Lymantria*) *dispar*	reduced diapause duration	Hoy 1977
	DIPTERA		
L	*Calliphora vicina*	increased inclination to diapause	Vinogradova 1975b
P	*Delia antiqua*	increased incidence of diapause (inadvertent)	Robinson *et al.* 1980
A	*Drosophila littoralis*	reduced incidence of diapause	Oikarinen & Lumme 1979 Lumme & Pohjola 1980

L	*Lucilia caesar*	low & high incidence of diapause	Ring 1971
P	*Poecilometopa spilogaster*	nondiapause	Delinger 1979
P	*Pseudosarcophaga affinis*	nondiapause (inadvertent)	House 1967
P	*Rhagoletis pomonella*	nondiapause	Baerwald & Boush 1967
P	*Sarcophaga bullata*	decreased incidence of diapause (inadvertent)	Denlinger 1972b
		increased diapause incidence and duration	Henrich & Denlinger 1982b
L	*Wyeomyia smithii*	increased and decreased incidence of diapause	Istock *et al.* 1976
	HYMENOPTERA		
L	*Apanteles melanoscelus* (French strain)	low incidence of diapause (inadvertent)	Hoy 1975b
L	*Gilpinia polytoma*	reduced incidence of diapause	Prebble 1941
L	*Megachile pacifica*	univoltinism	Hobbs & Richards 1976
	ACARINA		
A ♀	*Tetranychus urticae*	nondiapause	Helle 1962, 1968; Geyspitz 1968; Geyspitz *et al.* 1972

without any change in their diapause characteristics. Dingle *et al.* (1977) addressed this problem by examining responses to directional selection (under LD 12:12, a diapause-inducing day length) along three different lines: one for short duration of diapause, another for long duration, and two for intermediate duration. As expected, the line selected for a diapause of short duration resulted in the almost complete elimination of diapause under the experimental photoperiod. However, despite selection to the contrary, the other lines also exhibited a drastic loss of diapause that approximated that of the line selected for a short diapause. These results suggest either a physiological (nongenetic) response to the laboratory rearing conditions or a change in genotype frequencies that is associated with an advantage for early reproduction under laboratory conditions. Further experiments comparing parental and offspring responses indicated that, indeed, there is sufficient additive genetic variance (~ .70) to account for the overall decline in diapause duration.

This study points to an important omission in most experiments dealing with the artificial selection of diapause characters, and that is the use of proper controls consisting of nonselected lines for comparison. Most studies merely establish selected lines and compare results across generations. As shown by the experiments of Dingle *et al.* (1977), such a technique does not necessarily indicate a response to the selection pressure that was applied. Physiological effects and unavoidable genetic changes could account for the observed changes in many cases. Moreover, circaannual rhythms in sensitivity to diapause-controlling stimuli (see Sections 3.6.4 and 7.2) could be confused with a response to artificial stimuli. Thus, for all these reasons, we stress that proper controls are essential for selection experiments.

In addition to its direct effect on diapause, selection for specific diapause characteristics also influences nondiapause traits. In most, but not all, cases the change in nondiapause traits is deleterious. For example, high temperature (30°C) induces diapause in *Gryllus campestris*, and selection for low incidence of diapause at 30°C resulted in the simultaneous reduction in the ability to survive and develop at low temperature (Fuzeau-Braesch & Ismail 1976). Selection of a strain of false melon beetle, *Atrachya menetriesi*, without egg diapause at 26°C resulted in a shortened preoviposition period, but the strain also showed decreased fecundity and increased egg mortality (Ando & Miya 1968). Selection experiments with the pitcher-plant mosquito, *Wyeomia smithii*, demonstrated an association between developmental rates and the photoperiodic responses controlling diapause induction. In this case the association of characters appears to provide an increase in fitness (Istock *et al.* 1976). Similarly, nymphal developmental time is shorter in a strain of *Melanoplus sanguinipes* selected for nondiapause than it is in the diapause strain (Hollingsworth & Capinera 1983). Finally, in experiments with the gypsy moth, *Porthetria dispar*, male-female hatching rates were reversed in strains selected for nondiapause. Males tended to hatch from unchilled "nondiapause" eggs before females, whereas the reverse was true of chilled, wild egg masses (Knop *et al.* 1982). It is not known if the reversal is due to pleiotrophism with the nondiapause trait, linkage, or correlated selection.

Despite the rather rapid decrease in diapause incidence during the first several generations of selection, complete elimination of diapause can be very difficult. In the univoltine leaf beetle, *Atrachya menetriesi*, about 15 generations were sufficient to remove egg diapause from 80 to 90% of the population. After this, further response to selection was slight, and a pure nondiapause stock was not obtained even after 36 generations of stringent family selection (Ando 1978). Similar situations occurred in the selection for nondiapause (at 26°C) strains of the pink bollworm, *Pectinophora gossypiella* (Barry & Adkisson 1966), and three ichneumonid parasitoids (Aeschlimann 1974) and in experiments with the pitcher-plant mosquito, *Wyeomia smithii* (Istock *et al.* 1976).

Selection (under a diapause-maintaining photoperiod of LD 12:12) for reduced duration of diapause resulted in the virtual elimination of diapause in the milkweed bug, *Oncopeltus fasciatus*; however, the ability to respond to photoperiod was retained (Dingle *et al.* 1977). Transfer of the selected strain to a shorter day length (LD 11:13) resulted in a substantial incidence of diapause induction.

The rate of response to selection may vary among different local populations. Ando's studies with *Atrachya menetriesi* show that the saturation level of 80 to 90% nondiapause was reached after only four generations in a southern strain (from about 35°N), but more than 15 generations were required to obtain the same result in a northern strain (from about 40°N). Thus, it is not surprising that the propensity for diapause, though at a low level, may be retained in tropical and subtropical insects of temperate origin.

In comparison to selection for nondiapause, there are very few examples of selection for an increased tendency to enter diapause. However, available data show that selection is effective in both directions (see Istock *et al.* 1976; Masaki 1978a; Kikukawa & Chippendale 1983). From a strain of *Calliphora vicina* with variable diapause traits, Vinogradova (1975b) selected a line with a high inclination to enter diapause. Similarly, after 14 generations of selection for nondiapause, *Atrachya menetriesi* reverted to the original 100% level of diapause within seven additional generations of counterselection (Ando 1978). Other selection experiments have resulted in lengthened diapause duration (Langston & Watson 1975; Holtzer *et al.* 1976) and in a double-peaked response curve for diapause induction (Geyspitz *et al.* 1972).

From an evolutionary viewpoint, selection for diapause from predominantly nondiapause, tropical, or subtropical populations is especially interesting. In the temperate regions, most species of the ground crickets (*Pteronemobius*) enter autumnal-hibernal diapause in the egg stage, but the subtropical and tropical populations usually do not. Nevertheless, at least in *P. taprobanensis* and *P. fascipes*, a small proportion of eggs from subtropical and tropical populations show considerable delay in hatching. In the latter species, the incidence of delayed hatching may reach 10%. From individuals in this cohort, selection for egg diapause was started and continued for 15 generations; the result was an increase in percentage diapause to about 70% (Masaki 1978a; Masaki unpublished). Further exploration of possible selective changes in tropical insects is of utmost importance to our understanding the evolution of dormancy.

In summary, experiments with artificial selection have several important implications for the evolution of diapause. First, although they illustrate that insect seasonal cycles are characterized by considerable genetic variability that is responsive to selection pressure, they also show that this responsiveness is not without limitations. It is very difficult to eliminate completely the genetic propensity for diapause. This type of constraint, in conjunction with the responsiveness to reverse selection pressure, is consistent with the repeated and parallel or convergent evolution of diapause among insect species. Even under seemingly constant seasonal conditions, the genetic and physiological framework for diapause and associated seasonal adaptations persists. This suggests that under appropriate selection, apparently nondiapause stock can give rise to diapausing strains without the *de novo* evolution of any of the physiological processes involved in diapause. Future research comparing the physiology of diapause in naturally diapausing strains with that of strains with diapause produced artificially from predominantly nondiapause stock would be useful in verifying this idea.

Second, in some cases seasonal characteristics are correlated with each other in their response to selection; in other cases they can be altered independently. With regard to evolutionary models of diapause, future research could fruitfully determine at which level (memory system, neuroendocrine system, time-measuring system, other) the correlated traits are linked. Additional selection experiments could also determine to what degree correlated characteristics are amenable to decoupling—another feature that is of great importance to evolutionary models of both diapause and life histories.

8.2.2 Adaptations by colonizing species

The above reasoning predicts that insects have a high degree of evolutionary plasticity in their seasonal adaptations. This prediction is substantiated by studies of species, especially pest and beneficial species, that have been accidentally or intentionally introduced into new areas or that have expanded their geographical ranges in response to habitat disturbances by humans. Under these conditions insect seasonal cycles can respond to natural selection within relatively short periods of time, even if the immigrant stock is small in number and presumably of limited genetic variability. As a result, studies of the seasonal cycles of introduced species have very important implications not only to evolutionary biology but also to insect pest management and biological control programs and practices.

Gradual expansion of geographical distributions along climatic gradients is an important expression of successful colonization. Such range expansions can lead to the formation of phenological clines similar to those found in native species. Figure 8.3 compares the photoperiodic clines of two lepidopterans, *Chilo suppressalis* in its native Japanese islands (Kishino 1974), and *Cydia* (= *Laspeyresia*) *pomonella* in its Old World home and in its newly colonized area, North America (Riedl & Croft 1978). *Chilo* has a range of critical photoperiods from 13.5 to 15 hours of light per day over its 15° latitudal distribution. *C. pomonella* apparently has a similar range of

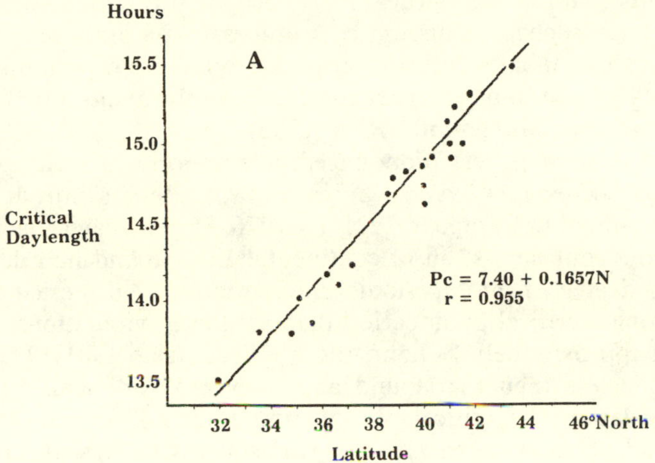

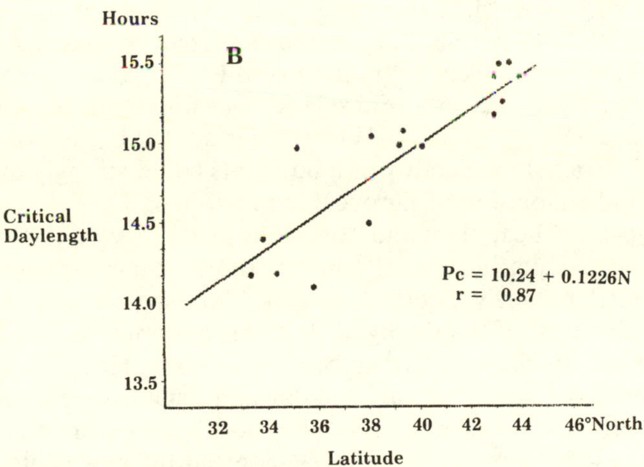

Fig. 8.3 Comparison of the geographical adaptation in the seasonal cycles of two lepidopteran species: a native species in its homeland (A. *Chilo suppressalis* in Japan; Kishino 1974) and an introduced species in its new homeland (B. *Cydia pomonella* in North America; Riedl & Croft 1978). Pc = critical photoperiod; N = °North latitude.

critical photoperiods over its original and new geographical distributions. This species was first recorded from North America in 1750 in New England, and its geographical expansion over 32 to 44°N during the last two centuries appears to have been accompanied by the evolution of variation in critical photoperiod similar to that which occurs in the original Old World populations over a similar range of latitudes. It also appears to have adapted to different hosts. For example, in southern California it thrives on a variety of fruit trees because of its variability in host preference for oviposition, phenological responses, and possibly in larval developmental requirements (Phillips & Barnes 1975). Apparently, adaptation to early maturing crops (such as plums, apricots, and early-maturing walnuts) occurs readily, but shifts to late-maturing crops (late varieties of walnuts) are rare. There appears to be some inherent constraint on the evolution of a late seasonal cycle in the codling moth (Riedl 1983).

Similar, but perhaps less pronounced, adaptation occurred in the Colorado potato beetle, *Leptinotarsa decemlineata*, after its introduction and first establishment in Europe in 1921. By 1976, 55 years later, the beetle had spread throughout almost all of continental Europe and had developed a substantial degree of photoperiodic differentiation. Northern populations exhibit a longer critical photoperiod than southern populations, the difference being approximately 1½ hours (de Wilde & Hsiao 1981). This range of variability is less than that found among North American populations; among the latter, the critical photoperiod ranges from about 15 hours (Logan, Utah) to fewer than 13 hours (Benson, Arizona) and a population from Roma-Los Saenz, Texas, shows very little response to photoperiod. However, comparisons between North American and European populations are complicated by the fact that food quality and species (de Wilde *et al.* 1969; Hsiao 1978; Hare 1983) play important roles in diapause induction. European beetles are largely restricted to cultivated potato, whereas North American beetles have several wild solanaceous hosts to which they are variously adapted (Hsiao 1978; de Wilde & Hsiao 1981; Jacobson & Hsiao 1983). The restricted availability of food plants could strongly influence the phenology and adaptation of European populations.

The influence of host plant and host plant phenology on the evolution of insect seasonal adaptations is well illustrated by another colonizing species, the swallowtail butterfly *Papilio zelicaon.* This species, which occurs in North America west of the Rocky Mountains, has a univoltine life cycle in areas where it feeds on its native umbelliferous hosts. However, the last 100 to 150 years have seen the introduction and establishment of new host plants that provide a source of larval food for a much longer period of the year. In areas where the papilionid has colonized the new hosts, it has several generations per year (Sims 1980). Artificial selection experiments confirm that univoltine populations have the genetic capacity for multivoltinism and that natural selection maintains univoltinism as an adaptation in those areas where only native, seasonally restricted host plants are available (Sims 1980, 1983). The introduction of new hosts during the early 1800s relaxed this selection pressure and populations colonizing the new hosts evolved a multivoltine life cycle (Sims 1980).

Similar types of changes—from univoltinism to bivoltinism—appear to have occurred in other introduced species as well. The mole cricket, *Scapteriscus acletus*, was introduced into the United States from South America, and between 1920 and 1960 it spread from northern Florida, where it has one generation per year, to southern Florida where two generations occur. In less than 40 years, it has made the transition from univoltinism to bivoltinism (Walker *et al.* 1983).

The corn rootworm, *Diabrotica virgifera*, shows a very different trend in the evolution of its diapause. This species probably originated in tropical or subtropical Mesoamerica and spread to the temperate regions of North America when the cultivation of corn was introduced into that region (Branson & Krysan 1981). The species is characterized by diapause in the egg stage and a univoltine life cycle even in tropical areas. Ecophysiological and genetic evidence indicates that the egg diapause is controlled by a single mechanism throughout the entire range of the species (Krysan & Branson 1977; Krysan 1982). Apparently, as the species spread into northern temperate areas there was a diminution in diapause intensity because the low temperatures associated with the consistently cold winters in the temperate zone prevent premature embryonic development (Krysan *et al.* 1977; Krysan 1978, 1982; Gustin 1983). Thus, the need for a deep diapause is eliminated. But in the tropics, where winter temperatures are not consistently low enough to prevent untimely development, natural selection appears to maintain a long diapause that synchronizes egg hatch with the seasonal occurrence of the larval host plant.

It is clear from work with several other tropical and subtropical species introduced into northern and southern temperate regions that diapause in tropical populations can form the basis for diapause in their temperate zone relatives (e.g., see Keeley *et al.* 1977). Japanese populations of the Indian meal moth, *Plodia interpunctella*, a pest that probably originated in southern India or another subtropical area, express a distinct larval diapause in response to short photoperiods and thermoperiods, low temperatures, and high levels of crowding (Tsuji 1963; Masaki & Kikukawa 1981). Many other populations in northern and southern temperate regions also are more or less sensitive to photoperiod, but the geographical pattern of their diapause response is probably obscured by frequent transport between locales (Bell *et al.* 1979). Other tropically derived pests of stored products that now occur in temperate regions show similarly variable diapause traits (Bell & Bowley 1980; see Howe 1962; Baker 1982), including perhaps a relic aestival diapause in *Ephestia kuehniella* (Cox *et al. 1981).*

Pectinophora gossypiella, the pink bollworm, which may have originated in India, has spread throughout almost all the cotton-producing regions of the world in large part due to its adaptable larval diapause (Ankersmit & Adkisson 1967). It was first reported in the New World in 1911, and since that time its seasonal cycle has adapted to local conditions over a very large geographical area. In a comprehensive study of this species in the New World, Ankersmit and Adkisson (1967) showed that populations from six localities ranging from El Paso, Texas (32°N), to Saenz Pena, Argentina (27°S), all enter diapause more frequently under 10 to 12 hours of light per day than under any other long or short day lengths. However, the incidence of diapause varies with latitude; populations close to the equator have a low, but significant, incidence of diapause in response to photoperiod, and populations further away from the equator have a high incidence. Populations also vary in their responsiveness to food and temperature; populations close to the equator are most affected by these nonphotoperiodic cues. Although it is claimed that the southern Indian population of this species does not have the ability to undergo diapause (Raina & Bell 1974; Raina *et al.* 1981), the evidence is meager and derived from experiments testing only photoperiod. It is highly unlikely that diapause evolved *de novo* in American populations, and it is probable that adaptation of the larval diapause to New World originated in characteristics already present in Old World tropical and temperate-zone populations.

The influence of selection on thermal responses is evident in *Diatraea grandiosella*, another species that has expanded its range into temperate regions. This species apparently originated in the neotropics and since 1900 has spread throughout midwestern and southern United States. The photoperiodic responses underlying diapause induction do not vary much among populations from Missouri, Kansas, Mississippi, and southern Mexico. However, the thermal requirements for the development of specific stages and the intensity of diapause differ among the geographical populations. These variations apparently result from independent and different selection pressures acting especially on thermally controlled traits (Takeda & Chippendale 1982b; Kikukawa & Chippendale 1983).

The ground cricket *Pteronemobius mikado* (formerly *taprobanensis*, see Masaki 1983) presents an interesting example in which diapause in the egg stage has virtually been lost in an introduced subtropical population. This species came to the Bonin Islands in the southeast of Japan approximately 100 years ago (Masaki 1978a). Although this island population reproduces all year round, laboratory experiments indicate that it retains a relic of the short-day-type regulation of nymphal development that is coadapted to egg diapause and hibernation at high latitudes, but that is not expressed at the low latitude of the Bonins (24-27°N). Acceleration of development by short day lengths characterizes the northern populations, but this characteristic is not adaptive under the warm climate of the south. Although phenotypic expression of this trait is environmentally suppressed in the Bonin Islands population, it has survived as a concealed trait. This example suggests that the success of an invading species may be enhanced by phenotypic modification of seasonal traits. The example also demonstrates that natural selection, like artificial selection, may result in the evolution of nondiapause strains from ancestral populations with diapause.

With some introduced species, adaptation may occur very slowly or may be limited, as illustrated by the Argentine stem weevil, *Hyperodes bonariensis*, after its introduction into New Zealand. Native South American weevils have a photoperiodically controlled hibernal diapause that adapts them to relatively severe winter conditions. In New Zealand populations, diapause appears to be maintained despite the mild winters that are typical of much of the beetle's new range. Apparently it persists as a relic response to photoperiodic cues that were significant in the native land, but that may not necessarily be adaptive to the new homeland (Goldson & Emberson 1980). Likewise, the tobacco budworm *Heliothis virescens*, which has recently become a major pest in Arizona, does not appear to be fully adapted to its new situation. A large percentage of the summer population does not enter diapause and apparently dies in autumn and winter because it is not sensitive to prevailing photoperiodic conditions (Potter & Watson 1980). An extended study of these species over the years would determine the adaptive changes made in their seasonal cycles.

It is important to stress that adaptation to new environmental conditions involves evolutionary changes in numerous diapause-mediated characters as well as life-history phenomena other than diapause. A population of the monarch butterfly, *Danaus plexippus*, recently introduced into Australia apparently has lost its propensity for long distance migration. Aggregated, overwintering populations are found in areas near Sydney and Adelaide, not far from where reproducing populations occur. Diapause in the Australian monarch also appears to be less intense than in the North American stock (James 1982, 1983; see Chapter 4). Thus, both diapause-mediated dormancy and migration behavior have been altered in this immigrant population.

The adaptation of numerous coordinated traits that accompanies the geographical spread of introduced species can lead to diversification of distinct biotypes with multiple differentiating characteristics. The European corn borer, *Ostrinia nubilalis*, is particularly interesting in this regard because it is a relatively recent immigrant that has been studied since its introduction into North America (Brindley *et al.* 1975; Gelman & Hayes 1980; Showers 1981). This species is among the most widespread and destructive pests of corn and other plants around the world. Near the beginning of the 20th century, it came into North America, probably through several independent introductions from various parts of Europe. Since that time it has spread throughout most of the eastern and central regions of the continent, from Quebec to Georgia and Alabama in the east and from South Dakota to Oklahoma in the west.

The European corn borer is characterized by an autumnal-hibernal diapause that is primarily controlled by photoperiod (Beck & Hanec 1960). Mature larvae enter diapause when day lengths fall below the critical photoperiod for diapause induction (actually when night lengths become long); low temperature enhances the diapause-inducing effects of short day lengths (Beck 1962). Diapause is maintained photoperiodically until sometime in December; subsequently, pupation is prevented until temperatures rise above approximately 10°C (McLeod & Beck 1963; Anderson *et al.* 1982a).

After adult emergence and oviposition, temperature and other factors such as the host plant (Anderson *et al.* 1982b), determine the rate of development in summer generations until day lengths and temperatures combine to induce diapause again.

Very soon after its introduction into North America, the European corn borer showed geographical variability in the number of generations per year (Arbuthnot 1944). In the central part of the continent, the variability increased as the geographical range of the species extended southward (Showers *et al.* 1971; Reed *et al.* 1978). Populations in some areas are restricted to one generation per year, whereas populations in others undergo two, three, or four life cycles per year. In a fine series of experiments that showed great prescience, Beck and Apple (1961) demonstrated that these differences are not only based on the physical conditions of the various regions, but that several life-history traits of the European corn borer populations themselves are adapted to local climatic conditions. Their studies showed that geographical populations ranging from 45°N (Wausau, Wisconsin) to 36°50′N (southeastern Missouri) differed in their responses to diapause-inducing photoperiods and temperatures and in their nondiapause growth rates. Later, Kim *et al.* (1967) also demonstrated geographical variability in morphometric characters.

In the midwest, therefore, several characteristics appear to have evolved along a latitudinal cline with northern populations exhibiting univoltine life cycles, intermediate (central) populations having bivoltine life cycles, and southern populations producing three or four generations per year (Beck & Apple 1961; Sparks *et al.* 1966; Showers *et al.* 1971, 1975; Reed *et al.* 1978; Showers 1981). In northeastern United States and eastern Canada, however, the variability appears to be more complex. In Ontario, Canada, univoltine and bivoltine populations occur along a stepped cline with interfacing populations (that are sometimes separated by no more than 25 miles) displaying distinctly different life-history patterns (e.g., McLeod 1976, 1978; McLeod *et al.* 1979). In other areas (e.g., in Massachusetts, Beck & Apple 1961; and in Quebec, McLeod *et al.* 1979) there are clear exceptions to the direction of the general geographical trends. And, finally, in certain areas of northern and western New York and in some parts of Ontario, distinct univoltine and multivoltine populations occur sympatrically or in adjacent areas. In these situations the relative abundance of each type may vary greatly from year to year and from locality to locality.

Such variability, as well as the distinct diapause characteristics and differential survival of individuals that have been moved to new areas (Chiang *et al.* 1968; Showers *et al.* 1972, 1975), illustrate the influence of local climatic conditions on population structure, and they raise some very interesting evolutionary questions: What is the source of the genetic variability that underlies such large seasonal differences among populations of an introduced species? How does natural selection act to maintain the genetic differences among sympatric or adjacent geographical populations? Are the differences in voltinism related to the evolution of reproductive isolation and speciation? Clues to the answers to some of these questions may reside in analyses of genetic differentiation among the geographical populations in North America and in the species' European homeland. For example, recent findings show that two different sex pheromones occur among the geographical populations of the corn borer in North America (see Cardé *et al.* 1978). Populations in Europe show responsiveness to the same pheromones (Klun *et al.* 1975; Cardé *et al.* 1978; Buechi *et al.* 1982), suggesting that the geographical variability that is evident in North America had its origin in several introductions from Europe. Other recent work in Europe also indicates that there is considerable variability in voltinism among native European populations (Danilevsky 1965; Khomyakova 1976; Robin 1980). Again, these data are consistent with the idea that the European corn borer was introduced into North America from more than one source in Europe.

The pheromone differences apparently can confer a relatively high degree of reproductive isolation, and in some areas they may be responsible for maintaining some degree of reproductive isolation among sympatric entities (Liebherr & Roelofs 1975; Harrison & Vawter 1977; Cardé *et al.* 1978; Buechi *et al.* 1982). Future work that analyzes the occurrence of the "pheromone strains" in relation to the geographical patterns in variation in voltinism may show that pheromone differences maintain isolation between sympatric or parapatric populations having different patterns of voltinism. Such appears to be the case in southern Ontario where the univoltine and bivoltine strains have clear differences in their pheromone responses (McLeod *et al.* 1979).

To summarize, studies of colonizing species in large part confirm and expand the findings from artificial selection experiments. Numerous experiments demonstrate the responsiveness of diapause-mediated seasonal cycles to natural selection, but there are no cases illustrating the *de novo* evolution of diapause from completely nondiapause populations. Tropical species that have been able to expand their geographical ranges with the introduction of their host plants into temperate areas illustrate this point well. In all known cases, diapause that was already present in the original tropical homeland formed the basis around which the life cycle adapted to the seasonal restrictions of the temperate zone.

Evolution of diapause is not restricted to a diapause-intensifying direction. Examples demonstrate that diapause can become less intense as populations expand their ranges out of the tropics. Other examples illustrate that diapause expression can be reduced when the ranges of northern populations expand to the south or when temperate zone species are introduced and established in tropical areas. Such findings have very important implications to our next section dealing with comparative studies, because they demonstrate that it cannot be assumed that the nondiapause condition is primitive to diapause.

8.2.3 Comparative studies

In the past, comparative biology has provided a very productive approach to evolutionary questions, whether at the morphological, genetic, biochemical, physiological, behavioral, or ecological levels. It is therefore reasonable to use comparative studies to verify, modify, or negate our evolutionary model of insect diapause. The first step in such an approach is to determine at what taxonomic level comparisons will be most meaningful.

In some cases, broad comparisons of interspecific differences and similarities among higher taxa have provided meaningful insight into evolutionary trends. This is especially true for relatively stable and conservative, but adaptively significant characteristics. However, this approach also leads to serious problems when it is applied to evolutionarily labile and yet environmentally and physiologically constrained traits such as seasonal cycles. The main problems stem from convergent and parallel evolution of traits that have limited alternative pathways.

Phylogenetically unrelated organisms (e.g., reptiles, birds, mammals, terrestrial and marine crustaceans, insects, and acari) exhibit some remarkable similarities both in their seasonal cycles and in the ecophysiological responses underlying these cycles (Tauber & Tauber 1978, 1981; Kurihara 1979; McQueen & Steel 1980; Marcus 1979, 1982a, b; various authors in Follett & Follett 1981 and in Flint *et al.* 1981). One striking example is provided by a comparison between two insects, the green lacewings *Chrysopa downesi* and *Meleoma signoretti*, and a bird, the white-crowned sparrow, *Zonotrichia leucophrys gambelii* (Fig. 8.4) (data on the white-crowned sparrow from Farner & Follett 1966; Farner & Lewis 1971; Farner 1975; data on the lacewing from Tauber & Tauber 1975a, 1976b, c, 1977c). Although the two lacewing species are both univoltine, with diapause expressed during late summer, autumn, and winter, the responses underlying their seasonal cycles are very different. *M. signoretti* develops continuously under long day conditions; short days induce diapause, and long days at about the time of the vernal equinox terminate it. There is no photorefractory period during which long day lengths do not stimulate development; short days are not important in diapause termination; and there is no quantitative response to day length. By contrast, the seasonal cycles and the mechanisms controlling the seasonal cycles in the white crowned sparrow and the lacewing *C. downesi* differ greatly from those in *M. signoretti*, but they show remarkable similarities to each other. The white-crowned sparrow is a single-brooded bird; migration and reproduction occur in the spring. *C. downesi* also produces only one generation per year, with reproduction in spring; summer, fall, and winter are spent in reproductive diapause. In both species, reproduction is followed by an aestival photorefractory phase during which an artificial increase in day length does not stimulate reproduction. Only after the animals have experienced the relatively short days of the autumnal equinox will transfer to long-day conditions initiate reproduction. And, after recovery from the photorefractory phase, the rate at which reproduction is initiated (i.e., the rate of gonadal growth, prenuptial molt, and certain premigratory events in the white-crowned sparrow, and the length of the preoviposition period in *C. downesi*) is quantitatively related to the absolute duration of day length. Similar adaptations are also found in fish, reptiles,

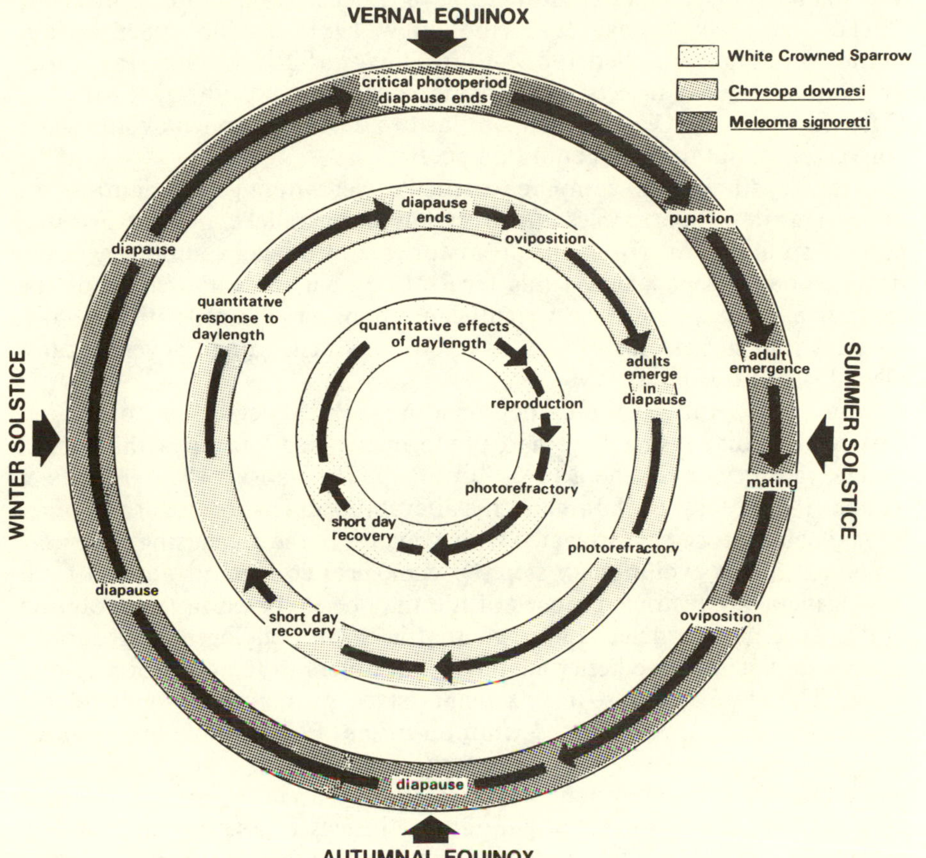

Fig. 8.4 The photoperiodic control of annual seasonal cycles in three species with varying degrees of phylogenetic relatedness. Each circle represents the annual cycle of one species. Although the periods of diapause in the two univoltine lacewing species, *Chrysopa downesi* and *Meleoma signoretti* (the two outside circles), overlap during late summer, autumn, and winter and end about the time of the vernal equinox, the underlying photoperiodic responses controlling the seasonal cycles differ greatly. By contrast, the white-crowned sparrow *(Zonotrichia leucophrys gambelii)* and the lacewing *C. downesi* (two inner circles) show remarkable overlap in (1) the features of their reproductive cycles (e.g., aestival photorefractory phases, short-day recovery periods, quantitative response to day lengths), (2) the annual timing of these events, and (3) the photoperiodic control of the events. Data from Farner and Follett (1966); Farner and Lewis (1971); Farner (1975); Tauber and Tauber (1975a, 1976b, 1977c).

and mammals (see Peter & Hontela 1978; Zucker *et al.* 1980; Johnston & Zucker 1980a, b; Kenagy & Bartholomew 1981; Marion 1982; various authors in Follett & Follett 1981 and in Flint *et al.* 1981); other similarities are found between insects and marine copepods and other zooplankton (Marcus 1982a, b). Clearly, such similarities develop through evolutionary convergence, not through common ancestry.

Likewise, studies that compare seasonal cycles within genera, families, or orders are rife with examples of convergent and parallel evolution. For that matter, so are comparisons of life histories and other evolutionary labile traits among higher taxa. At this level of comparison great care must be taken in assigning evolutionary significance to observed similarities and differences [the suggestions of Stearns 1980 notwithstanding]. Several examples illustrate this point.

Many comparative studies demonstrate a high degree of correlation between the diapausing stage and phylogenetic relationship within insect orders, families, and genera (e.g., Lumme 1978; Slansky 1974; Tauber & Tauber 1978; Masaki 1980). Thus, the alterations in the hormonal and other physiological processes associated with a change in the diapausing stage may involve a major evolutionary step. Or ecological constraints derived from phylogenetically restricted modes of life may be involved in the evolution of diapause in a particular life stage; that is, within phylogenetic groups of insects some stages are better adapted than others to the evolution of diapause. Therefore, diapause in a particular stage, even among closely related species, does not always imply common origin. For example, many Japanese species of crickets sharing the same diapause stage are more closely related to tropical, largely nondiapausing forms than to one another (Masaki 1978a, 1983). If the tropical origin of these insects is assumed, as is widely accepted (Alexander 1968), then they would have acquired diapause independently of each other in the course of their expansion from the tropics to the temperate region. Such appears to be the case in the two species-groups of ground crickets, *Pteronemobius taprobanensis* and *P. fascipes*. The temperate members of both species-groups enter egg diapause in Japan, but the eggs of tropical and subtropical populations of each species-group virtually lack all expression of diapause (Masaki 1978a, 1979, 1983, 1984).

Similarly, two Japanese species of *Teleogryllus* enter diapause as eggs; crosses between the two species produce hybrids with abbreviated diapause, probably because of differences between the parental species in the genetic background for diapause—a situation that suggests the independent evolution of their diapause (Ohmachi & Masaki 1964). Thus, egg diapause probably evolved independently, numerous times among cricket species in the same genus.

Several studies of insect families and orders have categorized species into ecological groups depending on the characteristics of the seasonal cycle. Although these classifications are useful in describing how life cycles adapt to tropical, temperate, and Arctic regions, they are of limited value in describing the evolutionary development of diapause and seasonal cycles. Corbet (Corbet *et al.* 1960; Corbet 1963) roughly divides the dragonflies into three ecological groups that he considers to represent an evolutionary trend in the adaptation to temperate conditions: univoltine summer species and semivoltine summer species in which adults emerge in a relatively unsynchronized manner late in the year, and spring species that emerge with a distinct peak, early in the season. Corbet proposes that the derived life cycles of northern species result from seasonal requirements that are quite different from those of their tropical relatives. Tropical species do not experience the drastic seasonal fluctuations in temperature that temperate species must endure, but their survival is highly dependent on the occurrence of patchily distributed, temporary pools of water to which larvae are restricted. Thus, dragonfly life cycles in the tropics are characterized by a long adult life that subserves dispersal among pools and a short larval life that is adaptive to survival in the ephemeral pools. Entry into the temperate zone resulted in a relaxation of the pressures associated with restricted larval resources, and seasonality assumed a major influence on the life cycle. Thus, adults became restricted to the warm season, and the stage most resistant to cold was restricted to the cold season. Diapause developed in the egg stage, and the life cycle essentially became univoltine. Such a seasonal cycle is exhibited by the summer species of dragon flies, for example, *Lestes sponsa.* Emergence is not synchronized in summer species, but adult life is still relatively long so that tight synchronization is not required.

As dragonflies acquired more northerly distributions, temperature conditions prevented the completion of the entire life cycle in one year. This resulted in the evolution of tolerance to cold in the final larval instar (the stage most likely to be overtaken by winter); later this characteristic extended to earlier instars. It is through this means that the semivoltine summer species with egg diapause evolved.

Spring species, according to Corbet, represent the final stage in the northerly extension of dragonflies. In these species, all larval stages are resistant to cold and able to tolerate winter conditions. Their adults can also tolerate low temperatures so that spring flight is common. Because of these factors, adult emergence must be synchronized; diapause that occurs in the final larval instar and that characterizes spring species serves this function.

Corbet supports his evolutionary scheme with data showing several trends in the life cycles of dragonflies from the tropics to northern temperate regions, specifically (1) a reversal in the relative lengths of larval and adult stages, (2) an increase in tolerance to low temperature, beginning with the last larval instar, extending through earlier instars, and reaching the reproductive stage last of all, and (3) an egg diapause in each generation which gives way to a final-instar diapause that can be averted under specific photoperiodic conditions. However, his studies do not compare the evolution of dragonfly seasonal cycles with independently derived dragonfly phylogenies. Thus, similarities among groups frequently represent convergent or parallel evolution. Elucidation of the specific steps in the evolution of seasonal cycles and diapause-controlling mechanisms will require detailed data from taxa within specific evolutionary lines.

Beetles in the family Carabidae constitute a particularly widespread group of insects whose seasonal cycles also have been the subject of broad comparative study. Because these insects occupy a vast variety of habitats, which include areas such as tropical lake shores and swamps without any apparent annual cycle, tropical riparian stream beds subject to periodic flooding or drying, and temperate, subarctic, and Arctic habitats characterized by large seasonal fluctuations in temperature and moisture, they offer a rich source of material for the study of the evolution of diapause (Thiele 1977b, 1979a, b; Paarmann 1979a). Among the carabids, the larval stage comprises the period of the life cycle that is most sensitive to extremes in temperature and moisture; thus, the seasonal cycle is timed so that larvae experience favorable conditions (see Thiele 1977b, 1979a; Paarmann 1979a). In summarizing the literature from many sources, Paarmann (1979a) proposes two main evolutionary pathways for the evolution of seasonal cycles in these beetles. Starting with seasonal cycles appropriate to specific tropical habitats with virtually no seasonal fluctuations, he postulates that the first line of descent goes through a sequence of changes that are adaptive to tropical habitats that remain humid but experience periodic inundation, to continuously humid habitats with and without periodic inundations in the subtropics. This evolutionary line ends in the temperate zone.

Paarmann's second line of descent has its beginnings in the same types of seasonally stable, tropical habitats as the first line. This line, however, sees the adaptation of life cycles to periodic dry seasons in the tropics and subtropics, and then the diversification of responses as the beetles expand in the temperate, subarctic, and perhaps Arctic regions.

Paarmann provides examples of seasonal cycles that fit nicely into these two evolutionary pathways. A few species that inhabit the seasonally constant habitats of tropical lakeshores and tropical swamps show continuous, uninterrupted development year round (Paarmann 1976a). In tropical areas that are subject to either periodic dry periods or periodic inundation. this type of life cycle is replaced by one with a restricted period of reproduction. The timing of this period is determined by a thermally or thermoperiodically controlled reproductive dormancy. Paarmann does not attribute this type of gonadal dormancy to diapause, but it is clear that the temperature changes that herald the arrival of unfavorable moisture conditions induce the onset of reproductive dormancy before the arrival of the adverse conditions (Paarmann 1976a, b, c, d, 1977, 1979b). Thus, the most simple form of dormancy in the Carabidae is one that is based on complex responses—the ability to perceive and respond to anticipatory (token) seasonal cues.

In this regard, and of special note, is the observation that even under certain conditions in which the monthly average change in atmospheric temperature amounts to no more than 0.9°C, there was a 5°C variance in the monthly average temperature of the soil surface where carabid beetles occur (Paarmann 1976a). No notice was made of the thermoperiod in the soil microhabitat, but considering recent work showing the diapause-controlling effect of thermoperiod (see, e.g., Beck 1982, 1983a, b), it is reasonable to suspect that, along with the seasonal fluctuation in mean temperature, changes in the daily rhythm of temperature fluctuations provide a significant seasonal cue.

Photoperiodic influence on diapause is found among Carabidae from the lower latitudes in species such as *Broscus laevigatus* and *Orthomus barbarus atlanticus* (Paarmann 1974, 1976a). Whether their photoperiodically controlled diapause evolved *de novo* or was derived from an ancestral thermal or thermoperiodic response is unknown. But, future work with geographical populations of these species, or closely related species and strains, could provide very interesting data regarding the evolution of photoperiodically controlled diapause.

In adapting to the temperate zone, the seasonal cycles of carabid beetles appear to have undergone a tremendous diversification appropriate to such a diverse taxon. Both larvae and adults are known to hibernate in the temperate regions, and thus far the beetles fall into five ecological categories with various forms of hibernation, aestivation, and development without dormancy (Thiele 1977b). As stated above, Paarmann proposed some possible trends in the evolution of seasonal cycles among temperate-zone carabid beetles, but as in the dragonflies and other groups, convergent and parallel evolution complicates the evolutionary picture. Elucidation of these pathways requires the comparative analysis of seasonal cycles and their controlling mechanisms within closely related groups of carabids with known or readily inferred phylogenies.

In reviewing the above, it is especially important to note that there is no evidence to support the supposition that the dormancy- and diapause-free life cycles of carabid beetles in the seasonally unvarying tropical habitats represent a primitive condition. It is well known (see Sections 8.2.1 and 8.2.2) that strains of insects with diapause-free development are readily selected from diapausing stock, and thus the uninterrupted life cycles of these tropical species could represent a derived adaptation to a seasonally nonfluctuating environment. Satisfactory analysis of the primitive nature of tropical seasonal cycles in the Carabidae, as in other insect groups, requires detailed comparative studies of the seasonal cycles of taxa with phylogenies that are known or reasonably well inferred from independent taxonomic study.

Tropical flesh flies (family Sarcophagidae) represent an insect group for which the tropical origin of diapause is reasonably inferred on the basis of relatively broad comparisons among taxa (Denlinger 1979). Among the flesh flies, tropical species are unique in that temperature and thermoperiod are the primary diapause-regulating cues (Denlinger 1979). Temperate-zone species rely almost exclusively on photoperiod. But, as in the Carabidae, studies of the seasonal cycles of tropical and temperate-zone flesh flies have not been related to taxonomic studies that derive phylogenetic lines. It is not known if the photoperiodic response evolved *de novo* as the flies expanded their distributions out of the tropics or if it evolved from existing thermal and thermoperiodic responses. Studies have shown that temperature and thermoperiod clearly act as anticipatory seasonal cues for tropical species (Denlinger 1979), as does photoperiod for temperate-zone species (Fraenkel & Hsiao 1968; Denlinger 1971, 1972b; Saunders 1971; Roberts & Warren 1975; Vinogradova 1976a, b; Kurahashi & Ohtaki 1977). These extensive studies form an invaluable beginning for the comprehensive analysis of the evolution of diapause—specifically with regard to the coupling of neuroendocrinologically controlled responses to photoperiodic regulation. To have a solid foundation, such an analysis requires comparative studies of seasonal cycles in conjunction with taxonomic treatment of the group and their relatives.

Comparative studies of primarily tropical and subtropical milkweed bugs in the *Oncopeltus* subgenus *Erythrichius* also suggest a tropical origin for diapause and the modification of diapause responses as populations extended into the temperate zone. Among the several species of this subgenus that occur in North America, only one, *Oncopeltus fasciatus*, extends into the temperate region. Apparently this species is able to do so because it has evolved a photoperiodically regulated diapause that delays reproduction and subserves long-distance migration. Under short day lengths, northern temperate populations of this species enter reproductive diapause and migrate southward, presumably to areas of relatively mild winter conditions. Tropical populations of *O. fasciatus* either do not respond to photoperiod by entering diapause or they show a very low incidence of diapause (Dingle 1978a; Dingle *et al.* 1980a). From the geographical distribution and diapause response of the species and its near relatives in North America, it is reasonable to assume that diapause evolved in tropical populations and that it became modified as the species extended its range into northern temperate regions.

Studies by several workers (summarized by Dingle 1978a) suggest a possible scenario for the evolution of *O. fasciatus* diapause. Nondiapause bugs can survive relatively long periods of starvation (at least 30 days) if they have access to water. During periods of starvation, metabolism is lowered and juvenile hormone production is shut down, producing a condition similar to diapause. Thus, it is proposed (Dingle 1978a) that the evolution of photoperiodically regulated diapause involves the linking of photoperiodic time-measurement to the neuroendocrine regulation of metabolism. Through this mechanism, reproduction is linked with photoperiodic control; subsequently, migration is also coupled to photoperiod. Further analysis of tropical populations of this species would be especially useful in elucidating the evolutionary relationships between diapause and migration.

Bradshaw and Lounibos's (1977) study of the evolution of seasonal cycles in the North American pitcher-plant mosquito *Wyeomyia smithii* considers traditional taxonomic characteristics of morphology and species distribution in deriving the evolutionary direction of dormancy in a geographically variable species. All evidence is consistent with the proposal that dormancy evolved from south to north. Thus, diapause in the fourth larval instar, which characterizes southern populations, is considered primitive, while that in the third instar, which is expressed only in northern populations, is derived. Furthermore, under certain circumstances northern populations can be induced to enter a second, fourth-instar diapause (Lounibos & Bradshaw 1975; Istock *et al.* 1975); presumably this characteristic is retained as a relic of southern ancestry. By contrast, southern populations never enter diapause as third instars, but in populations from south to north, the shorter day lengths increasingly prolong the third instar until diapause occurs in that stage. Thus, within this species the evolution of diapause has taken place through the progressively greater influence of photoperiod on the control of the developmental rate of a nondiapausing instar. Again, the diverse geographical populations of this species offer a valuable resource with which to analyze how changes in the photoperiodic, neuroendocrine, and memory systems contribute to the evolution of diapause.

Comparative study of seasonal cycles and their ecophysiological and genetic control, in conjunction with comparative morphological study has also been applied to green lacewings in the *Chrysopa carnea* species-complex. In North America this species-complex contains four basic life-cycle types—the multivoltine *carnea* type with long-day reproduction, the univoltine *downesi* and Alaska types, which have short-day/long-day requirements for reproduction, and the polymorphic *mohave* type with its photoperiodic and prey-mediated reproduction (Tauber & Tauber 1981b, 1982). A fifth type of life cycle, one without diapause, may occur in European populations near the Mediterranean (Alrouechdi & Canard 1979).

The *mohave*-type life cycle, with its strong dependence on prey as a diapause-regulating seasonal cue, is restricted to populations in relatively undisturbed habitats in western North America that are characterized by considerable between-year variability in prey availability. Both larval and adult morphological traits indicate that *carnea* is the ancestral form from which *mohave* was derived (Tauber & Tauber 1973a). From this it is concluded that response to prey, as a seasonal cue, is a derived condition in the *C. carnea* species-complex. All populations with *mohave* morphological characteristics and seasonal cycle retain the basic long-day requirement for reproduction, and to this they have added a prey-mediated aestival diapause. Adults of *carnea*-type populations in eastern North America show no seasonal response to prey. Genetic studies suggest that the response to prey is controlled by recessive alleles at more than one locus and that the trait is inherited independently of other seasonal traits in the species-complex. Other studies indicate that *mohave* genes are present in numerous populations throughout western North America and that natural selection acts to maintain them in populations that experience unpredictable seasonal abundance of prey. Ancestral or reintroduced *carnea* genes are predominant, especially in areas of western North America that are disturbed by horticultural and agricultural practices that promote the continuous presence of prey in the summer (Tauber & Tauber 1973a).

On the basis of biological, distributional, and morphological data (Tauber & Tauber 1977a), it was also concluded that the multivoltine *carnea* life cycle is ancestral to the univoltine *downesi* life cycle with its short-day/long-day requirement for reproduction. Populations with pure *downesi* characteristics are relatively restricted geographically (northern regions of North America) and in their plant associations. Unlike *C. carnea*, which occurs on a wide variety of deciduous, coniferous, and herbaceous plants, *C. downesi* populations are found only on evergreen conifers. *C. downesi* populations retain many aspects of the *carnea* response to photoperiod, specifically the quantitative response to day length that typifies the photoperiodic maintenance of diapause in *carnea* populations and the characteristic sensitive stages. Consequently, it is concluded that the *downesi* seasonal responses were derived from the ancestral *carnea* response (Tauber & Tauber 1976b). The difference between the two types lies primarily in the short-day/long-day requirement for avoidance of reproductive diapause. The presence of this requirement in *downesi* populations results in an aestival-autumnal-hibernal diapause; its absence in the *carnea*-type life cycle allows diapause-free reproduction during the summer. In eastern North America, populations with *downesi* and *carnea* life cycles are reproductively isolated from each other, primarily because of differences in the seasonal timing of reproduction. Hybridization and backcross tests indicate that recessive alleles at two autosomal loci control the *downesi* short-day/long-day requirement for reproduction. Thus, evolution of the univoltine *downesi* life cycle consists of the mutation of and selection for the two recessive alleles that result in a short-day/long-day requirement for reproduction.

Populations in western North America also possess the short-day/long-day requirement for reproduction, but in these populations the characteristic is expressed as a seasonal polymorphism (Tauber & Tauber 1981b, 1982). Preliminary genetic studies indicate that the trait is based on homologous genes in the eastern and western populations but that the mode of inheritance may be different among populations, suggesting that the alleles may have become linked in western populations (Tauber & Tauber in manuscript).

8.2.4 Summary

Evidence regarding the evolution of seasonal traits comes from three main sources: responses to artificial selection, adaptation by colonizing species, and comparative studies. All three types of studies have produced significant advances in the evolutionary analysis of insect seasonal cycles, but all are limited in the amount of information they have provided. Most studies have concentrated on the ecological or genetic aspects of evolutionary changes in seasonal cycles, thus omitting either important physiological aspects or essential phylogenetic considerations.

Insect seasonal cycles are very amenable to change by artificial selection over relatively few generations. Most such studies have aimed at selecting nondiapause strains from populations exhibiting diapause, and in doing so they have altered several characteristics of diapause, including the incidence of diapause, the length of diapause, the critical photoperiod, and responsiveness to temperature and population density. In some cases, selection in the opposite direction—that is, for an increased tendency to enter diapause—has been successful. Despite the ease with which seasonal traits are altered by artificial selection, most investigations have demonstrated a leveling-off in responsiveness after several generations. Thus, it is very difficult to eliminate completely the propensity to enter diapause. This apparent constraint, combined with the ability to respond to reverse selection, is consistent with the numerous examples of repeated or convergent evolution in seasonal cycles among insect taxa. It also suggests that predominantly nondiapause stock maintain genetic variability for diapause and in some cases can give rise to diapausing strains without the *de novo* evolution of diapause.

Introduced and native species that have expanded their geographical ranges in historical times offer a wealth of information on the evolution of seasonal adaptations. Adaptation of diapause and diapause-mediated characteristics can occur over relatively short periods of time, and it can result in the rapid evolution of distinct, locally adapted biotypes. All cases of seasonal adaptation by colonizing species involve modification of diapause that already existed in the original population; there are no reported cases of the *de novo* evolution of diapause from a truly nondiapause ancestor during historical times.

Despite the fact that diapause in tropical species is often considered to be primitive, it is highly adaptive. Thus, without supporting evidence it cannot be assumed that the nondiapause life cycles or that life cycles with subtle or variable diapause in tropical insects represent a primitive condition from which seasonally more restricted life cycles have evolved. Evolution in the reverse direction is possible. Studies of the evolutionary progression of diapause and its controlling mechanisms require comparative studies in conjunction with taxonomic studies that provide information on the direction of evolution along phylogenetic lines.

Broad, general comparisons of seasonal cycles and diapause characteristics among genera or higher taxa have not provided significant insight into the evolution of insect diapause. This is not to say that such studies have been without value to evolutionary biology—quite the contrary! Broad comparative studies are very useful in defining geographical and ecological trends in the adaptation of life histories and diapause and in deducing the significant ecological factors that determine the direction and extent of evolutionary change. But analysis of the steps involved in the evolution of diapause requires detailed comparisons of both the ecological and physiological mechanisms controlling diapause in groups of closely related species with known or readily inferred phylogenetic relationships. Comparative studies that meet such requirements of depth and breadth are lacking for all groups of insects, but several studies that have combined taxonomic analysis with comparative studies of seasonal adaptations illustrate that such an approach is a powerful tool to use in evolutionary studies of insect seasonal cycles.

Chapter 9

Seasonality, the Evolution of Life History, and Speciation

Seasonal cycles have multiple influences on the life histories of organisms. Periods of growth, development, and reproduction must be synchronized with the seasonal presence of energy resources, mates, and favorable physical conditions. The occurrence of vulnerable stages must be timed to avoid periods of physical extremes and the height of activity of natural enemies and competitors. Thus, periods of dormancy and migration intervene between periods of growth, development, and reproduction; this pattern simultaneously increases the chances of survival during unfavorable periods and the probability that growth, development, and reproduction will occur at an appropriate time and place.

Among insects the seasonal adaptations of diapause-mediated dormancy, migration, and polyphenism provide the framework for the entire life cycle. These adaptations involve, or are associated with, the postponement of growth and reproduction, sometimes under seemingly favorable conditions, for survival and increased reproductive success at a later date. Thus, seasonality entails distinct reproductive costs and benefits that can be measured in terms of their effect on some aspect of fitness (Istock 1982; Tauber & Tauber 1982). For example, the intervention of diapause during a particular life stage results in important demographic consequences that persist long after dormancy and migration have ended. Diapause traits interact in a dynamic way with the life-history parameters such as longevity, fecundity or fertility, viability, mating success, and so forth, that comprise fitness (see Istock 1983). In this context the genetic constraints and the genetic variation associated with seasonal characters both limit and provide the basis for the evolution of life histories and the diversification of taxa.

In this chapter we analyze the interaction between seasonality and life history. We begin by describing seasonality as a major life-history trait and then analyzing the role that seasonality plays in the evolution of life histories. First, we examine the interdependence between and among seasonal and other life-history traits; some interactions involve trade-offs, others coadaptation. Second, we consider variation in life-history traits in relation to seasonality, and we try to elucidate how life-history variation is affected and maintained by seasonal environments. Finally, we discuss the role of seasonality in speciation.

9.1 Seasonality as a Life-History Trait

The physical extremes of a locale set the ultimate constraints on the rates of growth and reproduction of a population. Within these limits, interspecific interactions (e.g., occurrence of hosts, natural enemies, and competitors) and intraspecific interactions (e.g., occurrence of mates) act as selective forces in molding life-history patterns. All of the above abiotic and biotic factors are characterized by seasonal cycles, and as a result, each species' life history evolves to take advantage of the availability of requisites and to avoid unfavorable conditions. As a result of the multiple facets of seasons, life-history traits (schedules of growth, reproduction, and longevity) and seasonal adaptations are intertwined so that selection pressure on one causes, to varying degrees, selection pressure on the other. Thus, when the evolution and expression of life histories are considered, seasonality emerges as a major consideration (see Istock 1982, 1983; Butler 1984).

Among insects, diapause is the primary means of achieving seasonal synchronization. This adaptation not only affects dormancy and survival during adverse conditions, it also determines, in large part, the timing and rates of growth, development, and reproduction during favorable periods. In terms of their direct impact on life history, perhaps two of the most important seasonal characteristics are the choice of a particular developmental stage for diapause and the timing of the onset and termination of dormancy. These characteristics affect the entire life-history pattern, and many interrelated and interdependent factors seem to be involved in their evolution.

9.1.1 The diapausing stage and survival during dormancy

At first sight, one of the most important factors in shaping life histories may seem to be the ability of a particular metamorphic stage to tolerate a long period of dormancy. However, this is probably not a decisive factor, because, among insects, diapause occurs in all stages from the egg to the adult (see Sections 3.1 and 4.1). Because insects share many basic neuroendocrine mechanisms and metabolic patterns, all species may ultimately have the capacity to evolve diapause at any stage. This is suggested by extirpation of the brain from pupae of the silkworm, *Bombyx mori*, whose diapausing stage is the egg; if such an operation is performed within four hours after pupation, the pupae remain in an "artificial diapause" for as long as 100 days or more at 25°C (Kobayashi 1955).

Another factor that could influence the evolution of a diapausing stage is the ability of that stage to tolerate extremes in weather conditions, such as cold, heat, and drought. Here, again, there seems to be no inherent advantage to one developmental stage over another. Hibernating eggs of certain species are as cold-hardy as hibernating larvae or pupae of other species showing similar magnitude of supercooling and accumulation of cryoprotective substances (see Section 4.4.4). Therefore, the ability to evolve cold-, heat-, or drought-hardiness does not appear to be restricted to particular stages.

All stages can thus evolve the ability to withstand long periods of dormancy and to tolerate harsh conditions. There is, however, a strongly biased distribution of diapausing stages among various taxa of insects, even though diapause has evolved repeatedly and independently within the taxa. Certain stages appear to be especially well adapted to dormancy, and this may be related to taxon-specific life-styles and developmental characteristics. For example, the egg is the predominant stage for diapause in crickets and grasshoppers in various parts of the world (Ohmachi & Matsuura 1951; Uvarov 1966; Alexander 1968); this is probably related to their common habit of laying eggs in the soil. The soil provides protection from extremes in temperature and moisture, and perhaps also from some parasitoids and predators (see Hubbell & Norton 1978). The occurrence of a few species of crickets that overwinter as nymphs might also be associated with a burrowing habit (Alexander & Bigelow 1960).

In general, adult beetles appear to be well adapted for dormancy in soil or leaf litter, perhaps because of their highly sclerotized integument. For example, most species of Chrysomelidae and Coccinellidae, which have relatively exposed eggs and larvae, hibernate as adults in the soil; in contrast, the common occurrence of larval hibernation in the Scarabaeidae can be ascribed to the subterranean life of their larvae. Among Lepidoptera many arboreal species enter diapause at the pupal stage—a stage that probably is better protected than any other because it is contained in a relatively hard cuticle and a tough cocoon. However, those Lepidoptera whose larvae bore into plant tissues frequently hibernate as larvae within the plant or soil.

9.1.2 Diapause as the life cycle timer

In addition to enhancing survival during harsh periods, diapause has an important effect on the seasonal arrangement of other parts of the life cycle. Selection acts on whole organisms, not on separate traits (Tuomi *et al.* 1983), to maintain or increase fitness. Increased survival of one stage is of no value if it puts a later stage into an unfavorable situation. Thus, the effects of selection must be considered in light of its influence on later stages, future offspring, and the full reproductive potential over time (i.e., over the entire seasonal cycle in some cases, longer in others) (see, e.g., Taylor 1980a, b; Southwood *et al.* 1983). As a result, the synchronization of the active life stages with seasonal requisites may be the primary selective factor during the evolution of the diapausing stage. Adaptation to extreme weather conditions evolves secondarily in the diapause stage thus selected. This is particularly true when acceptable food or habitats are limited to a very short period of the year.

Dung-inhabiting scarab beetles in the genus *Aphodius* illustrate well the constraints that the developmental requirements of nondiapausing stages put on the evolution of the diapause stage (Hanski 1980). In this genus, the adult is virtually the only stage that is mobile; larvae are largely restricted to their ephemeral habitat (primarily fresh animal droppings). In northern Europe, animal droppings are available year round, but since physical conditions may prevent year-round reproduction and development, it is not surprising that most northern European species of *Aphodius* overwinter as adults. Adults can locate suitable places for overwintering in autumn, and after dormancy they also can move and select a suitable habitat for oviposition and subsequent larval development the following spring. Other species of *Aphodius* in northern Europe overwinter as final instar larvae or prepupae, and they are synchronized with their larval habitat requirements in a manner somewhat similar to species that overwinter as adults. Late summer or autumnal adults find fresh animal droppings that are suitable for larval development; these adults produce offspring that overwinter as mature final-instar larvae or prepupae in their dung habitat. Postdiapause larvae resume development without feeding; the resulting adults emerge in the spring and seek new oviposition sites.

During the winter, animal droppings deteriorate and become unsuitable for larval development. It follows that no species overwinter as first- or second-instar larvae, apparently because these stages cannot disperse and find new habitats. Unexpectedly, however, a few species do overwinter in the egg stage. Newly hatched postdiapause larvae of these species find themselves in droppings that may not be suitable for larval development. However, there are indications that the larvae of these species can develop in habitats other than fresh animal droppings. Their occurrence illustrates that freedom from the restriction to a specific larval habitat or food offers latitude in the evolution of the diapausing stage.

The importance of the diapausing stage to the synchronization of postdiapause feeding stages that are restricted to a seasonal food supply is also illustrated by species of lepidopteran larvae that feed on *Quercus* or other perennial plants. Many of these species hibernate as eggs deposited close to dormant buds; larvae appear in the spring just after bud-break (Shirozu & Hara 1962). Young oak leaves provide food of acceptable texture and nutrition, whereas older leaves are less nutritious and tougher, and they tend to accumulate leaf tannins, which inhibit larval growth (Feeny 1970). Precise synchronization of hatching to bud-break is essential to ensure survival of the hatchlings and completion of larval development before leaves become unsuitable as food. If these lepidopterans were to overwinter in the pupal stage, the subtle timing required to synchronize the occurrence of newly hatched larvae with tender young leaves would be more difficult to achieve because of the intervening time required for pupal development and adult breeding activity after hibernation.

The importance of the seasonal occurrence of food as a selective force is illustrated by the close correlation between the proliferation-gemmulation pattern in freshwater sponges and the life cycles of the neuropteran, dipteran, and trichopteran predators that feed on them (Resh 1976a, b). Geographical variation in the timing of postdiapause emergence also illustrates the importance of food as a selective force in shaping life histories. In Japan, emergence by the rice stem borer, *Chilo suppressalis*, is closely correlated with the local planting date for rice plants (Masaki 1967b).

It is becoming clear that, in addition to influencing postdiapause events, the availability of food or habitats often shapes the overall seasonal cycle, especially in species that use a seasonally restricted resource. Indeed, many instances of summer diapause in nonarid areas occur under a range of weather conditions that would permit normal activities; the restrictions on periods of growth or reproduction are the result of seasonally recurring periods of food shortage (Tauber & Tauber 1973c, 1982; Slansky 1974; Shapiro 1979; Masaki 1980; Sims 1980; Niemelä *et al.* 1982). In the swallowtail butterfly, *Papilio zelicaon*, some populations are multivoltine and feed on introduced host plants that are continuously available throughout summer, whereas populations in areas which have only the native, spring and early summer food plant maintain a univoltine life cycle (Sims 1980). In the western North American *mohave* strain of the green lacewing *Chrysopa carnea*, aestival reproductive diapause is induced when populations of prey are low. The adults apparently do not require the nutrients provided by prey, but they use prey as a token stimulus to predict whether or not prey levels will be suitable for their larval offspring (Tauber & Tauber 1973c, 1982).

Variation in nondiapause developmental requirements and its effect on the evolution of the diapausing stage are well demonstrated in the seasonal cycles of two antlion species, *Myrmeleon formicarius* and *Hagenomyia micans*; in these cases the variation underlies the evolution of flexibility in the overwintering stage. The larvae of these species trap prey in funnel-shaped pitfalls in the sand or loose soil; their food supply can be intermittent and very variable. Two aspects of the antlions' life history allow them to cope with such a situation. First, larvae can survive weeks or even months of starvation, and second, they can overwinter in any instar. Flexibility in larval dormancy may have evolved in conjunction with dependence on an unpredictable food supply. If overwintering were restricted to a specific larval stage, survival and/or reproductive productivity would decline, fitness would decrease, and the food supply would be used less efficiently (Furunishi & Masaki 1981, 1982, 1983).

In some cases, it appears that temperature during the developmental stages exerts a major influence on the seasonal cycle. For example, optimum larval growth and optimum reproduction by certain black fly species occur over a distinct, narrow range of temperatures that exist for only a brief time during late winter and early spring. Development at higher temperatures after this time reduces final larval body size and possibly fecundity (Merritt *et al.* 1982). Simuliid species vary in their thermal optima; some are adapted to grow at higher temperatures and under conditions of greater temperature variation. Thus, the species in a stream may show temporal replacement along a complex thermal gradient (Merritt *et al.* 1982). Although temperature appears to exert a greater influence than food on the life histories of these species, the complex interactions between the two factors make it difficult to separate the relative contributions of each.

Studies with other aquatic species illustrate further that seasonal variability in habitat suitability and food availability can influence the reproductive cycle (see reviews by Berg & Knutson 1978; Butler 1984). Differential emergence by the sexes of the mosquito, *Aedes triseriatus*, serves as an example. In this species, emergence occurs throughout several seasons—with males predominant in spring and females in summer. The developmental differences between the sexes result from some aspect of postdiapause egg hatching requirements—probably differential responses to deoxygenated water (Shroyer & Craig 1981; Holzapfel & Bradshaw 1981). Although the evolutionary importance of this sexual difference is open to speculation, it is explainable, in part by the unpredictable nature of the larval habitat (treeholes that are subject to periodic desiccation) and the differential investment in reproduction by the two sexes (Shroyer & Craig 1981). Presumably, early hatching, emergence, and oviposition offer considerable risk to females because at the beginning of the season they have no way of discerning "good," relatively drought-resistant treeholes from those that will dry up before their offspring complete development. The loss of a single egg batch probably represents a large reproductive cost for females, and a very conservative seasonal strategy would reduce the risk of reproductive failure. In contrast, for males the loss of a batch of eggs does not represent a large loss in energy investment, and it may be adaptive for males to occur over the long season. The differential between the sexes would be especially significant if males survive for a long period of time and if summer males are very abundant relative to females.

In an interesting study of the relationship between the diapause stage and selected life-history traits in temperate-zone butterflies, Hayes (1982b) showed that diapause and voltinism are at the "hub of numerous intercorrelations" involving phagism, frost-free days, food plant form, and distribution. As expected, species without diapause are confined to warm habitats and are frequently associated with herbaceous host plants. However, species having an adult diapause are associated with warm habitats and woody host plants. Species with pupal diapause are relatively diverse with respect to the other factors but do show definite clustering around woody host plants and warm habitats. Species with larval diapause are also relatively diverse, although a large number are associated with cool habitats and herbaceous host plants. In contrast to species with larval and pupal diapause, species with egg diapause are less diverse and are associated with cool habitats and woody host plants. The causes and effects of these relationships are far from understood and are probably multifaceted (see also Slansky 1974), but they point out that the diapausing stage is strongly related to, and probably evolves in conjunction with, other life-history parameters.

Insect life histories are subject to the influence of many seasonal factors; as seen above, seasonal variations in physical conditions and food are the most obvious, but certainly not the only, factors. Seasonally varying interspecific interactions also play major roles. Although they have not received sufficient attention, they pose extremely intriguing problems, as the examples below illustrate.

In contrast to most other species of insects, Collembola often overwinter in many instars, but diapause is restricted to one or a few instars. Thus, their population structure makes Collembola very useful in studying the two separate functions of hibernal diapause: (1) a seasonal timer of life-history events and (2) an aid in increasing winter survival (Leinaas & Bleken 1983). The surface-dwelling collembolan *Lepidocyrtus lignorum* reproduces in Norway mainly during autumn, and most individuals overwinter as diapausing eggs. Hatching occurs synchronously in the spring, producing a distinct first cohort. Other individuals overwinter in stages other than the egg, primarily as adults. Some of these overwintered animals lay nondiapause eggs in spring and summer; these offspring constitute the second cohort, which is unsynchronized and widely distributed over time. The egg diapause therefore does not function as a requirement for overwintering survival, but it does ensure that hatching occurs when conditions are most favorable for juvenile survival. The first cohort of *L. lignorum* hatches when competition from other surface-dwelling Collembola is lowest, and presumably as a result of this phenomenon, the mortality of juveniles in the first cohort is much lower than that of the second cohort, which must contend with keen competition from other Collembola. It thus appears that egg diapause in the first cohort serves primarily to time the life cycle so that survival during the nondiapause period is enhanced (Leinaas & Bleken 1983).

Interspecific competition may also play an important role in determining the species composition of communities that use a single resource. Such a phenomenon may occur in dung-inhabiting beetles (Hanski & Koskela 1979) and algae-grazing stream insects (Georgian & Wallace 1983). These communities exhibit a seasonal progression of generally distantly related species—each of which is adapted to using the same resource but at a different stage in the seasonal development of the resource. Although it is intuitive to assume that such resource partitioning reduces interspecific competition, it is important to point out that few studies have tested the assumption, and interspecific competition has not been established as the evolutionary basis for such seasonal organization (see Bradshaw & Holzapfel 1983). Certainly, additional studies in this area are needed, but care must be taken not to draw evolutionary conclusions without taking into account the phylogenetic history of the species involved in the community interaction (see Section 8.2.3).

Similarly, aestival-hibernal diapause in the machilus gall midge, *Daphnephila machilicola*, appears to be largely influenced by the seasonal occurrence of parasitoids. Most species of gall midges undergo diapause as final instar larvae in the soil or in the gall, but *D. machilicola* and a few other species that attack broad-leaved evergreens enter diapause as first-instar larvae. The persistence of the leaves of *D. machilicola*'s host permits the long, two- to three-year, larval diapause. Parasitization of developing larvae can be very intense, and the gall midge's variable two- or three-year life cycle apparently serves to spread the risks of development and emergence over several years (Maeda *et al.* 1982).

Interaction with seasonally occurring predators may also have a strong influence on the evolution of the diapausing stage and the seasonal timing of diapause in prey species. Populations of the copepod *Diaptomus sanguineus* that inhabit permanent ponds in Rhode Island begin producing diapause eggs in March, before predaceous sunfish become active, whereas conspecific populations of the copepods in temporary ponds that lack predaceous fish continue to produce both diapause and nondiapause eggs until July (Hairston & Olds 1984).

The adaptive advantage of pupal summer diapause that results in the winter emergence of several species of moths is difficult to understand unless we hypothesize an adaptation involving escape from strong predation pressure in summer (Masaki 1980). Waldbauer and Sheldon (1971) and Waldbauer *et al.* (1977) reported a complicated interspecific interrelation involving seasonal variation in predation and dipteran mimics. The adult flies avoid emergence during the summer when newly fledged, insectivorous birds are abundant and have not yet learned to shun the aculeate hymenopteran models. Here we see that the evolution of life cycles is affected by a three-way interspecific interaction involving predators, models, and mimics.

Predation appears to be an important factor in determining the seasonal cycles and diapausing stages of several predator and prey species of mosquitoes that occur together in tree holes (Bradshaw & Holzapfel 1984). Both the predaceous and nonpredaceous species exhibit seasonal variability that is correlated with increased or decreased predation. However, as Bradshaw and Holzapfel (1984) are careful to point out, studies, such as theirs, that compare unrelated, interacting species are not intended to demonstrate the role of predation in the evolution of dormancy. Rather, such studies illustrate the interaction between predation and seasonal traits in the life cycles of both predator and prey species.

Another fascinating example of interspecific effects on seasonal synchrony occurs in the underwing moths, *Catocala* spp. (Sargent 1978). These moths have cryptic forewings and boldly colored hindwings that function as startle devices after attack by birds. The startle effect of the hindwings stems from their ability to introduce anomaly (an element of surprise) to the predator, and the repeated occurrence of any one type of hindwing pattern would result in the habituation of the bird. This situation appears to have a role in maintaining the local ratios of *Catocala* species with diversely colored hindwings. The local stability is based on very closely correlated seasonal fluctuations in the abundance of certain pairs of species having one member chromatic (brightly colored hindwings, with or without bands) and the other achromatic (hindwings black or white only). Probably, in these moths interspecific interactions influence the evolution of seasonal cycles.

Among the insects there are numerous examples that illustrate the role of seasonality in reproductive isolation between closely related, sympatric species (see Sections 9.2.2 and 9.3). In some instances, such as the case of the green lacewing, *Chrysopa downesi*, selection for reproductive isolation from a commonly occurring, closely related, sympatric species appears to play a major role in shaping the overall seasonal cycle (Tauber & Tauber 1982; see Section 9.2.1.2). In others, seasonal isolation may be achieved largely as a by-product of selection for other characteristics such as synchrony with diverse hosts that have different seasonal cycles (see Section 9.3.3).

An interesting example of an unusual role that mating seasons may play in the reproductive isolation of host-associated species occurs in the phytophagous ladybird beetles, *Henosepilachna pustulosa* and *H. vigintioctomaculata*. Although the pathways involved in their speciation have not been adequately reconstructed, differences in their host plant preference and the low hatchability of hybrid eggs are the primary reproductive isolating mechanisms. What is unusual about this situation is that long, overlapping periods of mating, rather than restricted and separated reproductive periods, appear to enhance reproductive isolation. This mechanism works because females mate often, and sperm from an intraspecific mating before or after an interspecific pairing appear to take precedence. Thus, a lengthened mating period increases the chances of a fertile, intraspecific pairing in areas where males of the opposite species are common (Katakura 1982).

The examples above show that the evolution of seasonal adaptation is influenced by many physiological, ecological, and behavioral aspects of the life history. It is also probable that seasonal strategies, in turn, affect the evolution of other adaptations, seemingly unrelated to seasonality. Such an instance occurs in the defense mechanisms of arctiid moths. Some species are silent while others produce a high-frequency sound that may serve as an aposematic cue that discourages nocturnal predators such as bats. The silent species fly mainly in the spring, and sound-producing species fly in the summer; this seasonal succession coincides with the seasonal levels of activity by bats (Fullard 1977). Since the silent species have vestigial tymbals, a seasonal cycle enabling them to escape from predation by bats might have allowed them to reduce their dependency on sound as a defense device.

9.1.3 Summary

Seasonality is a major life-history trait whose effects, both positive and negative, can be measured in terms of fitness. Among insects diapause-mediated dormancy, migration, and polyphenism influence the entire life cycle—not just survival during periods of environmental stress, but also the timing and rates of growth, development, and reproduction during nondiapause periods.

One of the most important factors in shaping life histories is the evolution of a particular metamorphic stage for diapause. Most evidence indicates that the developmental stages of insects do not differ in their inherent capabilities to survive long periods in a state of dormancy or in their abilities to withstand physical extremes. Thus, these attributes do not appear to play a leading role in the evolution of the diapausing stage. However, one life stage may be more suited to the evolution of diapause because of certain ecological, behavioral, or morphological characteristics.

The primary selective factor during the evolution of the diapausing stage may be the synchronization of active stages and generations with the seasonal occurrence of certain requisites. Comparative studies have been particularly useful in elucidating the important role of synchronization with food and other resources in the evolution of the diapausing stage. Interspecific competition, interspecific interactions with predators and parasitoids, and reproductive isolation can also have major roles.

9.2 Seasonality in the Evolution of Life History

Demographic studies of seasonal selection on life-history traits may predict the evolution of life histories, albeit on a local and somewhat static basis (Stearns 1983; Tuomi *et al.* 1983). In addition, life-history theory has recently become concerned with the interaction between genes and the locally varying environment. Thus, the goal of life-history studies has expanded to include the prediction of responses to selection in heterogeneous natural environments over long periods of time (Istock 1981; various authors in Dingle & Hegmann 1982; Stearns 1981, 1983). Again, seasonality, as it varies from year to year and among widespread areas, plays a central role in the dynamic interactions over time and space that determine fitness, and thus its study is crucial to our understanding of the evolution of life histories.

Much of our ability to predict life-history evolution is based on optimality models—that is, models in which reproductive effort and life-history traits in general are optimized by maximizing fitness under purely demographic forces of selection. Although useful in deriving predictions, this method of analysis has inherent problems for which alternative solutions have only recently been proposed (e.g., Tuomi *et al.* 1983). Of primary importance is the incorporation of an ecophysiological approach into evolutionary theory dealing with life histories. Such a step takes into account inherent (both genetic and physiological) constraints on the coevolution of life-history traits, and it also recognizes the importance of genetic and physiological couplings between life-history traits. Because selection acts on whole organisms, not on separate traits (see Stearns 1983; Tuomi *et al.* 1983), the approach should be both holistic and realistic. Within this context we analyze the role of seasonality in life-history evolution and speciation. Only through such analyses can the connection between genes and phenotypes be made (see Stearns 1983). Elucidation of this connection is necessary for a full understanding of the evolution of life histories—in fact, for a full synthesis of evolutionary theory itself.

9.2.1 Coevolution of life-history traits

Intuitively, it seems reasonable to expect some developmental, reproductive, or survival costs, as well as benefits, associated with a physiological trait such as diapause. Diapause simultaneously involves energy consumption (storage of metabolic reserves, complex pre- and postdiapause behavior patterns) and energy conservation (reduced metabolism and suppressed development and reproduction). Furthermore, the role that diapause plays in timing and coordinating numerous other life-history phenomena provides an added reason to anticipate costs as well as benefits from its evolution and expression.

Such costs can be direct, involving, for example, a reduced level of somatic investment in reproduction, or they can be indirect and occur as the result of increased vulnerability to some critical factor in the environment during diapause. It is also possible that diapause incurs no reproductive or other costs, as, for example, when input of resources is increased to compensate for energy consumption during diapause. The types of costs, or the lack of costs, associated with diapause are rarely measured, despite the fact that quantification of the relationship between seasonality and other life-history traits is essential to an understanding of the evolution of life histories.

In addition to expecting costs associated with diapause, it is also intuitively sound to expect that seasonal and other life-history traits coevolve in a positive manner, so that improvement in one trait simultaneously enhances another. Evolution along this line would presumably produce life histories comprising sets of coordinated traits that interact to adapt populations to their physical and biotic environment. Unfortunately, few studies have examined quantitatively the interaction between seasonal and other life-history traits, but those that have done so demonstrate some very interesting relationships. They provide solid groundwork for future, intensive studies in the area (see Dingle *et al.* 1982).

9.2.1.1 Costs of seasonal adaptations

The previous sections (9.1.1 to 9.1.3) clearly demonstrate that diapause is an important life-history trait with two major forms of benefits. First, it enables individuals to prepare for and endure rigorous seasonal conditions. Second, it subserves the timing of active stages and generations with favorable seasonal conditions. Thus, diapause contributes to fitness—the measure of life-history success. However, life-history theory predicts that the benefits associated with a life-history trait such as diapause often incur cost.

Other than the obvious delay in development and reproduction inherent in the diapause syndrome, several species exhibit conspicuous trade-offs between life-history traits before and after dormancy. Prediapause larvae of the milkweed leaf beetle, *Lapidomera clivicollis*, reared under short day lengths can successfully pupate at a significantly smaller body size than nondiapause larvae reared under long day lengths (Palmer 1982, 1983). Thus, the prediapause larvae do not require as much energy at the end of summer when the food supply is deteriorating, but it is possible that the reduced body size diminishes postdiapause fecundity. In this case, attainment of the diapause stage appears to be achieved at the cost of postdiapause reproduction.

Photoperiodic and thermal conditions also influence prediapause fecundity in the lepidopteran pests of apple, *Laspeyresia pomonella*, *Grapholitha funebrana*, and *G. molesta* (Deseö 1973; Deseö and Sáringer 1975b). These species overwinter as mature larvae in diapause, and we presume that reduced prediapause fecundity is associated with larger or "better" eggs that yield larger or more hardy prediapause larvae. Aphids, too, pay large reproductive costs for diapause by switching from parthenogenetic to bisexual reproduction in order to produce diapause eggs. Each female can produce only a few eggs, and it appears that reproductive potential is reduced in favor of sexual reproduction and increased survival during winter. As a result of the trade-offs, the timing of the end of parthenogenetic reproduction is closely related to the conditions of photoperiod and temperature that affect leaf fall and the completion of the final generation of aphids (Ward *et al.* 1984).

The speckled wood butterfly, *Pararge aegeria*, has univoltine and bivoltine populations in Sweden that illustrate well the long-term costs and benefits of diapause. Bivoltine individuals, that is, those that develop without aestival diapause, have the opportunity of producing a number of offspring that will overwinter and multiply the following spring. By contrast, univoltine individuals, those whose larval development is interrupted by aestival diapause, do not reproduce before winter; they therefore have to survive for a longer period before they reproduce the following spring. This increased risk, however, is not without its benefits. Second-generation, bivoltine larvae face greater risk in entering diapause before winter than do univoltine larvae; they are smaller and therefore presumably produce smaller, less fecund, and less competitive adults than do univoltine larvae. Thus, the occurrence of aestival diapause involves trade-offs in terms of survival and reproductive potential (Wiklund *et al.* 1983).

Overwintering itself may affect fecundity. In many cases the effect is negative, but this is not universal, and among insects diversity is the hallmark in such trade-offs. Although forewing length (and therefore body size) is largest in the overwintering generation of the pear leaf miner (*Bucculatrix pyrivorella*), overwintered females produce significantly fewer eggs than the females of any of the following three summer generations (Fujiie 1980). Thus, in one way or another, hibernation is associated with a reduction in postdiapause reproductive potential of the females. Similarly, in *Heliothis zea* egg production by postdiapause females is significantly reduced (compared to that of nondiapause females), and the offspring of postdiapause parents require more time for egg and larval development than do the offspring of nondiapause parents (Akkawi & Scott 1984).

A well-documented and interesting example of sex-biased loss of reproductive potential during dormancy occurs in the flesh fly *Sarcophaga bullata* (Denlinger 1981b). Nondiapause females oviposit about 40 eggs, whereas females that have undergone diapause for 100 days under laboratory conditions produce only about 18 eggs per female. An even more drastic reduction (to approximately 10 eggs per female) is noted among females that overwinter under natural winter conditions. A long diapause period does not cause similar reduction in male fertility. Denlinger considers that these data explain the occurrence of a male-biased sex ratio in overwintering pupae. Males, compared to females, enter diapause earlier in the season (during August) and in response to less intense diapause-inducing conditions. Many females do not enter diapause in August but squeeze in an additional generation before winter. Since diapause results in a large drop in fertility, the additional nondiapause generation represents a substantial increase in reproductive output by these females. It also may enhance the reproductive output of their diapausing F1 daughters by reducing the length of time spent in diapause. Although this conclusion requires tests comparing the fertility of early-diapause (August) and late-diapause females, it is clear that overwintering takes place at a higher cost to females than to males and that the selection pressures acting on the seasonal cycles of the two sexes may be quite different. A similar mechanism may account for seasonal changes in the sex ratio of other insects, for example, the nymphalid *Acraea quirina* (see Owen 1974). However, Adesiyun and Southwood (1979) showed that the apparent imbalance in overwintering by male and female frit flies, *Oscinella frit*, is actually a case of sex-biased propensity to migrate. In the spring, males tend to remain at the site of emergence, thus increasing the chances of inseminating females. By contrast, mated females have a higher propensity to disperse to new crops.

Among other species, such as the Colorado potato beetle, *Leptinotarsa decemlineata*, there appears to be no reduction in reproductive potential after overwintering (de Wilde & Hsiao 1981). Moreover, in the acridid *Tetrix undulata*, fecundity and fertility are higher for females after hibernation than for those that have not overwintered (Poras 1976). Perhaps these responses are related to the large food requirements of the adults of these species during their postdiapause preoviposition period. Thus, trade-offs in the allocation of food reserves between reproduction and dormancy may be influenced by adult feeding requirements.

Denlinger (1979) proposes that pupal diapause in tropical flesh flies formed the evolutionary basis for long periods of dormancy in temperate zone species. Long periods of dormancy such as those encountered in temperate species require a major supply of stored energy, and during pupal diapause this energy is reallocated from metabolic reserves that could otherwise be used by adults for egg maturation. Denlinger's suggestion is consistent with the observation that tropical flesh flies carry sufficient reserves through the pupal stage to enable them to mature a large batch of eggs without first taking a proteinaceous meal (autogenous reproduction). In contrast, their temperate relatives either completely lack the autogenous habit or have only limited capacity to produce eggs without a proteinaceous meal. Thus, it appears that in the flesh flies, dormancy of sufficient duration for adaptation to the temperate zone evolved at the expense of autogenous reproduction (Denlinger 1979).

An analogous situation may occur in temperate-zone mosquitoes that undergo diapause as adult females. In these insects, autogeny (the ability of adult females to undergo an initial gonotrophic cycle without a proteinaceous meal) occurs seasonally (Moore 1963; Spielman 1971; Spadoni *et al.* 1974). Autogenous females do not enter diapause; they either do not survive the winter or they breed continuously throughout the winter in sheltered sites (Spielman 1971; Nayar *et al.* 1979). Apparently only anautogenous females enter diapause (Spielman 1957, 1971; Spielman & Wong 1973a; Vinogradova 1960). Like diapause, autogeny in mosquitoes is under environmental and genetic control (Spielman 1957, 1979; Vinogradova 1960, 1961; Harwood 1966; Kalpage & Brust 1974; Nayar *et al.* 1979; Trpis 1978). It is tempting to speculate that diapause evolved in female mosquitoes at the expense of their ability to produce eggs without an external source of protein, or vice versa—autogeny evolving at the expense of diapause. Examination of these hypotheses requires further study involving comparisons of closely related species and strains of mosquitoes with known or readily inferred phylogenetic relationships (see Section 8.2.3).

The evolutionary pathway involving trade-offs in the allocation of resources between dormancy and reproduction, although possible for the flesh flies and some mosquitoes with adult diapause, is probably not universal among insects. In several species of mosquitoes that enter diapause as larvae, autogeny appears to be derived from anautogenous reproduction, and its evolution appears to involve primary trade-offs with fecundity rather than dormancy. However, even in these cases diapause may have an important influence on reproduction. For example, in *Wyeomyia smithii* the evolution of autogeny resulted in the commitment of almost all storable larval resources to adult reproduction, apparently at the expense of adult maintenance (Istock *et al.* 1975; Moeur & Istock 1980). The storage of nutrients for oögenesis occurs primarily in the fourth larval instar, which is also the stage that undergoes diapause. Thus, the evolution of autogeny in *W. smithii* involves interplay between the duration of the fourth larval stage, a variable larval food supply, the age at first reproduction, and fecundity.

An analogous situation appears in the salt-marsh sandfly *Culicoides furens*. The prediapause larval period of this species is prolonged, but the overwintered larvae produce adults that are larger and more fecund than those produced later in the year. In addition, the ability of females to develop eggs without a blood meal is greatest after winter, suggesting that high levels of nutrients are accumulated during the prolonged prediapause larval stage (Linley *et al.* 1970). Thus, prediapause larval growth rates and the storage of nutrients during diapause are important factors in determining the reproductive potential of postdiapause adults.

The cases cited above illustrate the types of trade-offs between diapause-mediated dormancy and various life-history traits. In addition, there are well-illustrated interactions between life-history traits and other seasonal adaptations, particularly diapause-mediated seasonal migration (see Dingle 1979, 1982). Among closely related migratory species, there is often a direct correlation between body size and the migratory habit. Large species are much more likely to be migrants than their small relatives, and even within species, populations of small individuals are less likely to exhibit migration than are populations of large individuals (see Dingle *et al.* 1980b).

Migration in insects appears to involve trade-offs in the allocation of energy for migration, on the one hand, and reproduction, development, and perhaps survival on the other. Flight polymorphisms illustrate the trade-offs well. Flightless morphs generally hold an advantage over winged morphs in two aspects of reproduction: fecundity and age at initiation of reproduction (Dingle 1982). Numerous cases of the association of reproductive advantage with flightlessness are known (Harrison 1980); noteworthy are studies with *Drepanosiphum dixoni* (Dixon 1972c) and *Pteronemobius taprobanensis* (Tanaka 1976). Apparently, energy that would otherwise be devoted to wing formation and flight is redirected by wingless morphs into other processes affecting reproduction and survival.

Interrelationships between diapause-mediated seasonal migration and other life-history traits were measured by Dingle (1981) for three geographical populations of *Oncopeltus fasciatus* milkweed bugs. Under short-day conditions, the population from the north-temperate region (Iowa) has a strong tendency to delay reproduction and migrate. Conversely, the population from Puerto Rico showed a high tendency for immediate reproduction, without diapause or large-scale migration. Low levels of short-range movement are maintained in the Puerto Rican population, so that the bugs can move between asynchronously available patches of food. In contrast to the other two populations that occur in areas of predictable climate, the Florida population is faced with an unpredictable environment. Variation occurs among years and among sites. In this case the trade-offs between reproduction and diapause-mediated migration are obvious. Even under long-day (diapause-averting) conditions, the Florida adults take almost twice as long as the Iowa and Puerto Rican adults to initiate oviposition, and a proportion of them enter reproductive diapause. Other measures of reproductive potential (e.g., number of egg clutches, total fecundity) are also lower in the Florida population, but this is correlated with greater adult survival. The results indicate that the effects of natural selection on life-history traits of local populations are inextricable from and are balanced against the effects on seasonal and other traits (e.g., behavioral responses).

Trade-offs between life-history, seasonal, and behavioral traits, similar to those described above, have recently been recognized as of major importance in the evolution of eusociality in haplodiploid insects (see Seger 1983; Brockmann 1984; Section 6.2). Voltinism (partial bivoltinism), sex-ratio biases, and the overwintering stage all appear to interact in the delicate balance of factors that encourage or discourage the evolution of social behavior. An interesting prediction comes from models of parental investment under seasonal conditions—that is, that eusocial behavior is more likely to evolve among haplodiploid species that overwinter as inseminated females than it will among species that overwinter as larvae or unmated adults (Seger 1983). This prediction holds quite well for most groups of Hymenoptera (see Brockmann 1984).

9.2.1.2 Coordination of life-history traits

The above examples give evidence that seasonal and other life-history traits are interdependent and that the benefits of diapause often are balanced by important costs that are met at the direct or indirect expense of other life-history traits. Conversely, reproductive and developmental potentials are enhanced by dormancy and other seasonal adaptations. Thus, given the costs and benefits of seasonal adaptations and the way in which seasonal traits are intertwined with other life-history traits, it is not at all surprising that life histories comprise sets of both antagonistic (e.g., see Rose 1983) and coordinated characters. Seasonal traits are balanced with each other to form adaptive but constrained seasonal cycles, and they, in turn, function in an interdependent, antagonistic manner with other life-history traits to form locally adaptive life cycles.

Several studies have demonstrated the interplay between seasonal and life-history traits as they evolve to form adaptive life cycles. One study examined two species-groups of Japanese crickets that have latitudinally correlated variation in voltinism and overwintering stage (Masaki 1978a). The transition between geographically varying populations involves stepped variation in the photoperiodic induction of diapause, the overwintering stage, diapause intensity, photoperiodic control of nymphal growth rates, and adult size (Kidokoro & Masaki 1978; Masaki 1978a, 1979). In each group, hybrids between egg and nymph overwintering forms are unlikely to survive in nature. By contrast, hybrids between the univoltine and bivoltine forms overwinter as eggs and appear to be well adapted to intermediate conditions. In this case, two alternate life cycles (univoltine and bivoltine) may occur in the transition zone. Apparently, reciprocal interaction between the various seasonal and life-history traits balances the life cycle in hybrids (Kidokoro & Masaki 1978).

A similar form of reciprocity has been demonstrated for traits underlying diapause in the grasshopper, *Oedipoda miniata* (Pener & Orshan 1980, 1983), and in the alfalfa blotch leaf miner, *Agromyza frontella* (Tauber *et al.* 1982). In these insects, thermal modification of the photoperiodic maintenance of diapause confers a high degree of synchrony between diapause termination and favorable local conditions. Such phenotypic flexibility in the interaction between diapause controlling traits may reduce selection pressure on individual diapause traits, and thus, geographical strains of *O. miniata* and *A. frontella* may not vary in their critical photoperiods for diapause maintenance (Pener & Orshan 1980, 1983; Tauber *et al.* 1982).

Extensive studies with the *Chrysopa carnea* species complex in North America also illustrate the coordination of seasonal and life-history traits in the evolution of geographically adapted populations (Tauber & Tauber 1982). The characters involved include the critical photoperiod for diapause induction, the depth of diapause, the thermal requirements for initiation of postdiapause oviposition, preimaginal developmental rates under a wide range of temperatures, influence of prey on the duration of the preoviposition period, and the photoperiodic and dietary requirements for nondiapause reproduction. The results are consistent with the conclusion that a variety of diverse, often opposing, selective forces cause the life-history traits to interact in forming adaptive life cycles.

Each *Chrysopa* population is characterized by a specific set or sets of life-history traits that adapt it to the physical and biotic factors in its particular locality and habitat. Selection for one character (e.g., photoperiodically controlled univoltinism) can strongly affect numerous other life-history traits (e.g., growth rate, fecundity, and oviposition schedule). In some instances, as in a population from Alaska, adaptation to a very harsh physical environment dominates the shaping of the life cycle and the underlying life history traits. Under the milder physical conditions of Strawberry Canyon, California, a locality where the availability of food varies considerably, reproduction is keyed to the seasonal occurrence of prey. In this case, the life-history traits subserve a polymorphic response to variable prey levels and constitute a "best-bet" strategy in an unpredictable environment. Similarly, seasonal occurrence of prey apparently influences the timing of vernal reproduction in the univoltine population from St. Ignatius, Montana.

In sympatric populations of *C. carnea* and *C. downesi* in the northeastern United States, diverse selection pressures appear to influence the evolution of life-history traits. In *C. downesi*, the life cycle is largely adapted to prevent interbreeding with the more common and dominant *C. carnea*. In *C. carnea*, the life cycle is not influenced by sympatry with *C. downesi* but is primarily adapted to local climatic conditions.

Several genetic mechanisms underlie the geographical variability in life-history traits of the *Chrysopa carnea* species-complex. In one situation, the photoperiodic responses that severely limit reproduction (*C. downesi*'s univoltinism) result from a simple Mendelian system, involving recessive alleles at two autosomal loci (see Section 7.2.2.2). More complicated genetic systems, involving several genes with dominance and recessiveness, regulate other traits, such as the prey-mediated aestival diapause in the Strawberry Canyon, California, population. Still other traits, such as the geographically variable critical photoperiod for diapause induction in *C. carnea*, apparently come under the control of multiple genes having additive effects.

Life cycles involving variation in sets of elaborate life history and seasonal traits are known for many other species. Among populations of the mosquito, *Aedes aegypti*, differences in oviposition behavior are associated with the degree of delayed hatching (Gillett 1955); in *Aedes triseriatus* embryonic and larval diapause are part of an integrated, adaptive, and finely tuned developmental continuum (Holzapfel & Bradshaw 1981; Sims 1982); in *Papilio polyxenes*, body size, thermoregulatory ability, reproduction rates, and the ability to enter photoperiodically controlled diapause are all intertwined in forming locally adaptive life histories (Blau 1981). Aphid life histories involve numerous interactions between traits such as fecundity, body size, migratory ability, sexual and asexual reproduction, and the ability to produce diapause eggs (e.g., see Dixon 1976a; Dixon & Dharma 1980; Wellings *et al.* 1980). The above and other studies by Tamura *et al.* (1959), Kishino (1970a, b), Shapiro (1976), Bradshaw and Lounibos (1977), McLeod *et al.* (1979), and Jordan (1980a, b) indicate that the various characteristics comprising seasonal adaptations evolve in an interrelated way with other life-history traits to form coordinated but constrained systems. Alterations in any one character cause greater or lesser changes in the others. The result is an overall adaptation to the physical and biotic characteristics of the environment.

Although some studies have shown genetic correlation between characters comprising the coordinated sets of traits (Fontana & Hogan 1969; Istock *et al.* 1976; Istock 1978, 1981; Hegmann & Dingle 1982), some characteristics appear to be genetically independent of each other (Honek 1979a; Kidokoro & Masaki 1978; Masaki 1978a, b, 1979; Tauber & Tauber 1982, unpublished). In all cases studied, natural selection appears to play a key role in maintaining the adaptive systems of coordinated traits in local populations (Dingle *et al.* 1980a, b; Istock 1981; Tauber & Tauber 1982; see also Ehrlich 1965; Ehrlich & White 1980).

9.2.2 Maintenance of genetic variability

Studies of life-history traits, in general, demonstrate abundant additive genetic variance. For example, prediapause developmental rates, the duration of diapause, and the critical photoperiod, among other traits, are readily altered by artificial selection and show high parent-offspring regression (Danilevsky 1965; Istock *et al.* 1975; Dingle *et al.* 1977; Hoy 1978a; see Chapter 7). These high levels of additive genetic variance are contrary to those expected for characters that determine fitness because, theoretically, under evolutionary equilibrium directional selection would lead to fixation of such critical life-history features (see Istock 1978, 1982; Rose 1983). Thus, evolutionary biologists are faced with the problem of examining the types of variability inherent in life histories (see Smith-Gill 1983) and explaining how this variability is maintained under natural conditions (see Giesel 1976; Istock 1983). Moreover, because the extent of genetic variation in these characters defines the limits to future patterns of evolution, high heritabilities in life-history traits indicate that life histories are subject to considerable alteration through natural selection. Such variation comprises the "raw material" for the evolutionary process (Istock 1981). Thus, analysis of life-history variation and the mechanisms involved in its maintenance are essential for the synthesis of Mendelian, quantitative, and ecological genetics into a comprehensive evolutionary theory.

A previous chapter (Chapter 7) discussed the types of variability found in seasonal cycles, including both environmentally controlled and nongenetic phenotypic flexibility in seasonal cycles, as well as genetically based variation. Here we ask how is the genetic variation maintained, and what role does seasonality play in this process? Several mechanisms have been proposed to account for the maintenance of genetic variation in life-history traits (e.g., Bradley 1982), and significant among them is variable selection in a seasonally uncertain environment. Such a situation, in some cases, favors the evolution of "bet-hedging" tactics—tactics that require and provide for the maintenance of sufficient genetic variation to allow expression of alternate developmental and reproductive pathways within a single population. Examples of such polymorphic life histories were given in Chapter 7, and we discuss a few in detail below.

In other cases, variable selection appears to favor the maintenance of genetic variability through diversification of subpopulations. The extent of diversification varies among species, and below (Section 9.2.2.2) we compare several insects that illustrate the range.

9.2.2.1 Bet-hedging in seasonally unpredictable environments

Several studies have correlated dispersal ability and the degree of instability of the habitat with the risk of population or subpopulation extinction (e.g., den Boer 1981). However, it should be emphasized that such correlations may be limited because many insects have a prolonged diapause in which a relatively small proportion of the population remains inactive for periods greater than a year (c.f. prolonged diapause in arid-region insects, Section 6.4.2). This characteristic is an excellent, but sometimes ignored, example of the bet-hedging or risk-spreading tactic. Individuals undergoing prolonged diapause have greatly reduced fitness in normal years (i.e., they produce no progeny), but in the event of a catastrophe to the active population, they alone, or in combination with immigrants, may serve as the source of population restoration (e.g., Takahashi 1977; Wise 1980; Danks 1983). In such a situation, prolonged diapause is probably maintained by selection during the occasional unfavorable years, but the genetic basis for the characteristic and the mechanisms involved in determining the incidence of its occurrence are poorly understood.

A relatively well-studied case of polymorphism for prolonged diapause involves the cecidomiid midge, *Hasegawia sasacola*, which forms galls in bamboo buds. The availability of these buds varies greatly from year to year, and thus local populations are faced with a high degree of seasonal uncertainty. Moreover, after flowering (which occurs at unpredictable intervals of about 50 years), the bamboo plants die synchronously, leaving local populations of the midge without a food source for a long period. Survival of the population depends on those individuals that have entered prolonged diapause and thus are able to withstand the host-free period (Sunose 1978).

An interesting trade-off appears to occur in the Swaine jack pine sawfly, *Neodiprion swainei*, in which prolonged diapause is associated with reduced susceptibility to attack by parasitoids. In a study in the Gatineau River area of Quebec, a small percentage (about 4%) of the sawfly cocoons were found to enter prolonged diapause. Unlike other cocoons, none of these cocoons were parasitized, indicating that the parasitoids were unable to find and oviposit in them. It is possible that the cocoons in prolonged diapause were buried more deeply in the forest litter than other cocoons and were therefore hidden from the parasitoids (Price & Tripp 1972; see also Minder 1973). In any case, the disadvantages associated with foregoing reproduction for a year or more may be offset by increased survival during periods of heavy parasitoid activity.

Prolonged diapause is not the only way in which insects utilize diapause as a bet-hedging device. Genetic variability for the induction of diapause over the season is another means. Many insects exhibit a partial generation at the end of their growing season. In some years this extra generation may be successfully completed, but in those years in which it is not, the part of the population in diapause serves as the sole source of the next year's generation (e.g., Tyshchenko *et al.* 1983).

The milkweed bug, *Oncopeltus fasciatus*, which is a long-distance migrant, has been shown to have considerable additive genetic variance in a number of life-history traits including diapause induction and wing length (Dingle *et al.* 1977, 1982; Hegmann & Dingle 1982). High additive variance in the traits of parental populations results in the intermediate expression of the traits in offspring, and variable selective regimes apparently prevent the traits from becoming fixed. Such a situation seems to occur in the absence of consistent selection for a particular adaptive response (e.g., when there is between-year variation in selection pressure) or when a species must scan an array of shifting environments (e.g., over a geographical range).

In some parts of Florida, *O. fasciatus* can reproduce all year during some, but not all, years. Additive genetic variance in these Florida populations underlies a bet-hedging strategy in which reproduction, with its potential risks and benefits, is balanced against delayed reproduction and survival during harsh periods. The genetic variance underlying the polymorphic life cycle apparently is maintained by seasonally variable selection emanating from the unpredictability of the environment and by interbreeding between geographically diverse populations that migrate into the region (Dingle *et al.* 1977; Miller & Dingle 1982).

The nonbiting mosquito *Wyeomyia smithii* presents an especially well-studied case in which both the genetic and physiological bases for intra-and interpopulation variation are well known (see summary by Istock 1981). Populations of this species exhibit mixed voltinism—from a single generation to three generations per year; diapause induction is determined by the environmental stimuli of photoperiod and food concentration and by the genetic makeup of the individual. Those individuals with an "extreme" diapause response enter diapause soon after the summer solstice and produce only one generation per year. Those individuals at the other end of the spectrum not only have a lower propensity to enter diapause but also are much faster developers, and thus they can produce three generations per year. Individuals between the two extremes exhibit mixed responses and produce two or three generations per year. Such variation in nature results in a bet-hedging life-history strategy. The extreme diapausers provide a margin of safety by entering diapause early in the season and remaining prepared for environmental extremes all through the summer.

Laboratory tests with field-collected individuals indicate that the full range of variation is expressed in the progeny of single females. However, the proportion of highly diapause-prone individuals among families declines through the season because individuals with high gene dosage for diapause enter diapause and leave the reproductive pool early. Both genetic segregation and recombination, mediated by sexuality, are essential in maintaining this type of genetic variation. Within such a framework, fluctuating selection for a shifting and intermediate amount of diapause appears to prevent the loss of genetic variation (Istock 1981).

Geographic populations of *W. smithii* have evolved life-history specializations encompassing the same diapause and developmental traits that vary within populations. Hybrids formed by crossing individuals from latitudinally diverse areas (e.g., Ontario and Florida) do not express deleterious effects in their primary fitness characters. From this, it is concluded that latitudinal diversification and specialization are not leading to reproductive isolation or speciation in *W. smithii* (Istock 1981).

9.2.2.2 Genetic diversification in seasonal environments

In contrast to the above examples in which fluctuating selection has resulted in bet-hedging and the polymorphic expression of diapause within an interbreeding population, some species appear to maintain variation along diversified lines, with varying degrees of reproductive isolation among the lines. The effects of such differences in mating systems (i.e., between panmictic populations and populations with reproductively separate lines are illustrated by a comparison among diverse host-specific insects in which synchronization with seasonally variable resources is achieved. The seasonal cycle of each of these insects is closely synchronized with that of the host plant, but in each case the seasonal occurrence of the host is only partially predictable. Thus, the insects are faced with the problem of maintaining synchrony with a seasonally variable resource, and each of the species responds in a very different way.

Our first example involves a panmictic breeder, the sycamore aphid, *Drepanosiphum platanoides*, in which hatching of overwintering eggs is highly correlated with bud-break. In autumn the sexual adults mate and deposit diapausing eggs on the bark of sycamore trees. The eggs hatch in the spring in conjunction with renewed growth of the trees. The most nutritive leaves are the youngest. Aphids on unfurling leaves achieve higher adult weight and maintain a greater reproductive output than those on older leaves, and their offspring can mature faster and achieve greater size than those on older leaves. Aphids that hatch out of synchrony with bud-break suffer a high degree of mortality. Thus, synchronization of egg hatch to bud-break is clearly advantageous (Dixon 1976b).

However, not all sycamore trees renew growth at the same time in the spring, and aphids must cope with the seasonal variability in the occurrence of their food source. Sycamore aphids appear to have overcome this dilemma by maintaining a relatively large amount of variability in the timing of egg hatch. Most aphid eggs hatch at the average time of bud burst, but variability in the timing of hatch encompasses both the earliest and latest budding trees. Early egg hatch is advantageous if the eggs are on early-opening sycamore trees; offspring of early maturing aphids can colonize a series of trees whose leaves are just unfurling, thus enabling the early aphid and her offspring to exploit a favorable food source for a long period and to produce a large number of offspring. However, early aphids that find themselves on late-opening trees succumb because of lack of food. Eggs that hatch late do well on late-opening sycamores because these trees are usually only very lightly infested (early-hatching aphids having succumbed). As a result aphids on late-budding trees often reach a larger size and are more fecund than aphids on heavily infested, early-budding trees. Thus, the high levels of variability in the seasonal cycle of the aphid *D. platanoides* appears to be maintained in response to fluctuating selection in an uncertain environment.

Like the sycamore aphid, the fall cankerworm, *Alsophila pometaria*, which attacks several species of trees in North America, also is adapted to the differences in the seasonal timing of bud-burst among its hosts. However, variability in the fall cankerworm is based on a very different genetic structure. Rather than maintaining large variability in hatching time within individual lines, this species is largely comprised of parthenogenetic, host-specific clones that are synchronized with the phenology of their specific hosts (Mitter *et al.* 1979). Disruptive selection, female winglessness, and absence of recombination in the parthenogenetic forms may aid in maintaining the diversified host-adapted clones, but this has not led to speciation.

In both of the above cases, seasonal synchrony with the host is achieved through the maintenance of high levels of variability in hatching time. In the sexually reproducing sycamore aphid, the variability apparently is expressed within the progeny of individual females and therefore probably is associated with relatively high levels of mortality. In the fall cankerworm, seasonal variation is largely expressed among host-adapted clones—a situation that presumably reduces the levels of mortality associated with maintaining seasonal variability.

Variation in seasonal life histories has reached an extreme in aphid life cycles. As in the fall cankerworm, asexual reproduction appears to play a role in maintaining the variation, but in a different way. The life cycles of most aphid species consist of parthenogenetic clones, each initiated in the spring by a fundatrix (stem mother). During spring and summer there are several generations that differ morphologically and physiologically from one another. Each morph has a particular role to perform and usually develops in response to specific environmental cues (Dixon 1976a). Seasonal changes in food quality (host plants) can be critical in determining morph composition and the quality of successive generations. In autumn, sexual forms are produced, and mated females oviposit diapausing eggs that overwinter and give rise in the spring to the new stem mothers.

Considerable variation in fecundity occurs both within and between generations. Individuals from the spring generations have the highest number of ovarioles and the highest fecundity, and during subsequent generations the number of ovarioles varies genetically over a particular range (Dixon 1976a; Dixon & Dharma 1980; Wellings *et al.* 1980). Within this structure there appear to be trade-offs among fecundity, speed of reproduction, and survival during successive seasonal generations; the future condition of host plants is anticipated and fecundity is adjusted genetically to compensate for these seasonal changes. The successful clones are those that produce egg-laying females in autumn. Selection therefore acts on the clone as a *whole* through its short-lived units; variability among clones appears to be maintained by both temporally and spatially varying selection pressures through sexual reproduction (Dixon & Dharma 1980).

Seasonal variability along intraspecific lines also occurs in the mosquito *Culex pipiens*. Among populations of this species, autogenous and anautogenous cohorts are, at least in part, reproductively isolated from each other because of differences in their seasonal cycles and breeding areas (Spielman 1971, 1979). Autogenous females do not enter diapause, whereas anautogenous females do. The two cohorts also respond very differently to inbreeding: anautogenous *C. pipiens* normally outbreed in nature, are much less adapted to brother-sister matings, and presumably are much more variable generally than are the autogenous mosquitoes. Occasional episodes of intermating between the two types may occur, thus resulting in the transmission of genetic variation to the autogenous cohorts. This genetic variation then is subject to intense selection pressure in the normally inbred autogenous lines.

Strains of autogenous/anautogenous *Aedes aegypti* mosquitoes in Mombasa, Kenya, exhibit a somewhat similar form of intraspecific variation; as in *C. pipiens* gene flow between the anautogenous and autogenous forms is restricted (Tabachnick *et al.* 1979). Apparently, seasonal differences in reproductive period and the degree of habitat preference influence the degree of reproductive isolation between the cohorts.

In contrast to the above cases in which seasonal variability involves two types of intraspecific genetic variability, *Rhagoletis* fruit flies and *Enchen-*

opa treehoppers express their seasonal variability among host-specific species (Bush 1969, 1974; Wood 1980; Wood & Guttman 1982). The utilization of diverse hosts requires the evolution of seasonal synchrony with each host. While the fall cankerworm overcomes this problem through parthenogenetic reproduction of host-adapted lines, *Rhagoletis* fruit flies and *Enchenopa* treehoppers do so through the evolution of reproductive isolation among host-associated races. These examples are discussed more fully in the next section.

9.2.3 Summary

Studies that combine ecophysiological and genetic analyses of life histories provide the basis for an evolutionary synthesis of life-history theory. Quantification of the relationships between seasonality and other life-history traits, as well as an understanding of the limits of genetic variation in life-history traits, is essential to this goal.

Life-history traits coevolve in a variety of ways. In some cases, there appears to be an antagonistic relationship between characters, with one evolving at the expense of another. In these instances, diapause entails large costs in reproductive potential. Such trade-offs between seasonal and other life-history traits occur in the evolution of diapause-mediated dormancy and migration, and they are evident in the evolution of autogenous reproduction. In other cases, seasonal traits appear to evolve relatively independently of other major life-history traits. Examples include species with large postdiapause food requirements.

Finally, life-history traits evolve in a coordinated fashion to form adaptive, but constrained, life cycles. Studies of the life histories of species with geographically diverse populations indicate that life-history traits are often free to vary independently but that natural selection causes them to coevolve and form locally adaptive life histories. In some cases the coordination involves simple quantitative adjustments of a few characters. In others, it involves elaborate seasonal, morphological, and other life-history traits.

The patterns of life-history evolution are defined by the extent of genetic variation in the life-history traits. Several studies have demonstrated abundant additive genetic variance in life-history characters, such as the critical photoperiod for diapause induction, the duration of diapause, and prediapause developmental rates, but the factors that maintain the variability have scarcely been examined. In a few cases, variable or fluctuating-stabilizing selection in a seasonally uncertain environment favors the evolution of "bet-hedging" tactics that allow the expression of alternative developmental and reproductive pathways within a single population. In other cases, gene flow between geographically diverse populations may play a role. Finally, in some populations variability is spread among diversified subpopulations—with or without the evolution of reproductive isolation.

9.3 Seasonal Cycles and Speciation

The previous section illustrated that the earth's seasonal cycles have an all-encompassing effect on the evolution of life histories. They can exert directional selection, as evidenced by geographical differentiation of life histories, and fluctuating selection, whose action is detectable among populations that are polymorphic in their life histories. They might also exert disruptive selection and thus cause divergence and speciation.

In considering the role of seasonal cycles in speciation, we analyze the problem in the context of three different models of speciation that involve natural selection as a driving force: allopatric, parapatric, and sympatric speciation (Bush 1975a; Endler 1977; White 1978; Barton & Charlesworth 1984). In so limiting our discussion, we exclude several models of speciation that are based largely on genetic drift or on the nature of the genetic system itself (White 1978; Templeton 1980); we also exclude sexual selection from discussion (see Nei *et al.* 1983; Wu 1985). However, we include an additional consideration—allochronic speciation—a form of sympatric speciation that is especially pertinent to our discussion of the role of seasonal cycles in speciation.

9.3.1 Allopatric speciation

Allopatric speciation entails the evolutionary divergence of physically (usually geographically) separated populations through adaptation to their respective environments. Under these circumstances, diversification is not a direct result of natural selection for differentiation; it occurs secondarily as the geographically separated populations adapt to their diverse environments. Likewise, reproductive isolation may evolve as a secondary result of the genetic divergence of the separate populations. In general, allopatric populations are considered reproductively isolated when divergence has been sufficient to produce deleterious effects on the primary fitness characteristics of hybrid populations (see Mayr 1963a; Istock 1981).

The seasonal cycle is only one of many traits that undergoes divergence in geographically separated populations. Nevertheless, because of its pervasive effects, it probably contributes in a major way to the divergence and speciation of allopatric populations. It has been proposed that allopatric speciation involves evolutionary change in polygenic life-history characters with relatively high additive genetic variance (Mayr 1963a; Lande 1980). This statement implies that quantitative changes in seasonal traits, such as the polygenically controlled critical photoperiod for diapause induction and the duration of diapause, are involved in this mode of speciation (Tauber & Tauber 1978, 1982). However, it does not exclude interacting, qualitatively varying traits from also having an important role.

The rate of diversification during allopatric speciation depends on (a) the strength of natural selection affecting adaptation (e.g., the degree of climatic differences between localities), (b) the pattern of genetic variation within the populations, and (c) the correlation among characters (see Lande 1980). It is likely that reciprocal interaction among coordinated life-history traits, including seasonal traits (such as demonstrated by Kidokoro and Masaki 1978 and Pener and Orshan 1983), or phenotypic modulation of life-history traits (such as described by Smith-Gill 1983), would reduce the impact of natural selection. These factors may contribute to the very slow rate of diversification frequently observed among allopatric populations (Ehrlich & Raven 1969).

As stated above, allopatric populations may evolve reproductive isolation as a result of their divergent adaptation to different geographical (climatic) conditions. Such appears to be the case in the autumnal and vernal biotypes of the weevil *Ceuthorrhynchus pleurostigma* in the Netherlands. The vernal biotype apparently originated in northern Europe and is associated with spring hosts, whereas the autumnal biotype presumably originated in southern Europe and oviposits on hosts that grow throughout autumn and winter. The autumnal biotype has been introduced into northern Europe, but the two biotypes have remained reproductively isolated largely because of differences in their seasonal cycles that were present before the two biotypes became sympatric (Ankersmit 1964).

Adaptation to diverse conditions is not the only means through which allopatric populations may evolve reproductive isolation. When two allopatrically differentiated populations come into secondary contact, they may experience strong selection pressure for the evolution of reproductive barriers. If the fitness of hybrids is sufficiently low, natural selection will favor the relatively quick evolution of seasonal or other premating reproductive isolating mechanisms (Mayr 1963a; Grant 1966). The evolution of seasonal asynchrony during secondary contact can involve qualitative changes in the life cycle over a relatively short time, depending on the strength of natural selection and on the type and amount of genetic variation in the populations (Tauber & Tauber 1978, 1982). Secondary contact of allopatrically differentiated populations may also result in character displacement (Brown & Wilson 1956); in such a situation allochrony is a possible mechanism for reducing competition among the populations in secondary contact. In addition, frequency- or density-dependent selection is likely to accelerate the rate of divergence and to act in the maintenance of differences in such cases (Rosenzwieg 1978; Gibbons 1979; Mani 1981).

The evolutionary scenario of allopatric speciation is difficult, if not impossible, to distinguish *post facto* from parapatric speciation (Endler 1977). In fact, because allopatric speciation is characterized by relatively slow rates of evolutionary change in many interacting and independent traits—both morphological and physiological—it is difficult to demonstrate the importance of any one facet (e.g., seasonal adaptation) to the process. This area of evolutionary biology is in need of bold, imaginative research.

9.3.2 Parapatric speciation

Parapatric speciation entails the evolution of clinal diversification and reproductive isolation among populations along a geographical gradient (Endler 1977). Because the seasonal environment is one of the most important geographically variable factors that affect life-history and other traits, adaptation to local seasonal conditions along a cline may play a leading role in parapatric speciation. The effect of the seasons on diversification is so striking that the term "climatic speciation" has been used to describe parapatric speciation in certain geographically widespread crickets that have considerable clinal variation in their seasonal adaptations (Masaki 1978a, 1983). Numerous other examples of seasonal variation along geographical clines are found in Section 7.1.2.

In parapatric speciation, reproductive isolation may evolve as a by-product of microevolutionary adaptation to local conditions (usually postmating isolation) and/or as a direct result of natural selection (usually premating isolation). Thus, both relatively slow changes in quantitative, polygenically controlled seasonal traits and relatively fast, qualitative changes in seasonal cycles may contribute to the evolution of reproductive isolation among parapatric populations. The rate of diversification under parapatric conditions depends on (a) the strength of natural selection (including the degree of the climatic differences among local areas and the relative fitness of hybrids between individuals from seasonally diversified populations), (b) the degree of movement (gene flow) between localities, and (c) the pattern of genetic variation and correlation between characters (Endler 1977; Lande 1980). As in allopatric speciation, reciprocal interaction among seasonal traits is expected to reduce selection for reproductive isolation (Kidokoro & Masaki 1978; Masaki 1978a; Pener & Orshan 1980), and density-dependent selection would increase it (Rosenzweig 1978). Genetic linkage of coordinated traits could also reduce selection for reproductive isolation (see Endler 1977) and could lead to the evolution of geographically variable polymorphisms [or geographically variable supergenic traits as demonstrated by Lumme (1978) and Lumme & Oikarinen (1977)], instead of speciation. However, the evolution of coadapted sets of life-history (and morphological) traits may promote or accelerate the speciation process.

Parapatric speciation probably constitutes the most common form of speciation (Endler 1977) and, particularly because of its previous neglect, it warrants further investigation. Of special importance is the analysis of the genetic control of geographically variable, coordinated seasonal traits. Particular emphasis should be placed on determining (a) the degree and type of genetic variability found in the locally adaptive, seasonal, and other life-history traits of adjacent and distant populations, (b) the phenotypic and genetic relationship between coordinated seasonal traits, (c) the relationship between clinally variable seasonal and life-history traits, and (d) the fitness of hybrids under natural conditions. Although reproductive isolation has not evolved between geographical populations of *Wyeomyia smithii*, the studies by Istock (1981, 1982, 1983) on this species serve as a paradigm. Studies on the Japanese crickets (e.g., Masaki 1978a) and the green lacewings of North America (Tauber & Tauber 1982) are also addressing significant questions concerning the role of seasonal cycles in the diversification of parapatric populations.

9.3.3 Sympatric speciation

During sympatric speciation, the two processes comprising speciation—diversification and the evolution of reproductive isolation—are somewhat more distinct than they are during either allopatric or parapatric speciation. That is, reproductive isolation does not evolve primarily as a by-product of adaptation to environmental conditions, but it evolves as a direct result of natural or sexual selection. Consequently, it has been proposed that sympatric speciation occurs relatively quickly (Maynard Smith 1966; Bush 1969; Tauber & Tauber 1977a, b) and that it may involve a form of frequency- or density-dependent selection (see Rosenzweig 1978; Mani 1981).

The first step in sympatric speciation is the establishment of a stable, two-niche polymorphism through the action of disruptive selection on relatively simple genetic differences (Maynard Smith 1966). The second step involves the evolution of premating reproductive isolation between the sympatrically differentiated morphs. Theoretically, diversification of seasonal cycles may contribute to both steps; however, evidence is strong only for its role in the second step.

Among animals, the first step in sympatric speciation (the establishment of a two-niche polymorphism) has been shown to involve diversification in host association (Bush 1969, 1974, 1975b; Phillips & Barnes 1975; Wood 1980; Wood & Guttman 1981; Tauber & Tauber 1985), spatial partitioning of a host (Gibbons 1979), and a shift in habitat association (Tauber & Tauber 1977a, b). Temporal (seasonal) partitioning of the niche or habitat has also been suggested as an initial phase of sympatric speciation (Alexander & Bigelow 1960; Alexander 1968; Southwood 1978). However, the evidence supporting this hypothesis has been challenged (see next section).

Evidence for the role of seasonal diversification in the second step of sympatric speciation (the evolution of reproductive isolation) is quite substantial. Seasonally asynchronous periods of sexual activity are important to reproductive isolation in host-specific fruit flies (Bush 1969, 1975b; Boller & Bush 1974), sawflies (Knerer & Atwood 1973), moths (Phillips & Barnes 1975; see Riedl 1983; Day 1984), treehoppers (Wood 1980; Wood & Guttman 1981), and lacewings (Tauber & Tauber 1985), and in habitat-specific lacewings (Tauber & Tauber 1977a, b). They may also contribute to reproductive isolation between habitat-specific scorpionflies (Sauer & Hensle 1977) and between detritus- and sponge-feeding cohorts of the caddisfly, *Ceraclea transversa* (Resh 1976a). In addition, they are partially responsible for the reproductive isolation between autogenous and anautogenous populations of *Culex pipiens* (Spielman 1971, 1979).

Both quantitative and qualitative changes in the seasonal cycle may contribute to the evolution of reproductive isolation between sympatric species. In the host-specific fruit flies, codling moth, and treehoppers, relatively small differences in the diapause characteristics appear to be sufficient to keep the divergent host races in synchrony with the different phenologies of their host plants and out of synchrony with each other (Bush 1969, 1974; Phillips & Barnes 1975; Wood 1980; Wood & Guttman 1982). Presumably, these changes result from natural selection acting on polygenic variation in diapause (perhaps diapause duration or the critical photoperiod for diapause induction). In contrast, reproductive isolation between *Chrysopa carnea* and *C. downesi*, two sympatric green lacewings in northeastern United States, is based on both qualitative and quantitative differences in their seasonal cycles. *C. carnea* is multivoltine and can reproduce continuously under the long day conditions of spring and summer in northeastern United States. *C. downesi* is univoltine—reproducing only once in late spring when *C. carnea* is not sexually active. *C. downesi*'s univoltine cycle is based on a short-day/long-day requirement for reproduction. These primary differences in the seasonal cycles of the two species result from allelic substitutions at two autosomal loci (Tauber *et al.* 1977). Other seasonal traits that are presumably polygenically controlled aid in fine-tuning the asynchrony between the two species (Tauber & Tauber 1977a, b).

Essentially, the evolution of seasonal isolation in sympatrically derived forms is similar to that in allopatric forms undergoing secondary contact. In both cases, the evolution of seasonal isolation goes hand in hand with seasonal adaptations to different hosts or habitats. In addition, both also involve direct natural selection for reproductive isolation between the differentiated forms and/or selection for reduced competition between the diverged populations (as discussed above). Distinction between these forms of speciation is difficult because they may occur through very similar mechanisms.

9.3.4 "Allochronic speciation"

The term "allochronic speciation" was originally proposed by Alexander and Bigelow (1960) to designate speciation based on the temporal diversification of a population without the intervention of geographic barriers. This form of speciation has been the subject of considerable debate (e.g., Alexander 1963; Mayr 1963a, b), and recently it has been proposed as a means of "species packing," through temporal partitioning of resources (Southwood 1978).

To differentiate it from other forms of sympatric speciation, we restrict the term "allochronic speciation" to sympatric speciation that is based primarily on the temporal partitioning of a resource or habitat. Therefore, allochronic speciation (1) does not include sympatric speciation in which a shift in temporal activity is secondary to host or habitat diversification [e.g., as in the cases of *Rhagoletis*, *Chrysopa*, *Enchenopa*, and others (see above)] and (2) includes only those cases in which temporal diversification occurs as the first (and perhaps only) step in sympatric speciation.

Given our restrictive definition of allochronic speciation, the only examples that appear to apply are certain North American crickets studied primarily by Alexander, Bigelow, and Walker (Alexander & Bigelow 1960; Alexander 1963; Walker 1974; Lloyd & Pace 1975) and certain sawflies discussed by Knerer and Atwood (1973). Even here there are problems. The species of crickets under question occur as sympatric pairs of very similar, but seasonally distinct, species. Originally, it appeared that speciation had occurred through shifts in the seasonal characteristics (the overwintering stages and timing of reproduction), but recent allozymic studies do not support the phylogenetic relationships proposed by the allochronic speciation model (Harrison 1979). Similar types of variation among pairs of cricket species in Japan have been attributed to a form of parapatric speciation involving selection along a climatic gradient (Masaki 1978a, 1983). Clearly, the speciation patterns in this group of organisms need additional study.

Evidence for allochronic speciation in the sawflies includes data showing seasonally diversified biotypes of a population adapted to balsam (Knerer & Atwood 1973). However, no evidence exists concerning the degree of reproductive isolation between the strains. Further work is obviously needed here also.

Thus, although allochronic speciation is an attractive model for speciation and species packing, its occurrence and importance are still open to question. Support for this theory will require the following types of evidence: (1) demonstration of relatively simple genetic differences in the seasonal responses of allochronic species and (2) genetic (or other) data demonstrating a common ancestry for the allochronic species.

9.3.5 Summary

In addition to their role in directional and stabilizing selection, the seasons can have a very important function in disruptive selection. This is evidenced by their numerous influences on speciation.

During allopatric speciation, the seasonal cycle is just one of many traits that undergo diversification as populations adapt to different environments. However, because of their pervasive effects, seasonal cycles may contribute in a major way to the process of allopatric speciation. Allopatric speciation is characterized by alterations in quantitative, interacting life-history traits including seasonal traits. The process is generally slow but can be accelerated if the allopatric populations come into secondary contact. In this case, seasonal cycles, as well as other traits, may diverge rapidly in response to natural selection for reproductive isolation.

Parapatric speciation, which entails the clinal diversification and reproductive isolation of populations along a geographical gradient, is probably the most common form of speciation. The numerous examples of geographic clines in seasonal cycles point to the importance of seasonality in this form of speciation. It can have a role in the initial adaptive divergence and in the evolution of reproductive isolation between geographically contiguous populations. Because parapatrically differentiated populations remain in contact, it is expected that the evolution of reproductive isolation, once begun, will proceed relatively quickly—in a manner similar to that in allopatric populations under secondary contact.

Theoretically, the diversification of seasonal cycles may play a role in the two major steps that comprise sympatric speciation—the evolution of a stable, two-niche polymorphism and the evolution of reproductive isolation. However, evidence is strong only for its role in the latter, in which it has evolved as a means of maintaining reproductive isolation between populations that are adapted to different hosts or habitats.

Temporal partitioning of resources as the first or only step in sympatric speciation (allochronic speciation in a restricted sense) has been proposed and is an attractive model, but it is in need of verification.

Chapter 10

Seasonality and Insect Pest Management

The fundamental unit of agricultural pest management is the crop ecosystem (Smith & van den Bosch 1967; Huffaker *et al.* 1984; Price 1984). In order to construct the required predictive models of agroecosystems and before one can take rational preventive or remedial action to solve pest problems, it is often necessary to quantify the phenologies of the important organisms involved within the ecosystem. This is because agroecosystems involve complexes of interacting and coevolving seasonal cycles (e.g., those of the crop, the pests, the natural enemies of pests, and any natural enemies of the beneficial species).

Effective manipulation of an agroecosystem requires an understanding of the ecophysiological and genetic mechanisms that underlie seasonal cycles. Such information is essential for (a) predicting accurately the timing of dormancy, migration, development, and reproduction in the field, (b) manipulating environmental or genetic factors that aid in suppressing insect pest populations, and (c) choosing reliable, well-adapted species or biotypes for use in biological control. In the sections below we examine the role of seasonality in each of these three areas of pest management (prediction, manipulation, and biological control).

Huffaker (1982) succinctly stated that the foremost requirement for integrated pest management is to understand the agroecosystem and how the pests and other components of the agroecosystem interact. This view applies not just to ecological interactions but also to evolutionary relationships (see Dingle 1979). We must understand the genetic variability present in the biological units of the agricultural system, and we must know how this variability responds to natural selection under various agronomic practices if we are to manage pest populations in a long-term, rational manner.

One recent example of this line of thinking comes from the work of Branson and Krysan (1981) who have taken the view that an evolutionary approach is useful in the development of management systems for corn rootworms (also see Gutierrez *et al.* 1979 for cotton boll weevil). From their analysis, it becomes apparent that rather than broadening the problem to an untractable, unrealistic level, such an evolutionary approach pinpoints crucial areas for emphasis and investigation. An important facet of the studies by Krysan and Branson is the coevolution of the seasonal cycles of the host plant and the pests, and it is within this context that we analyze the role of seasonality studies in integrated pest management.

10.1 Predictive Capability as the Key to Pest Management

Phenological studies can provide predictive capabilities that are important to insect pest management in two major ways. First, the theory and practice of integrated pest management are based on the concept that ecologically, environmentally, and economically appropriate *preventive* measures should be initiated *before* pest populations reach damage-causing levels. Integrated pest management, and even the effective use of chemicals or other control tactics, thus require predictive phenological models of crop, pest, and natural enemy life cycles (Haynes *et al.* 1973; Lieth 1974; Tummala *et al.* 1976; Norton & Holling 1977; Giese *et al.* 1975; Carter *et al.* 1982). Through the accurate forecasting of life-history events for the organisms within these three components, appropriate pest-suppression measures can be applied at times when they are most effective.

Second, many insect pests are opportunists that have taken advantage of a disturbed ecosystem—one whose equilibrium is artificially maintained by man's activities. Many of these pests are introduced species that have become established and are adapting to new geographical areas. Others are native species that have adapted to an introduced host and/or that have expanded their ranges in association with the extended range of a cultivated host. Knowledge of the seasonal adaptations and the seasonal variability of pests, potential pests, and beneficial species can provide a basis for predicting whether or not introduced species can survive in a new area, and, if so, their number of generations and seasonal prevalence. It will also allow the prediction of the climatic limits of range extension for both native and introduced species and their susceptibility to cultural or other pest suppression measures aimed at disrupting the seasonal synchrony between hosts and pests.

10.1.1 Predicting seasonal events

The forecasting of seasonal events in insect populations and the prediction of adaptation by colonizing species require detailed knowledge of seasonal responses and the genetic variability underlying them. Such information is scanty for most economically important pests and beneficial species. In addition, construction of phenological models presents many problems that are beyond the scope of this book. Thus, our analysis here is intended to provide only a general overview of the current state of predictive capabilities. We refer readers to the cited literature for other important considerations.

Prediction of seasonal events entails the modeling of two major components: (1) the onset and termination of diapause and (2) the continuous nondiapause development that occurs in between (Bradshaw 1974b). Nondiapause development is largely governed by temperature, food, and humidity—the dominant and most easily monitored of which is temperature (see, e.g., Gordon 1984). Changes in food and moisture are more difficult to measure in nature, and their seasonal fluctuations are dependent in large part on temperature conditions (see Davidson & Andrewartha 1948 for an example in the rose thrips). Thus, in many instances a reasonable amount of information on humidity and food is sufficient to predict growth rates during nondiapause periods if temperature conditions are known (Bradshaw 1974b; however, see Gage & Mukerji 1977; Sprenkel & Rabb 1981; Tauber & Tauber 1983; Wellington & Trimble 1984).

The primary approach to phenological prediction in agricultural systems has involved the physiological-time or degree-day approach[1]. For example, calculation of thermal thresholds and heat-degree days has found widespread use in relating plant growth, development, and maturation to local climatic conditions (e.g., Dethier & Vittum 1963), for predicting harvest dates and timing successive plantings (e.g., Hopp *et al.* 1972), and for identifying well-adapted, nonlocal stock in reforestation programs (e.g., Campbell 1974). Similarly, to predict the activity and appearance of a pest or beneficial species, it is necessary to know the heat requirements (e.g., thermal thresholds and heat-degree days) for growth and development. Specifically, this information is useful for forecasting damage (e.g., Apple 1952; Gutierrez *et al.* 1977), for predicting insect growth and development at different localities under differing conditions (e.g., Beck 1963; de Wilde 1969; Campbell *et al.* 1974; Anderson *et al.* 1982b; Tamaki *et al.* 1982b; Ring *et al.* 1983b; Nealis *et al.* 1984), for timing control applications (e.g., Strong & Apple 1958, Eckenrode & Chapman 1971; Rock *et al.* 1971; Sevacherian *et al.* 1977; Jorgensen *et al.* 1981), for determining the climatic limits to the distribution of a species (e.g., Messenger 1959, 1972; Greenbank 1970; Messenger & van den Bosch 1971; Campbell *et al.* 1974), for analyzing population dynamics and life-history interactions (e.g., Helgesen & Haynes 1972; Taylor 1982), for rearing and mass-producing beneficial species (e.g., van den Bosch & Messenger 1973; Obrycki & Tauber 1978, 1981, 1982; Obrycki *et al.* 1983a), and for evaluating beneficial species in the field (e.g., Force & Messenger 1968; Messenger 1972; Syme 1972).

The day-degree method is based on the concept that the rates of development, reproduction, and other biological functions are largely temperature-dependent. Usually the number of days needed for insects to complete development is measured under a series of constant temperatures. Developmental rate (100/days) is plotted against temperature, and the lower threshold is estimated by extrapolating the plot to the zero-development

[1]For a discussion of the effect of seasonality and diapause on sampling with food and pheromone traps, see Section 4.2.

level. In some cases, the rates of biological functions are assumed to be linear (up to a maximum temperature); in others the rates are assumed to be nonlinear and the developmental rate/temperature function is given the shape of a normal curve truncated to the right of the mean (see Laudien 1973; Taylor 1981). In both methods the developmental times of the various life stages are calculated in terms of physiological time, not calendar time.

From the relationships described above, it is possible to use temperature data to predict various phenological events in the field. The approach is particularly useful in predicting the completion of postdiapause development by overwintered insects and in forecasting within-generation rates of development by nondiapause insects. A number of methods are used to derive and interpret predictions from field temperature data (see Messenger & Flitters 1958, 1959; Logan *et al.* 1976; Baskerville & Emin 1969; Stinner *et al.* 1974, 1975; Watanabe 1978; Baker 1980; Taylor 1980a, b, 1981, 1982; Arnold 1960; Morris & Fulton 1970a; Wigglesworth 1972; Mack *et al.* 1981; Van Kirk & AliNiazee 1981; Stark & AliNiazee 1982; Logan & Boyland 1983). Each of the methods has limitations, but all of them result in predictions based on physiological time, instead of chronological time (calendar date) (see review by Wagner *et al.* 1984).

The occurrence of diapause and seasonally variable developmental rates precludes the universal application of purely temperature-related data to predict year-round development. Diapause usually intervenes when environmental conditions are still suitable for growth and reproduction, and it results in the curtailment of development and a substantial altering of the temperature-developmental rate relationship. Thus, it is essential to incorporate diapause into predictive phenological models.

The photoperiodic parameters for diapause induction can be incorporated into a photothermograph that relates the heat accumulation with the seasonal change of day length and thus predicts the number of summer generations and the initiation of diapause for multivoltine species. This method was first used extensively to predict voltinism for the European corn borer, *Ostrinia nubilalis*, in North America (Beck & Apple 1961; Beck 1963) and later for the fall webworm, *Hyphantria cunea* (Ito 1972; Masaki 1977; see also Bradshaw 1974b). In these species, the total heat units, accumulated from the time dormancy ends until the critical photoperiod is reached, determine the number of generations at each locality (Fig. 10.1). By graphing day-degrees against photoperiod, the possible number of generations can be estimated by projecting the occurrence of the life stages sensitive to photoperiod on the heat-unit axis in relation to the intercept of the date of the critical photoperiod (see Fig. 10.2). Between-year variance in temperature can be used to estimate the percentage of the larval population entering diapause or continuing development to produce another generation (see Riedl 1983).

Such a method requires the following information: (1) temperature/developmental rates, (2) critical photoperiod, (3) sensitive stage(s), (4) field temperatures, and (5) natural photoperiod. It does not take into account the

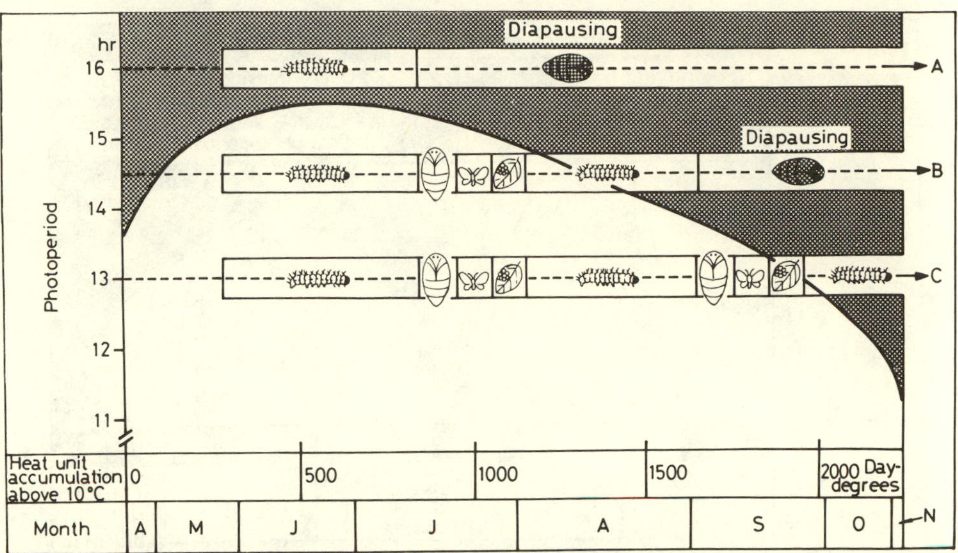

Fig. 10.1 Scheme showing the relationship between critical photoperiod and the timing of diapause. The critical photoperiod for diapause induction is assumed to be 16(A), 14.5(B), and 13(C) hours of light per day; the sensitive stage is assumed to be the large larva; and diapause occurs in the pupal stage. The life cycle would be univoltine in A, bivoltine in B, and in case C, diapause would not be induced.

effect of temperature on the critical photoperiod nor any possible influences of changing day lengths or food quality and quantity on growth and/or diapause induction. Verification of the accuracy of the predictions requires seasonal sampling over several years.

A photothermograph developed for the southwestern corn borer, *Diatraea grandiosella*, overcomes some of the problems associated with a photoperiod-temperature interaction by dividing the growing season into three periods—an early period in which long daylengths prevent diapause, a mid-season period during which temperatures are too high for diapause induction under any photoperiod, and a late-summer, short-day, low-temperature period during which diapause is induced (Takeda & Chippendale 1982b). The models show good agreement with the field situation for some populations, but unknown factors come into play for others.

Riedl and Croft (1978) used a similar method to predict the number and relative size of summer generations of the codling moth, *Cydia* (= *Laspeyresia*) *pomonella*. Although their predictions were tested against two years' field data and proved to be very accurate for the codling moth, the general applicability of their method is open to question. The main problem with their approach is in the use of the term "critical photoperiod." Generally, the critical photoperiod is defined as the day length (night length) that *induces* diapause in 50% of a population; it is the photoperiod that the sen-

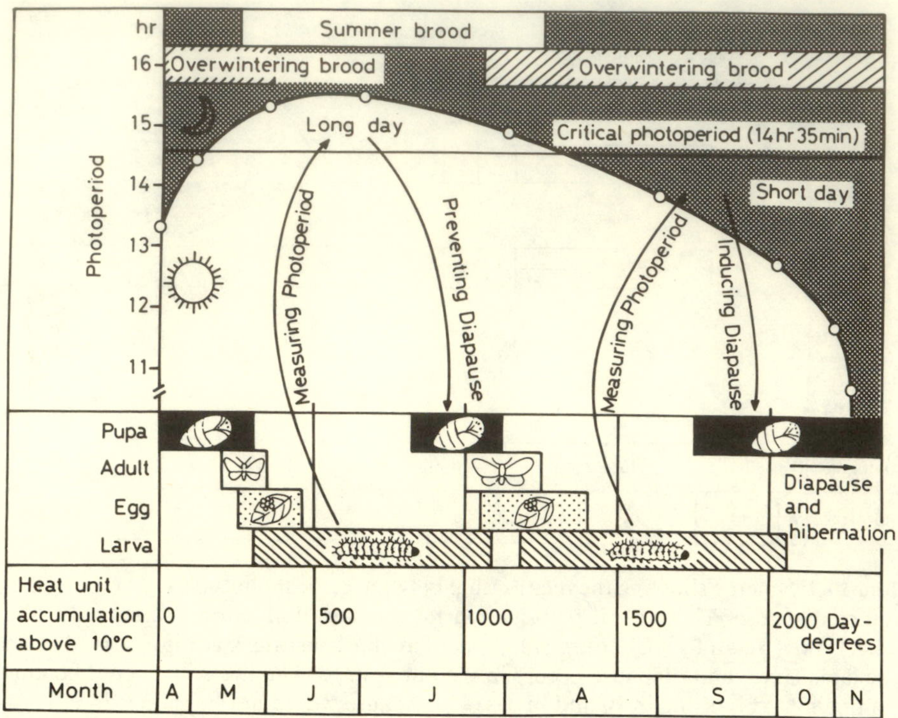

Fig. 10.2 Seasonal progress and programming of the life cycle of the fall webworm, *Hyphantria cunea,* in relation to heat accumulation and photoperiod (after Masaki *et al.* 1968).

sitive stage perceives, even though diapause may not be expressed until a much later life stage. However, Riedl and Croft regarded the critical photoperiod as the photoperiod occurring when 50% of the population *expresses* diapause. This may be justified only when the sensitive stage is very close to the diapausing stage, as may be the case in the codling moth (see Riedl 1983) but not for many other species whose sensitive and diapausing stages are widely separated.

In some cases, prediction of postdiapause development by the simple use of the heat-degree method also presents problems. After diapause ends, temperature is usually the primary factor governing the rate of growth and development. Because of this, the thermal thresholds and temperature-growth rates specific to the postdiapause period have been studied for many species to predict postdiapause biological events in the field (see Section 5.2.8). However, this procedure is often difficult because in many instances it is not known when diapause development ends and when the postdiapause period begins in nature. For some temperate-zone species that end diapause by midwinter and that also have relatively high thermal thresholds for development, there is a break between diapause and the resumption of growth, and therefore it may not be necessary to determine exactly when diapause ends in order to know when to begin summing heat units. However, this approach is inappropriate for species in which the diapause period merges into active growth, that is, those species that end diapause in late winter or spring or those with postdiapause developmental thresholds lower than the temperatures prevailing after diapause. For example, *Chrysopa carnea* from the northeastern United States has a relatively low temperature threshold for postdiapause development (~ 4°C) and ends diapause in mid- to late winter. Temperatures often exceed this threshold at the end of diapause, and thus postdiapause development can follow almost immediately (Tauber & Tauber 1973d). Similarly, overlap in the thermal range for diapause and postdiapause development occur in the tettigoniid *Ephippiger cruciger* and the Douglas-fir cone moth *Barbara colfaxiana* (Dean & Hartley 1977b; Sahota *et al.* 1982). In the pea moth, *Laspeyresia nigricana*, the change from the diapause to the postdiapause period is gradual, and reactions to temperature during the transitional period are variable (Wheatley & Dunn 1962). Therefore, the prediction of postdiapause activity in the field generally should be based on knowledge of the temperature reactions during diapause and during any transitional period between diapause and postdiapause, as well as knowledge of when diapause ends in nature (see Sections 5.2.7 and 5.2.8).

Comparison of field emergence data with laboratory-derived models is another method that can be used to estimate the timing of diapause and postdiapause events in the field. Prediction of postdiapause development in the pecan nut casebearer, *Acrobasis nuxvorella*, shows the least amount of error if the 72nd day (College Station, Texas) or the 80th day (Shreveport, Louisiana) of the year are used as the beginning dates for summation of heat units (Ring & Harris 1983; Ring *et al.* 1983a). Such data imply that diapause ends around these dates, but specific tests are needed to verify this conclusion.

Prediction of postdiapause migration is another area that is essential to phenological models for insect pest management (Rabb & Kennedy 1979; Pedgley 1982; Wellington 1983). This subject also needs considerable ecophysiological study, especially with regard to the proximal factors eliciting migratory behavior. In some cases, specific conditions are associated with the initiation of migration; for example, high daily maximum temperatures are closely correlated with an increase in migration in the rice leaf beetle, *Oulema oryzae* (Kidokoro 1983). Seasonal changes in host plant quality, including the emission of volatiles, may also influence migration (e.g., Hedin 1976). It is essential to include these factors in phenological models.

10.1.2 Predicting the colonization of new areas

Many of our most important insect pests are introduced species or native species whose geographical ranges have expanded. Moreover, classical biological control involves the introduction and colonization of exotic natural enemies. Therefore, questions inevitably arise concerning the phenological adaptation of colonizing species in agroecosystems and the climatic limits to their spread.

Colonization of new areas involves three, and sometimes four, stages: (1) introduction and survival, (2) establishment, (3) range expansion and adaptation to the new environment, and (4) divergence (Fig. 10.3). The seasonal features of the new locality exert a decisive influence on these steps. Thus, the immigrants' seasonal responses and the underlying genetic variability and physiological plasticity are of utmost importance to the success of the species.

Many factors impinge on the first step, introduction/survival, not the least of which is the insect's ability to tolerate the new seasonal conditions (Remington 1968; van den Bosch & Messenger 1973; Parsons 1982). Initial survival is primarily determined by the immigrants' ability to withstand the extremes of the seasonal physical factors, such as the temperature and humidity, that they encounter in the new area (Tauber & Tauber 1983). The

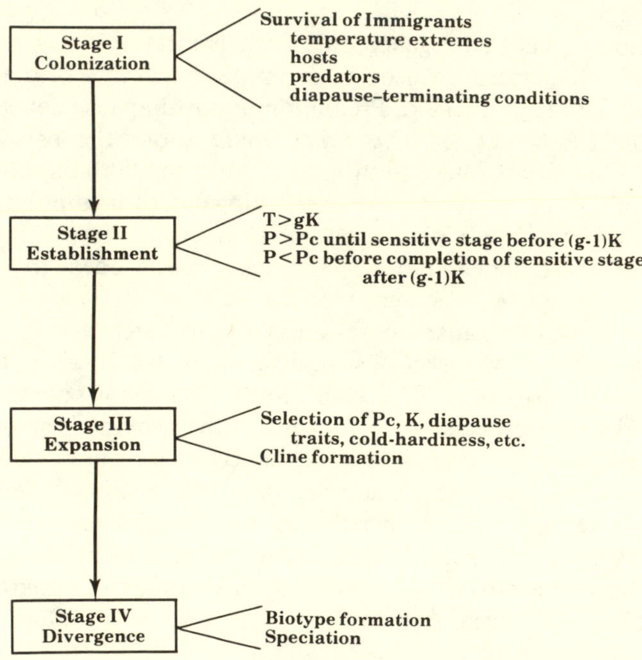

Fig. 10.3 Four stages in the introduction and adaptation of insects to new geographical areas. T = total number of heat units available during the growing season; K = heat units, above the lower thermal threshold, for completing one generation; P = effective day length; Pc = critical photoperiod for diapause induction; g = number of generations per year.

ability to find and utilize food sources and the ability to avoid predators, parasitoids, and pathogens also come into play during colonization. If diapausing individuals are introduced, the appropriate conditions for the timely end of diapause must be present to ensure survival.

After the introduction, the next step is establishment. Successful establishment depends on such factors as the presence of competitors (Geier & Springett 1976; Ehler & Hall 1982) and synchronization of the life cycle with the seasonal cycles of the biotic and abiotic characteristics of the new area (Tauber & Tauber 1973b; Masaki 1977). At least three conditions must be fulfilled for a species with a photoperiodically induced hibernal diapause to become established (Masaki 1981): (1) Conditions must be sufficiently warm to allow completion of a full generation ($T > gK$). (2) The critical photoperiod for diapause induction must be sufficiently short to allow development to continue without diapause induction until the sensitive stage of penultimate generation is completed [$P_c < P$, until the end of the sensitive stage before $(g - 1) K$]. (3) The critical photoperiod must be long enough to allow for diapause induction when insufficient thermal units remain for a complete generation [$P_c >$ P, before completion of the sensitive stage after $(g - 1) K$], where T = total number of heat units available during the growing season, g = number of generations per year, K = number of heat units required per generation, P = effective day length, and P_c = the critical photoperiod. This situation is schematically depicted in Figure 10.3. In some cases, factors such as a long interval between the occurrence of the sensitive stage and the initiation of diapause, a response to photoperiodic change, or thermal modification of the critical photoperiod may complicate prediction. But this photothermographic analysis gives a quick method for determining whether or not the immigrants have the possibility of establishment. Using the above method within 20 to 30 years of its introduction, researchers in Japan predicted the limit of expansion of the fall webworm, *Hyphantria cunea*, invading the Japanese islands and also deduced the latitude in North America from which the Japanese population of the webworm originated (Masaki *et al.* 1968; Masaki 1977).

Figure 10.3 A makes clear that the critical photoperiods of populations are adapted to their particular localities. Although this increases the fitness of the population *in situ*, the possibility of tolerating long latitudinal movement becomes restricted. The experimental transfer of individuals of southern populations of *Acronycta rumicis* to Leningrad illustrates this point (Danilevsky 1965). All the individuals from the southern populations perished with the arrival of winter conditions because they had failed to enter diapause under the long day lengths in the North. Therefore, in most cases only those immigrants coming from latitudinally and climatically similar areas ("phenological homologues") will be successful colonizers. Practitioners of classical biological control have begun to recognize the importance of such findings when importation of natural enemies is planned (see Messenger 1970; van den Bosch *et al.* 1979; Flint 1980; Obrycki *et al.* 1984; however, see also Levins 1969).

Figure 10.3 and work by Beck and Apple (1961), Masaki (1977), and others also indicate that not only the critical photoperiod (P_c) but also the amount of heat accumulation necessary for a complete generation (*K*) are important in determining the success of immigrants. Those species with a small *K* or a short generation time have a smaller risk than those with a large *K* or a long generation time, because the addition of a short generation can provide flexibility in the timing of photoperiodic induction of diapause at the end of the growing season, whereas the initiation of a long generation can result in asynchrony and, thus, mortality.

Flexibility and variability in seasonal responses are other characteristics that enhance the establishment of insects in new areas. Some species [e.g., the carrot fly, *Psila rosae* (Metcalf *et al.* 1951; Stadler 1970) and the satyrid butterfly, *Pararge aegeria* (Lees & Tilley 1980)], overwinter in more than one stage; this flexibility apparently adapts the species to colonization. *Agromyza frontella*, the alfalfa blotch leaf miner, which was introduced into North America from Europe, overwinters only in the pupal stage, but it has a thermally malleable diapause that allows modulation in the timing of diapause. High temperatures (> 21°C) prevent or even terminate autumnal diapause, allowing an additional generation in years or at localities with warm autumns (Tauber *et al.* 1982; Nechols *et al.* 1983). Such a flexible characteristic may subserve the geographic spread of a species, but it is adaptive only if *K* is small enough to allow a complete generation before the onset of winter conditions.

The Colorado potato beetle, *Leptinotarsa decemlineata*, exhibits another form of variability in its seasonal cycle that subserves colonization. With the cultivation of the potato, this native American species has spread to the north, east, and west in North America. Its ability to adapt to new hosts and its large genetic variability in phenological responses subserve rapid colonization, establishment, and adaptation to new areas. Adult Colorado potato beetles enter diapause primarily in response to short day lengths. However, low temperatures, senescent foliage, or a nonpreferred host greatly enhance the diapause-inducing effects of photoperiod (de Wilde & Ferket 1967; Hsiao 1978; Hare 1983). This phenotypic variability allows the beetles to adapt to local conditions even under day lengths that would consistently prevent diapause.

Genetic variation also underlies the colonizing ability of the Colorado potato beetle. In some areas of the USSR, approximately one-half of the females oviposit, one-fifth enter diapause without ovipositing, and another one-fifth oviposit some eggs and then enter diapause (Ushatinskaya 1966, 1976b). In addition, prolonged diapause (of two or more years) occurs in a small but significant number of Colorado potato beetles. Such a wide range of responses is typical of North American populations as well (Tauber *et al.* unpublished), and it virtually ensures that at least part of the population will be in a protected state of diapause at all times of the year. This variability has a strong genetic component, and it may be a major factor in the Colorado potato beetle's success in colonizing new areas (e.g., see Jacobson & Hsiao 1983).

In view of the relatively rapid response of insect seasonal cycles to artificial selection (Section 8.1), it is not surprising that immigrant species can form adaptive geographic clines once they have successfully colonized and become established. Thus, an established population may enter Stage 3, which is characterized by range expansion and adaptation to the climatic and other conditions of the new area (see Chapter 8). As selection proceeds under the new environmental conditions, the immigrants may undergo considerable diversification, and thus Stage 3 may lead to formation of biotypes and/or speciation (Fig. 10.3, Stage 4). Studies of introduced pests (see Section 8.2.2) and natural enemies (DeBach 1965; Callan 1969; Parsons 1982) indicate that under some circumstances this can occur after establishment.

10.1.3 Summary

Modern insect pest management is based on the premise that appropriate *preventive* measures should be initiated *before* pest populations reach damage-causing levels. Such an approach requires phenological prediction of both recurring seasonal events and the establishment and spread of colonizing species.

Prediction of recurring events involves the modeling of both nondiapause development and diapause induction and termination. Current predictive models are based largely on temperature, the most easily monitored and most reliable indicator of physiological time during the growing season. Several methods are used to describe the temperature/developmental rate relationship in insects and to verify laboratory-derived predictions with observations of growth and development in field populations.

Photoperiodic parameters for diapause induction can be combined with temperature/developmental rate data to predict the overall seasonal cycle—that is, the number of generations per year, diapause induction, and diapause termination, as well as nondiapause development. Models intended for such purposes incorporate a substantial amount of information, including, laboratory-derived temperature/developmental rates, critical photoperiod, sensitive stage, field temperatures, and natural photoperiod.

Many of the most important insect pests and beneficial insects are either introduced species or native species that are extending their geographical ranges in association with the spread of a host. Three, and sometime four, steps are involved in the introduction and colonization of immigrant species: (1) initial colonization, (2) establishment, (3) adaptation to the new area, and (4) diversification. Phenological data are essential to predicting the success of immigrants in all four of these steps.

Initial colonization involves the ability to survive the seasonal biotic and abiotic extremes of the new area. Establishment, the second step, requires the fulfillment of a number of conditions regarding the synchronization of the insects' life cycle with the biotic and abiotic seasonal characteristics of the new area. Once established, immigrant species may begin the third step, which involves range expansion and adaptation to climatic and other conditions of the new area. Such a process leads to the formation of clines, diversification of biotypes, and speciation.

10.2 Seasonal Considerations in Implementing Management Tactics

Climate affects many aspects of pest management, and all methods of suppressing insect pest populations entail seasonal considerations to one degree or another (Messenger 1970). In some cases, the seasonal synchrony between pest and host may be disrupted directly through cultural, hormonal, or genetic means. In other cases, the implications of seasonality are less direct, but nevertheless of primary importance. For example, seasonal cycles determine the most effective times for the application of management tactics, such as the application of chemical pesticides.

An important, but sometimes difficult, aspect of pest management is the selection of tactics for a particular pest problem. Recent studies suggest that an evolutionary, or comparative, approach involving the host-pest interaction can be especially efficient in making sound choices (Branson & Krysan 1981). Studies on the seasonal synchrony between hosts and pests, or between pests and natural enemies, can demonstrate life stages of the pest that are especially vulnerable to disruption. For example, the responses by pests and their near relatives to host plants and seasonal conditions can be very useful in prescribing which cultural or host resistance mechanisms should be developed as management tactics.

Numerous other such examples exist. In the section below we provide a brief review of the applications of phenological studies to the development of pest management tactics. Because biological control by predators, parasitoids, and pathogens involves at least three levels of seasonal interaction (host, pest, and natural enemies), we will treat it separately in its own section (Section 10.3).

10.2.1 Cultural methods

Seasonal synchrony with a host is essential for the success of a pest. Not only must the host be available when the pest requires it, but it must be in the proper stage of development. Thus, disruption of the insect pest-host plant (or animal) seasonal synchrony constitutes an important tactic in the overall pest management strategy (Eidt & Little 1968; Stern *et al.* 1976; Kogan 1982; Chippendale 1979, 1982). Pertinent cultural methods include the appropriate seasonal timing of planting, harvesting, and sanitation practices, and the rotation of crops.

Synchrony between host plant and insect pest is most frequently disrupted by altering planting dates. For example, before organophosphorous insecticides came into use, late planting of rice in southwestern parts of Japan, without any other method of pest suppression, resulted in a great decrease in the population density of the yellow rice stem borer, *Scirpophaga incertulas* (Ishikura & Nakatsuka 1955). Unfortunately, farmers now rely on chemicals rather than ecological methods of suppressing this pest. In Illinois late-planted bean fields escape colonization by overwintered bean leaf beetles, *Cerotoma trifurcata.* Moreover, the late planting of large areas can result in significant decreases in bean leaf beetle populations because without the host plant, most overwintered beetles die, either without ovipositing or at least without ovipositing on a suitable larval host (Waldbauer & Kogan 1976). The planting date also affects other insect pests—the Hessian fly, *Mayetiola* (= *Phytophaga*) *destructor*, on winter wheat (Metcalf *et al.* 1951); the European corn borer, *Ostrinia* (= *Pyrausta*) *nubilalis*, corn earworm, *Heliothis zea*, fall armyworm, *Spodoptera frugiperda*, and southwestern corn borer, *Diatraea grandiosella*, on corn (Everett *et al.* 1958; Chippendale 1982); the pink bollworm, *Pectinophora gossypiella*, on cotton (Adkisson 1972); the smaller rice leaf miner, *Hydrellia griseola* (Kidokoro *et al.* 1982); and several pests of cowpeas (Akingbohungbe 1982).

It should be pointed out that alteration of planting dates may invoke a selective response by the pest species, so that seasonal synchrony would be restored after some years. The close correlation between the eclosion time of *Chilo suppressalis* after hibernation and the planting date of rice suggests this possibility. A trivoltine ecotype occurs in a very restricted area on the Pacific Coast of Shikoku, where double cropping, which extends the period of food supply, has been practiced.

On a larger scale, crop rotation may be used to disrupt the synchrony between insect pests and their hosts. Through this practice, hosts are removed for an entire growing season so that overwintered pests emerge but are unable to find suitable food. Again, phenology, in the form of prolonged diapause, can play a large role in determining the success or failure of such a control tactic. As discussed in an earlier section (Section 7.1.1.2), some individuals have the ability or propensity to remain in diapause for more than one annual cycle. Both environmental and genetic factors determine the expression of prolonged diapause. Thus, a schedule of crop rotation should be based on knowledge of both the genetic variability and the environmental conditions that determine the incidence and duration of prolonged diapause. As with other methods of control, the amenability of prolonged diapause to manipulation through natural selection suggests that, although pest population levels may be reduced by crop rotation, the phenology of the pest's life cycle may become adapted to a variety of crop rotation schedules.

The management of preharvest vegetative growth and the timely disposal of crop residue are other important cultural practices that can be used to disrupt pest synchrony with the host plant. The tobacco hornworm, *Man-*

duca sexta, is an important pest of tobacco in North Carolina. A high percentage of the population does not enter diapause until after the second week in August—midway through the tobacco harvesting period. Reducing the amount of preferred food (especially preharvest suckers and postharvest crop residue) available from this time onward greatly reduces the number of overwintering pupae. Early transplant of tobacco also results in the avoidance of plant injury by the tobacco hornworm. However, this procedure is recommended only if the other cultural practices are adhered to; relaxation of compliance to the crop residue disposal and other late-season cultural recommendations would be very detrimental if early transplanting were initiated (Rabb *et al.* 1964; Reagan *et al.* 1979). Postharvest sanitation is also very important in the management of stalk boring and root pests of other annual crops, such as cotton and corn (Reynolds *et al.* 1982; Chippendale 1982).

10.2.2 Host resistance

Host, insect pest, and natural enemy phenology are important considerations in the development and use of pest-resistant hosts. The best-known examples involve the use of early- or late-maturing plant varieties—a practice that directly disrupts host plant-insect pest synchrony. For example, incorporation of early maturing cotton varieties into integrated pest management programs in Texas directly resulted in reduced damage by the boll weevil, *Anthonomus grandis*, and the pink bollworm, *Pectinophora gossypiella* (Phillips *et al.* 1980; Reynolds *et al.* 1982).

Phenology also has important indirect effects on the development of plant resistance that involves mechanisms other than early maturation. This includes potato plants bred for high densities of glandular trichomes that confer resistance to insects. Such a mechanism interacts in a complementary fashion over the season with the natural enemies of potato pests. The trichomes are highly effective in suppressing pest populations early in the growing season, whereas natural enemies are effective from midseason onward. The phenologies of both the resistant potato plants and the beneficial and pest insects influence the degree of pest suppression (Obrycki *et al.* 1983b, 1984; Obrycki & Tauber 1984).

Seasonal effects are not restricted to plant-insect interactions. Resistance to attack by the tick *Boophilus microplus* varies seasonally among steers and heifers of certain cattle in Australia and is subject to considerable alteration through seasonal changes in host nutrition (Sutherst *et al.* 1983).

10.2.3 Chemical methods

The goal of much research in pest management is to reduce the amount and persistence of the chemical pesticides used in suppressing pest populations.

Thus, seasonal considerations come into play in the use of insecticides. The aim of these procedures is to apply the chemical at a time during the insect's phenology when it is most effective and at a time in the crop's phenology when it is least disruptive both to the crop and to the environment.

Monitoring and predictive models of pest development aid in achieving these goals; the importance of considering diapause in the development of phenological models has been discussed (Section 10.1.1). It is also important to note that monitoring of insect populations with light, food, or pheromone traps can also be affected by diapause. The responses of diapausing insects to food and mates are usually much reduced from those in nondiapause insects. Such alterations in response may be manifest in light and bait trap catches (e.g., Mitchell & Hardee 1974; see Section 4.2). Thus, the use of such traps for insects that undergo diapause in the adult stage should be made with caution.

The occurrence of diapause can also have a profound effect on the response of the pest (and its natural enemies) to insecticide. Very often diapausing insects either migrate away from the crop or they hide in protected areas, thus escaping contact with the insecticide (e.g., Surgeoner & Wallner 1978; see Section 4.2.3). In addition, insects in diapause may show more resistance to insecticide than do nondiapausing insects (Sardesai 1972; Bartell *et al.* 1976; Matsumoto & Tsuji 1979).

Pest management of the boll weevil (oddly termed "diapause control" or "reproductive diapause control") takes these problems into account. Insecticidal applications are timed so that late-season reproductive and prediapause adults are killed before they migrate away from their feeding area. Such an approach greatly reduces the size of the overwintering population and limits late-season movement, which is largely responsible for dispersal of the pest. It also reduces the number of insecticidal applications required during the growing season when natural enemies of other cotton pests are present (Adkisson *et al.* 1966; Sterling & Adkisson 1966; Rummel *et al.* 1975).

10.2.4 Photoperiodic and hormonal methods

Disruption of the seasonal synchrony between insect pests and their hosts may be achieved through the use of factors that directly affect diapause—that is, the introduction of artificial light cycles or the application of diapause-altering hormones or antihormones. Research on the feasibility of these approaches to pest suppression is far from conclusive, especially with regard to the use of diapause-altering hormones or antihormones where field experiments have yet to be conducted. No data are available on the economic value of these approaches.

Field experiments have been conducted to determine if diapause (and therefore survival) can be altered through the use of artificial light. The effects on diapause of both (a) timed light breaks during the night and (b) artificial extension of natural day length have been determined for several species. Included are the tortricid moth *Adoxophyes orana* (Ankersmit 1968; Berlinger & Ankersmit 1976), the European corn borer, *Ostrinia nubilalis* (Hayes *et al.* 1970, 1974b, 1979), the codling moth *Cydia* (*Laspeyresia*) *pomonella* (Hayes *et al.* 1970, 1974a; Sáringer 1982), the tobacco budworm, *Heliothis virescens* (Hayes *et al.* 1974a), the pink bollworm, *Pectinophora gossypiella* (Sullivan *et al.* 1970; Hayes *et al.* 1974a), the oak silk moth, *Antheraea pernyi* (Hayes *et al.* 1974a), the rock pool mosquito, *Aedes atropalpus* (Beach & Craig 1979), and the turnip sawfly, *Athalia rosae* (Sáringer 1983). For all these species, photoperiod had some effect on the overwintering population. Artificial illumination prevented diapause induction in a substantial proportion of the European corn borer and codling moth populations under study. In addition, these and other studies show that the incidence of postdiapause emergence and the level of nondiapause oviposition can also be affected by the manipulated photoperiods (Nemec 1969; Schechter *et al.* 1974).

Temperature has an important effect on the photoperiodic induction of diapause; in some cases it alters the level of the critical photoperiod, and in others it may reduce or completely eliminate response to photoperiod (see Chapter 5). Thus, it is not surprising that temperature conditions in the field affect the reliability of photoperiodic manipulation as a control tactic (Hayes *et al.* 1974a, b). The intensity of the light provided and the heaviness of the leaf cover are two other factors that are crucial to the successful application of this control technique (Berlinger & Ankersmit 1976).

In considering the overall and long-term prospects for the successful utilization of photoperiodic manipulation as a tool in pest control, most researchers emphasize its application under certain restricted conditions. Also emphasized is its integration into pest management programs that are based on a variety of pest-regulating tactics. The above practical problems, economic considerations (Chippendale 1982), and possibility of the evolution of resistance to artificially altered light cycles are of major concern. Such an approach should be viewed only as a supplement to a broadly based pest management strategy.

The need for better understanding of insect endocrinology and biochemistry, the lack of field studies, and initial reports of the evolution of resistance are major factors for consideration in the development and assessment of diapause-altering hormones and antihormones as pest-suppressing tactics (Bowers 1971, 1982; Schneiderman 1972; Bowers *et al.* 1976; Morallo-Rejesus 1980; Chippendale 1982). Currently, juvenile hormone mimics are the only hormonal chemicals available for use in insect population management. Because of the involvement of juvenile hormone in adult and larval diapause (see Sections 4.1.1 and 4.1.3), diapause during these stages seems to be most susceptible to disruption by juvenile hormone mimics or antijuvenile hormones (Chippendale 1982). Evaluation of their potential value awaits field tests (Staal 1975; Morallo-Rejesus 1980).

10.2.5 Genetic control

Disruption of seasonal cycles by genetic means has been suggested as a means of suppressing pest populations (Klassen *et al.* 1970b; Hogan 1974). This method proposes to take advantage of the naturally occurring geographical variation in the seasonal adaptations of widespread insect pests by inundating pest populations with mass-reared individuals carrying genes for seasonal traits that are maladapted for the release area. Such traits are called "conditional lethal traits" because they are expressed only under certain conditions; under other conditions individuals carrying the conditional lethal genes may be indistinguishable from other members of the species. This method of pest suppression, like other forms of "genetic control" (e.g., the sterile male technique), does not require that the lethal trait persist in natural populations, but it does require the mass-rearing and periodic release of large numbers of genetically altered individuals that can compete with naturally occurring individuals for mates and whose offspring will succumb under natural conditions.

Klassen *et al.* (1970b) provide numerous examples of seasonal traits that might be useful as conditional lethal traits. An example is the inability to diapause, a trait expressed in some individuals from populations in the lower latitudes. Others include the delayed onset of diapause, premature termination of diapause, and changes in thermal requirements for development. Schemes for release of individuals carrying conditional lethal genes and possible scenarios of pest suppression are presented for conditional lethal traits controlled by one to four genes (Klassen *et al.* 1970a, b).

Although of considerable theoretical interest and possible practical usefulness under certain conditions in the future, such a method of insect management does not show immediate practicability. As Klassen *et al.* (1970b) and Hogan (1974) pointed out, there are three basic requirements before this approach can be used successfully: (1) natural populations must be reduced, by alternative methods, to levels low enough that they can be adequately inundated with genetically deficient insects, (2) methods of mass-rearing hundreds of millions of insects at reasonable cost must be perfected, and (3) vigorous, competitive insects bearing the lethal genetic trait or traits will have to be produced and released.

The first condition may be met by conventional pest suppression methods. However, the second two requirements present considerable economic and biological problems (see Whitten & Foster 1975; Bush 1979; Boller 1979). On a biological level, these conditions will be fulfilled only if we add a fourth and a fifth requirement: (4) the genetic basis for the seasonal traits to be used as conditional lethals should be understood and the traits should be amenable to genetic manipulation; (5) the effect of the conditional lethal traits and the effect of mass-rearing on the expression of major life-history traits must be elucidated. These are major challenges for future research. In addition, studies by Hoy and co-workers (Hoy 1977, 1978b; Hoy & Knop

1978; Lynch & Hoy 1978) on the genetics of diapause in the gypsy moth, *Porthetria* (= *Lymantria*) *dispar*, warn of a potential environmental hazard associated with the release of pests with genetically altered seasonal traits. Such practices could lead to the introduction of multivoltinism into normally univoltine pests—a situation that potentially could increase, rather than diminish, damage. Thus, although genetic disruption of pest seasonal synchrony with the host is an interesting and inviting alternative to conventional pest-suppressing methods, its successful application awaits considerably more biological study, and the investment of large sums of money for the development, maintenance, and staffing of mass-rearing facilities.

10.2.6 Summary

Seasonality is an essential component at all levels in the implementation of insect pest management. Broadly based comparative (evolutionary) studies of pest seasonal cycles and host relations offer an effective means for indicating specific areas in the pest-host interaction that may be vulnerable to disruption. Such studies can also elucidate the potentially most effective methods for achieving the desired disruption.

All methods of suppressing insect pest populations involve seasonal considerations. In some cases, seasonal synchrony between the host and pest is disrupted directly. Cultural methods, such as alterations in dates of planting, harvesting, and debris removal, are notable in this regard. Also, crop rotation is affected by phenology because prolonged diapause can result in extended emergence of the pest over several years.

The development of host resistance as a pest management tactic requires phenological considerations at several levels. An effective form of host resistance involves plant varieties with early or late maturation dates—a mechanism that directly disrupts host-pest seasonal synchrony. Phenology may also have an important indirect effect on resistance, in that pest-natural enemy interactions undergo seasonal changes in relation to various resistance mechanisms.

Phenological considerations are essential in the economically and environmentally prudent application of chemical pesticides. A clear understanding of the pest's biology and the ability to predict phenological events are keys to the judicious use of pesticides.

Photoperiodic, hormonal, and genetic methods of directly disrupting the pest's seasonal cycle present interesting considerations for alternatives to conventional pest-suppressing methods. At present, the full value of these three methods is generally undetermined; field tests especially are needed. Although each of these methods presents advantages under certain circumstances, each also has some potentially severe disadvantages that must be considered *before* widespread implementation.

10.3 Biological Control and Seasonality

Whether alone or as part of integrated pest management, biological control can be a significant, relatively low-cost, and environmentally safe method of pest suppression. Both native and imported predators, parasitoids, and pathogens of insects can be useful in biological control programs. An essential feature is the degree to which they are amenable to rearing and manipulation under laboratory and field conditions. Natural enemies are variable in this significant feature. Thus, biological control is particularly dependent on precise knowledge of key elements in the biology of the species involved. Such information is essential to all aspects of biological control—selection, release, conservation, and augmentation—and it is necessary for developing efficient mass-rearing techniques and for fulfilling the potential of natural enemies as suppressors of pests in the field.

The occurrence of diapause remains one of the major, sometimes unrecognized, problems limiting the successful use of natural enemies (see Messenger 1970). It affects the seasonal synchrony between pests and natural enemies and the susceptibility of pests to the effects of natural enemies, including microbial pathogens and nematodes (see Galloway & Brust 1977; Sharpe & Detroy 1979). It has prevented the continuous rearing of hosts and their parasitoids or predators for both basic study and field release (e.g., van den Bosch & Messenger 1973; Hoy 1977b). The problem is especially acute when the failure to avert or terminate diapause prevents the continuous, efficient rearing of univoltine species (van den Bosch & Messenger 1973) or the use of diapausing natural enemies from the opposite hemisphere (DeBach 1974).

The degree of the seasonal synchrony between host and parasitoid is very important in determining the permanence of host-parasitoid relations and the efficacy of biological control (van den Bosch *et al.* 1959, 1964; Schlinger 1960; Griffiths 1969b; Doutt *et al.* 1976; see also Andres & Bennett 1975 for biological control of aquatic weeds and Summy & Gilstrap 1983 for the relationship between host migration and natural enemies). Within this context, however, there is considerable controversy concerning the methods for establishing well-adapted and effective natural enemies. Some researchers recommend the release of multiple species, leaving the selection of the best adapted to natural forces (e.g., van den Bosch 1968). Other researchers (e.g., Levins 1969; Ehler & Hall 1982) propose a more empirical approach on the premise that factors such as environmental heterogeneity and competition with species already established may cause the introduction to fail even if the newly introduced species is better adapted to the pest than incumbent species.

Given a choice of releasing two closely related biological control agents, the following questions often arise. Which species is best suited for a particular situation? Should both species be released, conserved, and augmented? Are the two species subject to interspecific competition, and would release of both species into a new area likely lead to competitive exclusion of either species (see Ehler & Hall 1982)? Answers to these questions require, among other studies, a comparative and quantitative analysis of the natural enemies' life-history and seasonal traits and an assessment of the ecophysiological responses. In many cases, specific, simple tests of natural enemies' responses to selected environmental factors, such as day length, temperature, and humidity, can provide useful knowledge for (a) efficient mass rearing and (b) choosing well-adapted species or biotypes for release in either classical biological control programs or augmentation of native species (Force & Messenger 1968; Bartlett 1974; Tauber & Tauber 1975b, 1983).

Phenology is especially important in the conservation of natural enemies. For example, mortality rates are frequently high during or just after dormancy, and for this reason knowledge of the habits and requirements of hibernating natural enemies can be crucial. Provision of overwintering sites, alternative food, or supplemental food (e.g., Doutt & Nakata 1973; Hodek 1973; Hagen *et al.* 1970, 1976a) can improve survival during dormancy, as well as the subsequent effectiveness of natural enemies. Because such practices involve costs in material and labor, to be efficient, they should be done on a sound phenological basis.

The environmental control and genetics of parasitoid and predator seasonal cycles were discussed in previous chapters (Chapters 5, 6, 7). Unfortunately, very little is known about the seasonal cycles of insect pathogens (e.g., Fetter-Lasko & Washino 1983; Alves *et al.* 1984). Here, we discuss two aspects of seasonality that are of current interest and long-standing importance to biological control: the selection of seasonally well-adapted biotypes for release and the genetic improvement of the seasonal traits of predators and parasitoids.

10.3.1 Seasonally well-adapted biotypes

To a great extent, the success of biological control programs depends on which species or biotype of natural enemies is chosen for importation, release, conservation, or manipulation (van den Bosch & Messenger 1973). Only those biotypes that are well adapted to their environment can achieve the desired effect. Therefore, to increase the efficiency of biological control procedures, and to minimize the all too prevalent trial-and-error aspects of biological control, it is important to develop objective criteria for choosing well-adapted biotypes. This requires knowledge of (a) intraspecific variation in the ecological and physiological traits, (b) the mode of inheritance of geographically variable, adaptive characteristics, and (c) the processes underlying the evolution of the desired characteristics and their maintenance in mass-reared and natural populations (see Bush *et al.* 1976; Bush 1978; Bush and others in Hoy & McKelvey 1979).

In choosing "well-adapted" biotypes of natural enemies, it is important to keep in mind the objective of the proposed releases (Mackauer 1972). If parasitoids and/or predators are to be mass reared for inundative releases, important phenological questions may arise during the mass-rearing program and in the timing of releases. Specifically, for inundative releases, diapause must be easily averted or terminated under rearing and release conditions. Releases must be well timed with the phenology of the pest population, and in some cases diapausing insects may have to be stored for later release. However, the released natural enemies are not expected to persist indefinitely, and no particular emphasis is placed on the degree of long-term synchrony between the pests' and the introduced natural enemies' seasonal cycles.

By contrast, if inoculative releases of natural enemies are made for colonization and long-term interaction with the pest (as is the case in classical biological control), genetic variation in the seasonal traits of natural enemies becomes very important. Natural enemies exhibit various forms of genetic variability in their diapause characteristics (see Tauber *et al.* 1983 and Chapters 6 & 7). A striking example of the importance of this genetic variability is found in the attempts at biological control of the walnut aphid in California. Efforts to establish a French strain of the parasitoid, *Trioxys pallidus*, in the hot, dry regions of California were initially unsuccessful, although they succeeded in some coastal areas with a mild climate. Subsequent introduction of a second phenological strain from the central plateau of Iran, which has a hot, dry climate similar to that in California's major walnut-growing regions, resulted in establishment of the parasitoid and successful control of the walnut aphid (van den Bosch *et al.* 1979).

The *Chrysopa carnea* species-complex provides another example of an agriculturally important group of natural enemies (Ridgway & Jones 1969; Hagen *et al.* 1970, 1976b; Tauber & Tauber 1983) in which variability in seasonal traits is of major importance to biological control. The species-complex occurs throughout the Holarctic region where it is characterized by numerous, seasonally diverse species and biotypes (see summary in Tauber & Tauber 1982). This variability has two major effects on the usefulness of the group in biological control. In the short run, that is, until the variability among and within geographical populations is characterized and quantified, it leads to unreliable and unrepeatable results and impedes effective biological control. Thus, *C. carnea* and its relatives have not fulfilled their potential as biological control agents (Tauber & Tauber 1975b, 1983). However, once the variability is understood, it can enhance the effectiveness of the *C. carnea* species-complex because well-adapted biotypes can be selected for use in specific situations.

The two examples above and others reviewed by Kelleher (1969) and van den Bosch and Messenger (1973) illustrate several points that need emphasis: (1) Classical biological control requires natural enemies that are well adapted to the climatic and seasonal conditions of the new area (e.g., Messenger & van den Bosch 1971; Price 1972; DeBach 1965). (2) Genetic variability in the phenological characteristics of natural enemies should be examined as part of biological control programs. (3) Search for exotic natural enemies should focus on places that are ecologically and climatically similar to those of the proposed release areas. Implementation of these recommendations could increase the chances of importing biotypes of natural enemies that are well adapted to the new area and to suppressing pest populations.

10.3.2 Genetic improvement of natural enemies

Because seasonal cycles are of major importance to the effective use of natural enemies, they are prime characteristics for genetic improvement (see Simmonds 1963; Hoy 1979; Roush 1979). Relatively little research has been done in the area of genetic improvement of natural enemies, primarily because very little is known about insect genetics other than for a few well-studied groups such as *Drosophila*. As a result, this section is short, but we include it because "genetic engineering" is becoming an area of immense interest.

The seasonal cycles of natural enemies, especially parasitoids, are usually timed to follow those of their hosts; for example, parasitoid postdiapause emergence usually does not occur until after the host is in the proper stage for attack. Although such a seasonal pattern is adaptive, it may limit the parasitoid's effectiveness in suppressing the pest population (Mackauer 1972; Roush 1979). For example, some predators (e.g., certain indigenous coccinellids in the San Joaquin Valley, California) retain their original adaptation to the anticipated natural reduction in prey populations during summer, and they enter aestival diapause and cease to be effective even though modern agricultural practices, such as irrigation, fertilization, and introduced host plants, now consistently keep prey populations high during this period (Hoy 1979). These and other seasonal characteristics may be amenable to genetic alteration.

Two lines of evidence suggest that the genetic improvement of natural enemy seasonal cycles is feasible. First is the successful artificial selection of *Aphytis lingnanensis*, a parasitoid of the California red scale, *Aonidiella aurantii*, for adaptation to temperature extremes (White *et al.* 1970). This was a difficult, carefully planned, long-term study that required over 100 generations, but it was successful in that tolerance to temperature extremes was enhanced considerably and there were no observed detrimental traits associated with the selection process.

Other evidence in support of the amenability of natural enemy seasonal cycles to genetic improvement comes from observations that parasitoid and predator seasonal cycles are subject to the same high degree of variation that characterizes the seasonal cycles of phytophagous and other nonparasitic insects. Parasitoids that are reared under controlled laboratory conditions exhibit inter- and intrapopulation variation in their ability to enter and terminate diapause. Strains of the braconid *Trioxys complanatus* from California, Iran, and Italy differ in the proportion of individuals that enter diapause under various conditions of temperature, photoperiod, and relative humidity (Flint 1980). This type of variation can serve as the source of individuals from which to select nondiapause strains of parasitoids for use in mass rearing or inundative releases. Such a case occurred in laboratory rearings of the dipteran parasitoid, *Pseudosarcophaga affinis*, in which inadvertent selection of nondiapausing individuals as breeding stock resulted in the virtual elimination of the ability to enter diapause (House 1967).

Differences in critical photoperiods may underlie the intraspecific differences in diapause induction in both parasitoids and predators (Maslennikova & Mustafayeva 1971; Mustafayeva 1974; Eichhorn 1977b, 1978). For example, Hoy (1975b) and Weseloh (1982) demonstrated differences in photoperiodic responses among European, Indian, and North American strains of the gypsy moth parasitoid, *Apanteles melanoscelus*, and Tauber and Tauber (1972a, 1982) showed geographical variation in the critical photoperiod, diapause depth, and postdiapause developmental requirements of the predator, *Chrysopa carnea*. The results of these studies suggest polygenic control; for example, hybrids have photoperiodic reactions that are intermediate to those of the parental forms.

Once initiated, diapause in natural enemies often shows considerable interpopulation variation in the timing of termination. Variation can be expressed as within-year variation, as in the tachinid *Bessa harveyi*, where individuals with short diapause emerge in fall, whereas individuals with long diapause overwinter and do not emerge until the following spring (Turnock 1973). In contrast, the variation may be expressed as prolonged diapause of more than one year, as in *Platygaster* sp., a parasitoid of the cecidomyiid, *Hasegawia sassacola* (Sunose 1978). All of the above forms of variation can provide the basis for genetic selection and improvement of natural enemies.

10.3.3 Summary

Effective biological control, an entirely biologically based approach to pest population suppression, depends upon a sound knowledge of insect seasonality. Diapause can be a major impediment to the efficient mass rearing of hosts and natural enemies. It can also be important in the timing of natural enemy releases and in the provision of supplemental food or overwintering shelters that conserve or augment natural enemy populations.

One of the most crucial aspects of biological control programs is the use of biotypes that are well adapted to particular pest situations. A rational choice requires knowledge of (a) the purpose of the releases (inundative release for immediate, short-term pest suppression versus inoculative release for intermediate or longer term pest suppression), (b) intraspecific variation in the seasonal cycles of natural enemies, and (c) the mode of inheritance of the key seasonal traits. Such knowledge can provide the basis for the selection of well-adapted biotypes (whether naturally occurring or artificially produced) for use in specific pest-control situations.

Bibliography

Abo-Ghalia, A. and Thibout, E. (1983). Action du poireau *(Allium porrum)* sue l'activité reproductrice après la diapause imaginale chez le teigne du poireau, *Acrolepiopsis assectella. Entomol. Exp. Appl.* 33, 188–94.

Abou-Elela, R. and Hilmy, N. (1977). Wirkungen der Fotoperiode und Temperatur auf die Entwicklungsstadien von *Acrotylus insubricus* Scop. (Orthopt., Acrididae). *Anz. Schädlingskde., Pflanzenschutz, Umweltschutz* 50, 25–28.

Ackerman, J.D. (1983). Diversity and seasonality of male euglossine bees (Hymenoptera: Apidae) in central Panamá. *Ecology* 64, 274–83.

Ackerman, T.L. and Bamberg, S.A. (1974). Phenological studies in the Mojave Desert at Rock Valley (Nevada Test Site). In *Phenology and seasonality modeling* (ed. E.H. Lieth) pp. 215–26. Springer-Verlag, Berlin.

Adedokun, T.A. and Denlinger, D.L. (1984). Cold-hardiness: A component of the diapause syndrome in pupae of the flesh flies, *Sarcophaga crassipalpis* and *S. bullata. Physiol. Entomol.* 9, 361–64.

Adesiyun, A.A. and Southwood, T.R.E. (1979). Differential migration of the sexes in *Oscinella frit* (Diptera: Chloropidae). *Entomol. Exp. Appl.* 25, 59–63.

Adkisson, P.L. (1964). Action of the photoperiod in controlling insect diapause. *Amer. Natur.* 98, 357–74.

Adkisson, P.L. (1972). Use of cultural practices in insect pest management. In *Implementing pest management strategies* (eds. P.W. Bergman and D.P. Sanders) pp. 37–50. Proceedings of a National Extension Insect-Pest Management Workshop, Purdue University, Lafayette, Indiana, March 14–16, 1972.

Adkisson, P.L., Bell, R.A., and Wellso, S.G. (1963). Environmental factors controlling the induction of diapause in the pink bollworm, *Pectinophora gossypiella* (Saunders). *J. Insect Physiol.* 9, 299–310.

Adkisson, P.L., Rummel, D.R., Sterling, W.L., and Owen, W.L., Jr. (1966). Diapause boll weevil control: A comparison of two methods. *Texas A&M Univ. Bull.* 1054, 1–11.

Ae, S.A. (1978). A study on the immature stages and diapause of papilionid and other butterflies of the Philippines. Tyô to GA. (*Trans Lepid. Soc. Japan*) 29, 227–35.

Aeschlimann, J. (1974). Hibernation chez trois espèces de Metopiines: Hymenoptera, Ichneumonidae. *Entomol. Exp. Appl.* 17, 487–92.

Akhmedov, R.M. and Abdinbekova, A.A. (1977). Factors controlling the development and diapause of *Mamestra genistae* (Lepidoptera, Noctuidae). *Entomol. Rev.* 56, 1–5.

Akingbohungbe, A.E. (1982). Seasonal variation in cowpea crop performance at Ile-ife, Nigeria and the relationship to insect damage. *Insect Sci. Appl.* 3, 287–96.

Akinlosotu, T.A. (1982). Seasonal trend of green spider mite, *Mononychellus tanajoa* population on cassava, *Manihot esculenta* and its relationship with weather factors at Moor Plantation. *Insect Sci. Appl.* 3, 251–54.

Akkawi, M.M. and Scott, D.R. (1984). The effect of age of parents on the progeny of diapaused and nondiapaused *Heliothis zea. Entomol. Exp. Appl.* 35, 235–39.

Albrecht, F.O. (1973a). The decline of locust plagues: An essay in the ecology of photoperiodic regulation. *Acrida* 2, 97–107.

Albrecht, F.O. (1973b). The photoperiodic regulation of temperature requirements in crowded African migratory locusts (*Locusta migratoria migratorioides*, R. & F.) I. Constant photoperiods and mortality. *Acrida* 2, 9–24.

Albrecht, F.O. (1973c). The photoperiodic regulation of temperature requirements in crowded

African migratory locusts (*Locusta migratoria migratorioides*, R. & F.) II. Increasing and decreasing photoperiods and mortality. *Acrida* 2, 25–52.

Albrecht, F.O. and Lauga, J. (1979). Effets de la photopériode et de l'humidité sur le polymorphisme de *Locusta migratoria migratorioides* (R. et F.) (Orthopteres, Acridiens) eleve en isolement: Etude morphometrique des solitaires verts et bruns. *C. R. Acad. Sci.*, Paris, Series D, 289, 753–55.

Albrecht, F.O., Michel, R., and Casanova, D. (1978). The temperature and photoperiodic control of flight activity in crowded desert locusts, *Schistocerca gregaria* (Forsk.). II. Changing photoperiods. General discussion on the acquired flight ability of locusts. *Acrida* 7, 289–98.

Alexander, R.D. (1963). Animal species, evolution, and geographic isolation. *Syst. Zool.* 12, 202–4.

Alexander, R.D. (1968). Life cycle origins, speciation, and related phenomena in crickets. *Q. Rev. Biol.* 43, 1–41.

Alexander, R.D. and Bigelow, R.S. (1960). Allochronic speciation in field crickets, and a new species, *Acheta veletis*. *Evolution* 14, 334–46.

Ali, M.A. and El-Saedy, A.H.A. (1981). Vergleichende Untersuchungen über die Reservestoffe überwinternder und nicht überwinternder Adulter des Melonenmarienkäfers, *Epilachna chrysomelina* (F.) (Col., Coccinellidae). *Anz. Schädlingskde., Pflanzenschutz, Umweltschutz* 54, 5–8.

Ali, M. and Sáringer, G. (1975). Factors regulating diapause in alfalfa ladybird, *Subcoccinella 24-punctata* L. (Col., Coccinellidae). *Acta Phytopathol. Acad. Sci. Hung.* 10, 407–15.

Allemand, R. (1977). Le potentiel reproducteur des adultes de *Drosophila melanogaster*: Variations génétiques en reponse aux régimes lumineux. *Genetica* 47, 1–7.

Alrouechdi, K. and Canard, M. (1979). Mise en évidence d'un biotype sans diapause photopériodique dans une population méditerraneenne de *Chrysoperla carnea* (Stephens) (Insectes, Neuroptera). *C. R. Acad. Sci.*, Paris, Series D, 289, 553–55.

Alves, S.B., Risco, S.H., and Almeida, L.C. (1984). Influence of photoperiod and temperature on the development and sporulation of *Metarhizium anisopliae* (Metsch.) Sorok. *Z. Ang. Entomol.* 97, 127–29.

Anderson, J.F. (1968). Influence of photoperiod and temperature on the induction of diapause in *Aedes atropalpus* (Diptera: Culicidae). *Entomol. Exp. Appl.* 11, 321–30.

Anderson, J.F. (1970). Induction and termination of embryonic diapause in the salt marsh mosquito, *Aedes sollicitans* (Diptera: Culicidae). *Conn. Agr. Exp. Sta. Bull.* 711, 1–22.

Anderson, J.F. and Kaya, H.K. (1974). Diapause induction by photoperiod and temperature in the elm spanworm egg parasitoid, *Ooencyrtus* sp. *Ann. Entomol. Soc. Amer.* 67, 845–49.

Anderson, J.F. and Kaya, H.K. (1975). Influence of temperature on diapause termination in *Ooencyrtus ennomus*, an elm spanworm egg parasitoid. *Ann. Entomol. Soc. Amer.* 68, 671–72.

Anderson, T.E., Kennedy, G.G., and Stinner, R.E. (1982a). Temperature- dependent model for postdiapause development and spring emergence of the European corn borer, *Ostrinia nubilalis* (Hübner) (Lepidoptera: Pyralidae), in North Carolina. *Environ. Entomol.* 11, 1307–11.

Anderson, T.E., Kennedy, G.G., and Stinner, R.E. (1982b). Temperature- dependent models of European corn borer (Lepidoptera: Pyralidae) development in North Carolina. *Environ. Entomol.* 11, 1145–50.

Ando, Y. (1972). Egg diapause and water absorption in the false melon beetle, *Atrachya menetriesi* Faldermann (Coleoptera: Chrysomelidae). *Appl. Entomol. Zool.* 7, 142–54.

Ando, Y. (1978). Studies on egg diapause in the false melon beetle, *Atrachya menetriesi* Faldermann (Coleoptera: Chrysomelidae). *Bull. Fac. Agr. Hirosaki Univ.* 30, 131–215 (in Japanese, English summary).

Ando, Y. (1979). Geographic variation in the incidence of non-diapause eggs of the false melon beetle, *Atrachya menetriesi* Faldermann (Coleoptera: Chrysomelidae). *Appl. Entomol. Zool.* 14, 193–202.

Ando, Y. and Hartley, J.C. (1982). Occurrence and biology of a long-winged form of *Conocephalus discolor*. *Entomol. Exp. Appl.* 32, 238–41.

Ando, Y. and Miya, K. (1968). Diapause character in the false melon beetle, *Atrachya menetriesi* Faldermann, produced by crossing between diapause and non-diapause strains. *Bull. Fac. Agr. Iwate Univ.* 9, 87–96 (in Japanese, English summary).

Andres, L.A. and Bennett, F.A. (1975). Biological control of aquatic weeds. *Annu. Rev. Entomol.* 20, 31–46.

Andrewartha, H.G. (1943). Diapause in the eggs of *Austroicetes cruciata*, Sauss. (Acrididae) with particular reference to the influence of temperature on the elimination of diapause. *Bull. Entomol. Res.* 34, 1–17.

Andrewartha, H.G. (1944). The influence of temperature on the elimination of diapause from the eggs of the race of *Austroicetes cruciata* Sauss. (Acrididae) occurring in Western Australia. *Aust. J. Exp. Biol. & M. Sci.* 22, 17–20.

Andrewartha, H.G. (1952). Diapause in relation to the ecology of insects. *Biol. Rev.* 27, 50–107.

Andrewartha, H.G. and Birch, L.C. (1954). *The distribution and abundance of animals.* University of Chicago Press, Chicago.

Andrewartha, H.G. and Birch, L.C. (1973). The history of insect physiology. In *History of entomology* (eds. R.F. Smith, T.E. Mittler, and C.N. Smith) pp. 229–66. Annual Reviews, Palo Alto, CA.

Andrewartha, H.G., Miethke, P.M., and Wells, A. (1974). Induction of diapause in the pupa of *Phalaenoides glycinae* by a hormone from the suboesophageal ganglion. *J. Insect Physiol.* 20, 679–701.

Ankersmit, G.W. (1964). Voltinism and its determination in some beetles of cruciferous crops. *Meded. Landbouwhogesch.* Wageningen, 64, 1–62.

Ankersmit, G.W. (1968). The photoperiod as a control agent against *Adoxophyes reticulana* (Lepidoptera; Tortricidae). *Entomol. Exp. Appl.* 11, 231–40.

Ankersmit, G.W. and Adkisson, P.L. (1967). Photoperiodic responses of certain geographical strains of *Pectinophora gossypiella* (Lepidoptera). *J. Insect Physiol.* 13, 553–64.

Annila, E. (1982). Diapause and population fluctuations in *Megastigmus specularis* Walley and *Megastigmus spermotrophus* Wachtl. (Hymenoptera, Torymidae). *Ann. Entomol. Fenn.* 48, 33–36.

Annis, B., Tamaki, G., and Berry, R.E. (1981). Seasonal occurrence of wild secondary hosts of the green peach aphid, *Myzus persicae* (Sulzer), in agricultural systems in the Yakima Valley. *Environ. Entomol.* 10, 307–12.

Apple, J.W. (1952). Corn borer development and control on canning corn in relation to temperature accumulation. *J. Econ. Entomol.* 45, 877–79.

Arai, T. (1977). Effects of the daily cycle of light and temperature on hatchability and hatching time in *Metrioptera hime* Furukawa (Orthoptera, Tettigonidae). *Kontyû* 45, 107–20.

Arbuthnot, K.D. (1944). Strains of the European corn borer in the United States. *U.S.D.A. Tech. Bull.* 869, 1–20.

Arnold, C.Y. (1960). Maximum-minimum temperatures as a basis for computing heat units. *Amer. Soc. Hort. Sci.* 76, 682–92.

Arntfield, P.W., Gallaway, W.J., and Brust, R.A. (1982). Blood-feeding in overwintering *Culex tarsalis* (Diptera: Culicidae) from Manitoba. *Can. Entomol.* 114, 85–86.

Asahina, E. (1959). Prefreezing as a method enabling animals to survive freezing at an extremely low temperature. *Nature* 184, 1003–4.

Asahina, E. (1966). Freezing and frost resistance in insects. In *Cryobiology* (ed. H.T. Meryman) pp. 451–85. Academic Press, New York.

Asahina, E. (1969). Frost resistance in insects. In *Advances in insect physiology*, Vol. 6 (eds. J.W.L. Beament, J.E. Treherne, and V.B. Wigglesworth) pp. 1–49. Academic Press, London.

Ascerno, M.E., Hower, A.A., Jr., and Smilowitz, Z. (1978). Gonadal development of laboratory-reared male alfalfa weevils, *Hypera postica. Ann. Entomol. Soc. Amer.* 71, 239–42.

Ascerno, M.E., Smilowitz, Z., and Hower, A.A., Jr. (1980). Effects of the insect growth regulator hydroprene on diapausing *Microctonus aethiopoides*, a parasite of the alfalfa weevil. *Environ. Entomol.* 9, 262–64.

Ascerno, M.E., Smilowitz, Z., and Hower, A.A., Jr. (1981). Effects of the insect growth regulatory hydroprene on diapausing alfalfa weevils. *Environ. Entomol.* 10, 501–5.

Askew, R.R. (1971). *Parasitic insects.* Elsevier, New York.

Awiti, L.R.S. and Hidaka, T. (1982). Neuroendocrine mechanism involved in pupal colour dimorphism in swallowtail *Papilio xuthus L. Insect Sci. Appl.* 3, 181–91.

Azaryan, A.G. (1966). Some features of intraspecific geographic adaptations in *Dysaphis anthrisci* C.B. (Homoptera, Aphidinea). *Entomol. Rev.* 45, 278–83.

Baerwald, R.J. and Boush, G.M. (1967). Selection of a nondiapausing race of apple maggot. *J. Econ. Entomol.* 60, 682–84.

Bährmann, R. (1977). On the development of dormance-forms and non-dormance-forms in

Aleyrodes asari (Homoptera, Aleyrodina) in the same experimental conditions. *Zool. Jb. Syst.* 104, 80–97.

Baker, C.R.B. (1971). Egg and pupal development of *Spilosoma lubricipeda* in controlled temperatures. *Entomol. Exp. Appl.* 14, 15–22.

Baker, C.R.B. (1980). Some problems in using meteorological data to forecast the timing of insect life cycles. *EPPO Bull.* 10, 83–91.

Baker, C.R.B. and Miller, G.W. (1978). The effect of temperature on the post-diapause development of four geographical populations of the European cherry fruit fly (*Rhagoletis cerasi*). *Entomol. Exp. Appl.* 23, 1–13.

Baker, G.L. (1976). The seasonal life cycle of *Anoplolepis longipes* (Jerdon) (Hymenoptera: Formicidae) in a cacao plantation and under brushed rain forest in the northern district of Papua New Guinea. *Insectes Sociaux* 23, 253–62.

Baker, J.E. (1982). Termination of larval diapause-like condition in *Attagenus megatoma* (Coleoptera: Dermestidae) by low temperature. *Environ. Entomol.* 11, 506–8.

Bale, J.S. (1979). The occurrence of an adult reproductive diapause in the univoltine life cycle of the beech leaf mining weevil, *Rhynchaenus fagi* L. *Int. J. Invert. Reprod.* 1, 57–66.

Balling, S.S. and Resh, V.H. (1984). Life history variability in the water boatman, *Trichorixa reticulata* (Hemiptera: Corixidae), in San Francisco Bay salt marsh ponds. *Ann. Entomol. Soc. Amer.* 77, 14–19.

Barker, J.F. and Herman, W.S. (1976). Effect of photoperiod and temperature on reproduction of the monarch butterfly, *Danaus plexippus. J. Insect Physiol.* 22, 1565–68.

Barnard, D.R. and Mulla, M.S. (1977). Effects of photoperiod and temperature on blood feeding, oögenesis and fat body development in the mosquito, *Culiseta inornata. J. Insect Physiol.* 23, 1261–66.

Barnard, D.R. and Mulla, M.S. (1978). The ecology of *Culiseta inornata* in the Colorado Desert of California: Seasonal abundance, gonotrophic status, and oviparity of adult mosquitoes. *Ann. Entomol. Soc. Amer.* 71, 397–400.

Barnes, H.F. (1952). Studies of fluctuations in insect populations. XII. Further evidence of prolonged larval life in the wheat-blossom midges. *Ann. Appl. Biol.* 39, 370–73.

Barnes, J.K. (1976). Effect of temperature on development, survival, oviposition, and diapause in laboratory populations of *Sepedon fuscipennis* (Diptera: Sciomyzidae). *Environ. Entomol.* 5, 1089–98.

Baronio, P. and Sehnal, F. (1980). Dependence of the parasitoid *Gonia cinerascens* on the hormones of its lepidopterous hosts. *J. Insect Physiol.* 26, 619–26.

Barry, B.D. and Adkisson, P.L. (1966). Certain aspects of the genetic factors involved in the control of the larval diapause of the pink bollworm. *Ann. Entomol. Soc. Amer.* 59, 122–25.

Bartell, D.P., Sanborn, J.R., and Wood, K.A. (1976). Insecticide penetration of cocoons containing diapausing and nondiapausing *Bathyplectes curculionis*, an endoparasite of the alfalfa weevil. *Environ. Entomol.* 5, 659–61.

Bartelt, R.J., Kulman, H.M., and Jones, R.L. (1981). Effects of temperature on diapausing cocoons of the yellowheaded spruce sawfly, *Pikonema alaskensis. Ann. Entomol. Soc. Amer.* 74, 472–77.

Bartlett, B.R. (1974). Introduction into California of cold-tolerant biotypes of the mealybug predator, *Cryptolaemus montrouzieri*, and laboratory procedures for testing natural enemies for cold-hardiness. *Environ. Entomol.* 3, 553–56.

Barton, N.H. and Charlesworth, B. (1984). Genetic revolutions, founder effects, and speciation. *Annu. Rev. Ecol. Syst.* 15, 133–64.

Basedow, T. (1977). Der Einfluss von Temperature und Niederschlägen auf Diapause und Phänologie die Weizengallmücken *Contarinia tritici* (Kirby) und *Sitodiplosis mosellana* (Géhin) (Dipt., Cecidomyidae). *Zool. Jb. Syst. Bd.* 104, 302–26.

Baskerville, G.L. and Emin, P. (1969). Rapid estimation of heat accumulation from maximum and minimum temperatures. *Ecology* 50, 514–17.

Batten, A. (1967). Seasonal movements of swarms of *Locusta migratoria migratorioides* (R. & F.) in western Africa in 1928 to 1931. *Bull. Entomol. Res.* 57, 357–80.

Baust, J.G. (1982). Environmental triggers to cold hardening. *Comp. Biochem. Physiol.* 73A, 563–70.

Baust, J.G. and Lee, R.E., Jr. (1981). Divergent mechanisms of frost-hardiness in two populations of the gall fly, *Eurosta solidaginsis. J. Insect Physiol.* 27, 485–90.

Baust, J.G. and Miller, L.K. (1970). Seasonal variations in glycerol content and its influence on cold-hardiness in the Alaskan carabid beetle, *Pterostichus brevicornis*. *J. Insect Physiol.* 16, 979–90.

Baust, J.G. and Morrissey, R.E. (1975). Supercooling phenomenon and water content independence in the overwintering beetle, *Coleomegilla maculata*. *J. Insect Physiol.* 21, 1751–54.

Baxendale, F.P. and Teetes, G.L. (1983a). Factors influencing adult emergence from diapausing sorghum midge, *Contarinia sorghicola* (Diptera: Cecidomyiidae). *Environ. Entomol.* 12, 1064–67.

Baxendale, F.P. and Teetes, G.L. (1983b). Thermal requirements for emergence of overwintered sorghum midge (Diptera: Cecidomyiidae). *Environ. Entomol.* 12, 1078–82.

Beach, R. (1978). The required day number and timely induction of diapause in geographic strains of the mosquito *Aedes atropalpus*. *J. Insect Physiol.* 24, 449–55.

Beach, R.F. and Craig, G.B., Jr. (1979). Photoinhibition of diapause in field populations of *Aedes atropalpus*. *Environ. Entomol.* 8, 392–96.

Bean, D.W. and Beck, S.D. (1980). The role of juvenile hormone in the larval diapause of the European corn borer, *Ostrinia nubilalis*. *J. Insect Physiol.* 26, 579–84.

Bean, D.W. and Beck, S.D. (1983). Haemolymph ecdysteroid titres in diapause and nondiapause larvae of the European corn borer, *Ostrinia nubilalis*. *J. Insect Physiol.* 29, 687–93.

Bean, D.W., Beck, S.D., and Goodman, W.G. (1982). Juvenile hormone esterases in diapause and nondiapause larvae of the European corn borer, *Ostrinia nubilalis*. *J. Insect Physiol.* 28, 485–92.

Beards, G.W. and Strong, F.E. (1966). Photoperiod in relation to diapause in *Lygus hesperus* Knight. *Hilgardia* 37, 345–62.

Beatley, J.C. (1974). Phenological events and their environmental triggers in Mojave desert ecosystems. *Ecology* 55, 856–63.

Beck, S.D. (1962). Photoperiodic induction of diapause in an insect. *Biol. Bull.* 122, 1–12.

Beck, S.D. (1963). Physiology and ecology of photoperiodism. *Bull. Entomol. Soc. Amer.* 9, 8–16.

Beck, S.D. (1964). Time-measurement in insect photoperiodism. *Amer. Natur.* 48, 329–46.

Beck, S.D. (1967). Water intake and the termination of diapause in the European corn borer, *Ostrinia nubilalis*. *J. Insect Physiol.* 13, 739–50.

Beck, S.D. (1974a). Photoperiodic determination of insect development and diapause I. Oscillators, hourglasses, and a determination model. *J. Comp. Physiol.* 90, 275–95.

Beck, S.D. (1974b). Photoperiodic determination of insect development and diapause. II. The determination gate in a theoretical model. *J. Comp. Physiol.* 90, 297–310.

Beck, S.D. (1975). Photoperiodic determination of insect development and diapause. III. Effects of nondiel photoperiods. *J. Comp. Physiol.* 103, 227–45.

Beck, S.D. (1976a). Photoperiodic determination of insect development and diapause. IV. Effects of skeleton photoperiods. *J. Comp. Physiol.* 105, 267–77.

Beck, S.D. (1976b). Photoperiodic determination of insect development and diapause. V. Diapause, circadian rhythms, and phase response curves, according to the dual system theory. *J. Comp. Physiol.* 107, 97–111.

Beck, S.D. (1977). Dual system theory of the biological clock: Effects of photoperiod, temperature, and thermoperiod on the determination of diapause. *J. Insect Physiol.* 23, 1363–72.

Beck, S.D. (1980). *Insect photoperiodism* (2nd ed). Academic Press, New York.

Beck, S.D. (1982). Thermoperiodic induction of larval diapause in the European corn borer, *Ostrinia nubilalis*. *J. Insect Physiol.* 28, 273–77.

Beck, S.D. (1983a). Insect thermoperiodism. *Annu. Rev. Entomol.* 28, 91–108.

Beck, S.D. (1983b). Thermal and thermoperiodic effects on larval development and diapause in the European corn borer, *Ostrinia nubilalis*. *J. Insect Physiol.* 29, 107–12.

Beck, S.D. (1984). Effect of temperature on thermoperiodic determination of diapause. *J. Insect Physiol.* 30, 383–86.

Beck, S.D. and Apple, J.W. (1961). Effects of temperature and photoperiod on voltinism of geographical populations of the European corn borer, *Pyrausta nubilalis*. *J. Econ. Entomol.* 54, 550–58.

Beck, S.D. and Hanec, W. (1960). Diapause in the European corn borer, *Pyrausta nubilalis* (Hübn.). *J. Insect Physiol.* 4, 304–18.

Beckage, N.E. (1985). Endocrine interactions between endoparasitic insects and their hosts. *Annu. Rev. Entomol.* 30, 371–413.

Begon, M. (1976). Temporal variations in the reproductive condition of *Drosophila obscura* Fallén and *D. subobscura* Collin. *Oecologia* 23, 31–47.

Behrendt, K. (1963). Über die Eidiapause von *Aphis fabae* Scop. (Homoptera, Aphididae). *Zool. Jb. Physiol.* 70, 309–98.

Bell, C.H. (1976a). Factors governing the induction of diapause in *Ephestia elutella* and *Plodia interpunctella* (Lepidoptera). *Physiol. Entomol.* 1, 83–91.

Bell, C.H. (1976b). Factors influencing the duration and termination of diapause in the warehouse moth *Ephestia elutella. Physiol. Entomol.* 1, 169–78.

Bell, C.H. (1976c). Factors influencing the duration and termination of diapause in the Indian-meal moth, *Plodia interpunctella. Physiol. Entomol.* 1, 93–101.

Bell, C.H. (1982). Observations on the intensity of diapause and cold tolerance in larvae from twelve populations and two reciprocal crosses of the Indian meal moth, *Plodia interpunctella. Physiol. Entomol.* 7, 371–77.

Bell, C.H. (1983). The regulation of development during diapause in *Ephestia elutella* (Hübner) by temperature and photoperiod. *J. Insect Physiol.* 29, 485–90.

Bell, C.H. and Bowley, C.R. (1980). Effect of photoperiod and temperature on diapause in a Florida strain of the tropical warehouse moth *Ephestia cautella. J. Insect Physiol.* 26, 533–38.

Bell, C.H., Bowley, C.R., Cogan, P.M., and Sharma, S. (1979). Diapause in twenty-three populations of *Plodia interpunctella* (Hübner) (Lep., Pyralidae) from different parts of the world. *Ecol. Entomol.* 4, 193–97.

Bell, R.A. and Adkisson, P.L. (1964). Photoperiodic reversibility of diapause induction in an insect. *Science* 144, 1149–51.

Bell, R.A., Nelson, D.R., Borg, T.K., and Cardwell, D.L. (1975). Wax secretion in non-diapausing and diapausing pupae of the tobacco hornworm *Manduca sexta. J. Insect Physiol.* 21, 1725–29.

Bellamy, R.E. and Reeves, W.C. (1963). The winter biology of *Culex tarsalis* (Diptera: Culicidae) in Kern County, California. *Ann. Entomol. Soc. Amer.* 56, 314–23.

Belozerov, V.N. (1975). Photoperiodic regulation of behavioral reactions of ixodid ticks (Acarina, Ixodidae). In *Insect behavior as a basis for developing control measures against pests of field crops and forests* (ed. V.P. Pristavko) pp. 1–5. Naukova Dumka Publ., Kiev (English translation 1981, Oxonian Press Pvt. Ltd., New Delhi).

Belozerov, V.N. (1982). Diapause and biological rhythms in ticks. In *Physiology of ticks* (eds. F.D. Obenchain and R. Galun) pp. 469–500. Pergamon Press, Oxford.

Belozerov, V.N. and Galyal'murad, M. (1977). Photoperiodic regulation of nymphal diapause in *Hyalomma anatolicum* (Acarina, Ixodidae). *Entomol. Rev.* 3, 1–7.

Bengston, M. (1965). Overwintering behaviour of *Tetranychus telarius* (L.) in the Stanthorpe district, Queensland. *Queensland J. Agric. Anim. Sci.* 22, 170–76.

Benschoter, C.A. (1970). Culturing *Heliothis* species (Lepidoptera: Noctuidae) for investigation of photoperiod and diapause relationships. *Ann. Entomol. Soc. Amer.* 63, 699–701.

Berg, C.O. and Knutson, L. (1978). Biology and systematics of the Sciomyzidae. *Annu. Rev. Entomol.* 23, 239–58.

Berlinger, M.J. and Ankersmit, G.W. (1976). Manipulation with the photoperiod as a method of control of *Adoxophyes orana* (Lepidoptera, Tortricidae). *Entomol. Exp. Appl.* 19, 96–107.

Berry, S.J. (1981). Hormones and metabolism in the pupal diapause of silkmoths (Lepidoptera: Saturniidae). *Entomol. Gen.* 7, 233–43.

Bevan, D. and Carter, C.I. (1980). Frost proofed aphids. *Antenna* 4, 6–7.

Bewley, J.D. and Black, M. (1982). *Physiology and biochemistry of seeds*, Vol. 2. Springer-Verlag, Berlin.

Birch, L.C. (1942). The influence of temperatures above the developmental zero on the development of the eggs of *Austroicetes cruciata* Sauss. (Orthoptera). *Aust. J. Exp. Biol. & M. Sci.* 20, 17–25.

Birch, L.C. (1945). Diapause in *Scelio chortoicetes* Frogg. (Scelionidae), a parasite of the eggs of *Austroicetes cruciata* Sauss. *J. Aust. Inst. Agr. Sci.* 11, 189–90.

Birch, M.C. (1974). Seasonal variation in pheromone-associated behavior and physiology of *Ips pini. Ann. Entomol. Soc. Amer.* 67, 58–60.

Blackman, R.L. (1971). Variation in the photoperiodic response within natural populations of *Myzus persicae* (Sulz.) *Bull. Entomol. Res.* 60, 533–46.

Blackman, R.L. (1972). The inheritance of life-cycle differences in *Myzus persicae* (Sulz.) (Hem., Aphididae). *Bull. Entomol. Res.* 62, 281–94.

Blackman, R.L. (1974a). *Aphids.* Ginn & Co., Ltd., London.

Blackman, R.L. (1974b). Life cycle variation of *Myzus persicae* (Sulz.) (Hom., Aphididae) in different parts of the world, in relation to genotype and environment. *Bull. Entomol. Res.* 63, 595–607.

Blackman, R.L. (1975). Photoperiodic determination of the male and female sexual morphs of *Myzus persicae. J. Insect Physiol.* 21, 435–53.

Blake, G.M. (1958). Diapause and the regulation of development in *Anthrenus verbasci* (L.) (Col., Dermestidae). *Bull. Entomol. Res.* 49, 751–75.

Blake, G.M. (1959). Control of diapause by an "internal clock" in *Anthrenus verbasci* (L.) (Col., Dermestidae). *Nature* 183, 126–27.

Blake, G.M. (1960). Decreasing photoperiod inhibiting metamorphosis in an insect. *Nature* 188, 168–69.

Blake, G.M. (1963). Shortening of a diapause-controlled life-cycle by means of increasing photoperiod. *Nature* 198, 462–63.

Blakley, N. (1980). Divergence in seed resource use among Neotropical milkweed bugs *(Oncopeltus). Oikos* 35, 8–15.

Blakley, N. (1981). Life history significance of size-triggered metamorphosis in milkweed bugs (*Oncopeltus*). *Ecology* 62, 57–64.

Blakley, N. and Goodner, S.R. (1978). Size-dependent timing of metamorphosis in milkweed bugs *(Oncopeltus)* and its life history implications. *Biol. Bull.* 155, 499–510.

Blau, W.S. (1981). Latitudinal variation in the life histories of insects occupying disturbed habitats: A case study. In *Insect life history patterns* (eds. R.F. Denno and H. Dingle) pp. 75–95. Springer-Verlag, New York.

Block, W. (1982). Cold hardiness in invertebrate poikilotherms. *Comp. Biochem. Physiol.* 73A, 581–93.

Block, W. and Sømme, L. (1982). Cold hardiness of terrestrial mites at Signy Island, maritime Antarctic. *Oikos* 38, 157–67.

Boctor, I.Z. (1981). Changes in the free amino acids of the haemolymph of diapause and nondiapause pupae of the cotton bollworm, *Heliothis armigera* Hbn. (Lepidoptera: Noctuidae). *Experientia* 37, 125–26.

Bodnaryk, R.P. (1977). Stages of diapause development in the pupa of *Mamestra configurata* based on the β-ecdysone sensitivity index. *J. Insect Physiol.* 23, 537–42.

Bodnaryk, R.P. (1978). Factors affecting diapause development and survival in the pupa of *Mamestra configurata* (Lepidoptera: Noctuidae). *Can. Entomol.* 110, 183–91.

Bohle, H.W. (1972). Die Temperturabhängigkeit der Embryogenese und der embryonalen Diapause von *Ephemerella ignita* (Poda). *Oecologia* 10, 253–68.

Bohm, M.K. (1972). Effects of environment and juvenile hormone on ovaries of the wasp, *Polistes metricus. J. Insect Physiol.* 18, 1875–83.

Boiteau, G., Bradley, J.R., Jr., and Van Duyn, J.W. (1979). Bean leaf beetle: Some seasonal anatomical changes and dormancy. *Ann. Entomol. Soc. Amer.* 72, 303–7.

Boller, E.F. (1979). Behavioral aspects of quality in insectary production. In *Genetics in relation to insect management* (eds. M.A. Hoy and J.J. McKelvey, Jr.) pp. 153–60. Rockefeller Foundation, New York.

Boller, E.F. and Bush, G.L. (1974). Evidence for genetic variation in populations of the European cherry fruit fly, *Rhagoletis cerasi* (Diptera: Tephritidae) based on physiological parameters and hybridization experiments. *Entomol. Exp. Appl.* 17, 279–93.

Bonnemaison, L. (1951). Contribution à l'étude des facteurs provoquant l'apparition des formes ailées et sexuées chez les Aphidinae (Part 3). *Annls. Épiphyt.* 2, 205–380.

Bonnemaison, L. (1978). Actions de l'obscurité et de la lumière sur l'induction de la diapause chez trois espèces de Lépidoptères. *Z. Ang. Entomol.* 86, 178–204.

Borden, J.H. (1977). Behavioral responses of Coleoptera to pheromones, allomones, and kairomones. In *Chemical control of insect behavior* (eds. H.H. Shorey and J.J. McKelvey, Jr.) pp. 169–98. Wiley-Interscience, New York.

Borror, D.J., Delong, D.M., and Triplehorn, C.A. (1981). *An introduction to the study of insects* (5th ed.). Holt, Rinehart and Winston, New York.

Bowden, J. (1976). Weather and the phenology of some African Tabanidae. *J. Entomol. Soc. S. Afr.* 39, 207–45.

Bowen, M.F., Bollenbacher, W.E., and Gilbert, L.I. (1984). *In vitro* studies on the role of the brain and prothoracic glands in the pupal diapause of *Manduca sexta. J. Exp. Biol.* 108, 9–24.

Bowers, W.S. (1971). Insect hormones and their derivatives as insecticides. *Bull. Org. Mond. Santé (Bull. Wld. Hlth. Org.)* 44, 381–89.

Bowers, W.S. (1982). Endocrine strategies for insect control. *Entomol. Exp. Appl.* 31, 3–14.

Bowers, W.S. and Blickenstaff, C.C. (1966). Hormonal termination of diapause in the alfalfa weevil. *Science* 154, 1673–74.

Bowers, W.S., Ohta, T., Cleere, J.S. and Marsella, P.A. (1976). Discovery of insect anti-juvenile hormones in plants. *Science* 193, 542–47.

Bradfield, J.Y., IV, and Denlinger, D.L. (1980). Diapause development in the tobacco hornworm: A role for ecdysone or juvenile hormone? *Gen. Comp. Endocrinol.* 41, 101–7.

Bradley, B.P. (1982). Models for physiological and genetic adaptation to variable environments. In *Evolution and genetics of life histories* (eds. H. Dingle and J.P. Hegmann) pp. 33–50. Springer-Verlag, New York.

Bradshaw, W.E. (1969). Major environmental factors inducing the termination of larval diapause in *Chaoborus americanus* Johannsen (Diptera: Culicidae). *Biol. Bull.* 136, 2–8.

Bradshaw, W.E. (1970). Interaction of food and photoperiod in the termination of larval diapause in *Chaoborus americanus. Biol. Bull.* 139, 476–84.

Bradshaw, W.E. (1972). Action spectra for photoperiodic response in a diapausing mosquito. *Science* 175, 1361–62.

Bradshaw, W.E. (1973). Homeostasis and polymorphism in vernal development of *Chaoborus americanus. Ecology* 54, 1247–59.

Bradshaw, W.E. (1974a). Photoperiodic control of development in *Chaoborus americanus* with special reference to photoperiodic action spectra. *Biol. Bull.* 146, 11–19.

Bradshaw, W.E. (1974b). Phenology and seasonal modeling in insects. In *Phenology and seasonality modeling* (ed. H. Lieth) pp. 127–37. Springer-Verlag, Berlin.

Bradshaw, W.E. (1976). Geography of photoperiodic responses in diapausing mosquito. *Nature* 262, 384–85.

Bradshaw, W.E. and Holzapfel, C.M. (1975). Biology of tree-hole mosquitoes: Photoperiodic control of development in northern *Toxorhynchites rutilus* (Coq.). *Can. J. Zool.* 53, 889–93.

Bradshaw, W.E. and Holzapfel, C.M. (1983). Predator-mediated, non-equilibrium coexistence of tree-hole mosquitoes in southeastern North America. *Oecologia* 57, 239–56.

Bradshaw, W.E. and Holzapfel, C.M. (1984). Seasonal development of tree-hole mosquitoes (Diptera: Culicidae) and chaoborids in relation to weather and predation. *J. Med. Entomol.* 21, 366–78.

Bradshaw, W.E. and Lounibos, L.P. (1977). Evolution of dormancy and its photoperiodic control in pitcher-plant mosquitoes. *Evolution* 31, 546–67.

Bradshaw, W.E. and Phillips, D.L. (1980). Photoperiodism and the photic environment of the pitcher-plant mosquito, *Wyeomyia smithii. Oecologia* 44, 311–16.

Brakefield, P.M. and Larsen, T.B. (1984). The evolutionary significance of dry and wet season forms in some tropical butterflies. *Biol. J. Linnean Soc.* 22, 1–12.

Branson, T.F. (1976). The selection of a non-diapause strain of *Diabrotica virgifera* (Coleoptera: Chrysomelidae). *Entomol. Exp. Appl.* 19, 148–54.

Branson, T.F. and Krysan, J.L. (1981). Feeding and oviposition behavior and life cycle strategies of *Diabrotica*: An evolutionary view with implications for pest management. *Environ. Entomol.* 10, 826–31.

Branson, T.F., Guss, P.L., and Krysan, J.L. (1978). Winter populations of some *Diabrotica* in central Mexico. *Ann. Entomol. Soc. Amer.* 71, 165–66.

Branson, T.F., Reyes R., J., and Valdes M., H. (1982). Field biology of Mexican corn rootworm, *Diabrotica virgifera zeae* (Coleoptera: Chrysomelidae), in Central Mexico. *Environ. Entomol.* 11, 1078–83.

Braune, H.J. (1983). The influence of environmental factors on wing polymorphism in females of *Leptopterna dolobrata* (Heteroptera, Miridae). *Oecologia* 60, 340–47.

Brazzel, J.R. and Newsom, L.D. (1959). Diapause in *Anthonomus grandis* Boh. *J. Econ. Entomol.* 52, 603–11.

Breed, M.D. (1975). Sociality and seasonal size variation in halictine bees. *Insectes Sociaux* 22, 375–80.

Brelje, N.N. and Blickenstaff, C.C. (1974). Refrigeration of egg pods of a nondiapausing strain of the migratory grasshopper. *J. Econ. Entomol.* 67, 134–35.

Brian, M.V. (1955). Studies of caste differentiation in *Myrmica rubra* L. 3. Larval dormancy, winter size and vernalisation. *Insectes Sociaux* 2, 85–114.

Brian, M.V. (1963). Studies of caste differentiation in *Myrmica rubra* L. 6. Factors influencing the course of female development in the early third instar. *Insectes Sociaux* 10, 91–102.

Brian, M.V. (1965a). *Social insect populations.* Academic Press, London.

Brian, M.V. (1965b). Studies of caste differentiation in *Myrmica rubra* L. 8. Larval developmental sequences. *Insectes Sociaux* 12, 347–62.

Brian, M.V. (1975). Caste determination through a queen influence on diapause in larvae of the ant *Myrmica rubra. Entomol. Exp. Appl.* 18, 429–42.

Brian, M.V. (1977). The synchronisation of colony and climatic cycles. In *Proc. VIII Internat'l. Cong.*, pp. 202–6. International Union Study Social Insects, Wageningen, The Netherlands.

Brian, M.V. (1979). Caste differentiation and division of labor. In *Social insects*, Vol. 1 (ed. H.R. Hermann) pp. 121–222. Academic Press, New York.

Brian, M.V. and Kelly, A.F. (1967). Studies of caste differentiation in *Myrmica rubra* L. 9. Maternal environment and the caste bias of larvae. *Insectes Sociaux* 14, 13–24.

Briers, T., Peferoen, M., and de Loof, A. (1982). Ecdysteroids and adult diapause in the Colorado potato beetle, *Leptinotarsa decemlineata. Physiol. Entomol.* 7, 379–86.

Brindley, T.A., Sparks, A.N., Showers, W.B., and Guthrie, W.D. (1975). Recent research advances on the European corn borer in North America. *Annu. Rev. Entomol.* 20, 221–39.

Brittain, J.E. and Nagell, B. (1981). Overwintering at low oxygen concentrations in the mayfly *Leptophlebia vespertina. Oikos* 36, 45–50.

Brockmann, H.J. (1984). The evolution of social behaviour in insects. In *Behavioural ecology*, 2nd ed. (ed. J.R. Krebs and N.B. Davies) pp. 340–61. Sinauer Assoc., Sunderland, MA.

Broodryk, S.W. (1969). The biology of *Chelonus* (*Microchelonus*) *curvimaculatus* Cameron (Hymenoptera: Braconidae). *J. Entomol. Soc. S. Afr.* 32, 169–89.

Brower, J.H. (1980). Irradiation of diapausing and nondiapausing larvae of *Plodia interpunctella*: Effects on larval and pupal mortality and adult fertility. *Ann. Entomol. Soc. Amer.* 73, 420–26.

Brower, L.P. (1977). Monarch migration. *Natural History*, June/July, pp. 41–53.

Brower, L.P., Calvert, W.H., Hedrick, L.E., and Christian, J. (1977). Biological observations on an overwintering colony of monarch butterflies *(Danaus plexippus*, Danaidae) in Mexico. *J. Lepid. Soc.* 31, 232–42.

Brown, G.C., Berryman, A.A., and Bogyo, T.P. (1979). Density-dependent induction of diapause in the codling moth, *Laspeyresia pomonella* (Lepidoptera: Olethreutidae). *Can. Entomol.* 111, 431–33.

Brown, V.K. (1973). The overwintering stages of *Ectobius lapponicus* (L.) (Dictyoptera: Blattidae). *J. Entomol.* (A), 48, 11–24.

Brown, V.K. and Hodek, I. (eds.) (1983). *Diapause and life cycle strategies in insects.* Junk, The Hague, The Netherlands.

Brown, W.L., Jr., and Wilson, E.O. (1956). Character displacement. *Syst. Zool.* 5, 49–64.

Browning, T.O. (1952). On the rate of completion of diapause development at constant temperatures in the eggs of *Gryllulus commodus* Walker. *Aust. J. Sci. Res.*, Series B, *Biol. Sci.* 5, 344–53.

Browning, T.O. (1965). Observations on the absorption of water, diapause and embryogenesis in the eggs of the cricket *Teleogryllus commodus* (Walker). *J. Exp. Biol.* 43, 433–39.

Browning, T.O. (1979). Timing of the action of photoperiod and temperature on events leading to diapause and development in pupae of *Heliothis punctigera* (Lepidoptera: Noctuidae). *J. Exp. Biol.* 83, 261–69.

Browning, T.O. (1981). Ecdysteroids and diapause in pupae of *Heliothis punctiger. J. Insect Physiol.* 27, 715–19.

Brunel, E. (1968). Etude du développement nymphal de *Psila rosae* Fab. (Diptères Psilidés) en conditions naturelles et expérimentals: quiescence et diapause. *Société de Biologie de Rennes. Comptes Rendus* 162, 2223–28.

Bryan, G. (1983). Seasonal biological variation in some leaf-miner parasites in the genus *Achrysocharoides* (Hymenoptera, Eulophidae). *Ecol. Entomol.* 8, 259–70.

Buechi, R., Priesner, E., and Brunetti, R. (1982). Das sympatrische Vorkommen von zwei Pheromonstämmen des Maiszünslers, *Ostrinia nubilalis* Hbn., in der Sudschweiz. *Mitt. Schweiz. Entomol. Ges. Bull. Soc. Entomol. Suisse* 55, 33–53.

Buffington, J.D. (1972). Hibernaculum choice in *Culex pipiens. J. Med. Entomol.* 9, 128–32.

Bünning, E. (1973). *The physiological clock* (3rd ed). English University Press Ltd., London. Springer-Verlag, New York.

Burbutis, P.P., Curl, G.D., and Davis, C.P. (1976). Overwintering of *Trichogramma nubilale* in Delaware. *Environ. Entomol.* 5, 888–90.

Burn, A.J. and Coaker, T.H. (1981). Diapause and overwintering of the carrot fly, *Psila rosae* (F.) (Diptera: Psilidae). *Bull. Entomol. Res.* 71, 583–90.

Bush, G.L. (1969). Sympatric host race formation and speciation in frugivorous flies of the genus *Rhagoletis* (Diptera, Tephritidae). *Evolution* 23, 237–51.

Bush, G.L. (1974). The mechanism of sympatric host race formation in the true fruit flies. In *Genetic mechanisms of speciation in insects* (ed. M.J.D. White) pp. 3–23. D. Reidel, Boston.

Bush, G.L. (1975a). Modes of animal speciation. *Annu. Rev. Ecol. Syst.* 6, 339–64.

Bush, G.L. (1975b). Sympatric speciation in phytophagous parasitic insects. In *Evolutionary strategies of parasitic insects* (ed. P.W. Price) pp. 187–206. Plenum, London.

Bush, G.L. (1978). Planning a rational quality control program for the screwworm fly. In *The screwworm problem: Evolution of resistance to biological control* (ed. R.H. Richardson) pp. 37–47. University of Texas Press, Austin.

Bush, G.L. (1979). Ecological genetics and quality control. In *Genetics in relation to insect management* (eds. M.A. Hoy and J.J. McKelvey, Jr.) pp. 145–52. Rockefeller Foundation, New York.

Bush, G.L., Neck, R.W., and Kitto, G.B. (1976). Screwworm eradication: Inadvertent selection for non-competitive ecotypes during mass rearing. *Science* 193, 491–93.

Butler, M.G. (1982). A 7-year life cycle for two *Chironomus* species in arctic Alaskan tundra ponds (Diptera: Chironomidae). *Can. J. Zool.* 60, 58–70.

Butler, M.G. (1984). Life histories of aquatic insects. In *The ecology of aquatic insects* (eds. V.H. Resh and D.M. Rosenberg) pp. 24–55. Praeger Scientific, New York.

Butterfield, J. (1976). Effect of photoperiod on a winter and on a summer diapause in two species of cranefly (Tipulidae). *J. Insect Physiol.* 22, 1443–46.

Callan, E. McC. (1969). Ecology and insect colonization for biological control. *Proc. Ecol. Soc. Aust.* 4, 17–31.

Calvert, W.H. and Brower, L.P. (1981). The importance of forest cover for the survival of overwintering monarch butterflies (*Danaus plexippus*, Danaidae). *J. Lepid. Soc.* 35, 216–25.

Calvert, W.H. and Cohen, J.A. (1983). The adaptive significance of crawling up onto foliage for the survival of grounded overwintering monarch butterflies *(Danaus plexippus)* in Mexico. *Ecol. Entomol.* 8, 471–74.

Campbell, A., Frazer, B.D., Gilbert, N., Gutierrez, A.P., and Mackauer, M. (1974). Temperature requirements of some aphids and their parasites. *J. Appl. Ecol.* 11, 431–38.

Campbell, R.K. (1974). Use of phenology for examining provenance transfers in reforestation of Douglas fir. *J. Appl. Ecol.* 11, 1069–80.

Campbell, R.W. (1978). Some effects of gypsy moth density on rate of development, pupation time, and fecundity. *Ann. Entomol. Soc. Amer.* 71, 442–48.

Canard, M. (1982). Diapause reproductrice photoperiodique chez les adultes de *Nineta flava* (Scopoli) (Neuroptera, Chrysopidae). *Neuroptera International* 2, 59–68.

Cantelo, W.W. (1974). Diapause in a tropical strain of the tobacco hornworm. *Ann. Entomol. Soc. Amer.* 67, 828–30.

Cardé, R.T., Shapiro, A.M., and Clench, H.K. (1970). Sibling species in the *eurydice* group of *Lethe* (Lepidoptera: Satyridae). *Psyche* 77, 70–103.

Cardé, R.T., Roelofs, W.L., Harrison, R.G., Vawter, A.T., Brussard, P.F., Mutuura, A., and Munroe, E. (1978). European corn borer: Pheromone polymorphism or sibling species? *Science* 199, 555–56.

Carmona, A.S. and Barbosa, P. (1983). Overwintering egg mass adaptations of the eastern tent caterpillar, *Malacosoma americanum* (Fab.) (Lepidoptera: Lasiocampidae). *N.Y. Entomol. Soc.* 9, 68–74.

Carson, H.L. and Stalker, H.D. (1948). Reproductive diapause in *Drosophila robusta. Proc. Nat. Acad. Sci.* 34, 124–29.

Carter, N., Dixon, A.F.G., and Rabbinge, R. (1982). Cereal aphid populations: Biology, simulation and prediction. Wageningen Centre for Agricultural Publishing and Documentation, Wageningen, The Netherlands.

Carton, Y. and Claret, J. (1982). Adaptative significance of a temperature induced diapause in a cosmopolitan parasitoid of *Drosophila. Ecol. Entomol.* 7, 239–47.

Casagrande, R.A., Ruesink, W.G., and Haynes, D.L. (1977). The behavior and survival of adult cereal leaf beetles. *Ann. Entomol. Soc. Amer.* 70, 19–30.

Case, T.J., Washino, R.K., and Dunn, R.L. (1977). Diapause termination in *Anopheles freeborni* with juvenile hormone mimics. *Entomol. Exp. Appl.* 21, 155–62.

Cassier, P. (1966). Variabilité des effets de groupe (Effets immédiats et transmis) sur *Locusta migratoria migratorioides* (R. et F.). *Extrait du Bulletin Biologique de la France et de la Belgique*, Tome C 4, 520–22.

Cassier, P. (1967a). La variabilité des caractéristiques phasaires chez le Criquet migrateur africain, *Locusta migratoria migratorioides* (R. et F.), phase gregaria. Les effets cumulatifs des photopériodes longues. *Insectes Sociaux* 14, 193–228.

Cassier, P. (1967b). Variabilité des caractères phasaires chez le Criquet migrateur africain, *Locusta migratoria migratorioides* (R. et F.) phase grégaire. III. Les variations liés au poids a la naissance des femelles reproductrices. *Bulletin Biologique de la France et de la Belgique* 101, 219–46.

Cassier, P. (1967c). Sommation des effets photopériodiques et déterminisme des caractères phasaires chez *Locusta migratoria migratorioides*, phase grégaire. *Ann. Soc. Entomol. Fr.* 3, 873–83.

Cassier, P. (1968). Détermination de la période photosensible chez le Criquet migrateur africain, *Locusta migratoria migratorioides* (R. et F.), phase grégaire. *Insectes Sociaux* 15, 3–30.

Chambers, R.J. (1982). Maternal experience of crowding and duration of aestivation in the sycamore aphid. *Oikos* 39, 100–2.

Chantarasa-ard, S. (1984). Preliminary study of the overwintering of *Anagrus incarnatus* Haliday (Hymenoptera: Mymaridae), an egg parasitoid of the rice planthoppers. *Esakia* 22, 159–62.

Chaplin, S.B. and Wells, P.H. (1982). Energy reserves and metabolic expenditures of monarch butterflies overwintering in southern California. *Ecol. Entomol.* 7, 249–56.

Chapman, R.F. and Page, W.W. (1978). Embryonic development and water relations of the eggs of *Zonocerus variegatus* (L.) (Acridoidea: Pyrgmorphidae). *Acrida* 7, 243–52.

Chappell, M.A. (1983). Metabolism and thermoregulation in desert and montane grasshoppers. *Oecologia* 56, 126–31.

Charlesworth, P. and Shorrocks, B. (1980). The reproductive biology and diapause of the British fungal-breeding *Drosophila*. *Ecol. Entomol.* 5, 315–26.

Cherednikov, A.V. (1967). Photoperiodism in the honey bee, *Apis mellifera* L. (Hymenoptera, Apidae). *Entomol. Rev.* 46, 33–37.

Chetverikov, S.S. (1940). The selection of the Chinese oak silkworm (*Antheraea pernyi* Guer.) for univoltinism. In *Selection and acclimatization of oak silkworms* (ed. M. Sel'khozgiz) pp. 16–22. (through Hoy 1978a).

Chiang, H.C., Keaster, A.J., and Reed, G.L. (1968). Differences in ecological responses of three biotypes of *Ostrinia nubilalis* from the north central United States. *Ann. Entomol. Soc. Amer.* 61, 140–46.

Chino, H. (1957). Carbohydrate metabolism in diapause egg of the silkworm, *Bombyx mori*. I. Diapause and the change in glycogen content. *Embryologia* 3, 295–316.

Chino, H. (1958). Carbohydrate metabolism in the diapause egg of the silkworm, *Bombyx mori*—II. Conversion of glycogen into sorbitol and glycerol during diapause. *J. Insect Physiol.* 2, 1–12.

Chippendale, G.M. (1977). Hormonal regulation of larval diapause. *Annu. Rev. Entomol.* 22, 121–38.

Chippendale, G.M. (1978). Behavior associated with the larval diapause of the southwestern corn borer, *Diatraea grandiosella*: Probable involvement of juvenile hormone. *Ann. Entomol. Soc. Amer.* 71, 901–5.

Chippendale, G.M. (1979). The southwestern corn borer, *Diatraea grandiosella*: Case history of an invading insect. *Univ. Missouri-Columbia Agr. Exp. Sta. Res. Bull.* 1031, 4–52.

Chippendale, G.M. (1982). Insect diapause, the seasonal synchronization of life cycles, and management strategies. *Entomol. Exp. Appl.* 31, 24–35.

Chippendale, G.M. (1983). Larval and pupal diapause. In *Endocrinology of insects* (eds. R.G.H. Downer and H. Laufer) pp. 343–56. Alan R. Liss, New York.

Chippendale, G.M. and Kikukawa, S. (1983). Effect of daylength and temperature on the larval diapause of the sunflower moth, *Homoeosoma electellum*. *J. Insect Physiol.* 29, 643–49.

Chippendale, G.M. and Reddy, A.S. (1972). Diapause of the southwestern corn borer, *Diatraea grandiosella*: Transition from spotted to immaculate mature larvae. *Ann. Entomol. Soc. Amer.* 65, 882–87.

Chippendale, G.M. and Reddy, A.S. (1973). Temperature and photoperiodic regulation of diapause of the southwestern corn borer, *Diatraea grandiosella. J. Insect Physiol.* 19, 1397–1408.

Chippendale, G.M. and Turunen, S. (1981). Hormonal and metabolic aspects of the larval diapause of the southwestern corn borer, *Diatraea grandiosella* (Lepidoptera: Pyralidae). *Entomol. Gen.* 7, 223–31.

Chippendale, G.M. and Yin, C.-M. (1976). Endocrine interactions controlling the larval diapause of the southwestern corn borer, *Diatraea grandiosella. J. Insect Physiol.* 22, 989–95.

Chippendale, G.M. and Yin, C.-M. (1979). Larval diapause of the European corn borer, *Ostrinia nubilalis*: Further experiments examining its hormonal control. *J. Insect Physiol.* 25, 53–58.

Chippendale, G.M., Reddy, A.S., and Catt, C.L. (1976). Photoperiodic and thermoperiodic interactions in the regulation of the larval diapause of *Diatraea grandiosella. J. Insect Physiol.* 22, 823–28.

Church, N.S. (1955). Hormones and the termination and reinduction of diapause in *Cephus cinctus* Nort. *Can. J. Zool.* 33, 339–69.

Church, N.S. and Salt, R.W. (1952). Some effects of temperature on development and diapause in eggs of *Melanoplus bivittatus* (Say) (Orthoptera: Acrididae). *Can. J. Zool.* 30, 173–84.

Claret, J. (1966). Mise en évidence du rôle photorécepteur du cerveau dans l'induction de la diapause, chez *Pieris brassicae* (Lepido.). *Ann. d'Endocrinol.* (Paris) 27 (suppl.) 311–20.

Claret, J. (1973). La diapause facultative de *Pimpla instigator* (Hymenoptera, Ichneumonidae) I. Role de la photopériode. *Entomophaga* 18, 409–18.

Claret, J. (1978). La diapause facultative de *Pimpla instigator* (Hym.: Ichneumonidae) II. Rôle de la température. *Entomophaga* 233, 411–15.

Claret, J. (1982). Modification du signal photopériodique par la cuticle de l'hôte pour un endoparasite. *Comp. Rend. Soc. Biol.* 176, 834–38.

Claret, J. and Carton, Y. (1980). Diapause in a tropical species, *Cothonaspis boulardi* (Parasitic Hymenoptera). *Oecologia* 45, 32–34.

Claret, J., Porcheron, P., and Dray, F. (1978). La teneur en ecdysones circulants au cours du dernier stade larvaire de l'Hyménoptère endoparasite *Pimpla instigator* et l'entrée en diapause. *C. R. Acad. Sci.*, Paris, Series D, 286, 639–41.

Claridge, M.F. and Wilson, M.R. (1978). Seasonal changes and alternation of food plant preference in some mesophyll-feeding leafhoppers. *Oecologia* 37, 247–55.

Clay, M.E. and Venard, C.E. (1972). Larval diapause in the mosquito *Aedes triseriatus*: Effects of diet and temperature on photoperiodic induction. *J. Insect Physiol.* 18, 1441–46.

Clements, A.N. (1963). *The physiology of mosquitoes.* Pergamon Press, Oxford.

Cloutier, E.J. and Beck, S.D. (1963). Spermatogenesis and diapause in the European corn borer, *Ostrinia nubilalis. Ann. Entomol. Soc. Amer.* 56, 253–55.

Cohen, A.C. and Cohen, J.L. (1981). Microclimate, temperature and water relations of two species of desert cockroaches. *Comp. Biochem. Physiol.* 69A, 165–67.

Cohet, Y., Vouidibio, J., and David, J.R. (1980). Thermal tolerance and geographic distribution: A comparison of cosmopolitan and tropical endemic *Drosophila* species. *J. Therm. Biol.* 5, 69–74.

Collier, R.H. and Finch, S. (1983a). Completion of diapause in field populations of the cabbage root fly *(Delia radicum). Entomol. Exp. Appl.* 34, 186–92.

Collier, R.H. and Finch, S. (1983b). Effects of intensity and duration of low temperatures in regulating diapause development of the cabbage root fly *(Delia radicum). Entomol. Exp. Appl.* 34, 193–200.

Connin, R.V. and Hoopingarner, R.A. (1971). Sexual behavior and diapause of the cereal leaf beetle, *Oulema melanopus. Ann. Entomol. Soc. Amer.* 64, 655–60.

Conradi-Larsen, E.-M. and Sømme, L. (1973). Anaerobiosis in the overwintering beetle *Pelophila borealis. Nature* 245, 388–90.

Conradi-Larsen, E.-M. and Sømme, L. (1978). The effect of photoperiod and temperature on imaginal diapause in *Dolycoris baccarum* from southern Norway. *J. Insect Physiol.* 24, 243–49.

Corbet, P.S. (1956). The influence of temperature on diapause development in the dragonfly *Lestes sponsa* (Hansemann) (Odonata: Lestidae). *Proc. Roy. Entomol. Soc. Lond.* (A) 31, 45–48.

Corbet, P.S. (1963). *A biology of dragonflies.* Quadrangle Books, Chicago.

Corbet, P.S., Longfield, C., and Moore, N.W. (1960). *Dragonflies.* Collins, St. James Place, London.

Coulson, J.C., Horobin, J.C., Butterfield, J., and Smith, G.R.J. (1976). The maintenance of annual life-cycles in two species of Tipulidae (Diptera); A field study relating development, temperature and altitude. *J. Anim. Ecol.* 45, 215–33.

Cousin, G. (1961). Essai d'analyse de la spéciation chez quelques gryllides du continent Américain. *Bull. Biol.* 95, 155–74.

Cox, P.D., Mfon, M., Parkin, S., and Seaman, J.E. (1981). Diapause in a Glasgow strain of the flour moth, *Ephestia kuehniella. Physiol. Entomol.* 6, 349–56.

Cranham, J.E. (1972). Influence of temperature on hatching of winter eggs of fruit-tree red spider mite, *Panonychus ulmi* (Koch). *Ann. Appl. Biol.* 70, 119–37.

Cranham, J.E. (1973). Variation in the intensity of diapause in winter eggs of fruit tree red spider mite, *Panonychus ulmi. Ann. Appl. Biol.* 75, 173–82.

Crawford, C.S. (1981). *Biology of desert invertebrates.* Springer-Verlag, Berlin.

Croft, B.A. (1971). Comparative studies on four strains of *Typhlodromus occidentalis* (Acarina: Phytoseiidae). V. Photoperiodic induction of diapause. *Ann. Entomol. Soc. Amer.* 64, 962–64.

Crozier, A.J.G. (1979a). Supradian and infradian cycles in oxygen uptake of diapausing pupae of *Pieris brassicae. J. Insect Physiol.* 25, 575–82.

Crozier, A.J.G. (1979b). Diel oxygen uptake rhythms in diapausing pupae of *Pieris brassicae* and *Papilio machaon. J. Insect Physiol.* 25, 647–52.

Cullen, J.M. and Browning, T.O. (1978). The influence of photoperiod and temperature on the induction of diapause in pupae of *Heliothis punctigera. J. Insect Physiol.* 24, 595–601.

Curl, G.D. and Burbutis, P.P. (1977). The mode of overwintering of *Trichogramma nubilale* Ertle and Davis. *Environ. Entomol.* 6, 629–32.

Dallmann, S.H. and Herman, W.S. (1978). Hormonal regulation of hemolymph lipid concentration in the Monarch butterfly, *Danaus plexippus. Gen. Comp. Endocrinol.* 36, 142–50.

Danilevsky, A.S. (1965). *Photoperiodism and seasonal development of insects.* Oliver & Boyd, London (English translation).

Danilevsky, A.S., Goryshin, N.I., and Tyshchenko, V.P. (1970). Biological rhythms in terrestrial arthropods. *Annu. Rev. Entomol.* 15, 201–44.

Danks, H.V. (1971a). Overwintering of some north temperate and arctic Chironomidae. I. The winter environment. *Can. Entomol.* 103, 589–604.

Danks, H.V. (1971b). Overwintering of some north temperate and arctic Chironomidae. II. Chironomid biology. *Can. Entomol.* 103, 1875–1910.

Danks, H.V. (1978). Modes of seasonal adaptation in the insects I. Winter survival. *Can. Entomol.* 110, 1167–1205.

Danks, H.V. (1981). *Arctic arthropods.* Entomological Society of Canada, Ottawa.

Danks, H.V. (1983). Extreme individuals in natural populations. *Bull. Entomol. Soc. Amer.* 29, 41–46.

Danks, H.V. and Oliver, D.R. (1972). Seasonal emergence of some high arctic Chironomidae (Diptera). *Can. Entomol.* 104, 661–86.

Davey, K.G. (1956). The physiology of dormancy in the sweetclover weevil. *Can. J. Zool.* 34, 86–98.

Davidson, J. and Andrewartha, H.G. (1948). The influence of rainfall, evaporation and atmospheric temperature on fluctuations in the size of a natural population of *Thrips imaginis* (Thysanoptera). *J. Anim. Ecol.* 17, 200–22.

Davies, D.E. (1952). Seasonal breeding and migrations of the desert locust (*Schistocerca gregaria* Forskål) in north-eastern Africa and the Middle East. *Anti-Locust Memoir* 4, 1–57.

Davis, J.R. and Kirkland, R.L. (1982). Physiological and environmental factors related to the dispersal flight of the convergent lady beetle *Hippodamia convergens* (Guerin-Meneville). *J. Kans. Entomol. Soc.* 55, 187–96.

Dawe, A.R. (1983). Book review: *Hibernation and torpor in mammals and birds. Q. Rev. Biol.* 58, 449.

Day, K. (1984). Phenology, polymorphism and insect-plant relationships of the larch budmoth, *Zeiraphera diniana* (Guenée) (Lepidoptera: Tortricidae), on alternative conifer hosts in Britain. *Bull. Entomol. Res.* 74, 47–64.

Dean, J.M. (1982). Control of diapause induction by a change in photoperiod in *Melanoplus sanguinipes. J. Insect Physiol.* 28, 1035–40.

Dean, R.L. and Hartley, J.C. (1977a). Egg diapause in *Ephippiger cruciger* (Orthoptera: Tettigoniidae). I. The incidence, variable duration and elimination of the initial diapause. *J. Exp. Biol.* 66, 173–83.

Dean, R.L. and Hartley, J.C. (1977b). Egg diapause in *Ephippiger cruciger* (Orthoptera: Tettigoniidae). II. The intensity and elimination of the final egg diapause. *J. Exp. Biol.* 66, 185–95.

DeBach, P. (1965). Some biological and ecological phenomena associated with colonizing entomophagous insects. In *The genetics of colonizing species* (eds. H.G. Baker and G.L. Stebbins) pp. 287–306. Academic Press, New York.

DeBach, P. (1974). *Biological control by natural enemies.* Cambridge University Press, Cambridge.

de Kort, C.A.D. (1981). Hormonal and metabolic regulation of adult diapause in the Colorado beetle, *Leptinotarsa decemlineata* (Coleoptera: Chrysomelidae). *Entomol. Gen.* 7, 261–71.

de Kort, C.A.D. and Granger, N.A. (1981). Regulation of the juvenile hormone titer. *Annu. Rev. Entomol.* 26, 1–28.

de Kort, C.A.D., Schooneveld, H., and de Wilde, J. (1980). Endocrine regulation of seasonal states in the Colorado potato beetle, *Leptinotarsa decemlineata.* In *Integrated control of insect pests in the Netherlands* (ed. A.K. Minks and P. Gruys) pp. 233–40. Wageningen Cent. Agr. Publ. Documentation, Wageningen, The Netherlands.

de Kort, C.A.D., Khan, M.A., Bergot, B.J., and Schooley, D.A. (1981). The JH titre in the Colorado beetle in relation to reproduction and diapause. In *Juvenile hormone biochemistry* (eds. G.E. Pratt and G.T. Brooks) pp. 125–34. Elsevier-North-Holland Biomedical Press, Amsterdam.

de Kort, C.A.D., Bergot, B.J., and Schooley, D.A. (1982). The nature and titre of juvenile hormone in the Colorado potato beetle, *Leptinotarsa decemlineata. J. Insect Physiol.* 28, 471–74.

de Loof, A., Van Loon, J., and Vanderroost, C. (1979). Influence of ecdysterone, precocene, and compounds with juvenile hormone activity on induction, termination and maintenance of diapause in the parasitoid wasp, *Nasonia vitripennis. Physiol. Entomol.* 4, 319–28.

den Boer, P.J. (1981). On the survival of populations in a heterogeneous and variable environment. *Oecologia* 50, 39–53.

Denlinger, D.L. (1971). Embryonic determination of pupal diapause in the flesh fly *Sarcophaga crassipalpis. J. Insect Physiol.* 17, 1815–22.

Denlinger, D.L. (1972a). Seasonal phenology of diapause in the flesh fly *Sarcophaga bullata. Ann. Entomol. Soc. Amer.* 65, 410–14.

Denlinger, D.L. (1972b). Induction and termination of pupal diapause in *Sarcophaga* (Diptera: Sarcophagidae). *Biol. Bull.* 142, 11–24.

Denlinger, D.L. (1974). Diapause potential in tropical flesh flies. *Nature* 252, 223–24.

Denlinger, D.L. (1978). The developmental response of flesh flies (Diptera: Sarcophagidae) to tropical seasons. *Oecologia* 35, 105–7.

Denlinger, D.L. (1979). Pupal diapause in tropical flesh flies: Environmental and endocrine regulation, metabolic rate and genetic selection. *Biol. Bull.* 156, 31–46.

Denlinger, D.L. (1981a). Hormonal and metabolic aspects of pupal diapause in Diptera. *Entomol. Gen.* 7, 245–59.

Denlinger, D.L. (1981b). Basis for a skewed sex ratio in diapause-destined flesh flies. *Evolution* 35, 1247–48.

Denlinger, D.L. and Bradfield, J.Y., IV (1981). Duration of pupal diapause in the tobacco hornworm is determined by number of short days received by the larva. *J. Exp. Biol.* 91, 331–7.

Denlinger, D.L. and Shukla, M. (1984). Increased length and variability of the life cycle in tropical flesh flies (Diptera: Sarcophagidae) that lack pupal diapause. *Ann. Entomol. Soc. Amer.* 77, 46–9.

Denlinger, D.L., Willis, J.H., and Fraenkel, G. (1972). Rates and cycles of oxygen consumption during pupal diapause in *Sarcophaga* flesh flies. *J. Insect Physiol.* 18, 871–82.

Denlinger, D.L., Shukla, M., and Faustini, D.L. (1984). Juvenile hormone involvement in pupal diapause of the flesh fly *Sarcophaga crassipalpis*: Regulation of infradian cycles of O_2 consumption. *J. Exp. Biol.* 109, 191–99.

Denno, R.F. and Grissell, E.E. (1979). The adaptiveness of wing-dimorphism in the salt marsh-inhabiting planthopper, *Prokelisia marginata* (Homoptera: Delphacidae). *Ecology* 60, 221–36.

Depner, K.R. (1962). Effects of photoperiod and of ultraviolet radiation on the incidence of diapause in the horn fly, " *Haematobia irritans* (L.) (Diptera: Muscidae)." *Int. J. Biometeor.* 5, 68–71.

Derr, J.A. (1980). Coevolution of the life history of a tropical seed-feeding insect and its food plants. *Ecology* 6, 881–92.

Derr, J.A., Alden, B., and Dingle, H. (1981). Insect life histories in relation to migration, body size, and host plant array: A comparative study of *Dysdercus. J. Anim. Ecol.* 50, 181–93.

Deseö, K.V. (1973). Side-effect of diapause inducing factors on the reproductive activity of some lepidopterous species. *Nature New Biology* 242, 126–27.

Deseö, K.V. and Sáringer, G. (1975a). Photoperiodic regulation in the population dynamics of certain lepidopterous species. *Acta Phytopathologica Acad. Sci. Hung.* 10, 131–39.

Deseö, K.V. and Sáringer, G. (1975b). Photoperiodic effect on fecundity of *Laspeyresia pomonella, Grapholitha funebrana* and *G. molesta*: The sensitive period. *Entomol. Exp. Appl.* 18, 187–93.

Dethier, B.E. and Vittum, M.T. (1963). Growing degree days. *Bull. N.Y. State Agr. Exp. Sta. Geneva* 80, 1–84.

Deura, K. and Hartley, J.C. (1982). Initial diapause and embryonic development in the speckled bush-cricket, *Leptophyes punctatissima. Physiol. Entomol.* 7, 253–62.

de Wilde, J. (1954). Aspects of diapause in adult insects, with special reference to the Colorado potato beetle, *Leptinotarsa decemlineata* Say. *Arch. Neerl. Zool.* 10, 375–78.

de Wilde, J. (1962a). Photoperiodism in insects and mites. *Annu. Rev. Entomol.* 7, 1-26.

de Wilde, J. (1962b). Analysis of the diapause syndrome in the Colorado potato beetle (*Leptinotarsa decemlineata* Say); behavior and reproduction. *Acta Physiol. Pharmacol. Neerlandica* 11, 525.

de Wilde, J. (1969). Diapause and seasonal synchronization in the adult Colorado beetle (*Leptinotarsa decemlineata* Say). In *Dormancy and survival, 23rd Symp. Soc. Exp. Biol.* (ed. H.W. Woolhouse) pp. 263–84. Academic Press, New York.

de Wilde, J. (1970). Hormones and insect diapause. *Mem. Soc. Endocrinol.* 18, 487–514.

de Wilde, J. (1975). An endocrine view of metamorphosis, polymorphism, and diapause in insects. *Amer. Zool.* 15 (suppl. 1), 13–27.

de Wilde, J. (1983). Endocrine aspects of diapause in the adult stage. In *Endocrinology of insects* (eds. R.G.H. Downer and H. Laufer) pp. 357–67. Alan R. Liss, New York.

de Wilde, J. and de Boer, J.A. (1969). Humoral and nervous pathways in photoperiodic induction of diapause in *Leptinotarsa decemlineata. J. Insect Physiol.* 15, 661-75.

de Wilde, J. and de Loof, A. (1973). Reproduction—endocrine control. In *The physiology of Insecta,* Vol. I (2nd ed.) (ed. M. Rockstein) pp. 97–157. Academic Press, New York.

de Wilde, J. and Ferket, P. (1967). The hostplant as a source of seasonal information. *Med. Rijksfac. Landbouwwet. Gent* 32, 387–92.

de Wilde, J. and Hsiao, T. (1981). Geographic diversity of the Colorado potato beetle and its infestation in Eurasia. In *Advances in potato pest management* (eds. J.H. Lashomb and R. Casagrande) pp. 47–68. Hutchinson Ross, Stroudsburg, PA.

de Wilde, J., Duintjer, C.S., and Mook, L. (1959). Physiology of diapause in the adult Colorado beetle (*Leptinotarsa decemlineata* Say)—I. The photoperiod as a controlling factor. *J. Insect Physiol.* 3, 75–85.

de Wilde, J., Staal, G.B., de Kort, C.A.D., de Loof, A., and Baard, G. (1968). Juvenile hormone titre in the haemolymph as a function of photoperiodic treatment in the adult Colorado beetle (*Leptinotarsa decemlineata* Say). *Proc. Koninkl. Nederl. Akad. Wetenschappen,* Series C, 71, 321–26.

de Wilde, J., Bongers, W., and Schooneveld, H. (1969). Effects of hostplant age on phytophagous insects. *Entomol. Exp. Appl.* 12, 714–20.

de Wilde, J., de Kort, C.A.D., and de Loof, A. (1971). The significance of juvenile hormone titers. *Mitt. Schweiz. Entomol. Ges.* 44, 79–86.

Dickson, R.C. (1949). Factors governing the induction of diapause in the Oriental fruit moth. *Ann. Entomol. Soc. Amer.* 42, 511-37.

Dillwith, J.W. and Chippendale, G.M. (1984). Purification and properties of a protein that accumulates in the fat body of pre-diapausing larvae of the southwestern corn borer, *Diatraea grandiosella. Insect Biochem.* 14, 369–81.

Dingle, H. (1972). Migration strategies of insects. *Science* 175, 1327–35.

Dingle, H. (1974a). The experimental analysis of migration and life-history strategies in insects. In *Experimental analysis of insect behaviour* (ed. L. Barton Browne) pp. 329–42. Springer-Verlag, New York.

Dingle, H. (1974b). Diapause in a migrant insect, the milkweed bug *Oncopeltus fasciatus* (Dallas) (Hemiptera:Lygaeidae). *Oecologia* 17, 1–10.

Dingle, H. (1978a). Migration and diapause in tropical, temperate, and island milkweed bugs. In *Evolution of insect migration and diapause* (ed. H. Dingle) pp. 254–76. Springer-Verlag, New York.

Dingle, H. (ed.) (1978b). *Evolution of insect migration and diapause.* Springer-Verlag, New York.

Dingle, H. (1979). Adaptive variation in the evolution of insect migration. In *Movement of highly mobile insects: Concepts and methodology in research* (eds. R.L. Rabb and G.G. Kennedy) pp. 64–87. University Graphics, North Carolina State University, Raleigh.

Dingle, H. (1980). Ecology and evolution of migration. In *Animal migration, orientation, and navigation* (ed. S.A. Gauthreaux, Jr.) pp. 1–101. Academic Press, New York.

Dingle, H. (1981). Geographic variation and behavioral flexibility in milkweed bug life histories. In *Insect life history patterns* (eds. R.F. Denno and H. Dingle) pp. 57–73. Springer-Verlag, New York.

Dingle, H. (1982). Function of migration in the seasonal synchronization of insects. *Entomol. Exp. Appl.* 31, 36–48.

Dingle, H. and Arora, G. (1973). Experimental studies of migration in bugs of the genus *Dysdercus. Oecologia* 12, 119–40.

Dingle, H. and Baldwin, D. (1983). Geographic variation in life histories: A comparison of tropical and temperate milkweed bugs *(Oncopeltus).* In *Diapause and life cycle strategies in insects* (eds. V.K. Brown and I. Hodek) pp. 143–65. Junk, The Hague, The Netherlands.

Dingle, H. and Hegmann, J.P. (1982). *Evolution and genetics of life histories.* Springer-Verlag, New York.

Dingle, H., Brown, C.K., and Hegmann, J.P. (1977). The nature of genetic variance influencing photoperiodic diapause in a migrant insect, *Oncopeltus fasciatus. Amer. Natur.* 111, 1047–59.

Dingle, H., Alden, B.M., Blakley, N.R., Kopec, D., and Miller, E.R. (1980a). Variation in photoperiodic response within and among species of milkweed bugs (*Oncopeltus*). *Evolution* 34, 356–70.

Dingle, H., Blakley, N.R., and Miller, E.R. (1980b). Variation in body size and flight performance in milkweed bugs (*Oncopeltus*). *Evolution* 34, 371–85.

Dingle, H., Blau, W.S., Brown, C.K., and Hegmann, J.P. (1982). Population crosses and the genetic structure of milkweed bug life histories. In *Evolution and genetics of life histories* (eds. H. Dingle and J.P. Hegmann) pp. 209–29. Springer-Verlag, New York.

Dixon, A.F.G. (1969). Population dynamics of the sycamore aphid *Drepanosiphum platanoides* (Schr.) (Hemiptera: Aphididae): Migratory and trivial flight activity. *J. Anim. Ecol.* 38, 585–606.

Dixon, A.F.G. (1971a). The "interval timer" and photoperiod in the determination of parthenogenetic and sexual morphs in the aphid, *Drepanosiphum platanoides. J. Insect Physiol.* 17, 251–60.

Dixon, A.F.G. (1971b). Migration in aphids. *Sci. Prog., Oxf.* 59, 41–53.

Dixon, A.F.G. (1971c). The life-cycle and host preferences of the bird cherry-oat aphid, *Rhopalosiphum padi* L., and their bearing on the theories of host alternation in aphids. *Ann. Appl. Biol.* 68, 135–47.

Dixon, A.F.G. (1972a). Control and significance of the seasonal development of colour forms in the sycamore aphid, *Drepanosiphum platanoides* (Schr.). *J. Anim. Ecol.* 41, 689–97.

Dixon, A.F.G. (1972b). The "interval timer", photoperiod and temperature in the seasonal development of parthenogenetic and sexual morphs in the lime aphid, *Eucallipterus tiliae* L. *Oecologia* 9, 301–10.

Dixon, A.F.G. (1972c). Fecundity of brachypterous and macropterous alatae in *Drepanosiphum dixoni. Entomol. Exp. Appl.* 15, 335–40.

Dixon, A.F.G. (1973). Metabolic acclimatization to seasonal changes in temperature in the sycamore aphid, *Drepanosiphum platanoides* (Schr.), and lime aphid, *Eucallipterus tiliae* L. *Oecologia* 13, 205–10.

Dixon, A.F.G. (1975). Seasonal changes in fat content, form, state of gonads and length of adult life in the sycamore aphid, *Drepanosiphum platanoides* (Schr.). *Trans. Roy. Entomol. Soc. Lond.* 127, 87–99.

Dixon, A.F.G. (1976a). Reproductive strategies of the alate morphs of the bird cherry-oat aphid *Rhopalosiphum padi* L. *J. Anim. Ecol.* 45, 817–30.

Dixon, A.F.G. (1976b). Timing of egg hatch and viability of the sycamore aphid, *Drepanosiphum platanoidis* (Schr.), at bud burst of sycamore, *Acer pseudoplatanus* L. *J. Anim. Ecol.* 45, 593–603.

Dixon, A.F.G. (1977). Aphid ecology: Life cycles, polymorphism, and population regulation. *Annu. Rev. Ecol. Syst.* 8, 329–53.

Dixon, A.F.G. and Dharma, T.R. (1980). 'Spreading of the risk' in developmental mortality: Size, fecundity and reproductive rate in the black bean aphid. *Entomol. Exp. Appl.* 28, 301–12.

Dixon, A.F.G. and Glen, D.M. (1971). Morph determination in the bird cherry-oat aphid, *Rhopalosiphum padi* L. *Ann. Appl. Biol.* 68, 11–21.

Dixon, A.F.G., Burns, M.D., and Wangboonkong, S. (1968). Migration in aphids: Response to current adversity. *Nature* 220, 1337–38.

Dobzhansky, T. and Epling, C. (1944). Contributions to the genetics, taxonomy and ecology of *Drosophila pseudoobscura* and its relatives. *Publ. Carnegie Inst. Wash.* 554, 1–46.

Dortland, J.F. (1978). Synthesis of vitellogenins and diapause proteins by the fat body of *Leptinotarsa*, as a function of photoperiod. *Physiol. Entomol.* 3, 281–88.

Dortland, J.F. (1979). The hormonal control of vitellogenin synthesis in the fat body of the female Colorado potato beetle. *Gen. Comp. Endocrinol.* 38, 332–44.

Dortland, J.F. and de Kort, C.A.D. (1978). Protein synthesis and storage in the fat body of the Colorado potato beetle, *Leptinotarsa decemlineata. Insect Biochem.* 8, 93–98.

Douglas, M.M. and Grula, J.W. (1978). Thermoregulatory adaptations allowing ecological range expansion by the pierid butterfly, *Nathalis iole* Boisduval. *Evolution* 32, 776–83.

Doutt, R.L. (1959). The biology of parasitic Hymenoptera. *Annu. Rev. Entomol.* 4, 161-82.

Doutt, R.L. and Nakata, J. (1973). The *Rubus* leafhopper and its egg parasitoid: An endemic biotic system useful in grape-pest management. *Environ. Entomol.* 2, 381–86.

Doutt, R.L., Annecke, D.P., and Tremblay, E. (1976). Biology and host relationships of parasitoids. In *Theory and practice of biological control* (eds. C.B. Huffaker and P.S. Messenger) pp. 143–68. Academic Press, New York.

Downer, R.G.H. and Laufer, H. (eds.) (1983). *Endocrinology of insects.* Alan R. Liss, New York.

Downes, J.A. (1965). Adaptations of insects in the Arctic. *Annu. Rev. Entomol.* 10, 257–74.

Druzhelyubova, T.S. (1976). Temperature and light as factors affecting development and behavior in geographic populations of *Agrotus ypsilon* Rott. (Lepidoptera, Noctuidae). *Entomol. Rev.* 55, 9–13.

Dubynina, T.S. (1965). Onset of diapause and reactivation in *Tetranychus urticae* Koch (Acarina, Tetranychidae). *Entomol. Rev.* 44, 159–61.

Duman, J.G. (1980). Factors involved in overwintering survival of the freeze tolerant beetle, *Dendroides canadensis. J. Comp. Physiol.* 136, 53–59.

Duman, J.G. (1982). Insect antifreezes and ice-nucleating agents. *Cryobiology* 19, 613–27.

Duman, J.G. (1984). Change in overwintering mechanism of the cucujid beetle, *Cucujus clavipes. J. Insect Physiol.* 30, 235–39.

Duman, J.G., Horwarth, K.L., Tomchaney, A., and Patterson, J.L. (1982). Antifreeze agents of terrestrial arthropods. *Comp. Biochem. Physiol.* 73A, 545–55.

Dumortier, B. and Brunnarius, J. (1977). L'information thermopériodique et l'induction de la diapause chez *Pieris brassicae* L. *C.R. Acad. Sci.*, Paris, Series D, 284, 957–60.

Dumortier, B. and Brunnarius, J. (1981). Involvement of the circadian system in photoperiodism and thermoperiodism in *Pieris brassicae* (Lepidoptera). In *Biological clocks in seasonal reproductive cycles*, Proc. 32nd Symp. Colston Res. Soc., Bristol (1980) (eds. B.K. Follett and D.E. Follett) pp. 83–99. Halsted Press, New York.

Dyer, E.D.A. and Hall, P.M. (1977). Factors affecting larval diapause in *Dendroctonus rufipennis* (Coleoptera: Scolytidae). *Can. Entomol.* 109, 1485–90.

Earle, N.W. and Newsom, L.D. (1964). Initiation of diapause in the boll weevil. *J. Insect Physiol.* 10, 131–39.

Eckenrode, C.J. and Chapman, R.K. (1971). Effect of various temperatures upon rate of development of the cabbage maggot under artificial conditions. *Ann. Entomol. Soc. Amer.* 64, 1079–83.

Edmunds, M. (1978). Contrasting methods of survival in two sympatric cotton stainer bugs (Hem., Pyrrhocoridae) in Ghana during food shortage. *Entomologist's Monthly Mag.* 114, 241–44.

Edney, E.B. (1977). *Water balance in land arthropods.* Springer-Verlag, Berlin.

Edney, E. (1980). The components of water balance. In *Insect biology in the future "VBW 80"* (eds. M. Locke and D.S. Smith) pp. 39–58. Academic Press, London.

Eertmoed, G.E. (1978). Embryonic diapause in the psocid, *Peripsocus quadrifasciatus*: Photoperiod, temperature, ontogeny and geographic variation. *Physiol. Entomol.* 3, 197–206.

Eghtedar, E. (1970). Zur Biologie und Ökologie der Staphyliniden *Philonthus fuscipennis* Mannh. und *Oxytelus rugosus* Grav. *Pedobiologia* 10, 169–79.

Ehler, L.E. and Hall, R.W. (1982). Evidence for competitive exclusion of introduced natural enemies in biological control. *Environ. Entomol.* 11, 1–4.

Ehrlich, P.R. (1965). The population biology of the butterfly, *Euphydryas editha.* II. The structure of the Jasper Ridge colony. *Evolution* 19, 327–36.

Ehrlich, P.R. and Raven, P.H. (1969). Differentiation of populations. *Science* 165, 1228–32.

Ehrlich, P.R. and White, R.R. (1980). Colorado checkerspot butterflies: Isolation, neutrality, and the biospecies. *Amer. Natur.* 115, 328–41.

Eichhorn, O. (1976). Dauerzucht von *Diprion pini* L. (Hym.:Diprionidae) in Laboratorium unter Berücksichtigung der Fotoperiode. *Anz. Schädlingsk. Pflanzensch. Umweltsch.* 49, 38–41.

Eichhorn, O. (1977a). Autökologische Untersuchungen an Populationen der gemeinen Kiefern-Buschhornblattwespe *Diprion pini* (L.) (Hym.: Diprionidae). II. Zur Kenntnis der Larvenparasiten und ihrer Synchronisation mit dem Wirt. *Z. Ang. Entomol.* 83, 15–36.

Eichhorn, O. (1977b). Autökologische Untersuchungen an Population der gemeinen Kiefern-Buschhornblattwespe *Diprion pini* (L.) (Hym.: Diprionidae). III. Laborzuchten. *Z. Ang. Entomol.* 84, 264–82.

Eichhorn, O. (1978). Zur Prognose der Schlüpfwellen- und Generationfolge bei der gemeinen Kiefern-Buschhornblattwespe *Diprion pini* L. (Hymenopt.: Diprionidae). *Anz. Schädlingsk. Pflanzensch. Umweltsch.* 51, 65–69.

Eichhorn, O. (1979). Autökologische Untersuchungen an Populationen der gemeinen Kiefern-Buschhornblattwespe *Diprion pini* (L.) (Hym.: Diprioninidae). *Z. Ang. Entomol.* 88, 378–98.

Eichhorn, O. (1983). Dormanzverhalten der Gemeinen Kiefern-Buschhorn-blattwespe (*Diprion pini* L.) (Hym., Diprionidae) und ihrer Parasiten. *Z. Ang. Entomol.* 95, 482–98.

Eichhorn, O. and Pschorn-Walcher, H. (1976). Studies on the biology and ecology of the egg-parasites (Hym.: Calcidoidea) of the pine sawfly *Diprion pini* (L.) (Hym.: Diprionidae) in Central Europe. *Z. Ang. Entomol.* 80, 355–81.

Eichler, W. (1951). *Anopheles*—Beobachtungen in Südrussland. *Riv. Malariology* 30, 29–38.

Eickwort, G.C. (1981). Presocial insects. In *Social insects* Vol. II (ed. H.R. Hermann) pp. 199–280. Academic Press, New York.

Eidt, D.C. and Little, C.H.A. (1968). Insect control by artificially prolonging plant dormancy—A new approach. *Can. Entomol.* 100, 1278–79.

Elbert, A. (1979). Ökologische Untersuchungen zur Steuerung der Larvaldiapause von *Trogoderma variabile* Ballion (Col. Dermestidae). *Z. Ang. Entomol.* 88, 268–82.

Eldridge, B.F. (1963). The influence of daily photoperiod on blood-feeding activity of *Culex tritaeniorhynchus* Giles. *Amer. J. Hyg.* 77, 49–53.

Eldridge, B.F. (1966). Environmental control of ovarian development in mosquitoes of the *Culex pipiens* complex. *Science* 151, 826–28.

Eldridge, B.F. (1968). The effect of temperature and photoperiod on blood-feeding and ovarian development in mosquitoes of the *Culex pipiens* complex. *Amer. J. Trop. Med. Hyg.* 17, 133–40.

Eldridge, B.F. and Bailey, C.L. (1979). Experimental hibernation studies in *Culex pipiens* (Diptera: Culicidae): Reactivation of ovarian development and blood-feeding in prehibernating females. *J. Med. Entomol.* 15, 462–67.

El-Hariri, G. (1966). Studies of the physiology of hibernating Coccinellidae (Coleoptera): Changes in the metabolic reserves and gonads. *Proc. Roy. Entomol. Soc. Lond.* (A), 41, 133–44.

El-Hariri, G. (1970). Physiological studies on the fat and water contents of hibernating cucurbit beetles, *Epilachna chrysomelina* (F.) (Col., Coccinellidae). *Acta Phytopathol. Acad. Sci. Hung.* 5, 367–70.

Ellis, P.E. (1964a). Marching and colour in locust hoppers in relation to social factors. *Behaviour* 23, 177–92.

Ellis, P.E. (1964b). Changes in marching of locusts with rearing conditions. *Behaviour* 23, 193–202.

Ellis, P.E., Carlisle, D.B., and Osborne, D.J. (1965). Desert locusts: Sexual maturation delayed by feeding on senescent vegetation. *Science* 149, 546–47.

Embree, D.G. (1970). The diurnal and seasonal pattern of hatching of winter moth eggs, *Operophtera brumata. Can. Entomol.* 102, 759–68.

Emme, A.M. (1953). Some problems on the theory of insect diapause. *Usp. Sobr. Biol.* 35, 395–424 (in Russian).

Emmel, T.C. and Emmel, J.F. (1973). The butterflies of Southern California. *Natural History Museum of Los Angeles County*, Science series 26, 1–148.

Endler, J.A. (1977). *Geographic variation, speciation, and clines.* Princeton University Press, Princeton.

Endo, K. (1970). Relation between ovarian maturation and activity of the corpora allata in seasonal forms of the butterfly, *Polygonia c-aureum* L. Devel., *Growth, Differ.* 11, 297–304.

Endo, K. (1972). Activation of the corpora allata in relation to ovarian maturation in the seasonal forms of the butterfly, *Polygonia c-aureum* L. *Devel., Growth, Differ.* 14, 263–74.

Endo, K. (1973). Hormonal regulation of mating in the butterfly, *Polygonia c-aureum* L. *Devel., Growth, Differ.* 15, 1–10.

Englemann, F. (1970). *The physiology of insect reproduction.* Pergamon Press, Oxford.

Englemann, F. (1984). Reproduction in insects. In *Ecological entomology* (eds. C.B. Huffaker and R.L. Rabb) pp. 113–47. John Wiley & Sons, New York.

Englemann, W. and Shappirio, D.G. (1965). Photoperiodic control of the maintenance and termination of larval diapause in *Chironomus tentans. Nature* 207, 548–49.

Erpenbeck, A. and Kirchner, W. (1983). Zur Kälteresistenz der Kleinen Roten Waldameise *Formica polyctena* Foerst. (Hymenoptera, Formicidae). *Z. Ang. Entomol.* 96, 271–81.

Eskafi, F.M. and Legner, E.F. (1974). Fecundity, development, and diapause in *Hexacola* sp. near *websteri*, a parasite of *Hippelates* eye gnats. *Ann. Entomol. Soc. Amer.* 67, 769–71.

Etchegaray, J.B. and Nishida, T. (1975). Reproductive activity, seasonal abundance and parasitism of the monarch butterfly, *Danaus plexippus* (Lepidoptera: Danaidae) in Hawaii. *Proc. Hawaiian Entomol. Soc.* 22, 33–39.

Evans, A.A.F. and Perry, R.N. (1976). Survival strategies in nematodes. In *The organization of nematodes* (ed. N.A. Croll) pp. 383–424. Academic Press, London.

Evans, K.W. and Brust, R.A. (1972). Induction and termination of diapause in *Wyeomyia smithii* (Diptera: Culicidae), and larval survival studies at low and subzero temperatures. *Can. Entomol.* 104, 1937–50.

Everett, T.R., Chiang, H.C., and Hibbs, E.T. (1958). Some factors influencing populations of European corn borer [*Pyrausta nubilalis* (Hbn.)] in the North Central States. University of Minnesota Agricultural Experiment Station, North Central Regional Publ. No. 87 - *Tech. Bull.* 229, 1–63.

Ewen, A.B. (1966). A possible endocrine mechanism for inducing diapause in the eggs of *Adelphocoris lineolatus* (Goeze) (Hemiptera:Miridae). *Experientia* 22, 470.

Ewer, D.W. (1979). The pathways of water uptake by the eggs of *Locusta* and *Schistocerca* (Orthoptera: Acrididae). *Acrida* 8, 163–87.

Falkovich, M.I. (1979). Seasonal development of the desert Lepidoptera of Soviet Central Asia and a historical analysis of the development of the Lepidopteran fauna. *Entomol. Rev.* 58, 20–45.

Farner, D.S. (1975). Photoperiodic controls in the secretion of gonadotropins in birds. *Amer. Zool.* 15 (suppl.), 117–35.

Farner, D.S. and Follett, B.K. (1966). Light and other environmental factors affecting avian reproduction. *J. Anim. Sci.* 25 (suppl.), 90–115.

Farner, D.S. and Lewis, R.A. (1971). Photoperiodism and reproductive cycles in birds. In *Photophysiology* (ed. A.C. Giese) pp. 325–70. Academic Press, New York.

Farrow, R.A. (1979). Population dynamics of the Australian plague locust, *Chortoicetes terminifera* (Walker), in Central Western New South Wales. I. Reproduction and migration in relation to weather. *Aust. J. Zool.* 27, 717–45.

Farrow, R.A. and O'Neill, A. (1978). Differences in ovarian development between *Austracris proxima* (Walker) and *A. guttulosa* (Walker) (Orthoptera: Acrididae). *J. Aust. Entomol. Soc.* 17, 199–200.

Fasulati, S.R. (1979). Photoperiodic reaction and coloration of *Eurydema oleracea* (Heteroptera, Pentatomidae). *Entomol. Rev.* 58, 10–15.

Faure, J.C. (1932). The phases of locusts in South Africa. *Bull. Entomol. Res.* 23, 293–424.

Featherston, N.H. and Hays, S.B. (1971). Diapause in crosses of a laboratory and a wild strain of the plum curculio, *Conotrachelus nenuphar. J. Ga. Entomol. Soc.* 6, 95–101.

Feeny, P. (1970). Seasonal changes in oak leaf tannins and nutrients as a cause of spring feeding by winter moth caterpillars. *Ecology* 51, 565–81.

Ferenz, H.-J. (1975a). Anpassungen von *Pterostichus nigrita* F. (Col., Carab.) an subarktische Bedingungen. *Oecologia* 19, 49–57.

Ferenz, H.-J. (1975b). Photoperiodic and hormonal control of reproduction in male beetles, *Pterostichus nigrita. J. Insect Physiol.* 21, 331–41.

Ferenz, H.-J. (1977). Two-step photoperiodic and hormonal control of reproduction in the female beetle, *Pterostichus nigrita. J. Insect Physiol.* 23, 671–76.

Ferrari, D.C. and Hebert, P.D.N. (1982). The induction of sexual reproduction in *Daphnia magna*: genetic differences between arctic and temperate populations. *Can. J. Zool.* 60, 2143–48.

Ferris, G.F. (1919). A remarkable case of longevity in insects (Hem., Hom.). *Entomol. News* 30, 27 (through Lees 1955).

Fetter-Lasko, J.L. and Washino, R.K. (1983). In situ studies on seasonality and recycling pattern in California of *Lagenidium giganteum* Couch, an aquatic fungal pathogen of mosquitoes. *Environ. Entomol.* 12, 635–40.

Finch, S. and Collier, R.H. (1983). Emergence of flies from overwintering populations of cabbage root fly pupae. *Ecol. Entomol.* 8, 29–36.

Fisher, R.C. (1971). Aspects of the physiology of endoparasitic Hymenoptera. *Biol. Rev.* 46, 243–78.

Flanagan, T.R. and Hagedorn, H.H. (1977). Vitellogenin synthesis in the mosquito: The role of juvenile hormone in the development of responsiveness to ecdysone. *Physiol. Entomol.* 2, 173–78.

Flanders, S.E. (1944). Diapause in the parasitic Hymenoptera. *J. Econ. Entomol.* 37, 408–11.

Flint, A.P.F., Renfree, M.B., and Weir, B.J. (eds.) (1981). *Embryonic diapause in mammals. J. Reprod. and Fertil.*, Suppl. 29, 1–241.

Flint, M.L. (1980). Climatic ecotypes in *Trioxys complanatus*, a parasite of the spotted alfalfa aphid. *Environ. Entomol.* 9, 501–7.

Follett, B.K. and Follett, D.E. (eds.) (1981). *Biological clocks in seasonal reproductive cycles.* Halsted Press, John Wiley & Sons, New York.

Fontana, P.G. and Hogan, T.W. (1969). Cytogenetic and hybridization studies of geographic populations of *Teleogryllus commodus* (Walker) and *T. oceanicus* (Le Guillou) (Orthoptera: Gryllidae). *Aust. J. Zool.* 17, 13–35.

Force, D.C. and Messenger, P.S. (1968). The use of laboratory studies of three hymenopterous parasites to evaluate their field potential. *J. Econ. Entomol.* 61, 1374–78.

Fortescue-Foulkes, J. (1953). Seasonal breeding and migrations of the desert locust (*Schistocerca gregaria* Forskål) in southwestern Asia. *Anti-locust Memoir* 5, 1–35.

Foster, D.R. and Crowder, L.A. (1980). Diapause of the pink bollworm, *Pectinophora gossypiella* (Saunders), related to dietary lipids. *Comp. Biochem. Physiol.* 65B, 723–26.

Foster, W.A. (1978). Dispersal behaviour of an intertidal aphid. *J. Anim. Ecol.* 47, 653–59.

Foster, W.A. and Treherne, J.E. (1978). Dispersal mechanisms in an intertidal aphid. *J. Anim. Ecol.* 47, 205–17.

Fourche, J. (1977). The influence of temperature on respiration of diapausing pupae of *Pieris brassicae* (Lepidoptera). *J. Therm. Biol.* 2, 163–72.

Fraenkel, G.S. and Gunn, D.L. (1940). *The orientation of animals.* Oxford University Press, London.

Fraenkel, G.S. and Hsiao, C. (1968). Manifestations of a pupal diapause in two species of flies, *Sarcophaga argyrostoma* and *S. bullata. J. Insect Physiol.* 14, 689–705.

Frank, J.H. and Curtis, G.A. (1977). On the bionomics of bromeliad-inhabiting mosquitoes. III. The probable strategy of larval feeding in *Wyeomyia vanduzeei* and *Wy. medioalbipes. Mosq. News* 37, 200–6.

Frankie, G.W., Baker, H.G., and Opler, P.A. (1974). Comparative phenological studies of trees in tropical wet and dry forests in the lowlands of Costa Rica. *J. Ecol.* 62, 881–919.

Fraser, A. and Smith, W.F. (1963). Diapause in larvae of green blowflies (Diptera: Cyclorrhapha: *Lucilla* spp.). *Proc. Roy. Entomol. Soc. Lond.* A, 38, 90–97.

Free, J.B. (1955). Queen production in colonies of bumblebees. *Proc. Roy. Entomol. Soc. Lond.* (A), 30, 19–25.

Free, J.B. (1977). The seasonal regulation of drone brood and drone adults in a honeybee colony. *Proc. Eighth Inter. Cong. International Union for the Study of Social Insects*, pp. 207–10. Wageningen, The Netherlands.

Free, J.B. and Williams, I.H. (1975). Factors determining the rearing and rejection of drones by the honeybee colony. *Anim. Behav.* 23, 650–75.

Freeman, B.E. and Ittyeipe, K. (1982). Morph determination in *Melittobia*, a eulophid wasp. *Ecol. Entomol.* 7, 355–63.

Friedlander, M. (1982). Juvenile hormone and regulation of dichotomous spermatogenesis during the larval diapause of the codling moth. *J. Insect Physiol.* 28, 1009–12.

Fujiie, A. (1980). Ecological studies on the population of the pear leaf miner, *Bucculatrix pyrivorella* Kuroko (Lepidoptera: Lyonetiidae) III. Fecundity fluctuation from generation to generation within a year. *Appl. Entomol. Zool.* 15, 1–9.

Fujiyama, S. and Takahashi, F. (1973a). Studies on the self-regulation of life cycle in *Anomala cuprea* Hope (Coleoptera; Scarabaeidae). 1. The effects of constant temperature on the developmental stages. *Mem. Coll. Agr. Kyoto Univ.* 104, 23–30.

Fujiyama, S. and Takahashi, F. (1973b). Studies on the self-regulation of life cycle in *Anomala cuprea* Hope. (Coleoptera; Scarabaeidae). 2. The effects of low temperature and photoperiod on the induction and termination of the larval diapause. *Mem. Coll. Agr. Kyoto Univ.* 104, 31–39.

Fujiyama, S. and Takahashi, F. (1977). The survival rate and the duration of development in *Anomala cuprea* Hope (Coleoptera: Scarabaeidae) under the thermal changes along with its developmental stage. *Environ. Control Biol.* 15, 19–25.

Fukatami, A. (1984). Cold temperature resistance in *Drosophila lutescens* and *D. takahashii*. *Japan. J. Genet.* 59, 61–70.

Fukaya, M. and Mitsuhashi, J. (1961). Larval diapause in the rice stem borer with special reference to its hormonal mechanism. *Bull. Nat. Inst. Agr. Sci.*, Series C, No. 13, pp. 1–32.

Fukuda, S. (1951). The production of the diapause eggs by transplanting the suboesophageal ganglion in the silkworm. *Proc. Japan Acad.* 27, 672–77.

Fukuda, S. (1952). Function of the pupal brain and suboesophageal ganglion in the production of non-diapause and diapause eggs in the silkworm. *Annot. Zool. Japan* 25, 149–55.

Fukuda, S. (1953). Alternation of voltinism in the silkworm following transection of pupal oesophageal connectives. *Proc. Japan Acad.* 29, 389–91.

Fukuda, S. (1963). Déterminisme hormonal de la diapause chez le ver à soie. *Bull. Soc. Zoo. Fr.* 88, 151–79.

Fukuda, S. and Endo, K. (1966). Hormonal control of the development of seasonal forms in the butterfly, *Polygonia c-aureum* L. *Proc. Japan Acad.* 42, 1082–87.

Fukuda, S. and Takeuchi, S. (1967). Studies on the diapause factor-producing cells in the suboesophageal ganglion of the silkworm, *Bombyx mori* L. *Embryologia* 9, 333–53.

Fullard, J.H. (1977). Phenology of sound-producing arctiid moths and the activity of insectivorous bats. *Nature* 267, 42–43.

Furunishi, S. and Masaki, S. (1981). Photoperiodic response of the univoltine ant-lion *Myrmeleon formicarius* (Neuroptera, Myrmeleontidae). *Kontyû* 49, 653–67.

Furunishi, S. and Masaki, S. (1982). Seasonal life cycle in two species of ant-lion (Neuroptera: Myrmeleontidae). *Japan. J. Ecol.* 32, 7–13.

Furunishi, S. and Masaki, S. (1983). Photoperiodic control of development in the ant-lion *Hagenomyia micans* (Neuroptera: Myrmeleontidae). *Entomol. Gen.* 9, 51–62.

Furunishi, S., Masaki, S., Hashimoto, Y., and Suzuki, M. (1982). Diapause responses to photoperiod and night interruption in *Mamestra brassicae* (Lepidoptera: Noctuidae). *Appl. Entomol. Zool.* 17, 398–409.

Fuseini, B.A. and Kumar, R. (1975). Ecology of cotton stainers (Heteroptera: Pyrrhocoridae) in southern Ghana. *Biol. J. Linn. Soc.* 7, 113–46.

Fuzeau-Braesch, S. and Ismail, S. (1976). La diapause d'une espèce univoltine: *Gryllus campestris* L. (Orthop.): facteurs externes et programmation. *Ann. Zool. Ecol. Anim.* 8, 349–65.

Gage, S.H. and Mukerji, M.K. (1977). A perspective of grasshopper population distribution in Saskatchewan and interrelationship with weather. *Environ. Entomol.* 6, 469–79.

Gallaway, W.J. and Brust, R.A. (1982). Blood-feeding and gonotrophic dissociation in overwintering *Anopheles earlei* (Diptera: Culicidae) from southern Manitoba. *Can. Entomol.* 114, 1105–7.

Galloway, T.D. and Brust, R.A. (1977). Effects of temperature and photoperiod on the infection of two mosquito species by the mermithid *Romanomermis culicivorax*. *J. Nematol.* 9, 218–21.

Gambaro-Ivancich, P. (1958). L'azione del fotoperiodismo sullo suiluppo larvale di *Carpocapsa pomonella* L. *Mem. Acad. Patav.* 70, 1–7 (through Riedl 1983).

Garthe, W.A. (1970). Development of the female reproductive system and effect of males on oocyte production in *Sitona cylindricollis* (Coleoptera: Curculionidae). *Ann. Entomol. Soc. Amer.* 63, 367–70.

Geier, P.W. and Springett, B.P. (1976). Population characteristics of Australian leafrollers (*Epiphyas* spp., Lepidoptera) infesting orchards. *Aust. J. Ecol.* 1, 129–44.

Gelman, D.B. and Hayes, D.K. (1980). Physical and biochemical factors affecting diapause in insects; especially in the European corn borer, *Ostrinia nubilalis*: A review. *Physiol. Entomol.* 5, 367–83.

Gelman, D.B. and Woods, C.W. (1983). Haemolymph ecdysteroid titers of diapause- and non-diapause-bound fifth instars and pupae of the European corn borer, *Ostrinia nubilalis* (Hübner). *Comp. Biochem. Physiol.* 76A, 367–75.

Georgian, T. and Wallace, J.B. (1983). Seasonal production dynamics in a guild of periphyton-grazing insects in a southern Appalachian stream. *Ecology* 64, 1236–48.

Gerber, G.H. and Lamb, R.J. (1982). Phenology of egg hatching for the red turnip beetle, *Entomoscelis americana* (Coleoptera: Chrysomelidae). *Environ. Entomol.* 11, 1258–63.

Gerber, G.H., Neill, G.B. and Westdal, P.H. (1979). The reproductive cycles of the sunflower beetle, *Zygogramma exclamationis* (Coleoptera: Chrysomelidae), in Manitoba. *Can. J. Zool.* 57, 1934–43.

Geyspits, K.F. (1960). The effect of the conditions in which preceding generations were reared on the photoperiodic reaction of geographical forms of the cotton spider mite *(Tetranychus urticae* Koch.). *Trans. Peterhof. Biol. Inst., L.S.U.* 18, 169–77 (in Russian: through Danilevsky 1965).

Geyspits, K.F. (1968). Genetic aspects of variation of photoperiodic adaptations. In *Photoperiodic adaptations in insects and acari* (ed. A.C. Danilevsky) pp. 78–79. Leningrad University Press, Leningrad.

Geyspits, K.F. and Simonenko, N.P. (1970). An experimental analysis of seasonal changes in the photoperiodic reaction of *Drosophila phalerata* Meig. (Diptera, Drosophilidae). *Entomol. Rev.* 49, 46–54.

Geyspits, K.F. and Zarankina, A.I. (1963). Some features of the photoperiodic reaction of *Dasychira pudibunda* L. (Lepidoptera, Orgyidae). *Entomol. Rev.* 42, 14–19.

Geyspits, K.F., Sapozhnikova, F.D., and Taranets, M.N. (1971). Seasonal changes in the photoperiodic reaction and physiological state of the spider mite [*Tetranychus urticae* Kock] (Acarina, Tetranychidae). *Entomol. Rev.* 50, 156–62.

Geyspits, K.F., Glinyanaya, E.I., and Sapozhnikova, F.D. (1972). General principle of interrelationship of different manifestations of photoperiodic reactions of arthropods. *Dokl. Akad. Nauk SSSR* 202, 1229–32.

Geyspits, K.F., Glinyanaya, Y.I., Sapozhnikova, F.D., and Simonenko, N.P. (1974). The relationship between endogenous and exogenous factors in the regulation of seasonal changes in the photoperiodic reaction of arthropods. *Entomol. Rev.* 53, 27–34.

Geyspits, K.F., Glinyanaya, Y.I., Dubynina, T.S., Kvitko, N.V., Pidzhakova, T.V., Razumova, A.P., Sapozhnikova, F.D., Simonenko, N.P., and Taranets, M.N. (1978). The annual endogenous rhythm of changes in the photoperiodic reaction of arthropods and its relationship to exogenous factors. *Entomol. Rev.* 57, 495–505.

Gharib, B., Girardie, A., and De Reggi, M. (1981). Ecdysteroids and control of embryonic diapause: Changes in ecdysteroid levels and exogenous hormone effects in the eggs of cochineal *Lepidosaphes*. *Experientia* 37, 1107–8.

Gibbons, J.R.H. (1979). A model for sympatric speciation in *Megarhyssa* (Hymenoptera: Ichneumonidae): Competitive speciation. *Amer. Natur.* 114, 719–47.

Gibbs, D. (1975). Reversal of pupal diapause in *Sarcophaga argyrostoma* by temperature shifts after puparium formation. *J. Insect Physiol.* 21, 1179–86.

Giebultowicz, J.M. and Saunders, D.S. (1983). Evidence for the neurohormonal basis of commitment to pupal diapause in larvae of *Sarcophaga argyrostoma*. *Experientia* 39, 194–96.

Giese, R.L., Peart, R.M., and Huber, R.T. (1975). Pest management. *Science* 187, 1045–52.

Giesel, J.T. (1976). Reproductive strategies as adaptations to life in temporally heterogeneous environments. *Annu. Rev. Ecol. Syst.* 7, 57–79.

Gillett, J.D. (1955). The inherited basis of variation in the hatching-response of *Aedes* eggs (Diptera: Culicidae). *Bull. Entomol. Res.* 46, 255–65.

Gillett, J.D., Roman, E.A., and Phillips, V. (1977). Erratic hatching in *Aedes* eggs: A new interpretation. *Proc. Roy. Soc. Lond.* (B) 196, 223–32.

Gillett, S.D. (1978). Environmental determinants of phase polymorphism of the desert locust, *Schistocerca gregaria* (Forsk.), reared crowded. *Acrida* 7, 267–88.

Gillott, C. (1980). *Entomology*. Plenum Press, New York.

Girardie, A. and Granier, S. (1973a). Système endocrine et physiologie de la diapause imaginale chez le Criquet égyptien, *Anacridium aegyptium*. *J. Insect Physiol.* 19, 2341–58.

Girardie, A. and Granier, S. (1973b). Aspects ultrastructuraux des corps allates du Criquet égyptien femelle adulte (en diapause, actif et parasité). *J. Microsci.* 17, 60a-61a.

Glass, E.H. (1970). Changes in diapause response to photoperiod in laboratory strains of Oriental fruit moth. *Ann. Entomol. Soc. Amer.* 63, 74–76.

Goettel, M.S. and Philogène, B.J.R. (1980). Further studies on the biology of the banded woollybear, *Pyrrharctia* (*Isia*) *isabella* (J.E. Smith) (Lepidoptera: Arctiidae). IV. Diapause development as influenced by temperature. *Can. J. Zool.* 58, 317–20.

Golden, J.W. and Riddle, D.L. (1982). A pheromone influences larval development in the nematode *Caenorhabditis elegans*. *Science* 218, 578–80.

Golden, J.W. and Riddle, D.L. (1984). A pheromone-induced developmental switch in *Caenorhabditis elegans*: Temperature-sensitive mutants reveal a wild-type temperature-dependent process. *Proc. Natl. Acad. Sci. USA* 81, 819–23.

Goldschmidt, R. (1932). Untersuchungen zur Genetik der geographischen Variation. V. Analyse der Überwinterungszeit als Anpassungscharakter. *Arch. Entw. Mech. Org.* 126, 674–768.

Goldschmidt, R. (1934). *Lymantria. Bibliographia Genetica* 11, 1-186.

Goldson, S.L. and Emberson, R.M. (1980). Relict diapause in an introduced weevil in New Zealand. *Nature* 286, 489–90.

Gordon, H.T. (1984). Growth and development of insects. In *Ecological entomology* (eds. C.B. Huffaker and R.L. Rabb) pp. 53–77. John Wiley & Sons, New York.

Goryshin, N.I. and Tyshchenko, G.F. (1976). Parallel phenomena in the photoperiodic reaction and circadian rhythms of insects with light-dark cycles of various duration. *Entomol. Rev.* 55, 10–17.

Gösswald, K. and Bier, K. (1957). Untersuchungen zur Kastendetermination in der Gattung *Formica*. 5. Der Einfluss der Temperatur auf die Eiablage und Geschlechtsbestimmung. *Insectes Sociaux* 4, 335–48.

Granger, N.A. and Bollenbacher, W.E. (1981). Hormonal control of insect metamorphosis. In *Metamorphosis* (eds. L.I. Gilbert and E. Frieden) pp. 105–37. Plenum Press, New York.

Grant, V. (1966). The selective origin of incompatibility barriers in the plant genus *Gilia*. *Amer. Nat.* 100, 99–118.

Grechka, E.O. and Kipyatkov, V.E. (1983). Seasonal development cycle and caste differentiation in social wasp *Polistes gallicus* (Hymenoptera, Vespidae). I. Phenology and life cycle regulation. *Rev. Entomol. l'URSS* 62, 450–61 (in Russian, English summary).

Greenbank, D.O. (1970). Climate and the ecology of the balsam woolly aphid. *Can. Entomol.* 102, 546–78.

Griffiths, E. and Wratten, S.D. (1979). Intra- and inter-specific differences in cereal aphid low-temperature tolerance. *Entomol. Exp. Appl.* 26, 161–67.

Griffiths, K.J. (1969a). Development and diapause in *Pleolophus basizonus* (Hymenoptera: Ichneumonidae). *Can. Entomol.* 101, 907–14.

Griffiths, K.J. (1969b). The importance of coincidence in the functional and numerical responses of two parasites of the European pine sawfly, *Neodiprion sertifer*. *Can. Entomol.* 101, 673–713.

Grigo, F. and Topp, W. (1980). Einfluss von Adaptationstemperatur und Photoperiode auf den Sauerstoffverbrauch bei Staphyliniden (Col.) in Diapause und Non-Diapause. *Zool. Anz.* 204, 19–26.

Grinfeld, E.K. (1972). The influence of photoperiodism on the development cycle of the social wasp *Polistes gallicus* L. (Hymenoptera, Vespidae). *Vestn. Len. gos. Univ.* 21, 148–51 (in Russian, English summary).

Grinfeld, E.K. and Zakharova, L.V. (1971). The role of photoperiodism in the development cycle of bumble bees (preliminary communication). *Vestn. Len. gos. Univ.* 9, 149–50 (in Russian, English summary).

Grzelak, K., Szczesna, E., Krówczyńska, A., and Lassota, Z. (1981). Transcription in diapausing and developing pupae of *Celerio euphorbiae*. *Insect Biochem.* 11, 67–72.

Guerra, A.A., Garcia, R.F., Bodegas V., P.R., and de Coss F., M.E. (1984). The quiescent physiological status of boll weevils (Coleoptera: Curculionidae) during the noncotton season in the tropical zone of Soconusco in Chiapas, Mexico. *J. Econ. Entomol.* 77, 595–98.

Gupta, A.P. (1983). *Neurohemal organs of arthropods*. Charles C. Thomas, Springfield, IL.

Gurjanova, T.M. (1979). Importance of diapause of a parasite *Exenterus abruptorius* (Ichneumonidae) on population dynamics in *Neodiprion sertifer*. *Zool. Zh.* 58, 1339–49.

Gustin, R.D. (1983). *Diabrotica longicornis barberi* (Coleoptera: Chrysomelidae): Cold hardiness of eggs. *Environ. Entomol.* 12, 633–34.

Gutierrez, A.P., Leigh, T.F., Wang, Y., and Cave, R.D. (1977). An analysis of cotton production in California: *Lygus hesperus* (Heteroptera: Miridae) injury—an evaluation. *Can. Entomol.* 109, 1375–86.

Gutierrez, A.P., Wang, Y., and Daxl, R. (1979). The interaction of cotton and boll weevil (Lepidoptera: Gelichiidae)—A study of co-adaptation. *Can. Entomol.* 111, 357–66.

Habu, N. (1969). Life cycles of the pine moth, *Dendrolimus spectabilis* (Lepidoptera: Lasiocampidae) in Kyoto. *Japan. J. Appl. Entomol. Zool.* 13, 200–5 (in Japanese).

Hackett, D.S. and Gatehouse, A.G. (1982). Diapause in *Heliothis armigera* (Hübner) and *H. fletcheri* (Hardwick) (Lepidoptera: Noctuidae) in the Sudan Gezira. *Bull. Entomol. Res.* 72, 409–22.

Hadley, N.F. (1979). Wax secretion and color phases of the desert tenebrionid beetle *Cryptoglossa verrucosa* (LeConte). *Science* 203, 367–69.

Hagen, K.S. (1962). Biology and ecology of predaceous Coccinellidae. *Annu. Rev. Entomol.* 7, 289–326.

Hagen, K.S., Tassan, R.L., and Sawall, E.F., Jr. (1970). Some ecophysiological relationships between certain *Chrysopa*, honeydews, and yeasts. *Boll. Lab. Entomol. Agric. Filippo Silvestri Port.* 28, 113–14.

Hagen, K.S., Bombosch, S., and McMurtry, J.A. (1976a). The biology and impact of predators. In *Theory and practice of biological control* (eds. C.B. Huffaker and P.S. Messenger) pp. 93–142. Academic Press, New York.

Hagen, K.S., Greany, P., Sawall, E.F., Jr., and Tassan, R.L. (1976b). Tryptophan in artificial honeydews as a source of an attractant for adult *Chrysopa carnea*. *Environ. Entomol.* 5, 458–68.

Hagen, R.H. and Lederhouse, R.C. (1984). Polymodal emergence of the tiger swallowtail, *Papilio glaucus* (Lepidoptera: Papilionidae): source of a false second generation in central New York State. *Ecol. Entomol.* 10, 19–28.

Haglund, B.M. (1980). Proline and valine—cues which stimulate grasshopper herbivory during drought stress? *Nature* 288, 697–98.

Hagstrum, D.W. and Silhacek, D.L. (1980). Diapause induction in *Ephestia cautella*: An interaction between genotype and crowding. *Entomol. Exp. Appl.* 28, 29–37.

Hain, F.P. and Wallner, W.E. (1973). The life history, biology, and parasites of the pine candle moth, *Exoteleia nepheos*, on Scotch pine in Michigan. *Can. Entomol.* 105, 157–64.

Hairston, N.G., Jr., and Munns, W.R., Jr. (1984). The timing of copepod diapause as an evolutionarily stable strategy. *Amer. Natur.* 123, 733–51.

Hairston, N.G. and Olds, E.J. (1984). Population differences in the timing of diapause: adaptation in a spatially heterogeneous environment. *Oecologia* 61, 42–48.

Hamamura, T. (1982). Effects of photoperiod on nymphal development in *Philodromus subaureolus* Boesenberg et Strand (Araneae: Tomisidae). *Japan. J. Appl. Entomol. Zool.* 26, 131–37 (in Japanese, English summary).

Hand (née Smeeton), L. (1983). The effect of photoperiod by the red ant, *Myrmica rubra*. *Entomol. Exp. Appl.* 34, 169–73.

Hand, S.C. (1983). The effect of temperature and humidity on the duration of development and hatching success of eggs of the aphid, *Sitibion avenae*. *Entomol. Exp. Appl.* 33, 220–22.

Hans, H. (1961). Termination of diapause and continuous laboratory rearing of the sweet clover weevil, *Sitona cylindricollis* Fahr. *Entomol. Exp. Appl.* 4, 41–46.

Hansen, T.E., Viyk, M.O., and Luyk, A.K. (1980). Biochemical changes and cold-hardiness in overwintering bark-beetles *Ips typographus*. *Entomol. Rev.* 59, 9–12.

Hanski, I. (1980). Spatial variation in the timing of the seasonal occurrence in coprophagous beetles. *Oikos* 34, 311–21.

Hanski, I. and Koskela, H. (1979). Resource partitioning in six guilds of dung-inhabiting beetles (Coleoptera). *Ann. Entomol. Fenn.* 45, 1–12.

Harbo, J.R. and Kraft, K.J. (1969). A study of *Phanerotoma toreutae*, a parasite of the pine cone moth, *Laspeyresia toreuta*. *Ann. Entomol. Soc. Amer.* 62, 214–20.

Hardie, J. (1980a). Juvenile hormone mimics the photoperiodic apterization of the alate gynopara of aphid, *Aphis fabae*. *Nature* 286, 602–4.

Hardie, J. (1980b). Reproductive, morphological and behavioural affinities between the alate gynopara and virginipara of the aphid, *Aphis fabae*. *Physiol. Entomol.* 5, 385–96.

Hardie, J. (1981a). Juvenile hormone and photoperiodically controlled polymorphism in *Aphis fabae*: Prenatal effects on presumptive oviparae. *J. Insect Physiol.* 27, 257–65.

Hardie, J. (1981b). Juvenile hormone and photoperiodically controlled polymorphism in *Aphis fabae*: Postnatal effects on presumptive gynoparae. *J. Insect Physiol.* 27, 347–55.

Hardie, J. and Lees, A.D. (1983). Photoperiodic regulation of the development of winged gynoparae in the aphid, *Aphis fabae*. *Physiol. Entomol.* 8, 385–91.

Hare, J.D. (1983). Seasonal variation in plant-insect associations: Utilization of *Solanum dulcamara* by *Leptinotarsa decemlineata*. *Ecology* 64, 345–61.

Harper, A.M. and Lilly, C.E. (1982). Aggregations and winter survival in southern Alberta of *Hippodamia quinquesignata* (Coleoptera: Coccinellidae), a predator of the pea aphid (Homoptera: Aphididae). *Can. Entomol.* 114, 303–9.

Harrewijn, P. (1978). The role of plant substances in polymorphism of the aphid *Myzus persicae*. *Entomol. Exp. Appl.* 24, 198–214.

Harrison, R.G. (1979). Speciation in North American field crickets: Evidence from electrophoretic comparisons. *Evolution* 33, 1009–23.

Harrison, R.G. (1980). Dispersal polymorphisms in insects. *Annu. Rev. Ecol. Syst.* 11, 95–118.

Harrison, R.G. and Vawter, A.T. (1977). Allozyme differentiation between pheromone strains of the European corn borer, *Ostrinia nubilalis*. *Ann. Entomol. Soc. Amer.* 70, 717–20.

Hartley, J.C. and Warne, A.C. (1972). The developmental biology of the egg stage of Western European Tettigoniidae (Orthoptera). *J. Zool.*, London 168, 267–98.

Hartman, M.J. and Hynes, C.D. (1980). Embryonic diapause in *Tipula simplex* and the action of photoperiod in its termination (Diptera: Tipulidae). *Pan- Pac. Entomol.* 56, 207–12.

Harvey, A.W. (1983). *Schistocerca piceifrons* (Walker) (Orthoptera: Acrididae), the swarming locust of tropical America: A review. *Bull. Entomol. Res.* 73, 171–84.

Harvey, G.T. (1957). The occurrence and nature of diapause-free development in the spruce budworm, *Choristoneura fumiferana* (Clem.) (Lepidoptera: Tortricidae). *Can. J. Zool.* 35, 549–72.

Harvey, G.T. (1967). On coniferophagous species of *Choristoneura* (Lepidoptera: Tortricidae) in North America. V. Second diapause as a species character. *Can. Entomol.* 99, 486–503.

Harwood, R.F. (1966). The relationship between photoperiod and autogeny in *Culex tarsalis* (Diptera, Culicidae). *Entomol. Exp. Appl.* 9, 327–31.

Hasegawa, K. (1952). Studies on the voltinism in the silkworm, *Bombyx mori* L., with special reference to the organs concerning determination of voltinism. *J. Fac. Agr. Tottori Univ.* 1, 83–126.

Hasegawa, K. (1963). Studies on the mode of action of the diapause hormone in the silkworm, *Bombyx mori* L. I. The action of diapause hormone injected into pupae of different ages. *J. Exp. Biol.* 40, 517–29.

Hasegawa, K. and Yamashita, O. (1967). Control of metabolism in the silkworm pupal ovary by the diapause hormone. *J. Seric. Sci. Japan* 36, 297–300.

Hasegawa, K. and Yamashita, O. (1970). Mode d'action de l'hormone de diapause dans le métabolisme glucidique du ver à soie *Bombyx mori* L. *Anales d'Endocrinol.*, Paris 31, 631–36.

Hassell, M.P. (1969). A study of the mortality factors acting upon *Cyzenis albicans* (Fall.), a tachinid parasite of the winter moth (*Operophtera brumata* (L.)). *J. Anim. Ecol.* 38, 329–39.

Havelka, J. (1980). Photoperiodism of the carnivorous midge *Aphidoletes aphidimyza* (Diptera, Cecidomyiidae). *Entomol. Rev.* 59, 1–8.

Hawke, S.D. and Farley, R.D. (1973). Ecology and behavior of the desert burrowing cockroach, *Arenivaga* sp. (Dictyoptera, Polyphagidae). *Oecologia* 11, 263–79.

Hayakawa, Y. and Chino, H. (1981). Temperature-dependent interconversion between glycogen and trehalose in diapausing pupae of *Philosamia cynthia ricini* and *preyeri*. *Insect Biochem.* 11, 43–47.

Hayes, D.K., Sullivan, W.N., Oliver, M.Z., and Schechter, M.S. (1970). Photoperiod manipulation of insect diapause: A method of pest control? *Science* 169, 382–83.

Hayes, D.K., Horton, J., Schechter, M.S., and Halberg, F. (1972). Rhythm of oxygen uptake in diapausing larvae of the codling moth at several temperatures. *Ann. Entomol. Soc. Amer.* 65, 93–96.

Hayes, D.K., Cawley, B.M., Sullivan, W.N., Adler, V.E., and Schechter, M.S. (1974a). The effect of added light pulses on overwintering and diapause, under natural light and temperature conditions, of four species of Lepidoptera. *Environ. Entomol.* 3, 863–65.

Hayes, D.K., Hewing, A.N., Odesser, D.B., Sullivan, W.N., and Schechter, M.S. (1974b). The effect on diapause of photoperiod manipulation at different temperatures. In *Chronobiology* (eds. L.E. Scheving, F. Halberg, and J.E. Pauly) pp. 593–96. Igaky Shoin Ltd., Tokyo.

Hayes, D.K., Sullivan, W.N., Schechter, M.S., Cawley, B.M., Jr., and Campbell, L.E. (1979). European corn borer: Effect of manipulated photoperiods on survival in the field. *J. Econ. Entomol.* 72, 61–63.

Hayes, J.L. (1982a). A study of the relationships of diapause phenomena and other life history characters in temperate butterflies. *Amer. Natur.* 120, 160–70.

Hayes, J.L. (1982b). Diapause and diapause dynamics of *Colias alexandra* (Lepidoptera: Pieridae). *Oecologia* 53, 317–22.

Hayes, J.L. and Dingle, H. (1983). Male influence on the duration of reproductive diapause in the large milkweed bug, *Oncopeltus fasciatus*. *Physiol. Entomol.* 8, 251–56.

Haynes, D.L., Brandenburg, R.K., and Fisher, P.D. (1973). Environmental monitoring network for pest management systems. *Environ. Entomol.* 2, 889–99.

Hazel, W.N. and West, D.A. (1979). Environmental control of pupal colour in swallowtail butterflies (Lepidoptera: Papilioninae): *Battus philenor* (L.) and *Papilio polyxenes* Fabr. *Ecol. Entomol.* 4, 393–400.

Hazel, W.N. and West, D.A. (1983). The effect of larval photoperiod on pupal colour and diapause in swallowtail butterflies. *Ecol. Entomol.* 8, 37–42.

Hedin, P.A. (1976). Seasonal variations in the emission of volatiles by cotton plants growing in the field. *Environ. Entomol.* 5, 1234–38.

Hedlin, A.F., Miller, G.E., and Ruth, D.S. (1982). Induction of prolonged diapause in *Barbara colfaxiana* (Lepidoptera: Olethreutidae): Correlations with cone crops and weather. *Can. Entomol.* 114, 465–71.

Hegdekar, B.M. (1977). Photoperiodic and temperature regulation of diapause induction in the bertha armyworm, *Mamestra configurata* Walker. *Manitoba Entomol.* 11, 56–60.

Hegdekar, B.M. (1979). Epicuticular wax secretion in diapause and non-diapause pupae of the bertha armyworm. *Ann. Entomol. Soc. Amer.* 72, 13–15.

Hegmann, J.P. and Dingle, H. (1982). Phenotypic and genetic covariance structure in milkweed bug life history traits. In *Evolution and genetics of life histories* (eds. H. Dingle and J.P. Hegmann) pp. 177–85. Springer-Verlag, New York.

Heinrich, B. (1974). Themoregulation in endothermic insects. *Science* 185, 747–56.

Heinrich, B. (1975). Thermoregulation and flight energetics of desert insects. In *Environmental physiology of desert organisms* (ed. N.F. Hadley) pp. 90–105. Dowden, Hutchinson and Ross, London.

Heinrich, B. (ed.) (1981). *Insect thermoregulation*. Wiley-Interscience, New York.

Helfert, B. (1980). Die regulative Wirkung von Photoperiode und Temperatur auf den Lebenszyklus ökologisch unterschiedlicher Tettigoniiden-Arten (Orthoptera, Saltatoria). 2. Teil: Embryogenese und Dormanz der Filialgeneration. *Zool. Jb. Syst.* 107, 449–500.

Helgesen, R.G. and Haynes, D.L. (1972). Population dynamics of the cereal leaf beetle, *Oulema melanopus*: A model for age specific mortality. *Can. Entomol.* 104, 797–814.

Helle, W. (1962). Genetics of resistance to organophosphorus compounds and its relation to diapause in *Tetranychus urticae* Koch (Acari). *Tijdschr. Pl.- ziekten* 63, 155–95.

Helle, W. (1968). Genetic variability of photoperiodic response in an arrhenotokous mite (*Tetranychus urticae*). *Entomol. Exp. Appl.* 11, 101–13.

Helle, W. and Overmeer, W.P.J. (1973). Variability in tetranychid mites. *Annu. Rev. Entomol.* 18, 97–120.

Henderson, C.F. (1955). Overwintering, spring emergence and host synchronization of two egg parasites of the beet leafhopper in southern Idaho. U.S.D.A. Cir. No. 967.

Henneberry, T.J. and Clayton, T.E. (1983). Pink bollworm (Lepidoptera: Gelechiidae): Effects of soil moisture on behavior of diapausing larvae and adult emergence from bolls. *Environ. Entomol.* 12, 1490–95.

Henrich, V.C. and Denlinger, D.L. (1982a). A maternal effect that eliminates pupal diapause in progeny of the flesh fly, *Sarcophaga bullata*. *J. Insect Physiol.* 28, 881–84.

Henrich, V.C. and Denlinger, D.L. (1982b). Selection for late pupariation affects diapause incidence and duration in the flesh fly, *Sarcophaga bullata*. *Physiol. Entomol.* 7, 407–11.

Herbert, H.J. and McRae, K.B. (1982). Predicting eclosion of overwintering eggs of the European red mite, *Panonychus ulmi* (Acarina: Tetranychidae), in Nova Scotia. *Can. Entomol.* 114, 703–12.

Herman, W.S. (1973). The endocrine basis of reproductive inactivity in monarch butterflies overwintering in central California. *J. Insect Physiol.* 19, 1883–87.

Herman, W.S. (1981). Studies on the adult reproductive diapause of the monarch butterfly, *Danaus plexippus*. *Biol. Bull.* 160, 89–106.

Herman, W.S., Lessman, C.A., and Johnson, G.D. (1981). Correlation of juvenile hormone titer changes with reproductive tract development in the posteclosion monarch butterfly. *J. Exp. Zool.* 218, 387–95.

Heron, R.J. (1972). Differences in postdiapause development among geographically distinct populations of the larch sawfly, *Pristiphora erichsonii* (Hymenoptera: Tenthredinidae). *Can. Entomol.* 104, 1307–12.

Herron, J.C. (1953). Biology of the sweet clover weevil and notes on the biology of the clover root curculio. *Ohio J. Sci.* 53, 105–12.

Herzog, G.A. and Phillips, J.R. (1974). Selection for a nondiapause strain of the bollworm, *Heliothis zea* (Lepidoptera: Noctuidae). *Environ. Entomol.* 3, 525–27.

Hew, C.L., Kao, M.H., So, Y.P., and Lim, K.-P. (1983). Presence of cystine-containing antifreeze proteins in the spruce budworm *Choristoneura fumiferana. Can. J. Zool.* 61, 2324–28.

Hidaka, T. (1956). Recherches sur le déterminisme hormonal de la coloration pupale chez lépidoptères. I. Les effets de la ligature, de l'ablation des ganglions et de l'incision des nerfs chez prépupes et larves âgées de quelque papilionides. *Annot. Zool. Japon.* 29, 69–74.

Hidaka, T. (1977). Mating behaviour. In *Adaptation and speciation in the fall webworm* (ed. T. Hidaka) pp. 81–100. Kodansha Ltd., Tokyo.

Hidaka, T. and Aida, S. (1963). Day length as the main factor of seasonal form determination in *Polygonia c-aureum* (Lepidoptera, Nymphalidae). *Zool. Mag.* 72, 77–83.

Hidaka, T. and Takahashi, H. (1967). Temperature conditions and maternal effect as modifying factors in the photoperiodic control of the seasonal form in *Polygonia c-aureum* (Lepidoptera, Nymphalidae). *Annot. Zool. Japon.* 40, 200–4.

Hidaka, T., Ishizuka, Y., and Sakagami, Y. (1971). Control of pupal diapause and adult differentiation in a univoltine papilionid butterfly, *Luehdorfia japonica. J. Insect Physiol.* 17, 197–203.

Highnam, K.C. and Hill, L. (1977). *The comparative endocrinology of the invertebrates* (2nd ed). Edward Arnold, London.

Hill, H.F., Jr., Wenner, A.M., and Wells, P.H. (1976). Reproductive behavior in an overwintering aggregation of monarch butterflies. *Amer. Midl. Natur.* 95, 10–19.

Hille Ris Lambers, D. (1966). Polymorphism in Aphididae. *Annu. Rev. Entomol.* 11, 46–78.

Hillman, W.S. (1973). Non-circadian photoperiodic timing in the aphid *Megoura. Nature* 242, 128–29.

Hinton, H.E. (1951). A new Chironomid from Africa, the larva of which can be dehydrated without injury. *Proc. Zool. Soc. Lond.* 121, 371-80.

Hinton, H.E. (1960). Cryptobiosis in the larva of *Polypedilum vanderplanki* Hint. (Chironomidae). *J. Insect Physiol.* 5, 286–300.

Hiruma, K. (1979). Prevention of pupal diapause by the application of juvenile hormone analogue to the last instar larvae of *Mamestra brassicae. Appl. Entomol. Zool.* 14, 76–82.

Hobbs, G.A. and Richards, K.W. (1976). Selection for a univoltine strain of *Megachile* (Eutricharaea) *pacifica* (Hymenoptera: Megachilidae). *Can. Entomol.* 108, 165–67.

Hodek, I. (1962). Experimental influencing of the imaginal diapause in *Coccinella septempunctata* L. *Acta Soc. Entomol. Cesk.* 59, 297–313.

Hodek, I. (1971a). Termination of adult diapause in *Pyrrhocoris apterus* (Heteroptera: Pyrrhocoridae) in the field. *Entomol. Exp. Appl.* 14, 212–22.

Hodek, I. (1971b). Sensitivity of larvae to photoperiods controlling the adult diapause of two insects. *J. Insect Physiol.* 17, 205–16.

Hodek, I. (1971c). Sensitivity to photoperiod in *Aelia acuminata* (L.) after adult diapause. *Oecologia* 6, 152–55.

Hodek, I. (1973). *Biology of Coccinellidae.* Junk, The Hague; Academia, Prague.

Hodek, I. (1974). Development of diapause in *Pyrrhocoris apterus* females in the middle period of winter dormancy (Heteroptera). *Vestn. Cesk. Spol. Zool.* 3, 161–69.

Hodek, I. (1976). Two contrasting types of environmental regulation of adult diapause. *Atti della Accademia delle Scienze dell' Istituto di Bologna l3*, 3, 81-88.

Hodek, I. (1977). Photoperiodic response in spring in three Pentatomoidea (Heteroptera). *Acta Entomol. Bohemoslov.* 74, 209–18.

Hodek, I. (1979). Intermittent character of adult diapause in *Aelia acuminata* (Heteroptera). *J. Insect Physiol.* 25, 867–71.

Hodek, I. (1981). Le role des signaux de l'environnement et des processus endogènes dans la régulation de la reproduction par la diapause imaginale. *Bull. Soc. Zool. de France* 106, 317–25.

Hodek, I. (1983). Role of environmental factors and endogenous mechanisms in insects diapausing as adults. In *Diapause and life cycle strategies in insects* (eds. V.K. Brown and I. Hodek) pp. 9–33. Junk, The Hague.

Hodek, I. and Čerkasov, J. (1958). A study of the imaginal hibernation of *Semiadalia undecimnotata* Schneid. (Coccinellidae, Col.) in the open. I. (5th contribution to the study of ecology of Coccinellidae). *Acta Soc. Zool. Bohemoslov.* 22, 180–92.

Hodek, I. and Čerkasov, J. (1961). Experimental influencing of the imaginal diapause in *Coccinella septempunctata* L. (Coccinellidae, Col.). *Acta Soc. Zool. Bohemoslov.* 25, 70–90.

Hodek, I. and Čerkasov, J. (1963). Imaginal dormancy in *Semiadalia undecimnotata* Schneid. (Coccinellidae, Col.). II. Changes in water, fat and glycogen content. *Acta Soc. Zool. Bohemoslov.* 27, 298–318.

Hodek, I. and Hodková, M. (1981). Relationship between respiratory rate and diapause intensity in adults of two heteropteran species. *Vest. Cesk. Spol. Zool.* 45, 27–34.

Hodek, I. and Honek, A. (1970). Incidence of diapause in *Aelia acuminata* (L.) populations from southwest Slovakia (Heteroptera). *Vest. Cesk. Spol. Zool.* 33, 170–83.

Hodek, I. and Landa, V. (1971). Anatomical and histological changes during dormancy in two Coccinellidae. *Entomophaga* 16, 239–51.

Hodek, I. and Růžička, Z. (1977). Insensitivity to photoperiod after diapause in *Semiadalia undecimnotata* (Col. Coccinellidae). *Entomophaga* 22, 169–74.

Hodek, I. and Růžička, Z. (1979). Photoperiodic response in relation to diapause in *Coccinella septempunctata* (Coleoptera). *Acta. Entomol. Bohemoslov.* 76, 209–18.

Hodek, I., Iperti, G., and Rolley, F. (1977). Activation of hibernating *Coccinella septempunctata* (Coleoptera) and *Perilitus coccinellae* (Hymenoptera) and the photoperiodic response after diapause. *Entomol. Exp. Appl.* 21, 275–86.

Hodek, I., Bonet, A., and Hodková, M. (1981). Some ecological factors affecting diapause in adults of *Acanthoscelides obtectus* from Mexican mountains. In *The ecology of bruchids attacking legumes (Pulses)* (ed. V. Labeyrie) pp. 43–55. Junk, The Hague.

Hodková, M. (1976). Nervous inhibition of corpora allata by photoperiod in *Pyrrhocoris apterus*. *Nature* 263, 521–23.

Hodková, M. (1982). Interaction of feeding and photoperiod in regulation of the corpus allatum activity in females of *Pyrrhocoris apterus* L. (Hemiptera). *Zool. Jb. Physiol.* 86, 477–88.

Hoffmann, H.J. (1970). Neuro-endocrine control of diapause and oocyte maturation in the beetle *Pterostichus nigrata*. *J. Insect Physiol.* 16, 629–42.

Hoffmann, R.J. (1974). Environmental control of seasonal variation in the butterfly *Colias eurytheme*: Effects of photoperiod and temperature on pteridine pigmentation. *J. Insect Physiol.* 20, 1913–24.

Hoffmann, R.J. (1978). Environmental uncertainty and evolution of physiological adaptation in *Colias* butterflies. *Amer. Natur.* 112, 999–1015.

Hogan, T.W. (1960a). The onset and duration of diapause in eggs of *Acheta commodus* (Walk.) (Orthoptera). *Aust. J. Biol. Sci.* 13, 14–29.

Hogan, T.W. (1960b). The effects of subzero temperatures on the embryonic diapause of *Acheta commodus* (Walk.) (Orthoptera). *Aust. J. Biol. Sci.* 13, 527–40.

Hogan, T.W. (1967). Inter-racial mating of a non-diapausing and a diapausing race of *Teleogryllus commodus* (Walk.) (Orthoptera: Gryllidae). *Aust. J. Zool.* 15, 541–45.

Hogan, T.W. (1974). A genetic approach to the population suppression of the common field cricket *Teleogryllus commodus*. In *The use of genetics in insect control* (eds. R. Pal and M.J. Whitten) pp. 57–70. Elsevier-North-Holland, Amsterdam.

Hokyo, N. (1971). Applicability of the temperature-sum rule in estimating the emergence time of the overwintering population of the green rice leafhopper, *Nephotettix cincticeps* Uhler. *Appl. Entomol. Zool.* 6, 1–10.

Hokyo, N., Suzuki, H., and Murai, M. (1983). Egg diapause in the Oriental chinch bug, *Cavelerius saccharivorus* Okajima (Heteroptera: Lygaeidae) I. Incidence and intensity. *Appl. Entomol. Zool.* 18, 382–91.

Hollingsworth, C.S. and Capinera, J.L. (1983). Comparison of nymphal developmental times in diapausing and nondiapausing strains of *Melanoplus sanguinipes* (F.) (Orthoptera: Acrididae). *J. Kans. Entomol. Soc.* 56, 612–16.

Holtzer, T.O., Bradley, J.R., Jr., and Rabb, R.L. (1976). Geographic and genetic variation in time required for emergence of diapausing *Heliothis zea*. *Ann. Entomol. Soc. Amer.* 69, 261–65.

Holzapfel, C.M. and Bradshaw, W.E. (1981). Geography of larval dormancy in the tree-hole mosquito, *Aedes triseriatus* (Say). *Can. J. Zool.* 59, 1014–21.

Homma, K. (1966). Photoperiodic responses in two local populations of the smaller tea tortrix,

Adoxophyes orana Fischer von Röslerstamm (Lepidoptera: Tortricidae). *Appl. Entomol. Zool.* 1, 32–36.

Honda, K. (1981). Environmental factors affecting the pupal coloration in *Papilio protenor demetrius* Cr. (Lepidoptera: Papilionidae). II. Effect of physical stimuli. *Appl. Entomol. Zool.* 16, 467–71.

Honek, A. (1972). Selection for non-diapause in *Aelia acuminata* and *A. rostrata* (Heteroptera, Pentatomidae) under various selective pressures. *Acta. Entomol. Bohemoslov.* 69, 73–77.

Honek, A. (1973). Induction of a winter coloration in *Chrysopa carnea* Steph. (Neuroptera: Chrysopidae). *Věst. Česk. Spol. Zool.* 37, 253–57.

Honek, A. (1979a). Independent response of 2 characters to selection for insensitivity to photoperiod in *Pyrrhocoris apterus*. *Experientia* 35, 762–63.

Honek, A. (1979b). Regulation of diapause, number of instars, and body growth in the moth species *Amathes c-nigrum* (Lepidoptera: Noctuidae). *Entomol. Gen.* 5, 221–29.

Hong, J.W. and Platt, A.P. (1975). Critical photoperiod and daylength threshold differences between northern and southern populations of the butterfly *Limenitis archippus*. *J. Insect Phyisol.* 21, 1159–65.

Hopp, R.J., Vittum, M.T., Canfield, N.L., and Dethier, B.E. (1972). Regional phenological studies with Persian lilac (*Syringa persica*). *N.Y. Food Life Sci. Bull. Geneva* 17, 1–8.

Horsfall, W.R., Fowler, H.W., Jr., Moretti, L.J., and Larsen, J.R. (1973). *Bionomics and embryology of the inland floodwater mosquito Aedes vexans.* University Illinois Press, Urbana.

Horwath, K.L. and Duman, J.G. (1982). Involvement of the circadian system in photoperiodic regulation of insect antifreeze proteins. *J. Exp. Zool.* 219, 267–70.

Horwath, K.L. and Duman, J.G. (1983a). Preparatory adaptations for winter survival in the cold hardy beetles, *Dendroides canadensis* and *Dendroides concolor*. *J. Comp. Physiol.* 151, 225–32.

Horwath, K.L. and Duman, J.G. (1983b). Induction of antifreeze protein production by juvenile hormone in larvae of the beetle, *Dendroides canadensis*. *J. Comp. Physiol.* 151, 233–40.

Horwath, K.L. and Duman, J.G. (1983c). Photoperiodic and thermal regulation of antifreeze protein levels in the beetle *Dendroides canadensis*. *J. Insect Physiol.* 29, 907–17.

House, H.L. (1967). The decreasing occurrence of diapause in the fly *Pseudosarcophaga affinis* through laboratory-reared populations. *Can. J. Zool.* 45, 149–53.

Howe, R.W. (1962). The influence of diapause on the status as pests of insects found in houses and warehouses. *Ann. Appl. Biol.* 50, 611–14.

Hoy, M.A. (1975a). Diapause in the mite *Metaseiulus occidentalis*: Stages sensitive to photoperiodic induction. *J. Insect Physiol.* 21, 745–51.

Hoy, M.A. (1975b). Hybridization of strains of the gypsy moth parasitoid, *Apanteles melanoscelus*, and its influence upon diapause. *Ann. Entomol. Soc. Amer.* 68, 261–64.

Hoy, M.A. (1977). Rapid response to selection for a nondiapausing gypsy moth. *Science* 196, 1462–63.

Hoy, M.A. (1978a). Variability in diapause attributes of insects and mites: Some evolutionary and practical implications. In *Evolution of insect migration and diapause* (ed. H. Dingle) pp. 101–26. Springer-Verlag, New York.

Hoy, M.A. (1978b). Selection for a non-diapausing gypsy moth: Some biological attributes of a new laboratory strain. *Ann. Entomol. Soc. Amer.* 71, 75–80.

Hoy, M.A. (1979). The potential for genetic improvement of predators for pest management programs. In *Genetics in relation to insect management* (eds. M.A. Hoy and J.J. McKelvey, Jr.) pp. 106–15. Rockefeller Foundation, New York.

Hoy, M.A. and Flaherty, D.L. (1970). Photoperiodic induction of diapause in a predaceous mite, *Metaseiulus occidentalis*. *Ann. Entomol. Soc. Amer.* 63, 960–63.

Hoy, M.A. and Flaherty, D.L. (1975). Diapause induction and duration in vineyard-collected *Metaseiulus occidentalis*. *Environ. Entomol.* 4, 262–64.

Hoy, M.A. and Knop, N.F. (1978). Development, hatch dates, overwintering success, and spring emergence of a "non-diapausing" gypsy moth strain (Lepidoptera: Orgyiidae) in field cages. *Can. Entomol.* 110, 1003–8.

Hoy, M.A. and McKelvey, J.J., Jr. (eds.) (1979). *Genetics in relation to insect management.* A Rockefeller Foundation Conference in Bellagio, Italy, March 31–April 5, 1978. Rockefeller Foundation, New York.

Hsiao, T.H. (1978). Host plant adaptations among geographic populations of the Colorado potato beetle. *Entomol. Exp. Appl.* 24, 437–47.

Hsiao, T.H. (1981). Ecophysiological adaptations among geographic populations of the Colorado potato beetle in North America. In *Advances in potato pest management* (eds. J. Lashomb and R. Casagrande) pp. 69–85. Hutchinson Ross, Stroudsburg, PA.

Hubbell, T.H. and Norton, R.M. (1978). The systematics and biology of the cave-crickets of the North American tribe Habenoecini (Orthoptera Saltatoria: Ensifera: Rhaphidophoridae: Dolichopodinae). *Misc. Publ. Mus. Zool.*, University of Michigan 156, 1–124.

Hudson, A. (1966). Proteins in the haemolymph and other tissues of the developing tomato hornworm, *Protoparce quinquemaculata* Haworth. *Can. J. Zool.* 44, 541–55.

Hudson, J.E. (1977). Induction of diapause in female mosquitoes, *Culiseta inornata* by a decrease in daylength. *J. Insect Physiol.* 23, 1377–82.

Hudson, J.E. (1979). Follicle development, blood feeding, digestion and egg maturation in diapausing mosquitoes, *Culiseta inornata. Entomol. Exp. Appl.* 25, 136–45.

Huffaker, C.B. (1982). Overall approach to insect problems in agriculture. In *Biometeorology in integrated pest management* (eds. J.L. Hatfield and I.J. Thomason) pp. 171–92. Academic Press, New York.

Huffaker, C.B., Gordon, H.T., and Rabb, R.L. (1984). Meaning of ecological entomology—the ecosystem. In *Ecological entomology* (eds. C.B. Huffaker and R.L. Rabb) pp. 3–17. John Wiley & Sons, New York.

Huggans, J.L. and Blickenstaff, C.C. (1964). Effects of photoperiod on sexual development in the alfalfa weevil. *J. Econ. Entomol.* 57, 167–68.

Hummelen, P.J. (1974). Relations between two rice borers in Surinam, *Rupela albinella* (Cr.) and *Diatraea saccharalis* (F.), and their hymenopterous larval parasites. *Meded. Landbouwhogesch.*, Wageningen 74–1.

Hunter, D.M. (1980). Production of diapause eggs by the Australian plague locust after migration. *J. Aust. Entomol. Soc.* 19, 210.

Hunter, D.M. and Gregg, P.C. (1984). Variation in diapause potential and strength in eggs of the Australian plague locust, *Chortoicetes terminifera* (Walker) (Orthoptera: Acrididae). *J. Insect Physiol.* 30, 867–70.

Hussey, N.W. (1955). The life-histories of *Megastigmus spermotrophus* Wachtl (Hymenoptera: Chalcidoidea) and its principal parasite, with descriptions of the developmental stages. *Trans. Roy. Entomol. Soc. Lond.* 106, 133–51.

Ichijo, N., Kimura, M.T., and Minami, N. (1980). Eco-physiological aspects of reproductive diapause in *Drosophila sordidula* and *D. lacertosa* (Diptera: Drosophilidae). *Japan. J. Ecol.* 30, 221–28.

Ichinosé, T. and Iwasaki, N. (1979). Pupal diapause in some Japanese papilionid butterflies III. The difference in the termination of diapause between the two subspecies of *Papilio protenor* Cramer and their development. *Kontyû* 47, 272–80.

Ichinosé, T. and Negishi, H. (1979). Pupal diapause in some Japanese papilionid butterflies. II. The difference in the induction of diapause between the two subspecies of *Papilio protenor* Cramer. *Kontyû* 47, 89–98.

Iheagwam, E.U. (1977). Comparative flight performance of the seasonal morphs of the cabbage whitefly, *Aleyrodes brassicae* (Wlk.), in the laboratory. *Ecol. Entomol.* 2, 267–71.

Iheagwam, E.U. (1983). On the relationship between the so-called wet season and dry season Mendelian populations of the variegated grasshopper pest, *Zonocerus variegatus* L. (Orthopt., Pyrgomorphidae) at Nsukka, Nigeria. *Z. Ang. Entomol.* 96, 10–15.

Ikan, R., Gottlieb, R., Bergmann, E.D., and Ishay, J. (1969). The pheromone of the queen of the Oriental hornet, *Vespa orientalis. J. Insect Physiol.* 15, 1709–12.

Ilyinskaya, N.B. (1968). Diapause and cellular resistance to injury. *Proc. l3th Intl. Cong. Entomol.* 387–88.

Ilyinskaya, N.B. (1969). The diapause and the dynamics of seasonal changes in electrical resistivity of the muscles of the codling moth, *Laspeyresia pomonella* L. (Lepidoptera, Tortricidae). *Entomol. Rev.* 48, 285–92.

Ingram, B.R. (1975). Diapause termination in two species of damselflies. *J. Insect Physiol.* 21, 1909–16.

Injeyan, H.S. and Tobe, S.S. (1981). Phase polymorphism in *Schistocerca gregaria*: Assessment of juvenile hormone synthesis in relation to vitellogenesis. *J. Insect Physiol.* 27, 203–10.

Iperti, G. and Hodek, I. (1974). Induction alimentaire de la dormance imaginale chez *Semiadalia undecimnotata* Schn. (Coleop. Coccinellidae) pour aider a la conservation des coccinelles élevées au laboratoire avant une utilisation ultérieure. *Ann. Zool. - Ecol. Anim.* 6, 41–51.

Ishay, J. (1975). Caste determination by social wasps: Cell size and building behaviour. *Anim. Behav.* 23, 425–31.

Ishay, J.S., Iny, Y., and Rosenzweig, E. (1984). Longevity of hibernating queens in *Vespa orientalis* (Hymenoptera: Vespinae)—Effects of physical and chemical treatments. *J. Insect Physiol.* 30, 357–62.

Ishii, M. and Hidaka, T. (1979). Seasonal polymorphism of the adult rice-plant skipper, *Parnara guttata guttata* (Lepidoptera: Hesperiidae) and its control. *Appl. Entomol. Zool.* 14, 173–84.

Ishii, M. and Hidaka, T. (1982). Characteristics of pupal diapause in the univoltine papilionid, *Luehdorfia japonica* (Lepidoptera, Papilionidae). *Kontyû* 50, 610–20.

Ishii, M. and Hidaka, T. (1983). The second pupal diapause in the univoltine papilionid, *Luehdorfia japonica* (Lepidoptera: Papilionidae) and its terminating factor. *Appl. Entomol. Zool.* 18, 456–63.

Ishii, M., Johki, Y., and Hidaka, T. (1983). Studies on summer diapause in zygaenid moths (Lepidoptera, Zygaenidae) I. Factors affecting the pupal duration in *Pryeria sinica*. *Kontyû* 51, 122–27.

Ishikawa, K. (1936). *Silkworm pathology.* Meibundo, Tokyo.

Ishikura, H. and Nakatsuka, K. (1955). A review on prevalence and forecasting of the yellow stemborer. Plant Protection Association of Japan, Tokyo (in Japanese).

Ishitani, M. and Sato, N. (1981). The life cycle and a method of its control of the turnip maggot *Hylemya floralis* Fállen (Diptera: Anthomiidae) in Aomori Prefecture. *Bull. Aomori Field Crops Hortic. Exp. Sta.* No. 4, 1–16.

Ismail, S. and Fuzeau-Braesch, S. (1972). Analyse du déterminisme de la larvaire de *Gryllus campestris* (Orthoptères): Suppression de la diapause par sélection génétique. *C. R. Acad. Sci., Paris* 275, 1007–9.

Ismail, S. and Fuzeau-Braesch, S. (1976a). Programmation de la diapause chez *Gryllus campestris. J. Insect Physiol.* 22, 133–39.

Ismail, S. and Fuzeau-Braesch, S. (1976b). L'hormone de mue et la diapause chez *Gryllus campestris* L. (Orthop.): Dosage, injections et métabolisme. *Comp. Rend. Soc. Biol.* 170, 37–40.

Ismail, S., Baehr, J.-C., Porcheron, P., and Fuzeau-Braesch, S. (1984). The role of juvenile hormones in the larval diapause of the cricket *Gryllus campestris* (Orthoptera). *Comp. Biochem. Physiol.* 77A, 383–87.

Isobe, M. and Goto, T. (1980). Diapause hormones. In *Neurohormonal techniques in insects* (ed. T.A. Miller) pp. 216–43. Springer-Verlag, New York.

Istock, C.A. (1978). Fitness variation in a natural population. In *Evolution of insect migration and diapause* (ed. H. Dingle) pp. 171–90. Springer-Verlag, New York.

Istock, C.A. (1981). Natural selection and life history variation: Theory plus lessons from a mosquito. In *Insect life history patterns* (eds. R.F. Denno and H. Dingle) pp. 113–27. Springer-Verlag, New York.

Istock, C.A. (1982). Some theoretical considerations concerning life history evolution. In *Evolution and genetics of life histories* (eds. H. Dingle and J.P. Hegmann) pp. 7–20. Springer-Verlag, New York.

Istock, C.A. (1983). The extent and consequences of heritable variation for fitness characters. In *Population biology* (eds. C.E. King and P.S. Dawson) pp. 61–96. Columbia University Press, New York.

Istock, C.A., Wasserman, S.S., and Zimmer, H. (1975). Ecology and evolution of the pitcher-plant mosquito: 1. Population dynamics and laboratory responses to food and population density. *Evolution* 29, 296–312.

Istock, C.A., Zisfein, J., and Vavra, K.J. (1976). Ecology and evolution of the pitcher-plant mosquito. 2. The substructure of fitness. *Evolution* 30, 535–47.

Ito, Y. (1972). Prediction of seasonal occurrence of *Hyphantria cunea. Shokubutsu Boeki* 26, 139–43 (in Japanese).

Ivanovic, J., Jankovic-Hladni, M., Stanic, V., Milanovic, M., and Nenadovic, V. (1979). Possible role of neurohormones in the process of acclimatization and acclimation in *Morimus funereus* larvae (Insecta)—I. Changes in the neuroendocrine system and target organs (midgut, haemolymph) during the annual cycle. *Comp. Biochem. Physiol.* 63A, 95–102.

Ives, W.G.H. (1973). Heat units and outbreaks of the forest tent caterpillar, *Malacosoma disstria*. *Can. Entomol.* 105, 529–43.

Iwao, S. (1968). Some effects of grouping in Lepidopterous insects. In *l'Effet de groupe chez les animaux, Colloques Internationaux du Centre National de la Recherche Scientifique*, pp. 185–212. Centre National de la Recherche Scientifique, Paris.

Iwao, S. (1983). Phase variation in insects. *Res. Popul. Ecol.* 3, 3–14.

Iwao, Y., Kimura, M.T., Minami, N., and Watabe, H. (1980). Bionomics of Drosophilidae (Diptera) in Hokkaido. III. *Drosophila auraria* and *D. biauraria*. *Kontyû* 48, 160–68.

Izumiyama, S., Suzuki, K., and Miya, K. (1983). Glycerol in the eggs of the two-spotted cricket, *Gryllus bimaculatus* de Geer. *Appl. Entomol. Zool.* 18, 295–300.

Jackson, D.J. (1963). Diapause in *Caraphractus cinctus* Walker (Hymenoptera: Mymaridae), a parasitoid of the eggs of Dytiscidae (Coleoptera). *Parasitology* 53, 225–51.

Jacobson, J.W. and Hsiao, T.H. (1983). Isozyme variation between geographic populations of the Colorado potato beetle, *Leptinotarsa decemlineata* (Coleoptera: Chrysomelidae). *Ann. Entomol. Soc. Amer.* 76, 162–66.

Jacquemard, P. (1976a). La diapause de *Diparopsis watersi* (Roths.) (Lepidoptera Noctuidae) dans le Nord du Cameroun. *Cot. Fib. Trop.* 31, 297–311.

Jacquemard, P. (1976b). Relations entre la diapause de *Diparopsis watersi* (Roths.) (Lep. Noct.) et la diapause de son parasite *Eucarcelia* sp. [? *evolans* (Wied.)] (Dipt. Tachin.) dans le Nord du Cameroun. *Cot. Fib. Trop.* 31, 313–21.

James, B.D. and Luff, M.L. (1982). Cold-hardiness and development of eggs of *Rhopalosiphum insertum*. *Ecol. Entomol.* 7, 277–82.

James, D.G. (1982). Ovarian dormancy in *Danaus plexippus* (L.) (Lepidoptera: Nymphalidae)—oligopause not diapause. *J. Aust. Entomol. Soc.* 21, 31–35.

James, D.G. (1983). Induction of reproductive diapause in Australian monarch butterflies, *Danus plexippus* (L.). *Aust. J. Zool.* 31, 491–98.

James, D.G. and Hales, D.F. (1983). Sensitivity to juvenile hormone is not reduced in clustering monarch butterflies, *Danaus plexippus*, in Australia. *Physiol. Entomol.* 8, 273–76.

Janzen, D.H. (1973). Sweep samples of tropical foliage insects: Effects of seasons, vegetation types, elevation, time of day, and insularity. *Ecology* 54, 687–708.

Janzen, D.H. (1984). Weather-related color polymorphism of *Rothschildia lebeau* (Saturniidae). *Bull. Entomol. Soc. Amer.* 30, 16–20.

Janzen, D.H. and Waterman, P.G. (1984). A seasonal census of phenolics, fibre and alkaloids in foliage of forest trees in Costa Rica: Some factors influencing their distribution and relation to host selection by Sphingidae and Saturniidae. *Biol. J. Linnean Soc.* 21, 439–54.

Jeppson, L.R., Keifer, H.H., and Baker, E.W. (1975). *Mites injurious to economic plants*. University of California Press, Berkeley.

Jermy, T. (1967). Experiments on the factors governing diapause in the codling moth, *Cydia pomonella* L. (Lepidoptera, Tortricidae). *Acta Phytopath. Acad. Sci. Hung.* 2, 49–60.

Jervis, M.S. (1980). Ecological studies on the parasite complex associated with typhiocybine leafhoppers (Homoptera, Cicadellidae). *Ecol. Entomol.* 5, 123–36.

Johansson, T.S.K. and Johansson, M.P. (1979). The honeybee colony in winter. *Bee World* 60, 155–70.

Johnson, C.G. (1963). Physiological factors in insect migration by flight. *Nature* 198, 423–27.

Johnson, C.G. (1969). *Migration and dispersal of insects by flight*. Methuen, London.

Johnson, C.G., Taylor, L.R., and Southwood, T.R.E. (1962). High altitude migration of *Oscinella frit* L. (Diptera: Chloropidae). *J. Anim. Ecol.* 31, 373–83.

Johnston, P.G. and Zucker, I. (1980a). Photoperiodic regulation of reproductive development in white-footed mice *(Peromyscus leucopus)*. *Biol. Reprod.* 22, 983–89.

Johnston, P.G. and Zucker, I. (1980b). Photoperiodic regulation of the testes of adult white-footed mice *(Peromyscus leucopus)*. *Biol. Reprod.* 23, 859–66.

Joly, L., Hoffmann, J., and Joly, P. (1977). Controle humoral de la différenciation phasaire chez *Locusta migratoria migratorioides* (R. & F.) (Orthoptères). *Acrida* 6, 33–42.

Jordan, R.G. (1980a). Embryonic diapause in three populations of the western tree hole mosquito, *Aedes sierrensis*. *Ann. Entomol. Soc. Amer.* 73, 357–59.

Jordan, R.G. (1980b). Geographic differentiation in the development of *Aedes sierrensis* (Diptera: Culicidae) in nature. *Can. Entomol.* 112, 205–10.

Jordan, R.G. and Bradshaw, W.E. (1978). Geographic variation in the photoperiodic response of the western tree-hole mosquito, *Aedes sierrensis*. *Ann. Entomol. Soc. Amer.* 71, 487–90.

Jorgensen, C.D., Rice, R.E., Hoyt, S.C., and Westigard, P.H. (1981). Phenology of the San Jose scale (Homoptera: Diaspididae). *Can. Entomol.* 113, 149–59.

Jorgensen, J. (1976). Biological peculiarities of *Hylemya floralis* Fall. in Denmark. *Ann. Agr. Fenn.* 15, 16–23.

Judson, C.J., Hokama, Y., and Kliewer, J.W. (1966). Embryogeny and hatching of *Aedes sierrensis* eggs (Diptera: Culicidae). *Ann. Entomol. Soc. Amer.* 59, 1181–84.

Kai, H. and Haga, Y. (1978). Further studies on "Esterase A" in *Bombyx* eggs in relation to diapause development. *J. Seric. Sci. Japan* 47, 125–33.

Kai, H. and Hasegawa, K. (1973). An esterase in relation to yolk cell lysis at diapause termination in the silkworm, *Bombyx mori*. *J. Insect Physiol.* 19, 799–810.

Kai, H. and Kawai, T. (1981). Diapause hormone in *Bombyx* eggs and adult ovaries. *J. Insect Physiol.* 27, 623–27.

Kai, H. and Nishi, K. (1976). Diapause development in *Bombyx* eggs in relation to "Esterase A" activity. *J. Insect Physiol.* 22, 1315–20.

Kai, H., Kawai, T., and Kaneto, A. (1984). Esterase A_4 elevation mechanism in relation to *Bombyx* (Lepidoptera: Bombycidae) egg diapause development. *Appl. Entomol. Zool.* 19, 8–14.

Kalmes, R. (1975). Influence de certaines caractéristiques périodiques de l'écosystème sur l'utilisation alimentaire de *Pieris brassicae* (Lepidoptera: Pieridae) en diapause par un parasitoide (*Pimpla instigator* F.: Hymenoptera: Ichneumonidae). *C. R. Acad. Sci.*, Paris, Series D, 280, 1881–84.

Kalpage, K.S.P. and Brust, R.A. (1974). Studies on diapause and female fecundity in *Aedes atropalpus*. *Environ. Entomol.* 3, 139–45.

Kambysellis, M.P. and Heed, W.B. (1974). Juvenile hormone induces ovarian development in diapausing cave-dwelling *Drosophila* species. *J. Insect Physiol.* 20, 1779–86.

Kamm, J.A. (1972). Photoperiodic regulation of growth in an insect: response to progressive changes in daylength. *J. Insect Physiol.* 18, 1745–49.

Kappus, K.D. and Venard, C.E. (1967). The effects of photoperiod and temperature on the induction of diapause in *Aedes triseriatus* (Say). *J. Insect Physiol.* 13, 1007–19.

Katakura, H. (1982). Long mating season and its bearing on the reproductive isolation in a pair of sympatric phytophagous ladybirds (Coleoptera, Coccinellidae). *Kontyû* 50, 599–603.

Kato, Y. and Sakate, S. (1981). Studies on summer diapause in pupae of *Antheraea yamamai* (Lepidoptera: Saturniidae) III. Influence of photoperiod in the larval stage. *Appl. Entomol. Zool.* 16, 499–500.

Kato, Y., Yamauchi, M., Katsu, Y., and Sakate, S. (1979). Studies on summer diapause in pupae of *Antheraea yamamai* (Lepidoptera: Saturniidae) I. Shortening of the "pupal" duration under certain environmental conditions. *Appl. Entomol. Zool.* 14, 389–96.

Kats, T.S. (1982). A histological study of the neuroendocrine system in the aphid *Megoura viciae* Buckt. (Homoptera, Aphididae). *Entomol. Rev.* 61, 17–25.

Keeley, L.L., Moody, D.S., Lynn, D., Joiner, R.L., and Vinson, S.B. (1977). Succinate-cytochrome c reductase activity and lipids in diapause and non-diapause *Anthonomus grandis* from different latitudes. *J. Insect Physiol.* 23, 231–34.

Kefuss, J.A. (1978). Influence of photoperiod on the behaviour and brood-rearing activities of honeybees in a flight room. *J. Apicult. Res.* 17, 137–51.

Kelleher, J.S. (1969). Introduction practices—past and present. *Bull. Entomol. Soc. Amer.* 15, 235–36.

Kenagy, G.J. and Bartholomew, G.A. (1981). Effects of day length, temperature, and green food on testicular development in a desert pocket mouse *Perognathus formosus*. *Physiol. Zool.* 54, 62–73.

Kennedy, J.S. (1956). Phase transformation in locust biology. *Biol. Rev.* 31, 349–70.

Kennedy, J.S. (1961). A turning point in the study of insect migration. *Nature* 189, 785–91.

Kennedy, J.S. (1975). Insect dispersal. In *Insects, science, and society* (ed. D. Pimentel) pp. 103–19. Academic Press, New York.

Kennedy, J.S. and Booth, C.O. (1954). Host alternation in *Aphis fabae* Scop. II. Changes in the aphids. *Ann. Appl. Biol.* 41, 88–106.

Kennedy, J.S. and Booth, C.O. (1961). Host finding by aphids in the field. III. Visual attraction. *Ann. Appl. Biol.* 49, 1–21.

Kennedy, J.S. and Booth, C.O. (1963a). Free flight of aphids in the laboratory. *J. Exp. Biol.* 40, 67–85.

Kennedy, J.S. and Booth, C.O. (1963b). Co-ordination of successive activities in an aphid. The effect of flight on the settling responses. *J. Exp. Biol.* 40, 351–69.

Khaldey, Y.L. (1977). The biology, phenology and photoperiodic reaction of *Forficula tomis* (Dermaptera, Forficulidae). *Entomol. Rev.* 56, 6–12.

Khalil, G.M. (1976). The subgenus *Persicargas* (Ixodoidea: Argasidae: *Argas*). 26. *Argas (P.) arboreus*: Effect of photoperiod on diapause induction and termination. *Exp. Parasitol.* 40, 232–37.

Khelevin, N.V. (1958). The effect of environmental factors on the induction of embryonic diapause and on the number of generations in a season of *Aedes caspius dorsalis* Mg. (Diptera, Culicidae). Effect of temperature on the induction of embryonic diapause in *Aedes caspius dorsalis* Mg. *Entomol. Rev.* 37, 19–35.

Khomyakova, V.O. (1976). Photoperiodic and temperature reactions in the caterpillars of geographical populations of the European cornborer, *Ostrinia nubilalis* Hb. (Lepidoptera, Pyralidae). *Entomol. Rev.* 55, 1–5.

Khoo, S.G. (1968). Experimental studies on diapause in stoneflies. II. Eggs of *Diura bicaudata* (L.) *Proc. Roy. Entomol. Soc. Lond.* (A) 43, 49–56.

Kidd, N.A.C. (1979). The control of seasonal changes in the pigmentation of lime aphid nymphs, *Eucallipterus tiliae. Entomol. Exp. Appl.* 25, 31–38.

Kidokoro, T. (1983). Migration and dispersal after hibernation in the rice leaf beetle, *Oulema oryzae* Kuwayama (Coleoptera: Chrysomelidae). *Appl. Entomol. Zool.* 18, 211–19.

Kidokoro, T. and Masaki, S. (1978). Photoperiodic response in relation to variable voltinism in the ground cricket, *Pteronemobius fascipes* Walker (Orthoptera: Gryllidae). *Japan. J. Ecol.* 28, 291–98.

Kidokoro, T., Fujisaki, Y., and Takano, T. (1982). Transplanting time of rice as a factor in the occurrence of outbreaks of the smaller rice leaf miner, *Hydrellia griseola* Fallén (Diptera: Ephydridae). *Japan. J. Appl. Entomol. Zool.* 26, 306–8 (in Japanese, English abstract).

Kikukawa, S. and Chippendale, G.M. (1983). Seasonal adaptations of populations of the southwestern corn borer, *Diatraea grandiosella*, from tropical and temperate regions. *J. Insect Physiol.* 29, 561–67.

Kim, K.C., Chiang, H.C., and Brown, B.W., Jr. (1967). Morphometric differences among four biotypes of *Ostrinia nubilalis* (Lepidoptera: Pyralidae). *Ann. Entomol. Soc. Amer.* 60, 796–801.

Kimura, M.T. (1982). Effects of photoperiod and temperature on reproductive diapause in *Drosophila testacea. Experientia* 38, 371–72.

Kimura, M.T. (1983). Geographic variation and genetic aspects of reproductive diapause in *Drosophila triauraria* and *D. quadraria. Physiol. Entomol.* 8, 181–86.

Kimura, M.T. (1984). Geographic variation of reproductive diapause in the *Drosophila auraria* complex (Diptera: Drosophilidae). *Physiol. Entomol.* 9, 425–31.

Kimura, M.T. and Minami, N. (1980). Reproductive diapause in *Drosophila auraria*: Sensitivity of different developmental states to photoperiod and temperature. *Kontyû* 48, 458–63.

Kimura, M.T., Beppu, K., Ichijo, N., and Toda, M.J. (1978). Bionomics of Drosophilidae (Diptera) in Hokkaido. II. *Drosophila testacea. Kontyû* 46, 585–95.

Kimura, T. (1975). *Luehdorfia* butterflies waiting for emergence in spring. *Insectarium* 12, 76–79 (in Japanese).

Kimura, T. and Masaki, S. (1977). Brachypterism and seasonal adaptation in *Orgyia thyellina* Butler (Lepidoptera, Lymantriidae). *Kontyû* 45, 97–106.

Kimura, T., Takano, H., and Masaki, S. (1982). Photoperiodic programming of summer diapause after hibernation in *Spilarctia imparilis* Butler (Lepidoptera: Arctiidae). *Appl. Entomol. Zool.* 17, 218–26.

Kind, T.V. (1969). The role of neurosecretory cells of the subesophageal ganglion in determination of embryonic diapause in *Orgyia antiqua* L. *Dokl. Akad. Nauk. SSSR* 187, 517–20 (English translation).

Kind, T.V. (1977). A study of the reactivation of diapausing pupae of *Acronycta rumicis* L. (Lepidoptera, Noctuidae). I. Cold activation of the prothoracic glands in brainless pupae. *Entomol. Rev.* 56, 13–16.

Kind, T.V. (1978). A study of the reactivation of diapausing pupae of *Acronycta rumicis* (Lepidoptera, Noctuidae). 2. Reactivation of intact and brainless pupae at various temperatures. *Entomol. Rev.* 57, 163–67.

King, A.B.S. (1974). Photoperiodic induction and inheritance of diapause in *Pionea forficalis* (Lepidoptera: Pyralidae). *Entomol. Exp. Appl.* 17, 397–409.

King, J.R. and Farner, D.S. (1974). Biochronometry and bird migration: General perspective. In *Chronobiology* (ed. L.E. Scheving, F. Halberg, and J.E. Pauly) pp. 625–29. Igaku Shoin Ltd., Tokyo.

Kingsolver, J.G. (1979). Thermal and hydric aspects of environmental heterogeneity in the pitcher plant mosquito. *Ecol. Monogr.* 49, 357–76.

Kipyatkov, V.Y. (1973). Observations on the photoperiodic reaction in ants of the genus *Myrmica* (Hymenoptera, Formicidae). *Doklady Biological Sciences* 205, 470–72.

Kipyatkov, V.Y. (1974). A study of the photoperiodic reaction in the ant *Myrmica rubra* L. (Hymenoptera, Formicidae). Communication 1. Basic parameters of the reaction. *Entomol. Rev.* 53, 35–41.

Kipyatkov, V.Y. (1976). A study of the photoperiodic reaction in the ant *Myrmica rubra* (Hymenoptera, Formicidae). Communication 5. Perception of photoperiodic information by an ant colony. *Entomol. Rev.* 55, 27–34.

Kipyatkov, V.Y. (1979). The ecology of photoperiodism in the ant *Myrmica rubra* (Hymenoptera, Formicidae). 1. Seasonal changes in the photoperiodic reaction. *Entomol. Rev.* 58, 10–19.

Kishino, K. (1970a). Ecological studies on the local characteristics of seasonal development in the rice stem borer, *Chilo suppressalis* Walker. II. Local characteristics of diapause and development. *Japan. J. Appl. Entomol. Zool.* 14, 1–11.

Kishino, K. (1970b). Ecological studies on the local characteristics of development of the rice stem borer, *Chilo suppressalis* Walker. III. Seasonal development on the transitional zone from the univoltine to the bivoltine areas in the rice stem borer. *Japan. J. Appl. Entomol. Zool.* 14, 182–90 (in Japanese, English abstract).

Kishino, K. (1974). Ecological studies on the local characteristics of the seasonal development in the rice stem borer *Chilo suppressalis* Walker. *Bull. Tohoku Agr. Exp. Sta.* 47, 13–114 (in Japanese).

Kisimoto, R. (1956). Factors determining the wing-form of adult, with special reference to the effect of crowding during the larval period of the brown planthopper, *Nilaparvata lugens* Stal. *Oyo-Kontyû* 12, 105–11 (in Japanese, English summary).

Kisimoto, R. (1957). Studies on the polymorphism in the planthoppers (Homoptera, Araeopidae) III. Differences in several morphological and physiological characters between two wing-forms of the planthoppers. *Japan. J. Appl. Entomol. Zool.* 3, 164–73 (in Japanese, English summary).

Klassen, W., Creech, J.F., and Bell, R.A. (1970a). The potential for genetic suppression of insect populations by their adaptations to climate. *U.S. Dept. Agr. Misc. Publ.* 1178, 1–77.

Klassen, W., Knipling, E.F., and McGuire, J.U., Jr. (1970b). The potential for insect-population suppression by dominant conditional lethal traits. *Ann. Entomol. Soc. Amer.* 63, 238–55.

Klun, J.A. and Cooperators (1975). Insect sex pheromones: Intraspecific pheromonal variability of *Ostrinia nubilalis* in North America and Europe. *Environ. Entomol.* 4, 891–94.

Knerer, G. and Atwood, C.E. (1973). Diprionid sawflies: Polymorphism and speciation. *Science* 179, 1090–99.

Knop, N.F., Hoy, M.A. and Montgomery, M.E. (1982). Altered hatch sequence of males and females from unchilled eggs of a "non-diapause" gypsy moth strain (Lepidoptera: Lymantriidae). *N.Y. Entomol. Soc.* 90, 82–86.

Kobayashi, M. (1955). Relationship between the brain hormone and the imaginal differentiation of silkworm, *Bombyx mori. J. Seric. Sci. Japan* 24, 389–92 (in Japanese).

Kogan, M. (1982). Plant resistance in pest management. In *Introduction to insect pest management* (2nd ed.) (eds. R.L. Metcalf and W.H. Luckmann) pp. 93–134. Wiley-Interscience, New York.

Kogure, M. (1933). The influence of light and temperature on certain characters of the silkworm, *Bombyx mori. J. Dept. Agr. Kyushu Univ.* 4, 1-93.

Koidsumi, K. and Makino, K. (1958). Intake of food during hibernation of the rice stem borer, *Chilo suppressalis* Walker. *Japan. J. Appl. Entomol. Zool.* 2, 135–38.

Kolesova, D.A., Kuznetsova, V.G., and Shaposhnikov, G.K. (1980). Clonal variability in *Myzus persicae* Sulz. (Homoptera, Aphididae). *Entomol. Rev.* 59, 21–34.

Koller, D. (1969). The physiology of dormancy and survival of plants in desert environments. In *Dormancy and survival*, 23rd Symp. Soc. Exp. Biol. (ed. H.W. Woolhouse) pp. 449–69. Academic Press, New York.

Komazaki, S. (1983). Overwintering of the spirea aphid, *Aphis citricola* van der Goot (Homoptera:Aphididae) on citrus and spirea plants. *Appl. Entomol. Zool.* 18, 301–7.

Kono, Y. (1973). Difference of cuticular surface between diapause and non-diapause pupae of *Pieris rapae crucivora* (Lepidoptera: Pieridae). *Appl. Entomol. Zool.* 8, 50–52.

Kono, Y. (1979). Abnormal photoperiodic and phototactic reactions of the beetle, *Epilachna vigintioctopunctata*, reared on sliced potatoes. *Appl. Entomol. Zool.* 14, 185–92.

Kono, Y. (1980). Endocrine activities and photoperiodic sensitivity during prediapause period in the phytophagous lady beetle, *Epilachna vigintioctopunctata*. *Appl. Entomol. Zool.* 15, 73–80.

Kono, Y. (1982). Change of photoperiodic sensitivity with fat body development during prediapause period in the twenty-eight-spotted lady beetle, *Henosepilachna vigintioctopunctata* Fabricius (Coleoptera: Coccinellidae). *Appl. Entomol. Zool.* 17, 92–101.

Kono, Y., Kobayashi, M., and Claret, J. (1983). A putative photoreceptor organelle in insect brain glial cell. *Appl. Entomol. Zool.* 18, 116–21.

Kozhantshikov, I.W. (1948). Hibernation and diapause in lepidopterous insects of the family Orgyidae (Lepidoptera, Insecta). *Bull. Acad. Sci. URSS*, Biol. Ser., No. 6, 653 (in Russian; through Lees 1955).

Krehan, I. (1970). Die Steuerung von Jahresrhythmik und Diapause bei Larval- und Imagoüberwinterern der Gattung *Pterostichus* (Col., Carab.). *Oecologia* 6, 58–105.

Krüll, F. (1976a). The position of the sun is a possible Zeitgeber for Arctic animals. *Oecologia* 24, 141–48.

Krüll, F. (1976b). Zeitgebers for animals in the continuous daylight of high Arctic summer. *Oecologia* 24, 149–57.

Krüll, F. (1976c). The synchronizing effect of slight oscillations of light intensity on activity period of birds. *Oecologia* 25, 301–8.

Krysan, J.L. (1978). Diapause, quiescence, and moisture in the egg of the western corn rootworm, *Diabrotica virgifera*. *J. Insect Physiol.* 24, 535–40.

Krysan, J.L. (1982). Diapause in the Nearctic species of the *virgifera* group of *Diabrotica*: Evidence for tropical origin and temperate adaptations. *Ann. Entomol. Soc. Amer.* 75, 136–42.

Krysan, J.L. and Branson, T.F. (1977). Inheritance of diapause intensity in *Diabrotica virgifera*. *J. Hered.* 68, 415–17.

Krysan, J.L., Branson, T.F., and Diaz Castro, G. (1977). Diapause in *Diabrotica virgifera* (Coleoptera: Chrysomelidae): A comparison of eggs from temperate and subtropical climates. *Entomol. Exp. Appl.* 22, 81–89.

Kurahashi, H. and Ohtaki, T. (1977). Crossing between nondiapausing and diapausing races of *Sarcophaga peregrina*. *Experientia* 33, 186–87.

Kurata, S., Koga, K., and Sakaguchi, B. (1979a). Differential changes in nucleolar size and ribosomal RNA synthesis during diapause break by prolonged chilling in *Bombyx* eggs. *J. Insect Physiol.* 25, 115–18.

Kurata, S., Koga, K., and Sakaguchi, B. (1979b). RNA content and RNA synthesis in diapause and non-diapause eggs of *Bombyx mori*. *Insect Biochem.* 9, 107–9.

Kurihara, K. (1979). Photoperiodic regulation of winter diapause in the grass spider. *Experientia* 35, 1479–80.

Kuznetsova, I.A. and Tyshchenko, V.P. (1979). The part played by photoperiod and temperature in regulating cessation of diapause in *Drosophila transversa* Fll. and *D. phalerata* Mg. (Diptera, Drosophilidae). *Entomol. Rev.* 58, 8–15.

Lakovaara, S., Saura, A., Koref-Santibañez, S., and Ehrman, L. (1972). Aspects of diapause and its genetics in northern drosophilids. *Hereditas* 70, 89–96.

Lande, R. (1980). Genetic variation and phenotypic evolution during allopatric speciation. *Amer. Natur.* 116, 463–79.

Landis, B.J., Powell, D.M., and Fox, L. (1972). Overwintering and winter dispersal of the potato aphid in eastern Washington. *Environ. Entomol.* 1, 68–71.

Langston, D.T. and Watson, T.F. (1975). Influence of genetic selection on diapause termination of the pink bollworm. *Ann. Entomol. Soc. Amer.* 68, 1102–6.

Lankinen, P. and Lumme, J. (1984). Genetic analysis of geographical variation in photoperiodic diapause and pupal eclosion rhythm in *Drosophila littoralis*. In *Photoperiodic regulation of insect and molluscan hormones* (eds. R. Porter and G.M. Collins) pp. 97–114. Ciba Foundation Symposium 104, Pitman, London.

Laudien, H. (1973). Changing reaction systems. In *Temperature and life* (eds. H. Precht, J. Christophersen, H. Hensel, and W. Larcher) pp. 355–99. Springer-Verlag, New York.

Laugé, G. and Launois, M. (1980). Effets de deux conditions photothermopériodique ou apériodique sur le Criquet migrateur malgache *Locusta migratoria capito* (Saussure) [Orthopt. Acrididae]. *Ann. Soc. Entomol. Fr.* (N.S.) 16, 221–31.

Lea, A.O. (1963). Some relationships between environment, corpora allata, and egg maturation in aedine mosquitoes. *J. Insect Physiol.* 9, 793–809.

LeBerre, J.-R. (1953). Contribution a l'étude biologique du criquet migrateur des landes (*Locusta migratoria gallica* Remaudière). *Bull. Biol.* 87, 227–73.

Lee, R.E., Jr. (1980). Physiological adaptations of Coccinellidae to supranivean and subnivean hibernacula. *J. Insect Physiol.* 26, 135–38.

Lees, A.D. (1953). Environmental factors controlling the evocation and termination of diapause in the fruit tree red spider mite *Metatetranychus ulmi* Koch (Acarina: Tetranychidae). *Ann. Appl. Biol.* 40, 449–86.

Lees, A.D. (1955). *The physiology of diapause in arthropods.* Cambridge University Press, London.

Lees, A.D. (1956). The physiology and biochemistry of diapause. *Annu. Rev. Entomol.* 1, 1–16.

Lees, A.D. (1959). The role of photoperiod and temperature in the determination of parthenogenetic and sexual forms in the aphid *Megoura viciae* Buckton—I. The influence of these factors on apterous virginoparae and their progeny. *J. Insect Physiol.* 3, 92–117.

Lees, A.D. (1960). The role of photoperiod and temperature in the determination of parthenogenetic and sexual forms in the aphid *Megoura viciae* Buckton—II. The operation of the "interval timer" in young clones. *J. Insect Physiol.* 4, 154–75.

Lees, A.D. (1961). Clonal polymorphism in aphids. *Symp. Roy. Entomol. Soc. Lond.* 1, 68–79.

Lees, A.D. (1963). The role of photoperiod and temperature in the determination of parthenogenetic and sexual forms in the aphid *Megoura viciae* Buckton—III. Further properties of the maternal switching mechanism in apterous aphids. *J. Insect Physiol.* 9, 153–64.

Lees, A.D. (1964). The location of the photoperiodic receptors in the aphid *Megoura viciae* Buckton. *J. Exp. Biol.* 41, 119–33.

Lees, A.D. (1966). The control of polymorphism in aphids. In *Advances in insect physiology* (eds. J.W.L. Beament, J.E. Treherne, and V.B. Wigglesworth) pp. 207–77. Academic Press, London.

Lees, A.D. (1967). Direct and indirect effects of day length on the aphid *Megoura viciae* Buckton. *J. Insect Physiol.* 13, 1781–85.

Lees, A.D. (1968). Photoperiodism in insects. In *Photophysiology* (ed. A.C. Giese) pp. 47–137. Academic Press, New York.

Lees, A.D. (1970). Insect clocks and timers. Inaugural lecture, Imperial College of Science and Technology, 1970, pp. 1–15.

Lees, A.D. (1971). The relevance of action spectra in the study of insect photoperiodism. In *Biochronometry* (ed. M. Menaker) pp. 372–80. National Academy of Science, Washington, DC.

Lees, A.D. (1972). The role of circadian rhythmicity in photoperiodic induction in animals. In *Circadian rhythmicity, Proc. Intl. Symp. Circadian Rhythmicity*, pp. 87–110. Centre for Agricultural Publishing and Documentation, Wageningen.

Lees, A.D. (1973). Photoperiodic time measurement in the aphid *Megoura viciae. J. Insect Physiol.* 19, 2279–316.

Lees, A.D. (1978). Endocrine aspects of photoperiodism in aphids. In *Comparative endocrinology* (eds. P.J. Gaillard and H.H. Boer) pp. 165–68. Elsevier-North Holland Biomedical Press, Amsterdam.

Lees, A.D. (1980). The development of juvenile hormone sensitivity in alatae of the aphid *Megoura viciae. J. Insect Physiol.* 26, 143–51.

Lees, A.D. (1981). Action spectra for the photoperiodic control of polymorphism in the aphid *Megoura viciae. J. Insect Physiol.* 27, 761–71.

Lees, A.D. (1983). The endocrine control of polymorphism in aphids. In *Endocrinology of insects* (eds. R.G.H. Downer and H. Laufer) pp. 369–77. Alan R. Liss, New York.

Lees, A.D. and Hardie, J. (1981). The photoperiodic control of polymorphism in aphids: Neuroendocrine and endocrine components. In *Biological clocks in seasonal reproduction cycles* (eds. B.K. Follett and D.E. Follett) pp. 125–35. Proc. 32nd Symp. Colston Res. Soc., Bristol (1980). Halsted Press, New York.

Lees, E. and Tilley, R.J.D. (1980). Influence of photoperiod and temperature on larval development in *Pararge aegeria* (L.) (Lepidoptera: Satyridae). *Entomol. Gaz.* 31, 3–6.

Legner, E.F. (1979). Emergence patterns and dispersal in *Chelonus* spp. near *curvimaculatus* and *Pristomerus hawaiiensis*, parasitic on *Pectinophora gossypiella. Ann. Entomol. Soc. Amer.* 72, 681–86.

Legner, E.F. (1983). Patterns of field diapause in the navel orangeworm (Lepidoptera: Phycitidae) and three imported parasites. *Ann. Entomol. Soc. Amer.* 76, 503–6.

Legner, E.F., Olton, G.A., and Eskafi, F.M. (1966). Influence of physical factors on the developmental stages of *Hippelates collusor* in relation to the activities of its natural parasites. *Ann. Entomol. Soc. Amer.* 59, 851–61.

Leibee, G.L., Pass, B.C., and Yeargan, K.V. (1980). Effect of various temperature-photoperiod regimes on initiation of oviposition in *Sitona hispidulus* (Coleoptera: Curculionidae). *J. Kans. Entomol. Soc.* 53, 763–69.

Leigh, E.G., Jr., Rand, A.S., and Windsor, D.M. (eds.) (1982). *The ecology of a tropical forest.* Smithsonian Institution Press, Washington, DC.

Leinaas, H.P. and Bleken, E. (1983). Egg diapause and demographic strategy in *Lepidocyrtus lignorum* Fabricius (Collembola; Entomobrydiae). *Oecologia* 58, 194–99.

Lepage, M.G. (1983). Foraging of *Macrotermes* spp. (Isoptera: Macrotermitinae) in the tropics. In *Social insects in the tropics*, Vol. 2 (ed. P. Jaisson) pp. 205–18. Université Paris Nord.

Lessman, C.A. and Herman, W.S. (1981). Flight enhances juvenile hormone inactivation of *Danaus plexippus plexippus* L. (Lepidoptera: Danaidae). *Experientia* 37, 599–601.

Lessman, C.A. and Herman, W.S. (1983). Seasonal variation in hemolymph juvenile hormone of adult monarchs (*Danaus p. plexippus*: Lepidoptera). *Can. J. Zool.* 61, 88–94.

Levings, S.C. and Windsor, D.M. (1982). Seasonal and annual variation in litter arthropod populations. In *The ecology of a tropical forest* (eds. E.G. Leigh, Jr., A.S. Rand, and D.M. Windsor) pp. 355–87. Smithsonian Institution Press, Washington, DC.

Levins, R. (1968). *Evolution in changing environments.* Princeton University Press, Princeton.

Levins, R. (1969). Some demographic and genetic consequences of environmental heterogeneity for biological control. *Bull. Entomol. Soc. Amer.* 15, 237–40.

Liebherr, J. and Roelofs, W. (1975). Laboratory hybridization and mating period studies using two pheromone strains of *Ostrinia nubilalis. Ann. Entomol. Soc. Amer.* 68, 305–9.

Lieth, H. (1974). *Phenology and seasonality modeling.* Springer-Verlag, New York.

Lincoln, C. and Palm, C.E. (1941). Biology and ecology of the alfalfa snout beetle. *Cornell Univ. Agr. Exp. Sta. Mem.* 236, 1-44.

Linley, J.R., Evans, H.T., and Evans, F.D.S. (1970). A quantitative study of autogeny in a naturally occurring population of *Culicoides furens* (Poey) (Diptera: Ceratopogonidae). *J. Anim. Ecol.* 39, 169–83.

Litsinger, J.A. and Apple, J.W. (1973a). Oviposition of the alfalfa weevil in Wisconsin. *Ann. Entomol. Soc. Amer.* 66, 17–20.

Litsinger, J.A. and Apple, J.W. (1973b). Estival diapause of the alfalfa weevil in Wisconsin. *Ann. Entomol. Soc. Amer.* 66, 11–16.

Lloyd, E.P., Tingle, F.C., and Gast, R.T. (1967). Environmental stimuli inducing diapause in the boll weevil. *J. Econ. Entomol.* 60, 99–102.

Lloyd, J.E. and Pace, A.E. (1975). Seasonality in northern field crickets. *Fla. Entomol.* 58, 31–32.

Loeb, M.J. (1982). Diapause and development in the tobacco budworm, *Heliothis virescens*: A comparison of haemolymph ecdysteroid titres. *J. Insect Physiol.* 28, 667–73.

Loeb, M.J. and Hayes, D.K. (1980a). Neurosecretion during diapause and diapause development in brains of mature embryos of the gypsy moth, *Lymantria dispar. Ann. Entomol. Soc. Amer.* 73, 432–36.

Loeb, M.J. and Hayes, D.K. (1980b). Critical periods in the regulation of the pupal molt of the tobacco budworm, *Heliothis virescens. Ann. Entomol. Soc. Amer.* 73, 679–82.

Lofgren, C.S., Banks, W.A., and Glancey, B.M. (1975). Biology and control of imported fire ants. *Annu. Rev. Entomol.* 20, 1–30.

Logan, J.A., Wollkind, D.J., Hoyt, S.C., and Tanigoshi, L.K. (1976). An analytical model for description of temperature dependent rate phenomena in arthropods. *Environ. Entomol.* 5, 1133–40.

Logan, S.H. and Boyland, P.B. (1983). Calculating heat units via a sine function. *J. Amer. Soc. Hort. Sci.* 108, 977–80.

Lopez, J.D., Jr. (1982). Emergence pattern of an overwintering population of *Cardiochiles nigriceps* in Central Texas. *Environ. Entomol.* 11, 838–42.

Lopez, J.D., Jr., and Morrison, R.K. (1980). Overwintering of *Trichogramma pretiosum* in central Texas. *Environ. Entomol.* 9, 75–78.

Lopez, J.D., Hartstack, A.W., Jr., and Witz, J.A. (1983). Diapause development of *Heliothis* spp. *J. Ga. Entomol. Soc.* 18, 104–11.

Lounibos, L.P. and Bradshaw, W.E. (1975). A second diapause in *Wyeomyia smithii*: Seasonal incidence and maintenance by photoperiod. *Can. J. Zool.* 53, 215–21.

Lounibos, L.P., van Dover, C., and O'Meara, G.F. (1982). Fecundity, autogeny, and the larval environment of the pitcher-plant mosquito, *Wyeomyia smithii. Oecologia* 55, 160–64.

Lowman, M.D. (1982). Seasonal variation in insect abundance among three Australian rain forests, with particular reference to phytophagous types. *Aust. J. Ecol.* 7, 353–61.

Ludwig, D. (1932). The effect of temperature on the growth curves of the Japanese beetle *(Popillia japonica* Newnan). *Physiol. Zöol.* 5, 431–47.

Lumme, J. (1978). Phenology and photoperiodic diapause in northern populations of *Drosophila.* In *Evolution of insect migration and diapause* (ed. H. Dingle) pp. 145–70. Springer-Verlag, New York.

Lumme, J. (1981). Localization of the genetic unit controlling the photoperiodic adult diapause in *Drosophila littoralis. Hereditas* 94, 241–44.

Lumme, J. (1982). The genetic basis of the photoperiodic timing of the onset of winter dormancy in *Drosophila littoralis. Acta Univ. Ouluensis, Ser. A., Sci. Nat.* 129, *Biol.* 16, 1–42.

Lumme, J. and Keränen, L. (1978). Photoperiodic diapause in *Drosophila lummei* Hackman is controlled by an X-chromosomal factor. *Hereditas* 89, 261–62.

Lumme, J. and Lakovaara, S. (1983). Seasonality and diapause in drosophilids. In *The genetics and biology of Drosophila,* Vol. 3d. (eds. M. Ashburner, H.L. Carson, and J.N. Thompson, Jr.) pp. 171–220. Academic Press, London.

Lumme, J. and Oikarinen, A. (1977). The genetic basis of the geographically variable photoperiodic diapause in *Drosophila littoralis. Hereditas* 86, 129–42.

Lumme, J. and Pohjola, L. (1980). Selection against photoperiodic diapause started from monohybrid crosses in *Drosophila littoralis. Hereditas* 92, 377–78.

Lumme, J., Lakovaara, S., Oikarinen, A., and Lokki, J. (1975). Genetics of the photoperiodic diapause in *Drosphila littoralis. Hereditas* 79, 143–48.

Lumme, J., Muona, O., and Orell, M. (1978). Phenology of boreal drosophilids (Diptera). *Ann. Entomol. Fenn.* 44, 73–85.

Lumme, J., Lakovaara, S., Muona, O., and Järvinen, O. (1979). Structure of a boreal community of drosophilids (Diptera). *Aquilo Ser. Zool.* 20, 65–73.

Lutz, P.E. (1968). Effects of temperature and photoperiod on larval development in *Lestes eurinus* (Odonata: Lestidae). *Ecology* 49, 637–44.

Lutz, P.E. (1974). Effects of temperature and photoperiod on larval development in *Tetragoneura cynosura* (Odonata: Libellulidae). *Ecology* 55, 370–77.

Lutz, P.E. and Jenner, C.E. (1964). Life-history and photoperiodic responses of nymphs of *Tetrangoneura cynosura* (Say). *Biol. Bull.* 27, 304–16.

Lyman, C.P., Willis, J.S., Malan, A., and Wang, L.C.H. (1982). *Hibernation and torpor in mammals and birds.* Academic Press, New York.

Lynch, C.B. and Hoy, M.A. (1978). Diapause in the gypsy moth: Environment-specific mode of inheritance. *Genet. Res., Camb.* 32, 129–33.

Lyon, R.L., Richmond, C.E., Robertson, J.L., and Lucas, B.A. (1972). Rearing diapause and diapause-free western spruce budworm *(Choristoneura occidentalis)* (Lepidoptera: Tortricidae) on an artificial diet. *Can. Entomol.* 104, 417–26.

MacFarlane, J.R. and Drummond, F.H. (1970). Embryonic diapause in a hybrid between two Australian species of field crickets, *Teleogryllus* (Orthoptera: Gryllidae). *Aust. J. Zool.* 18, 265–72.

Machida, J. (1935). On the growth of *Sturmia sericariae.* Oyo-Dobutsugaku *Zasshi. J. Appl. Zool.* 7, 17–26 (in Japanese).

Mack, T.P., Bajusz, B.A., Nolan, E.S., and Smilowitz, Z. (1981). Development of a temperature-mediated functional response equation. *Environ. Entomol.* 10, 573–79.

Mackauer, M. (1972). Genetic aspects of insect production. *Entomophaga* 17, 27–48.

MacKay, P.A. (1977). Alata-production by an aphid: The "interval timer" concept and maternal age effects. *J. Insect Physiol.* 23, 889–93.

MacLeod, E.G. (1967). Experimental induction and elimination of adult diapause and autumnal coloration in *Chrysopa carnea* (Neuroptera). *J. Insect Physiol.* 13, 1343–49.

Maddrell, S. (1980). The control of water relations in insects. In *Insect Biology in the future "VBW" 80* (eds. M. Locke and D.S. Smith) pp. 179–99. Academic Press, London.

Madrid, F.J. and Stewart, R.K. (1981). Ecological significance of cold hardiness and winter mortality of eggs of the gypsy moth *Lymantria dispar* L., in Quebec. *Environ. Entomol.* 10, 586–89.

Maeda, N., Sato, S., and Yukawa, J. (1982). Polymodal emergence pattern of the Machilus leaf gall midge, *Daphnephila machilicola* Yukawa (Diptera, Cecidomyiidae). *Kontyû* 50, 44–50.

Magnarelli, L.A. (1979). Blood-feeding and gonotrophic dissociation in *Anopheles punctipennis* (Diptera: Culicidae) prior to hibernation in Connecticut. *J. Med. Entomol.* 15, 278–81.

Mane, S.D. and Chippendale, G.M. (1981). Hydrolysis of juvenile hormone in diapausing and non-diapausing larvae of the southwestern corn borer, *Diatraea grandiosella. J. Comp. Physiol.* 144, 205–14.

Mani, G.S. (1981). Conditions for balanced polymorphism in the presence of differential delay in developmental time. *Theor. Pop. Biol.* 20, 363–93.

Mansingh, A. (1971). Physiological classification of dormancies in insects. *Can. Entomol.* 103, 983–1009.

Mansingh, A. and Smallman, B.N. (1972). Variation in polyhydric alcohol in relation to diapause and cold-hardiness in the larvae of *Isia isabella. J. Insect Physiol.* 18, 1565–71.

Marcovitch, S. (1923). Plant lice and light exposure. *Science* 58, 537–38.

Marcus, N.H. (1979). On the population biology and nature of diapause of *Labidocera aestiva* (Copepoda: Calanoida). *Biol. Bull.* 157, 297–305.

Marcus, N.H. (1982a). The reversibility of subitaneous and diapause egg production by individual females of *Labidocera aestiva* (Copepoda: Calanoida). *Biol. Bull.* 162, 39–44.

Marcus, N.H. (1982b). Photoperiodic and temperature regulation of diapause in *Labidocera aestiva* (Copepoda: Calanoida). *Biol. Bull.* 162, 45–52.

Marcus, N.H. (1984). Variation in the diapause response of *Labidocera aestiva* (Copepoda: Calanoida) from different latitudes and its importance in the evolutionary process. *Biol. Bull.* 166, 127–39.

Marion, K.R. (1982). Reproductive cues for gonadal development in temperate reptiles: temperature and photoperiod effects on the testicular cycle of the lizard *Sceloporus undulatus. Herpetologica* 38, 26–39.

Masaki, S. (1956a). The effect of temperature on the termination of diapause in the egg of *Lymantria dispar* Linné. *Japan. J. Appl. Zool.* 21, 148–57.

Masaki, S. (1956b). The local variation in the diapause pattern of the cabbage moth, *Barathra brassicae* Linné, with particular reference to the aestival diapause (Lepidoptera: Noctuidae). *Bull. Fac. Agr. Mie Univ.* 13, 29–46.

Masaki, S. (1958). The response of a "short-day" insect to certain external factors: The induction of diapause in *Abraxas miranda* Butler. *Japan. J. Appl. Entomol. Zool.* 2, 285–94.

Masaki, S. (1959). Seasonal changes in the mode of diapause in the pupae of *Abraxas miranda* Butler. *Bull. Fac. Agric. Hirosaki Univ.* 5, 14–27.

Masaki, S. (1961). Geographic variation of diapause in insects. *Bull. Fac. Agr. Hirosaki Univ.* 7, 66–98.

Masaki, S. (1962). The influence of temperature on the intensity of diapause in the eggs of the Emma field cricket (Orthoptera: Gryllidae). *Kontyû* 30, 9–16.

Masaki, S. (1963). Adaptation to local climatic conditions in the Emma field cricket (Orthoptera: Gryllidae). *Kontyû* 31, 249–60.

Masaki, S. (1965). Geographic variation in the intrinsic incubation period: a physiological cline in the emma field cricket (Orthoptera: Gryllidae: *Teleogryllus*). *Bull. Fac. Agr. Hirosaki Univ.* 11, 59–90.

Masaki, S. (1966). Photoperiodism and geographic variation in the nymphal growth of *Teleogryllus yezoemma* (Ohmachi & Matsuura). *Kontyû* 34, 277–88.

Masaki, S. (1967a). Geographic variation and climatic adaptation in a field cricket (Orthoptera: Gryllidae). *Evolution* 21, 725–41.

Masaki, S. (1967b). Some aspects of adaptation in the life cycle of insects. *Kontyû* 35, 205–20 (in Japanese).

Masaki, S. (1972). Photoperiodism in the seasonal cycle of crickets. In *Problems of photoperiodism and diapause in insects* (ed. N.I. Goryshin) pp. 25–50. Leningrad State University, Leningrad (in Russian, English summary).

Masaki, S. (1973). Climatic adaptation and photoperiodic response in the band-legged ground cricket. *Evolution* 26, 587–600.

Masaki, S. (1977). Life cycle programming. In *Adaptation and speciation in the fall webworm* (ed. T. Hidaka) pp. 31–60. Kodansha Ltd., Tokyo.

Masaki, S. (1978a). Seasonal and latitudinal adaptations in life cycles of crickets. In *Evolution of insect migration and diapause* (ed. H. Dingle) pp. 72–100. Springer-Verlag, New York.

Masaki, S. (1978b). Climatic adaptation and species status in the lawn ground cricket. II. Body size. *Oecologia* 35, 343–56.

Masaki, S. (1979). Climatic adaptation and species status in the lawn ground cricket. I. Photoperiodic response. *Kontyû* 47, 48–65.

Masaki, S. (1980). Summer diapause. *Annu. Rev. Entomol.* 25, 1–25.

Masaki, S. (1981). Evolution of seasonal adaptation. In *Recent advances in entomology* (ed. S. Ihsii) pp. 143–56. Tokyo University Press, Tokyo (in Japanese).

Masaki, S. (1983). Climatic speciation in Japanese ground crickets. *GeoJournal* 7.6, 483–90.

Masaki, S. (1984). Unity and diversity in insect photoperiodism. In *Photoperiodic regulation of insect and molluscan hormones* (eds. R. Porter and G.M. Collins) pp. 7–25. Ciba Foundation Symposium 104, Pitman, London.

Masaki, S. and Kikukawa, S. (1981). The diapause clock in a moth: Response to temperature signals. In *Biological clocks in seasonal reproductive cycles* (eds. B.K. Follett and D.E. Follett) pp. 101–12. Proc. 32nd Symp. Colston Res. Soc., Bristol (1980). Halsted Press, New York.

Masaki, S. and Ohmachi, F. (1967). Divergence of photoperiodic response and hybrid development in *Teleogryllus* (Orthoptera: Gryllidae). *Kontyû* 35, 83–105.

Masaki, S. and Oyama, N. (1963). Photoperiodic control of growth and wing form in *Nemobius yezoensis*. *Kontyû* 31, 16–26.

Masaki, S. and Sakai, T. (1965). Summer diapause in the seasonal life cycle of *Mamestra brassicae* Linné (Lepidoptera: Noctuidae). *Japan. J. Appl. Entomol. Zool.* 9, 191-205.

Masaki, S., Umeya, K., Sekiguchi, Y., and Kawaski, R. (1968). Biology of *Hyphantria cunea* Drury (Lepidoptera: Artiidae) in Japan. III. Photoperiodic induction of diapause in relation to the seasonal life cycle. *Appl. Entomol. Zool.* 3, 55–66.

Masaki, S., Ando, Y., and Watanabe, A. (1979). High temperature and diapause termination in the eggs of *Teleogryllus commodus* (Orthoptera: Gryllidae). *Kontyû* 47, 493–504.

Maslennikova, V.A. (1958). The conditions determining diapause in the parasitic Hymenopteran *Apanteles glomeratus* L. (Hymenoptera, Braconidae) and *Pteromalus puparum* L. (Hymenoptera, Chalcididae). *Entomol. Rev.* 37, 466–72.

Maslennikova, V.A. (1959). On the diapause and hibernation in *Trichogramma evanescens. Vest. Leningr. Univ., Ser. Biol.* 3, 91–96.

Maslennikova, V.A. (1968). The regulation of seasonal development in parasitic insects. In *Photoperiodic adaptations in insects and acari* (ed. A.C. Danilevsky) pp. 129–52. Leningrad State University, Leningrad (in Russian, English abstract).

Maslennikova, V.A. and Chernysh, S.I. (1973). Effect of ecdysterone on diapause determination in *Pteromalus puparum* L. *Dokl. Akad. Nauk. SSSR* 213, 480–2 (English translation).

Maslennikova, V.A. and Mustafayeva, T.M. (1971). An analysis of photoperiodic adaptations in geographic populations of *Apanteles glomeratus* L. (Hymenoptera, Braconidae) and *Pieris brassicae* L. (Lepidoptera, Pieridae). *Entomol. Rev.* 50, 281–84.

Maslennikova, V.A., Chernysh, S.I., and Abdul Nabi, A.A. (1976). The ecdysone titer in induction of winter and summer diapause in the cabbage moth, *Barathra brassicae* (Lepidoptera, Noctuidae). *Entomol. Rev.* 55, 22–26.

Matsuka, M. and Mittler, T.E. (1979). Production of males and gynoparae by apterous viviparae of *Myzus persicae* continuously exposed to different scotoperiods. *J. Insect Physiol.* 25, 587–93.

Matsumoto, K. and Tsuji, H. (1979). Occurrence of two colour types in the green peach aphid, *Myzus persicae*, and their susceptibility to insecticides. *Japan J. Appl. Entomol. Zool.* 23, 92–99.

Matthée, J.J. (1951). The structure and physiology of the egg of *Locustana pardalina* (Walk.). *Union S. Afr. Dept. Agr. Sci. Bull.* 316, 1–83.

Matthée, J.J. (1978). The induction of diapause in eggs of *Locustana pardalina* (Walker) (Acrididae) by high temperature. *J. Entomol. Soc. S. Afr.* 41, 25–30.

Maynard Smith, J. (1966). Sympatric speciation. *Amer. Natur.* 100, 637–50.

Mayr, E. (1963a). *Animal species and evolution*. Belknap Press, Cambridge.

Mayr, E. (1963b). Reply to criticism by R.D. Alexander. *Syst. Zool.* 12, 204–6.

McCafferty, W.P. and Huff, B.L., Jr. (1978). The life cycle of the mayfly *Stenacron interpunctatum* (Ephemeroptera: Heptageniidae). *Great Lakes Entomol.* 11, 209–16.

McCaffery, A.R. and Page, W.W. (1978). Factors influencing the production of long-winged *Zonocerus variegatus. J. Insect Physiol.* 24, 465–72.

McHaffey, D.G. (1972). Photoperiod and temperature influences on diapause in eggs of the floodwater mosquito *Aedes vexans* (Meigen) (Diptera:Culicidae). *J. Med. Entomol.* 9, 564–71.

McHaffey, D.G. and Harwood, R.F. (1970). Photoperiod and temperature influences on diapause in eggs of the floodwater mosquito, *Aedes dorsalis* (Meigen) (Diptera: Culicidae). *J. Med. Entomol.* 7, 631–44.

McLachlan, A. (1983). Life-history tactics of rain-pool dwellers. *J. Anim. Ecol.* 52, 545–61.

McLeod, D.G.R. (1976). Geographical variation of diapause termination in the European corn borer, *Ostrinia nubilalis* (Lepidoptera: Pyralidae), in southwestern Ontario. *Can. Entomol.* 108, 1403–8.

McLeod, D.G.R. (1978). Genetics of diapause induction and termination in the European corn borer, *Ostrinia nubilalis* (Lepidoptera: Pyralidae), in southwestern Ontario. *Can. Entomol.* 110, 1351–53.

McLeod, D.G.R. and Beck, S.D. (1963). Photoperiodic termination of diapause in an insect. *Biol. Bull.* 124, 84–96.

McLeod, D.G.R., Ritchot, C., and Nagai, T. (1979). Occurrence of a two generation strain of the European corn borer, *Ostrinia nubilalis* (Lepidoptera: Pyralidae), in Quebec. *Can. Entomol.* 111, 233–36.

McMullen, R.D. (1967a). A field study of diapause in *Coccinella novemnotata* (Coleoptera: Coccinellidae). *Can. Entomol.* 99, 42–49.

McMullen, R.D. (1967b). The effects of photoperiod, temperature, and food supply on rate of development and diapause in *Coccinella novemnotata. Can. Entomol.* 99, 578–86.

McMurtry, J.A., Mahr, D.L., and Johnson, H.G. (1976). Geographic races in the predaceous mite, *Amblyseius potentillae* (Acari: Phytoseiidae). *Intl. J. Acar.* 2, 23–28.

McNeil, J.N. and Rabb, R.L. (1973). Physical and physiological factors in diapause initiation of two hyperparasites of the tobacco hornworm, *Manduca sexta. J. Insect Physiol.* 19, 2107–18.

McPherson, R.M. and Hensley, S.D. (1978). Response of the parasite *Lixophaga diatraeae* (Tachinidae) to photoperiod and temperature. *Environ. Entomol.* 7, 136–38.

McQueen, D.J. and Steel, C.G.H. (1980). The role of photoperiod and temperature in the initiation of reproduction in the terrestrial isopod *Oniscus asellus* Linnaeus. *Can. J. Zool.* 58, 235–40.

Meats, A. (1976). Seasonal trends in acclimatization to cold in the Queensland fruit fly (*Dacus tryoni*, Diptera) and their prediction by means of a physiological model fed with climatological data. *Oecologia* 26, 73–87.

Meats, A. (1983). Critical periods for developmental acclimation to cold in the Queensland fruit fly, *Dacus tryoni. J. Insect Physiol.* 29, 943–46.

Meidell, E.M. (1983). Diapause, aerobic and anaerobic metabolism in alpine, adult *Melasoma collaris* (Coleoptera). *Oikos* 41, 239–44.

Mellini, E. (1972). Studi sui Ditteri Larvevoridi. XXV. Sul determinismo ormonale delle influenze escercitate dagli ospiti sui loro parassiti. *Boll. Entomol. Bologna* 31, 165–203.

Menaker, M. and Gross, G. (1965). Effect of fluctuating temperature on diapause induction in the pink bollworm. *J. Insect Physiol.* 11, 911–14.

Meola, R.W. and Adkisson, P.L. (1977). Release of prothoracicotropic hormone and potentiation of developmental ability during diapause in the bollworm, *Heliothis zea. J. Insect Physiol.* 23, 683–88.

Meola, R. and Gray, R. (1984). Temperature-sensitive mechanism regulating diapause in *Heliothis zea. J. Insect Physiol.* 30, 743–49.

Merkl, M.E. and McCoy, J.R. (1978). Boll weevils: Seasonal response over five years to pheromone baited traps. *J. Econ. Entomol.* 71, 730–31.

Merritt, R.W., Ross, D.H., and Larson, G.J. (1982). Influence of stream temperature and seston on the growth and production of overwintering larval black flies (Diptera: Simuliidae). *Ecology* 63, 1322–31.

Messenger, P.S. (1959). Bioclimatic studies with insects. *Annu. Rev. Entomol.* 4, 183–206.

Messenger, P.S. (1969). Bioclimatic studies of the aphid parasite *Praon exsoletum.* 2. Thermal limits to development and effects of temperature on rate of development and occurrence of diapause. *Ann. Entomol. Soc. Amer.* 62, 1026–31.

Messenger, P.S. (1970). Bioclimatic inputs to biological control and pest management programs. In *Concepts of pest management* (eds. R.L. Rabb and F.E. Guthrie) pp. 84–102. Proc. of Conf. at North Carolina State University, Raleigh, March 25–27, 1970. North Carolina State University Press, Raleigh.

Messenger, P.S. (1972). Climatic limitations to biological controls. *Proc. Tall Timbers Conf. Ecol. Anim. Control Habitat Manage.* 3, 97–114.

Messenger, P.S. and Flitters, N.E. (1958). Effect of constant temperature environments on the egg stage of three species of Hawaiian fruit flies. *Ann. Entomol. Soc. Amer.* 51, 109–19.

Messenger, P.S. and Flitters, N.E. (1959). Effect of variable temperature environments on egg development of three species of fruit flies. *Ann. Entomol. Soc. Amer.* 52, 191–204.

Messenger, P.S. and van den Bosch, R. (1971). The adaptability of introduced biological control agents. In *Biological control* (ed. C.B. Huffaker) pp. 68–92. Plenum, New York.

Messina, F.J. (1982). Timing of dispersal and ovarian development in goldenrod leaf beetles *Trirhabda virgata* and *T. borealis*. *Ann. Entomol. Soc. Amer.* 74, 78–83.

Metcalf, C.L., Flint, W.P., and Metcalf, R.L. (1951). *Destructive and useful insects* (3rd ed.). McGraw-Hill, New York.

Meyer, A. (1979). Colour polymorphism in the grasshopper *Paulinia acuminata*. *Entomol. Exp. Appl.* 25, 21–30.

Meyer, R.P. (1977). Fall and winter populations of mosquitoes sampled by dry ice baited CDC miniature light traps in Central California. *Proc. Calif. Mosq. Vector. Contr. Assoc.* 45, 177–80.

Michener, C.D. and Amir, M. (1977). The seasonal cycle and habitat of a tropical bumble bee. *Pacific Insects* 17, 237–40.

Michener, C.D., Breed, M.D., and Bell, W.J. (1979). Seasonal cycles, nests, and social behavior of some Colombian halictine bees (Hymenoptera: Apoidea). *Rev. Biol. Trop.* 27, 13–34.

Michieli, S. (1968). Poraba kisika med imaginalnim spreminjanjem barve pri stenici *Nezara virdula* (L.) (Heteroptera, Pentatomidae). Ph.D. thesis, *Slovenska Akademija Znanosti in Umetnosti* 11, 231–43.

Michieli, S. and Žener, B. (1968). Der Sauerstoffverbrauch verschiedener Farbstadien bei der Wanze *Nezara viridula* (L). *Zeit. Vergl. Physiol.* 58, 223–24.

Micinski, S., Boethel, D.J., and Boudreaux, H.B. (1979). Influence of temperature and photoperiod on development and oviposition of the pecan leaf scorch mite, *Eotetranychus hicoriae*. *Ann. Entomol. Soc. Amer.* 72, 649–54.

Miller, E.R. and Dingle, H. (1982). The effect of host plant phenology on reproduction of the milkweed bug, *Oncopeltus fasciatus*, in tropical Florida. *Oecologia* 52, 97–103.

Miller, K. (1982). Cold-hardiness strategies of some adult and immature insects overwintering in interior Alaska. *Comp. Biochem. Physiol.* 73A, 595–604.

Miller, L.K. and Smith, J.S. (1975). Production of threitol and sorbitol by an adult insect: Association with freezing tolerance. *Nature* 258, 519–20.

Mills, N.J. (1981). The mortality and fat content of *Adalia bipunctata* during hibernation. *Entomol. Exp. Appl.* 30, 265–68.

Minami, N. and Kimura, M.T. (1980). Geographical variation of photoperiodic adult diapause in *Drosophila auraria*. *Japan. J. Genet.* 55, 319–24.

Minami, N., Kimura, M.T., and Ichijo, N. (1979). Physiology of reproductive diapause in *Drosophila auraria*: Photoperiod and temperature as controlling factors. *Kontyû* 47, 244–48.

Minder, I.F. (1966). Hibernation conditions and survival rate of the Colorado beetle in different types of soils. In *Ecology and physiology of diapause in the Colorado beetle* (ed. K.V. Arnoldi) pp. 28–58. (Izdatel'stvo "Nauka", Moscow). trans. (1976) Indian Nat'l. Sci. Doc. Cent., New Delhi.

Minder, I.F. (1973). Long-term diapause of the *Neodiprion sertifer* larvae. *Zool. Zh.* 52, 1661–70.

Minder, I.F. (1980). Photoperiodic reactivation of eonymphs of *Neodiprion sertifer* Geoffr. (Hym.: Diprionidae). *Doklady Biol. Sci.* 248, 1075–77.

Missonnier, J. (1963). Etude ecologique du developpement nymphal de deux Dipteres Muscides phytophages: *Pegomyia betae* Curtis et *Chortophila brassicae* Bouche. *Ann. Epiphyties* 14, 293–310.

Mitchell, C.J. (1981). Diapause termination, gonoactivity, and differentiation of host-seeking behavior from blood-feeding behavior in hibernating *Culex tarsalis* (Diptera: Culicidae). *J. Med. Entomol.* 18, 386–94.

Mitchell, C.J. (1983). Differentiation of host-seeking behavior from blood-feeding behavior in overwintering *Culex pipiens* (Diptera: Culicidae) and observations on gonotrophic dissociation. *J. Med. Entomol.* 20, 157–63.

Mitchell, E.B. and Hardee, D.D. (1974). Seasonal determination of sex ratios and condition of diapause of boll weevils in traps and in the field. *Environ. Entomol.* 3, 386–88.

Mitter, C., Futuyma, D.J., Schneider, J.C., and Hare, J.D. (1979). Genetic variation and host plant relations in a parthenogenetic moth. *Evolution* 33, 777–90.

Mittler, T.E. (1973). Aphid polymorphism as affected by diet. In *Perspectives in aphid biology* (ed. A.D. Lowe). *Entomol. Soc. New Zealand Bull. No.* 2, 65–75.

Mittler, T.E. and Dadd, R.H. (1966). Food and wing determination in *Myzus persicae* (Homoptera:Aphidae). *Ann. Entomol. Soc. Amer.* 59, 1162–66.

Mittler, T.E. and Sutherland, O.R.W. (1969). Dietary influences on aphid polymorphism. *Entomol. Exp. Appl.* 12, 703–13.

Mittler, T.E., Nassar, S.G., and Staal, G.B. (1976). Wing development and parthenogenesis induced in progenies of kinoprene-treated gynoparae of *Aphis fabae* and *Myzus persicae*. *J. Insect Physiol.* 22, 1717–25.

Mittler, T.E., Eisenbach, J., Searle, J.B., Matsuka, M., and Nassar, S.G. (1979). Inhibition by kinoprene of photoperiod-induced male production by apterous and alate viviparae of the aphid *Myzus persicae*. *J. Insect Physiol.* 25, 219–26.

Miyata, T. (1974). Studies on diapause in *Actias* moths (Lepidoptera, Saturniidae). I. Photoperiodic induction and termination. *Kontyû* 42, 51–63.

Mochida, O. and Yoshimeki, M. (1962). Relations with development of the gonads, dimensional changes of the corpora allata, and duration of postdiapause period in hibernating larvae of the rice stem borer. *Japan. J. Appl. Entomol. Zool.* 6, 114–23.

Modder, W.W.D. (1978). Respiratory and weight changes, and water uptake, during embryonic development and diapause in the African grasshopper *Zonocerus variegatus* (L.) (Acridoidea:Pyrgomorphidae). *Acrida* 7, 253–65.

Moeur, J.E. and Istock, C.A. (1980). Ecology and evolution of the pitcher-plant mosquito. IV. Larval influence over adult reproductive performance and longevity. *J. Anim. Ecol.* 49, 775–92.

Mogi, M. (1981). Studies on *Aedes togoi* (Diptera: Culicidae) 1. Alternative diapause in the Nagasaki strain. *J. Med. Entomol.* 18, 477–80.

Moore, C.G. (1963). Seasonal variation in autogeny in *Culex tarsalis* Coq. in northern California. *Mosq. News* 23, 238–41.

Morallo-Rejesus, B. (1980). Juvenile hormones and insect control. *Philipp. Entomol.* 4, 21–54.

Morant, V. (1947). Migrations and breeding of the red locust (*Nomadacris septemfasciata* Serville) in Africa, 1927–1945. *Anti-locust Memoir* 2, 1–60.

Morden, R.D. and Waldbauer, G.P. (1980). Diapause and its termination in the psychid moth, *Thyridopteryx ephemeraeformis*. *Entomol. Exp. Appl.* 28, 322–33.

Moreau, R., Olivier, D., Gourdoux, L., and Dutrieu, J. (1981). Carbohydrate metabolism in *Pieris brassicae* L. (Lepidoptera); variations during normal and diapausing development. *Comp. Biochem. Physiol.* 68B, 95–99.

Morgan, T.D. and Chippendale, G.M. (1983). Free amino acids of the haemolymph of the southwestern corn borer and the European corn borer in relation to their diapause. *J. Insect Physiol.* 29, 735–40.

Moriarty, F. (1969). Egg diapause and water absorption in the grasshopper *Chorthippus brunneus*. *J. Insect Physiol.* 15, 2069–74.

Morimoto, N. and Takafuji, A. (1983). Comparison of diapause attributes and host preference among three populations of the citrus red mite, *Panonychus citri* (McGregor) occurring in the southern part of Okayama Prefecture, Japan. *Japan. J. Appl. Entomol. Zool.* 27, 224–28.

Morohoshi, S. (1959). Hormonal studies on the diapause and non-diapause eggs of the silkworm, *Bombyx mori* L. *J. Insect Physiol.* 3, 28–40.

Morohoshi, S. and Oshiki, T. (1969). Effect of the brain on the suboesophageal ganglion and determination of voltinism in *Bombyx mori*. *J. Insect Physiol.* 15, 167–75.

Morris, R.F. (1967). Factors inducing diapause in *Hyphantria cunea*. *Can. Entomol.* 99, 522–28.

Morris, R.F. and Fulton, W.C. (1970a). Models for the development and survival of *Hyphantria cunea* in relation to temperature and humidity. *Mem. Entomol. Soc. Can.* 70, 1–60.

Morris, R.F. and Fulton, W.C. (1970b). Heritability of diapause intensity in *Hyphantria cunea* and correlated fitness responses. *Can. Entomol.* 102, 927–38.

Müller, F.P. (1971). Bisher unbekannte Überwinterungsformen bei anholozyklischen Aphiden. *Wiss. Zeits. Univ. Rostock* 20, *Math.- naturw. Reihe*, Heft 1, 91–6.

Müller, F.P. and Möller, F.W. (1973). Möglichkeiten der parthenogenetischen Überwinterung von Aphiden unter den klimatischen Bedingungen des Bezirkes Rostock. *Wiss. Zeit. Universität Rostock -Mathematisch- Naturwissenschaftliche Reihe, Heft* 6/7, 799–804.

Müller, H.J. (1957). Über die Diapause von *Stenocranus minutus* Fabr. (Homoptera: Auchenorrhyncha). *Beiträge zur Entomol.* 7, 203–26.

Müller, H.J. (1958a). Über den formbildenden Einfluss der Tageslänge bei Insekten. *Verh. Dtsch. Zool. Ges.*, Frankfurt, 76–84.

Müller, H.J. (1958b). Über den Einfluss der Photoperiode auf Diapause und Körpergrösse der

Delphacide *Stenocranus minutus* Fabr. (Homoptera Auchenorrhyncha). *Zool. Anz.* 160, 294–312.

Müller, H.J. (1960a). Über die Abhängigkeit der Oogenese von *Stenocranus minutus* Fabr. (Hom. Auchenorrhyncha) von Dauer und Art der Täglichen Beleuchtung. *Verh. XI Intl. Kongr. Entomol. Wien* l, 678–89.

Müller, H.J. (1960b). Die Bedeutung der Photoperiode im Lebensablauf der Insekten. *Z. Ang. Entomol.* 47, 7–24.

Müller, H.J. (1970). Formen der Dormanz bei Insekten. *Nova Acta Leopoldina* 35, 1–27.

Müller, H.J. (1974). Die regulative Wirkung der Tageslänge auf die Entwicklung der Insekten. In *Umweltbiophysik, Abhandl. Akad. Wissenschaften DDR* (eds. R. Glaser, K. Unger, and M. Koch) pp. 113–24.

Müller, H.J. (1976a). Formen der Dormanz bei Insekten als Mechanismen ökologischer Anpassung. *Verh. Dtsch. Zool. Ges.* 1976, 46–58.

Müller, H.J. (1976b). Über die Parapause als Dormanzform am Beispiel der Imaginal-Diapause von *Mocydia crocea* H.S. (Homoptera Auchenorrhyncha). *Zool. Jb. Physiol. Bd.* 80, 231–58.

Müller, H.J. (1981). The effects of dormancy pattern on population dynamics of *Euscelis incisus* (Kbm) (Homoptera Cicadellidae). *Zool. Jb. Syst.* 108, 314–34 (in German, English abstract).

Münster-Swendsen, M. and Nachman, G. (1978). Asynchrony in insect host-parasite interaction and its effect on stability, studied by a simulation model. *J. Anim. Ecol.* 47, 159–71.

Muona, O. and Lumme, J. (1981). Geographical variation in the reproductive cycle and photoperiodic diapause of *Drosophila phalerata* and *D. transversa* (Drosophilidae:Diptera). *Evolution* 35, 158–67.

Muroga, H. (1951). On the consumption coefficient of inhibitory substance in the silkworm eggs. *J. Seric. Sci. Japan* 20, 92–94 (in Japanese, English title).

Mustafa, T.M. and Hodgson, C.J. (1984). Observations on the effect of photoperiod on the control of polymorphism in *Psylla pyricola. Physiol. Entomol.* 9, 207–13.

Mustafayeva, T.M. (1974). The photoperiodic adaptations of geographical populations of the parasitic chalcid *Pteromalus puparum* L. (Hymenoptera, Chalcidoidea). *Entomol. Rev.* 53, 1–3.

Nagell, B. (1980). Overwintering strategy of *Cloeon dipterum* (L.) larvae. In *Advances in Ephemeroptera biology* (eds. J.F. Flannagan and K.E. Marshall) pp. 259–64. Plenum Press, New York.

Nair, K.S.S. (1974). Studies on the diapause of *Trogoderma granarium*: Effects of juvenile hormone analogues on growth and metamorphosis. *J. Insect Physiol.* 20, 231–44.

Nair, K.S.S. and Desai, A.K. (1973). Studies on the isolation of diapause and non-diapause strains of *Trogoderma granarium* Everts (Coleoptera, Dermestidae). *J. Stored Prod. Res.* 9, 181–88.

Nakamura, I. and Ae, S.A. (1977). Prolonged pupal diapause of *Papilio alexanor*: Arid zone adaptation directed by larval host plant. *Ann. Entomol. Soc. Amer.* 70, 481–84.

Nakasuji, F. and Kimura, M. (1984). Seasonal polymorphism of egg size in a migrant skipper, *Parnara guttata guttata* (Lepidoptera, Hesperiidae). *Kontyû* 52, 253–59.

Natuhara, Y. (1983). The influences of food and photoperiod on flight activity and reproduction of the bean bug, *Riptortus clavatus* Thunberg (Heteroptera: Coreidae). *Appl. Entomol. Zool.* 18, 392–400.

Nayar, J.K., Pierce, P.A., and Haeger, J.S. (1979). Autogeny in *Wyeomyia vanduzeei* in Florida. *Entomol. Exp. Appl.* 25, 311–16.

Nealis, V.G., Jones, R.E. and Wellington, W.G. (1984). Temperature and development in host-parasite relationships. *Oecologia* 61, 224–29.

Nechols, J.R. and Tauber, M.J. (1977). Age-specific interaction between the greenhouse whitefly and *Encarsia formosa*: Influence of host on the parasite's oviposition and development. *Environ. Entomol.* 6, 143–49.

Nechols, J.R. and Tauber, P.J. (1982). Thermal requirements for postdiapause development and survival in the giant silkworm, *Hyalophora cecropia* (Lepidoptera: Saturniidae). *N.Y. Entomol. Soc.* 90, 252–57.

Nechols, J.R., Tauber, M.J., and Helgesen, R.G. (1980). Environmental control of diapause and postdiapause development in *Tetrastichus julis* (Hymenoptera: Eulophidae), a parasite of the cereal leaf beetle, *Oulema melanopus* (Coleoptera: Chrysomelidae). *Can. Entomol.* 112, 1277–84.

Nechols, J.R., Tauber, M.J., Tauber, C.A., and Helgesen, R.G. (1983). Environmental regulation of dormancy in the alfalfa blotch leafminer, *Agromyza frontella* (Diptera: Agromyzidae). *Ann. Entomol. Soc. Amer.* 76, 116–19.

Nei, M., Maruyama, T., and Wu, C.-I. (1983). Models of evolution of reproductive isolation. *Genetics* 103, 557–79.

Nemec, S.J. (1969). Use of artificial lighting to reduce *Heliothis* spp. populations in cotton fields. *J. Econ. Entomol.* 62, 1138–40.

Neudecker, C. and Thiele, H.-U. (1974). Die jahreszeitliche Synchronisation der Gonadenreifung bei *Agonum assimile* Payk. (Coleopt. Carab.) durch Temperature und Photoperiode. *Oecologia* 17, 141–57.

Niehaus, M. (1982). Technique for rearing the small tortoiseshell *Aglais urticae* without diapause at different temperatures (Lepidoptera: Nymphalidae). *Entomol. Gen.* 7, 365–73.

Niemelä, P., Tahvanainen, J., Sorjonen, J., Hokkanen, T., and Neuvonen, S. (1982). The influence of host plant growth form and phenology on the life strategies of Finnish macrolepidopterous larvae. *Oikos* 39, 164–70.

Nijhout, H.F. and Wheeler, D.E. (1982). Juvenile hormone and the physiological basis of insect polymorphisms. *Q. Rev. Biol.* 57, 109–33.

Nijhout, H.F. and Williams, C.M. (1974). Control of moulting and metamorphosis in the tobacco hornworm, *Manduca sexta* (L.): Cessation of juvenile hormone secretion as a trigger for pupation. *J. Exp. Biol.* 61, 493–501.

Nilsen, E.T. and Muller, W.H. (1982). The influence of photoperiod on drought induction of dormancy in *Lotus scoparius*. *Oecologia* 53, 79–83.

Noda, T. (1984). Short day photoperiod accelerates the oviposition in the Oriental green stink bug, *Nezara antennata* Scott (Heteroptera: Pentatomidae). *Appl. Entomol. Zool.* 19, 119–20.

Nolte, D.J. (1974). The gregarization of locusts. *Biol. Rev.* 49, 1–14.

Nopp-Pammer, E. and Nopp, H. (1967). Zur hormonalen Steureung des Spinnverhaltens von *Philosamia cynthia*. *Naturwissenschaften* 54, 592–93.

Nordin, J.H., Cui, Z., and Yin, C.-M. (1984). Cold-induced glycerol accumulation by *Ostrinia nubilalis* larvae is developmentally regulated. *J. Insect. Physiol.* 30, 563–66.

Norling, U. (1971). The life history and seasonal regulation of *Aeshna viridis* Eversm. in southern Sweden (Odonata). *Entomol. Scand.* 2, 170–90.

Norling, U. (1975). Livscykler hos svenska Odonater. *Entomologen* 4, 1–14.

Norris, K.H., Howell, F., Hayes, D.K., Adler, V.E., Sullivan, W.N., and Schechter, M.S. (1969). The action spectrum for breaking diapause in the codling moth, *Laspeyresia pomonella* (L.), and the oak silkworm, *Antheraea pernyi* Guer. *Proc. Natl. Acad. Sci.* 63, 1120–27.

Norris, M.J. (1958). Influence of photoperiod on imaginal diapause in Acridids. *Nature* 181, 58.

Norris, M.J. (1962). Diapause induced by photoperiod in a tropical locust, *Nomadacris septemfasciata* (Serv.). *Ann. Appl. Biol.* 50, 600–3.

Norris, M.J. (1964). Environmental control of sexual maturation in insects. *Symp. Roy. Entomol. Soc. Lond.* 2, 56–65.

Norris, M.J. (1965a). The influence of constant and changing photoperiods on imaginal diapause in the red locust (*Nomadacris septemfasciata* Serv.). *J. Insect Physiol.* 11, 1105–19.

Norris, M.J. (1965b). Reproduction of the grasshopper *Anacridium aegyptium* L. in the laboratory. *Proc. Roy. Entomol. Soc. Lond.* (A) 40, 19–29.

Norris, M.J. (1970). Aggregation response in ovipositing females of the desert locust, with special reference to the chemical factor. *J. Insect Physiol.* 16, 1493–1515.

Norton, G.A. and Holling, C.S., eds. (1977). *Proc. Conf. on Pest Management*, International Institute for Applied Systems Analysis, A-2361, Laxenburg, Austria.

Numata, H. and Hidaka, T. (1980). Development of male sex cells in the swallowtail, *Papilio xuthus* L. (Lepidoptera: Papilionidae) in relation to pupal diapause. *Appl. Entomol. Zool.* 15, 151–58.

Numata, H. and Hidaka, T. (1981). Development of male sex cells in the swallowtail, *Papilio xuthus* L. (Lepidoptera: Papilionidae) after the termination of diapause. *Appl. Entomol. Zool.* 16, 313–14.

Numata, H. and Hidaka, T. (1982). Photoperiodic control of adult diapause in the bean bug, *Riptortus clavatus* Thunberg (Heteroptera: Coreidae). I. Reversible induction and termination of diapause. *Appl. Entomol. Zool.* 17, 530–38.

Numata, H. and Hidaka, T. (1983a). Compound eyes as the photoperiodic receptors in the bean bug. *Experientia* 39, 868–69.

Numata, H. and Hidaka, T. (1983b). Photoperiodic control of adult diapause in the bean bug, *Riptortus clavatus* Thunberg (Heteroptera: Coreidae) II. Termination of diapause induced under different photoperiods. *Appl. Entomol. Zool.* 18, 439–41.

Numata, H. and Hidaka, T. (1984). Role of the brain in post-diapause adult development in the swallowtail, *Papilio xuthus*. *J. Insect Physiol.* 30, 165–68.

Oatman, E.R. and Platner, G.R. (1972). Colonization of *Trichogramma evanescens* and *Apanteles rubecula* on the imported cabbageworm on cabbage in Southern California. *Environ. Entomol.* 1, 347–52.

Obrycki, J.J. and Tauber, M.J. (1978). Thermal requirements for development of *Coleomegilla maculata* (Coleoptera: Coccinellidae) and its parasite *Perilitus coccinellae* (Hymenoptera: Braconidae). *Can. Entomol.* 110, 407–12.

Obrycki, J.J. and Tauber, M.J. (1979). Seasonal synchrony of the parasite *Perilitus coccinellae* and its host, *Coleomegilla maculata*. *Environ. Entomol.* 8, 400–5.

Obrycki, J.J. and Tauber, M.J. (1981). Phenology of three coccinellid species: Thermal requirements for development. *Ann. Entomol. Soc. Amer.* 74, 31–6.

Obrycki, J.J. and Tauber, M.J. (1982). Thermal requirements for development of *Hippodamia convergens* (Coleoptera: Coccinellidae). *Ann. Entomol. Soc. Amer.* 75, 678–83.

Obrycki, J.J. and Tauber, M.J. (1985). Seasonal occurrence and relative abundance of aphid predators and parasitoids on pubescent potato plants. *Can. Entomol.* (in press).

Obrycki, J.J., Tauber, M.J., Tauber, C.A., and Gollands, B. (1983a). Environmental control of the seasonal life cycle of *Adalia bipunctata* (Coleoptera: Coccinellidae). *Environ. Entomol.* 12, 416–21.

Obrycki, J.J., Tauber, M.J., and Tingey, W.M. (1983b). Predator and parasitoid interaction with aphid-resistant potatoes to reduce aphid densities: A two-year field study. *J. Econ. Entomol.* 76, 456–62.

Obrycki, J.J., Tauber, M.J., Tauber, C.A., and Gollands, B. (1985). *Edovum puttleri* (Hymenoptera: Eulophidae), an exotic egg parasitoid of the Colorado potato beetle (Coleoptera: Chrysomelidae): Responses to temperate zone conditions and resistant potato plants. *Environ. Entomol.* 14, 48–54.

Ochando, D. (1980). Seasonal distribution of *Drosophila* species. *Experientia* 36, 163–64.

Odiyo, P.O. (1975). Seasonal distribution and migrations of *Agrotis ipsilon* (Hufnagel) (Lepidoptera, Noctuidae). Centre for Overseas Pest Research, *Tropical Pest Bull.* 4, 1–22.

Ogata, T. and Sasakawa, M. (1983). Effects of the aestivation on the feeding and reproductive activities of the Viburnum leaf beetle, *Pyrrhalta humeralis* Chen (Coleoptera: Chrysomelidae). *Japan. J. Appl. Entomol. Zool.* 27, 276–79.

Ogura, N. (1975). Hormonal control of larval coloration in the armyworm, *Leucania separata*. *J. Insect Physiol.* 21, 559–76.

Ogura, N. and Saito, T. (1973). Induction of embryonic diapause in the silkworm, *Bombyx mori* L. (Lepidoptera: Bombycidae), by implantation of ganglia of the common armyworm larvae, *Leucania separata* Walker (Lepidoptera: Noctuidae). *Appl. Entomol. Zool.* 8, 46–48.

Ohmachi, F. and Masaki, S. (1964). Intercrossing and development of hybrids between the Japanese species of *Teleogryllus* (Orthoptera: Gryllidae). *Evolution* 18, 405–16.

Ohmachi, F. and Matsuura, I. (1951). Observations and experiments on four types in the life history of Gryllodea. *J. Appl. Zool.* 16, 104–10 (in Japanese).

Ohnishi, E., Ohtaki, T., and Fukuda, S. (1971). Ecdysone in the eggs of *Bombyx* silkworm. *Proc. Japan Acad.* 47, 413–15.

Ohtaki, T. (1960). Hormonal control of pupal coloration in the cabbage butterfly, *Pieris rapae crucivora*. *Ann. Zool. Japon.* 33, 97–103.

Ohtaki, T. and Takahashi, M. (1972). Induction and termination of pupal diapause in relation to the change of ecdysone titer in the fleshfly, *Sarcophaga peregrina*. *Japan. J. Med. Sci. Biol.* 25, 369–76.

Oikarinen, A. and Lumme, J. (1979). Selection against photoperiodic reproductive diapause in *Drosophila littoralis*. *Hereditas* 90, 119–25.

Okada, M. (1971). Role of the chorion as a barrier to oxygen in the diapause of the silkworm, *Bombyx mori* L. *Experientia* 27, 658–60.

Oku, T. (1966). Diapause of some apple leaf-rollers at an earlier larval instar (Lepidoptera: Tortricidae). I. Pre-diapause behaviour and peculiarity of life cycle. *Kontyû* 34, 144–53 (in Japanese).

Oku, T. (1982). Overwintering of eggs in the Siberian cutworm, *Euxoa sibirica* Boisduval (Lepidoptera:Noctuidae). *Appl. Entomol. Zool.* 17, 244–52.

Oku, T. and Kobayashi, T. (1978). Migratory behaviors and life-cycles of noctuid moths (Insecta, Lepidoptera) with notes on the recent status of migrant species in northern Japan. *Bull. Tohoku Nat. Agr. Exp. Sta.* 58, 97–209 (in Japanese).

Okuda, T. and Hodek, I. (1983). Response to constant photoperiods in *Coccinella septempunctata bruckii* populations from central Japan (Coleoptera, Coccinellidae). *Acta Entomol. Bohemoslov.* 80, 74–75.

Oldfield, G.N. (1970). Diapause and polymorphism in California populations of *Psylla pyricola* (Homoptera: Psyllidae). *Ann. Entomol. Soc. Amer.* 63, 180–84.

Omata, K. (1984). Influence of the physiological condition of the host, *Papilio xuthus* L. (Lepidoptera: Papilionidae), on the development of the parasite, *Trogus mactator* Tosquinet (Hymenoptera: Ichneumonidae). *Appl. Entomol. Zool.* 19, 430–35.

O'Meara, G.F. (1972). Polygenic regulation of fecundity in autogenous *Aedes atropalpus. Entomol. Exp. Appl.* 15, 81–89.

O'Meara, G.F. and Craig, G.B., Jr. (1969). Monofactorial inheritance of autogeny in *Aedes atropalpus. Mosq. News* 29, 14–22.

O'Meara, G.F. and Krasnick, G.J. (1970). Dietary and genetic control of the expression of autogenous reproduction in *Aedes atropalpus* (Coq.)(Diptera: Culicidae). *J. Med. Entomol.* 7, 328–34.

Omura, S. (1950). Researches on the hibernation and ecological characteristics of the wild silkworm *Bombyx mandarina. Bull. Seric. Exp. Stn.* 13, 79–130 (in Japanese).

Orshan, L. and Pener, M.P. (1979a). Repeated reversal of the reproductive diapause by photoperiod and temperature in males of the grasshopper, *Oedipoda miniata. Entomol. Exp. Appl.* 25, 219–26.

Orshan, L. and Pener, M.P. (1979b). Termination and reinduction of reproductive diapause by photoperiod and temperature in males of the grasshopper, *Oedipoda miniata. Physiol. Entomol.* 4, 55–61.

Owen, D.F. (1971). *Tropical butterflies.* Clarendon Press, Oxford.

Owen, D.F. (1974). Seasonal change in sex ratio in *Acraea quirina* (F.) (Lep. Nymphalidae), and notes on the factors causing distortions of the sex ratio in butterflies. *Entomol. Scand.* 5, 110–14.

Paarmann, W. (1974). Der Einfluss von Temperatur und Lichtwechsel auf die Gonadenreifung des Laufkäkers *Broscus laevigatus* Dej. (Col., Carab.) aus Nordafrika. *Oecologia* 15, 87–92.

Paarmann, W. (1975). Frielanduntersuchungen in Marokko (Nordafrika) zur Jahresrhythmik von Carabiden (Col., Carab.) und zum Mikroklima im Lebensraum der Käfer. *Zool. Jb. Syst. Bd.* 102, 72–88.

Paarmann, W. (1976a). Die Bedeutung exogener Faktoren für die Gonadenreifung von *Orthomus barbarus atlanticus* (Coleoptera, Carabidae) aus Nordafrika. *Entomol. Exp. Appl.* 19, 23–36.

Paarmann, W. (1976b). The annual periodicity of the polyvoltine ground beetle *Pogonus chalceus* Marsh. (Col., Carabidae) and its control by environmental factors. *Zool. Anz. Jena* 196, 150–60.

Paarmann, W. (1976c). Die Trockenzeitdormanz der Carabiden des zentraafrikanischen Hochlands und ihre Steuerung durch Aussenfaktoren. *Verh. Dtsch. Zool. Ges.* 1976, 209.

Paarmann, W. (1976d). Jahreszeitliche Aktivatät und Fortpflanzungsrhythmik von Laufkäfern (Col., Carab.) im Kivugebiet (Ost-Zaire, Zentralafrika). *Zool. Jb. Syst. Bd.* 103, 311–54.

Paarmann, W. (1977). Propagation rhythm of subtropical and tropical Carabidae (Coleoptera) and its control by exogenous factors. In *Advances in invertebrate reproduction*, Vol. 1 (eds. K.G. Adiyodi and R.G. Adiyodi) pp. 49–60. Peralam-Kenoth, Karivellur, Kerala, India.

Paarmann, W. (1979a). Ideas about the evolution of the various annual reproduction rhythms in carabid beetles of the different climatic zones. In *On the evolution of behaviour in carabid beetles* (eds. P.J. den Boer, H.U. Thiele, and F. Weber) pp. 119–32. Misc. Papers 18, Agricultural University of Wageningen, The Netherlands.

Paarmann, W. (1979b). A reduced number of larval instars, as an adaptation of the desert carabid beetle *Thermophilum* (*Anthia*) *sexmaculatum* F. (Coleoptera, Carabidae) to its arid environment. In *On the evolution of behaviour in carabid beetles* (eds. P.J. den Boer, H.U. Thiele, and F. Weber) pp. 113–17. Misc. Papers 18, Agricultural University of Wageningen, The Netherlands.

Page, T.L. (1982). Extraretinal photoreception in entrainment and photoperiodism in invertebrates. *Experientia* 38, 1007–13.

Palmer, J.O. (1982). Photoperiodic effect on size-related metamorphosis in the milkweed leaf beetle, *Labidomera clivicollis. Physiol. Entomol.* 7, 37–41.

Palmer, J.O. (1983). Photoperiodic control of reproduction in the milkweed leaf beetle, *Labidomera clivicollis. Physiol. Entomol.* 8, 187–94.

Pammer, E. (1966). Auslösung und Steuerung des Spinnverhaltens und der Diapause bei *Philosamia cynthia* Dru. (Saturniidae, Lep.). *Z. vergl. Physiol.* 53, 99–113.

Papaj, D.R. and Rausher, M.D. (1983). Individual variation in host location by phytophagous insects. In *Herbivorous insects* (ed. S. Ahmad) pp. 77–124. Academic Press, New York.

Parker, F.D. and Tepedino, V.J. (1982). Maternal influence on diapause in the alfalfa leafcutting bee (Hymenoptera: Megachilidae). *Ann. Entomol. Soc. Amer.* 75, 407–10.

Parr, W.J. and Hussey, N.W. (1966). Diapause in the glasshouse red spider mite (*Tetranychus urticae* Koch): A synthesis of present knowledge. *Horticult. Res.* 6, 1–21.

Parrish, D.S. and Davis, D.W. (1978). Inhibition of diapause in *Bathyplectes curculionis*, a parasite of the alfalfa weevil. *Ann. Entomol. Soc. Amer.* 71, 103–7.

Parsons, P.A. (1982). Adaptive strategies of colonizing animal species. *Biol. Rev.* 57, 117–48.

Passera, L. (1977). Production des soldats dans les sociétés sortant d'hibernation chez la fourmi *Pheidole pallidula* (Nyl.) (Formicidae, Myrmicinae). *Insectes Sociaux* 24, 131–46.

Patterson, J.L. and Duman, J.G. (1978). The role of the thermal hysteresis factor in *Tenebrio molitor* larvae. *J. Exp. Biol.* 74, 37–45.

Pedgley, D. (1982). *Windborne pests and diseases.* Ellis Horwood Ltd., Chichester.

Pener, M.P. (1970). The corpus allatum in adult acridids: The inter-relation of its functions and possible correlations with the life cycle. In *Proc. Intl. Study Conf. Current and Future Problems of Acridology* (eds. C.F. Hemming and T.H.C. Taylor) pp. 135–47. Centre for Overseas Pest Research, London.

Pener, M.P. (1974). Neurosecretory and corpus allatum controlled effects on male sexual behaviour in acridids. In *Experimental analysis of insect behaviour* (ed. L. Barton Browne) pp. 264–77. Springer-Verlag, Berlin.

Pener, M.P. (1976). The differential effect of the corpora allata on yellow coloration in crowded and isolated *Locusta migratoria migratorioides* (R. & F.) males. *Acrida* 5, 269–85.

Pener, M.P. (1977). The effect of photoperiod on male mating behavior of acridids in relation to reproductive diapause. *Proc. XII Intl. Conf. Intl. Soc. Chronobiol.*, pp. 607–13. Publishing House Il Ponte, Milan.

Pener, M.P. (1983). Endocrine aspects of phase polymorphism in locusts. In *Endocrinology of insects* (eds. R.G.H. Downer and H. Laufer) pp. 379–94. Alan R. Liss, New York.

Pener, M.P. and Broza, M. (1971). The effect of implanted, active corpora allata on reproductive diapause in adult females of the grasshopper *Oedipoda miniata. Entomol. Exp. Appl.* 14, 190–202.

Pener, M.P. and Lazarovici, P. (1979). Effect of exogenous juvenile hormones on mating behaviour and yellow colour in allatectomized adult male desert locusts. *Physiol. Entomol.* 4, 251–61.

Pener, M.P. and Orshan, L. (1980). Reversible reproductive diapause and intermediate states between diapause and full reproductive activity in male *Oedipoda miniata* grasshoppers. *Physiol. Entomol.* 5, 417–26.

Pener, M.P. and Orshan, L. (1983). The reversibility and flexibility of the reproductive diapause in males of a "short day" grasshopper, *Oedipoda miniata.* In *Diapause and life cycle strategies in insects* (eds. V.K. Brown and I. Hodek) pp. 67–85. Junk, The Hague, The Netherlands.

Pener, M.P., Girardie, A., and Joly, P. (1972). Neurosecretory and corpus allatum controlled effects on mating behavior and color change in adult *Locusta migratoria migratorioides* males. *Gen. Comp. Endocrinol.* 19, 494–508.

Peter, R.E. and Hontela, A. (1978). 2. Annual gonadal cycles in teleosts: Environmental factors and gonadotropin levels in blood. In *Environmental endocrinology* (eds. I. Assenmacher and D.S. Farner) pp. 20–25. Springer-Verlag, Berlin.

Pfaender, S.L., Rabb, R.L., and Sprenkel, R.K. (1981). Physiological attributes of reproductively active and dormant Mexican bean beetles. *Environ. Entomol.* 10, 222–25.

Phillips, J.R. (1976). Diapause as it relates to the boll weevil *Anthonomus grandis* Boheman. In *Boll weevil suppression, management, and elimination technology*, pp. 10–11. Proc. Conf. Memphis, TN. USDA-ARS-S-71.

Phillips, J.R., Gutierrez, A.P., and Adkisson, P.L. (1980). General accomplishments toward better insect control in cotton. In *New technology of pest control* (ed. C.B. Huffaker) pp. 123–53. Wiley-Interscience, New York.

Phillips, P.A. and Barnes, M.M. (1975). Host race formation among sympatric apple, walnut and plum populations of the codling moth, *Laspeyresia pomonella*. *Ann. Entomol. Soc. Amer.* 68, 1053–60.

Phipps, J. (1968). The ecological distribution and life cycles of some tropical African grasshoppers (Acridoidea). *Bull. Entomol. Soc. Nigeria* 1, 71–97.

Pickford, R. and Randell, R.L. (1969). A non-diapause strain of the migratory grasshopper, *Melanoplus sanguinipes* (Orthoptera: Acrididae). *Can. Entomol.* 101, 894–96.

Pienkowski, R.L. (1976). Behavior of the adult alfalfa weevil in diapause. *Ann. Entomol. Soc. Amer.* 69, 155–57.

Pinger, R.R. and Eldridge, B.F. (1977). The effect of photoperiod on diapause induction in *Aedes canadensis* and *Psorophora ferox* (Diptera: Culicidae). *Ann. Entomol. Soc. Amer.* 70, 437–40.

Pittendrigh, C.S. (1981). Circadian organization and the photoperiodic phenomena. In *Biological clocks in seasonal reproductive cycles* (eds. B.K. Follett and D.E. Follett) pp. 1–35. Proc. 32nd Symp. Colston Res. Soc., Bristol (1980). Halsted Press, New York.

Pittendrigh, C.S. and Minis, D.H. (1971). The photoperiodic time measurement in *Pectinophora gossypiella* and its relation to the circadian system in that species. In *Biochronometry* (ed. M. Menaker) pp. 212–50. National Academy of Sciences, Washington, DC.

Poitout, S. and Bues, R. (1977a). Caractéristiques du développement de *Mamestra oleracea* L. (Noctuidae, Hadeninae) en fonction des conditions (température, temps d'éclairement) d'exposition des chenilles. Mise en évidence d'une diapause de type "estival" dans la population de la basse vallée du Rhône. *Ann. Zool. Écol. Anim.* 9, 211–23.

Poitout, S. and Bues, R. (1977b). Études comparées des diapauses nymphales estivales existant dans les populations de basse vallée du Rhône de deux Noctuidae Hadeninae (*Mamestra brassicae* L., *Mamestra oleracea* L.). *Ann. Zool. Écol. Anim.* 9, 225–34.

Poitout, S. and Bues, R. (1977c). Quelques aspects génétiques de l'hétérogénéité de manifestation de la diapause estivale dans les populations européennes de deux Lépidoptères Noctuidae Hadeninae (*Mamestra oleracea* L. et *Mamestra brassicae* L). *Ann. Zool. Écol. Anim.* 9, 235–39.

Polis, G.A. (1980). Seasonal patterns and age-specific variation in the surface activity of a population of desert scorpions in relation to environmental factors. *J. Anim. Ecol.* 49, 1–18.

Poras, M. (1973). Physiologie des insects. Effet de la photoperiode et de l'hibernation sur la diapause de *Tetrix undulata* (Sow.). *C. R. Acad. Sci. Paris*, Series D, 277, 205–8.

Poras, M. (1976). Influence de la photopériode et de la température sur quelques aspects de la diapause imaginale chez les femelles de *Tetrix undulata* (Sow.) (Orthoptère, Tetrigidae). *Ann. Zool. Ecol. Anim.* 8, 373–80.

Poras, M. (1978). Etude du cycle biologique et de son contròle écophysiologique chez un acridien (*Tetrix undulata*, Sowerby) hibernant à l'état imaginal et larvaire. [Ph.D. thesis, Université Pierre et Marie Curie, Paris]. Abstract in *Acrida* 9 (Suppl.) 1980.

Poras, M. (1981). La diapause larvaire de *Tetrix undulata* (Sowerby, 1806) (Orthoptera, Tetrigoidea). *Can. J. Zool.* 59, 422–27.

Poras, M. (1982). Le contrôle endocrinien de la diapause imaginale des femelles de *Tetrix undulata* (Sowerby, 1806) (Orthoptere, Tetrigidae). *Gen. Comp. Endocrinol.* 46, 200–10.

Potter, M.F., Huber, R.T., and Watson, T.F. (1981). Heat unit requirements for emergence of overwintering tobacco budworm, *Heliothis virescens* (F.) in Arizona. *Environ. Entomol.* 10, 543–45.

Potter, M.F. and Watson, T.F. (1980). Induction of diapause in the tobacco budworm in Arizona. *J. Econ. Entomol.* 73, 820–23.

Powell, J.A. (1974). Occurrence of prolonged diapause in ethmiid moths. *Pan- Pac. Entomol.* 50, 220–25.

Prebble, M.L. (1941). The diapause and related phenomena in *Gilpinia polytoma* (Hartig). *Can. J. Res.* 19, 295–322.

Price, P.W. (1972). Methods of sampling and analysis for predictive results in the introduction of entomophagous insects. *Entomophaga* 17, 211–22.

Price, P.W. (1980). *Evolutionary biology of parasites*. Monograph in Population Biology, No. 15. Princeton University Press, Princeton.

Price, P.W. (1984). The concept of the ecosystem. In *Ecological entomology* (eds. C.B. Huffaker and R.L. Rabb) pp. 19–50. John Wiley & Sons, New York.

Price, P.W. and Tripp, H.A. (1972). Activity patterns of parasitoids on the swaine jack pine sawfly, *Neodiprion swainei* (Hymenoptera: Diprionidae), and parasitoid impact on the host. *Can. Entomol.* 104, 1003–16.

Principi, M.M., Memmi, M., and Pasqualini, E. (1977). Induzione e mantenimento della oligopausa larvale in *Chrysopa flavifrons* Brauer (Neuroptera, Chrysopidae). *Bollettino del'Istituto di Entomologia della Universitá di Bologna* 33, 301–14.

Prokopy, R.J., Armbrust, E.J., Cothran, W.R., and Gyrisco, G.G. (1967). Migration of the alfalfa weevil, *Hypera postica* (Coleoptera: Curculionidae), to and from estivation sites. *Ann. Entomol. Soc. Amer.* 60, 26–31.

Propp, G.D., Tauber, M.J., and Tauber, C.A. (1969). Diapause in the neuropteran *Chrysopa oculata*. *J. Insect Physiol.* 15, 1749–57.

Proshold, F.I. and LaChance, L.E. (1974). Analysis of sterility in hybrids from interspecific crosses between *Heliothis virescens* and *H. subflexa*. *Ann. Entomol. Soc. Amer.* 67, 445–49.

Raabe, M. (1982). *Insect neurohormones.* Plenum Press, New York.

Rabb, R.L. (1966). Diapause in *Protoparce sexta*. *Ann. Entomol. Soc. Amer.* 59, 160–65.

Rabb, R.L. (1969). Diapause characteristics of two geographical strains of the tobacco hornworm and their reciprocal crosses. *Ann. Entomol. Soc. Amer.* 62, 1252–56.

Rabb, R.L. and Kennedy, G.G. (eds.) (1979). *Movement of highly mobile insects: Concepts and methodology in research.* Proc. Conf. Raleigh, N.C., 1979. University Graphics, North Carolina State University, Raleigh.

Rabb, R.L. and Stinner, R.E. (1978). The role of insect dispersal and migration in population processes. In *Radar, insect population ecology and pest management* (eds. C.R. Vaughn, W. Wolf, and W. Klassen) pp. 3–16. NASA Conf. Publ. No. 2070.

Rabb, R.L. and Thurston, R. (1969). Diapause in *Apanteles congregatus*. *Ann. Entomol. Soc. Amer.* 62, 125–28.

Rabb, R.L., Neunzig, H.H., and Marshall, H.V., Jr. (1964). Effect of certain cultural practices on the abundance of tobacco hornworms, tobacco budworms, and corn earworms on tobacco after harvest. *J. Econ. Entomol.* 57, 791–92.

Raignier, A., van Boven, J., and Ceusters, R. (1974). Der Polymorphismus der afrikanischen Wanderameisen unter biometrischen und biologischen Gesichtspunkten. In *Sozialpolymorphismus bei Insekten* (ed. G.H. Schmidt) pp. 668–93. Wissenschaftliche Verlagsgesellschaft MBH, Stuttgart.

Raina, A.K. and Bell, R.A. (1974). A nondiapausing strain of pink bollworm from southern India. *Ann. Entomol. Soc. Amer.* 67, 685–86.

Raina, A.K., Bell, R.A., and Klassen, W. (1981). Diapause in the pink bollworm: Preliminary genetic analysis. *Insect Sci. Appl.* 1, 231–35.

Rainey, R.C. (1963). Meteorology and the migration of desert locusts. *World Meteorological Organization Tech. Note* No. 54, 1–115. (*Anti-locust Memoir* 7, 115).

Rainey, R.C. (1974). Biometeorology and insect flight: Some aspects of energy exchange. *Annu. Rev. Entomol.* 19, 407–39.

Rainey, R.C. (1978). The evolution and ecology of flight: The "oceanographic" approach. In *Evolution of insect migration and diapause* (ed. H. Dingle) pp. 33–48. Springer-Verlag, New York.

Rains, T.D. and Dimock, R.V., Jr. (1978). Seasonal variation in cold hardiness of the beetle *Popilius disjunctus*. *J. Insect Physiol.* 24, 551–54.

Rakshpal, R. (1962). The effect of cold on pre- and post-diapause eggs of *Gryllus pennsylvanicus* Burmeister (Orthoptera: Gryllidae). *Proc. Roy. Entomol. Soc. Lond.* (A) 37, 117–20.

Rankin, M.A. (1974). The hormonal control of flight in the milkweed bug, *Oncopeltus fasciatus*. In *Experimental analysis of insect behavior* (ed. L. Barton Browne) pp. 317–28. Springer-Verlag, Berlin.

Rankin, M.A. (1978). Hormonal control of insect migratory behavior. In *Evolution of insect migration and diapause* (ed. H. Dingle) pp. 5–32. Springer-Verlag, New York.

Rankin, M.A. (1980). Effects of precocene I and II on flight behaviour in *Oncopeltus fasciatus*, the migratory milkweed bug. *J. Insect Physiol.* 26, 67–73.

Rankin, M.A. and Riddiford, L.M. (1978). Significance of haemolymph juvenile hormone titer changes in timing of migration and reproduction in adult *Oncopeltus fasciatus*. *J. Insect Physiol.* 24, 31–38.

Rankin, M.A. and Singer, M.C. (1984). Insect movement: Mechanisms and effects. In *Ecological entomology* (eds. C.B. Huffaker and R.L. Rabb) pp. 185–216. John Wiley & Sons, New York.

Rankin, S.M. and Rankin, M.A. (1980). The hormonal control of migratory flight behaviour in the convergent ladybird beetle, *Hippodamia convergens*. *Physiol. Entomol.* 5, 175–82.

Rausher, M.D. (1980). Host abundance, juvenile survival, and oviposition preference in *Battus philenor*. *Evolution* 34, 342–55.

Readio, P.A. (1931). Dormancy in *Reduvius personatus*. *Ann. Entomol. Soc. Amer.* 24, 19–39.

Ready, P.D. and Croset, H. (1980). Diapause and laboratory breeding of *Phlebotomus perniciosus* Newstead and *Phlebotomus ariasi* Tonnoir (Diptera: Psychodidae) from southern France. *Bull. Entomol. Res.* 70, 511–23.

Reagan, T.E., Stinner, R.E., Rabb, R.L., and Tuttle, C. (1979). A simulation model depicting the production of overwintering hornworms on tobacco. *Environ. Entomol.* 8, 268–73.

Reddy, A.S. and Chippendale, G.M. (1973). Water involvement in diapause and the resumption of morphogenesis of the southwestern corn borer, *Diatraea grandiosella*. *Entomol. Exp. Appl.* 16, 445–54.

Reed, G.L., Showers, W.B., Guthrie, W.D., and Lynch, R.E. (1978). Larval age and the diapause potential in northern and southern ecotypes of the European corn borer. *Ann. Entomol. Soc. Amer.* 71, 928–30.

Reed, G.L., Guthrie, W.D., Showers, W.B., Barry, B.D., and Cox, D.F. (1981). Sex-linked inheritance of diapause in the European corn borer: Its significance to diapause physiology and environmental response of the insect. *Ann. Entomol. Soc. Amer.* 74, 1–8.

Reeves, W.C. (1974). Overwintering of arboviruses. *Progr. Med. Virol.* 17, 193–220.

Régnière, J., Rabb, R.L., and Stinner, R.E. (1981). *Popillia japonica*: Simulation of temperature-dependent development of the immatures, and prediction of adult emergence. *Environ. Entomol.* 10, 290–96.

Remington, C.L. (1968). The population genetics of insect introduction. *Annu. Rev. Entomol.* 13, 415–26.

Remmert, H. (1980). *Arctic animal ecology*. Springer-Verlag, Berlin.

Remmert, H. and Wisniewski, W. (1970). Low resistance to cold of polar animals in summer. *Oecologia* 4, 111–12.

Renfer, A. (1975). Contribution à l'étude de la diapause larvaire de *Phytodietus griseanae* Kerrich (Hym., Ichneumonidae) parasitöide de la tordeuse grise du mélèze, *Zeiraphera diniana* Guénée (Lep., Tortricidae). *Mitt. Schweiz. Entomol. Ges.* 48, 59–67.

Resh, V.H. (1976a). Life histories of coexisting species of *Ceraclea* caddisflies (Trichoptera: Leptoceridae): The operation of independent functional units in a stream ecosystem. *Can. Entomol.* 108, 1303–18.

Resh, V.H. (1976b). Life cycles of invertebrate predators of freshwater sponge. In *Aspects of sponge biology* (eds. F.W. Harrison and R.R. Cowden) pp. 299–314. Academic Press, New York.

Rettenmeyer, C.W. (1963). Behavioral studies of army ants. *Univ. Kans. Sci. Bull.* 44, 281–465.

Reynolds, H.T., Adkisson, P.L., Smith, R.F., and Frisbie, R.E. (1982). Cotton insect pest management. In *Introduction to insect pest management* (2nd ed.) (eds. R.L. Metcalf and W.H. Luckmann) pp. 375–441. Wiley-Interscience, New York.

Riddiford, L.M. and Truman, J.W. (1978). Biochemistry of insect hormones and insect growth regulators. In *Biochemistry of insects* (ed. M. Rockstein) pp. 307–57. Academic Press, New York.

Ridgway, R.L. and Jones, S.L. (1969). Inundative releases of *Chrysopa carnea* for control of *Heliothis* on cotton. *J. Econ. Entomol.* 62, 177–80.

Riedl, H. (1983). Analysis of codling moth phenology in relation to latitude, climate and food availability. In *Diapause and life cycle strategies in insects* (eds. V.K. Brown and I. Hodek) pp. 233–52. Junk, The Hague, The Netherlands.

Riedl, H. and Croft, B.A. (1978). The effects of photoperiod and effective temperatures on the seasonal phenology of the codling moth (Lepidoptera: Tortricidae). *Can. Entomol.* 110, 455–70.

Ring, D.R. and Harris, M.K. (1983). Predicting pecan nut casebearer (Lepidoptera: Pyralidae) activity at College Station, Texas. *Environ. Entomol.* 12, 482–86.

Ring, D.R., Calcote, V.R., and Harris, M.K. (1983a). Verification and generalization of a degree-day model predicting pecan nut casebearer (Lepidoptera: Pyralidae) activity. *Environ. Entomol.* 12, 487–89.

Ring, D.R., Calcote, V.R., Cooper, J.N., Olszak, R., Begnaud, J.E., Fuchs, T.W., Neeb, C.W., Parker, R.D., Henson, J.L., Jackman, J.A., Flynn, M.S., and Harris, M.K. (1983b). Generalization and application of a degree-day model predicting pecan nut casebearer (Lepidoptera: Pyralidae) activity. *J. Econ. Entomol.* 76, 831–35.

Ring, R.A. (1967). Maternal induction of diapause in the larva of *Lucilia caesar* L. (Diptera: Calliphoridae). *J. Exp. Biol.* 46, 123–36.

Ring, R.A. (1971). Variations in the photoperiodic reaction controlling diapause induction in *Lucilia caesar* L. (Diptera: Calliphoridae). *Can. J. Zool.* 49, 137–42.

Ring, R.A. (1972). Relationship between diapause and supercooling in the blowfly, *Lucilia sericata* (Mg.) (Diptera: Calliphoridae). *Can. J. Zool.* 50, 1601–5.

Ring, R.A. (1977). Cold-hardiness of the bark beetle, *Scolytus ratzeburgi* Jans. (Col. Scolytidae). *Norw. J. Entomol.* 24, 125–36.

Ring, R.A. (1980). Insects and their cells. In *Low temperature preservation in medicine and biology* (eds. M.J. Ashwood-Smith and J. Farrant) pp. 187–217. Pitman Medical Ltd., Tumbridge Wells, Kent.

Ring, R.A. (1982). Freezing-tolerant insects with low supercooling points. *Comp. Biochem. Physiol.* 73A, 605–12.

Ring, R.A. and Tesar, D. (1980). Cold-hardiness of the arctic beetle, *Pytho americanus* Kirby Coleoptera, Pythidae (Salpingidae). *J. Insect. Physiol.* 23, 763–74.

Ring, R.A. and Tesar, D. (1981). Adaptations to cold in Canadian Arctic insects. *Cryobiology* 18, 199–211.

Rivnay, E. and Sobrio, G. (1967). The phenology and diapause of *Saturnia pyri* Schiff. in temperate and subtropical climates. *Z. Ang. Entomol.* 59, 59–63.

Roach, S.H. and Adkisson, P.L. (1970). Rôle of photoperiod and temperature in the induction of pupal diapause in the bollworm, *Heliothis zea*. *J. Insect Physiol.* 16, 1591-97.

Roberts, B. and Warren, M.A. (1975). Diapause in the Australian flesh fly *Tricholioproctia impatiens* (Diptera: Sarcophagidae). *Aust. J. Zool.* 23, 563–67.

Robin, J.C. (1980). Influence de la température sur l'induction de la diapause de Pyrale du Maïs *Ostrinia nubilalis* Hbn. (Lepidoptere Pyralidae) de differentes origines geographiques. *Rev. Zool. Agr. Pathol. Végétale* 79, 3–7.

Robinson, A.S., Herfst, M., and Vosselman, L. (1980). Genetic control of *Delia antiqua* (Meigen) (Diptera: Anthomyiidae). Sensitivity to diapause interfering with a field-cage experiment using a homozygous chromosomal translocation. *Bull. Entomol. Res.* 70, 103–11.

Rock, G.C. (1983). Thermoperiodic effects on the regulation of larval diapause in the tufted apple budmoth (Lepidoptera: Tortricidae). *Environ. Entomol.* 12, 1500–3.

Rock, G.C. and Shaffer, P.L. (1983). Tufted apple budmoth (Lepidoptera: Tortricidae): Effects of constant daylengths and temperatures on larval diapause development. *Environ. Entomol.* 12, 71–75.

Rock, G.C., Yeargan, D.R., and Rabb, R.L. (1971). Diapause in the phytoseiid mite, *Neoseiulus* (T.) *fallacis*. *J. Insect Physiol.* 17, 1651–59.

Rock, G.C., Shaffer, P.L., and Shaltout, A.D. (1983). Tufted apple budmoth (Lepidoptera: Tortricidae): Photoperiodic induction of larval diapause and stages sensitive to induction. *Environ. Entomol.* 12, 66–70.

Roe, R.M., Hammond, A.M., Jr., Douglas, E.E., and Philogène, B.J.R. (1984). Photoperiodically induced delayed metamorphosis in the sugarcane borer, *Diatraea saccharalis* (Lepidoptera: Pyralidae). *Ann. Entomol. Soc. Amer.* 77, 312–18.

Roff, D. (1980). Optimizing development time in a seasonal environment: The "ups and downs" of clinal variation. *Oecologia* 45, 202–8.

Rojas, R.R., Lee, R.E., Jr., Luu, T.A., and Baust, J.G. (1983). Temperature dependence-independence of antifreeze turnover in *Eurosta solidaginis* (Fitch). *J. Insect Physiol.* 29, 865–69.

Rolley, F., Hodek, I., and Iperti, G. (1974). Influence de la nourriture aphidienne (selon l'âge de la plante-hôte a partir de laquelle les pucerons se multiplient) sur l'induction de la dormance chez *Semiadalia undecimnotata* Schn. (Coleop., Coccinellidae). *Ann. Zool. Ecol. Anim.* 6, 53–60.

Root, R.B. and Chaplin, S.J. (1976). The life-styles of tropical milkweed bugs, *Oncopeltus* (Hemiptera: Lygaeidae) utilizing the same hosts. *Ecology* 57, 132–40.

Rose, M.R. (1983). Theories of life-history evolution. *Amer. Zool.* 23, 15–23.

Röseler, P.F. (1970). Unterschiede in der Kastendetermination zwischen den Hummelarten *Bombus hypnorum* und *Bombus terrestris*. *Z. naturforsch* 25 b, 543–48.

Rosenzweig, M.L. (1978). Competitive speciation. *Biol. J. Linn. Soc.* 10, 275–89.

Roush, R.T. (1979). Genetic improvement of parasites. In *Genetics in relation to insect management* (eds. M.A. Hoy and J.J. McKelvey, Jr.) pp. 97–105. Rockefeller Foundation, New York.

Rowell, C.H.F. (1971). The variable coloration of the acridoid grasshoppers. In *Advances in insect physiology*, Vol. 8 (eds. J.W.L. Beament, J.E. Treherne, and V.B. Wigglesworth) pp. 145–98. Academic Press, London.

Rummel, D.R. and Bottrell, D.G. (1976). Seasonally related decline in response of boll weevils to pheromone traps during mid-season. *Environ. Entomol.* 5, 783–87.

Rummel, D.R., Bottrell, D.G., Adkisson, P.L., and McIntyre, R.C. (1975). An appraisal of a 10–year effort to prevent the westward spread of the boll weevil. *Bull. Entomol. Soc. Amer.* 21, 6–11.

Ryan, R.B. (1965). Maternal influence on diapause in a parasitic insect, *Coeloides brunneri* Vier. (Hymenoptera: Braconidae). *J. Insect Physiol.* 11, 1331–36.

Saeki, H. (1966a). The effect of the population density on the occurrence of the macropterous form in a cricket, *Scapsipedus aspersus* Walker. *Japan. J. Ecol.* 16, 1–4 (in Japanese, English summary).

Saeki, H. (1966b). The effect of the day-length on the occurrence of the macropterous form in a cricket, *Scapsipedus aspersus* Walker. *Japan. J. Ecol.* 16, 49–52 (in Japanese, English summary).

Safranek, L., Cymborowski, B., and Williams, C.M. (1980). Effects of juvenile hormone on ecdysone-dependent development in the tobacco hornworm, *Manduca sexta. Biol. Bull.* 158, 248–56.

Sahota, T.S., Ruth, D.S., Ibaraki, A., Farris, S.H., and Peet, F.G. (1982). Diapause in the pharate adult stage of insect development. *Can. Entomol.* 114, 1179–83.

Sakagami, S.F. (1976). Specific differences in the bionomic characters of bumblebees. A comparative review. *J. Fac. Sci. Hokkaido Univ. Ser. VI, Zool.* 20, 390–447.

Sakagami, S.F. and Hayashida, K. (1968). Bionomics and sociology of the summer matrifilial phase in the social Halictine bee, *Lasioglossum duplex. J. Fac. Sci. Hokkaido Univ. Ser. VI, Zool.* 16, 413–513.

Salt, G. (1975). The fate of an internal parasitoid, *Nemeritis canescens*, in a variety of insects. *Trans. Roy. Entomol. Soc. Lond.* 127, 141–61.

Salt, R.W. (1947). Some effects of temperature on the production and elimination of diapause in the wheat stem sawfly, *Cephus cinctus* Nort. *Can. J. Res. Ser. D*, 25, 66–86.

Salt, R.W. (1961). Principles of insect cold-hardiness. *Annu. Rev. Entomol.* 6, 55–74.

Salt, R.W. (1969). The survival of insects at low temperatures. In *Dormancy and survival, 23rd Symp. Soc. Exp. Biol.* (ed. H.W. Woolhouse) pp. 331–50. Academic Press, New York.

Sanburg, L.L. and Larsen, J.R. (1973). Effect of photoperiod and temperature on ovarian development in *Culex pipiens pipiens. J. Insect Physiol.* 19, 1173–90.

Sardesai, J.B. (1972). Response of diapausing and nondiapausing larvae of *Plodia interpunctella* to hydrogen cyanide and methyl bromide. *J. Econ. Entomol.* 65, 1562–65.

Sargent, T.D. (1978). On the maintenance of stability in hindwing diversity among moths of the genus *Catocala* (Lepidoptera: Noctuidae). *Evolution* 32, 424–34.

Sáringer, G. (1982). Photoperiod as a potential control against *Cydia pomonella* Linne (Lep.; Tortricidae). *Proceedings of the 2nd Egyptian-Hungarian Conference of Plant Protection*, pp. 127–32. University of Alexandria, Alexandria, Egypt.

Sáringer, G. (1983). Illumination for half an hour at a time in autumn, in the scotophase of the photoperiod, as a possible ecological method of controlling the turnip sawfly *Athalia rosae* L. (Hym., Tenthredinidae). *Z. Ang. Entomol.* 96, 287–91.

Sáringer, G. and Szentkirályi, F. (1980). Contribution to the knowledge of the diapause of *Grapholitha funebrana* Treitschke (Lepid., Tortricidae). A study on the correlations of body weight, diapause and mortality. *Z. Ang. Entomol.* 90, 493–505.

Sauer, K.P. and Hensle, R. (1977). Reproduktive Isolation, ökologische Sonderung und morphologische Differenz der Zwillingsarten *Panorpa communis* L. und *P. vulgaris* Imhoff und Labram (Insecta, Mecoptera). *Z. Zool. Syst. Evol. Forsch.* 15, 169–207.

Saulich, A.K. (1975). The effects of photoperiod and density on the development of the noctuids *Cirphus unipuncta* Haw. and *Laphygma exigua* Hb. (Lepidoptera, Noctuidae). *Entomol. Rev.* 34, 52–59.

Saunders, D.S. (1965). Larval diapause of maternal origin: Induction of diapause in *Nasonia vitripennis* (Walk.) (Hymenoptera: Pteromalidae). *J. Exp. Biol.* 42, 495–508.

Saunders, D.S. (1966). Larval diapause of maternal origin. II. The effect of photoperiod and temperature on *Nasonia vitripennis. J. Insect Physiol.* 12, 569–81.

Saunders, D.S. (1971). The temperature-compensated photoperiodic clock "programming" development and pupal diapause in the flesh-fly, *Sarcophaga argyrostoma. J. Insect Physiol.* 17, 801-12.

Saunders, D.S. (1972). Circadian control of larval growth rate in *Sarcophaga argyrostoma*. *Proc. Natl. Acad. Sci. USA* 69, 2738–40.

Saunders, D.S. (1973). Thermoperiodic control of diapause in an insect: Theory of internal coincidence. *Science* 181, 358–60.

Saunders, D.S. (1974). Circadian rhythms and photoperiodism in insects. In *The physiology of Insecta* (2nd ed.) (ed. M. Rockstein) pp. 461–533. Academic Press, New York.

Saunders, D.S. (1978a). An experimental and theoretical analysis of photoperiodic induction in the flesh-fly, *Sarcophaga argyrostoma*. *J. Ccmp. Physiol.* 124, 75–95.

Saunders, D.S. (1978b). Internal and external coincidence and the apparent diversity of photoperiodic clocks in the insects. *J. Comp. Physiol.* 127, 197–207.

Saunders, D.S. (1981a). Insect photoperiodism—the clock and the counter: A review. *Physiol. Entomol.* 6, 99–116.

Saunders, D.S. (1981b). Insect photoperiodism: Entrainment within the circadian system as a basis for time measurement. In *Biological clocks in seasonal reproductive cycles* (eds. B.K. Follett and D.E. Follett) pp. 67–81. Proc. 32nd Symp. Colston Res. Soc., Bristol (1980). Halsted Press, New York.

Saunders, D.S. (1982). *Insect clocks* (2nd ed). Pergamon Press, Oxford.

Saunders, D.S. (1983). A diapause induction-termination asymmetry in the photoperiodic responses of the linden bug, *Pyrrhocoris apterus* and an effect of near-critical photoperiods on development. *J. Insect Physiol.* 29, 399–405.

Saunders, D.S., Sutton, D., and Jarvis, R.A. (1970). The effect of host species on diapause induction in *Nasonia vitripennis*. *J. Insect Physiol.* 16, 405–16.

Sawchyn, W.W. and Church, N.S. (1973). The effects of temperature and photoperiod on diapause development in the eggs of four species of *Lestes* (Odonata: Zygoptera). *Can. J. Zool.* 51, 1257–65.

Sawchyn, W.W. and Gillott, C. (1974). The life history of *Lestes congener* on the Canadian prairies. *Can. Entomol.* 106, 367–76.

Schad, G.A., Chowdhury, A.B., Dean, C.G., Kochar, V.K., Nawalinski, T.A., Thomas, J., and Tonascia, J.A. (1973). Arrested development in human hookworm infections: An adaptation to a seasonally unfavorable external environment. *Science* 180, 502–4.

Schaefer, C.H. and Washino, R.K. (1969). Changes in the composition of lipids and fatty acids in adult *Culex tarsalis* and *Anopheles freeborni* during the overwintering period. *J. Insect Physiol.* 15, 395–402.

Schaller, F. (1968). Action de la température sur la diapause et le développement de l'embryon d'*Aeshna mixta* (Odonata). *J. Insect Physiol.* 14, 1477–83.

Schaller, F. (1972). Action de la température sur la diapause embryonnaire et sur le type de développement d'*Aeshna mixta* Latreille (Anisoptera: Aeshnidae). *Odonatologica* 1, 143–53.

Schaller, F., Andries, J.C., Mouze, M., and Defossez, A. (1974). Nouveaux aspects du contrôle hormonal du cycle biologique des Odonates: Recherches sur la larve d'*Aeshna cyanea* (Müller) (Anisoptera: Aeshnidae). *Odonatologica* 3, 49–62.

Schechter, M.S., Sullivan, W.N., and Hayes, D.K. (1974). The use of artificial light outdoors to control agricultural insects. In *Chronobiology* (eds. L.E. Scheving, F. Halberg, and J.E. Pauly) pp. 617–21. Igaku Shoin Ltd., Tokyo.

Scheltes, P. (1976). The role of graminaceous host-plants in the induction of aestivation-diapause in the larvae of *Chilo zonellus* Swinhoe and *Chilo argyrolepia* Hamps. *Symp. Biol. Hung.* 16, 247–53.

Scheltes, P. (1978a). Ecological and physiological aspects of aestivation-diapause in the larvae of two pyralid stalk borers of maize in Kenya. Ph.D. thesis, Meded. Lanbouwhogeschool, Wageningen, The Netherlands.

Scheltes, P. (1978b). The condition of the host plant during aestivation-diapause of the stalk borers *Chilo partellus* and *Chilo orichalcociliella* (Lepidoptera, Pyralidae) in Kenya. *Entomol. Exp. Appl.* 24, 479–88.

Schipper, A.L. (1938). Some effects of hypertonic solution upon development and oxygen consumption. *Physiol. Zool.* 6, 40–53.

Schlinger, E.I. (1960). Diapause and secondary parasites nullify the effectiveness of rose-aphid parasites in Riverside, California, 1957–1958. *J. Econ. Entomol.* 53, 151–54.

Schlinger, E.I. and Hall, J.C. (1959). A synopsis of the biologies of three imported parasites of the spotted alfalfa aphid. *J. Econ. Entomol.* 52, 154–57.

Schlinger, E.I. and Hall, J.C. (1960). The biology, behavior, and morphology of *Praon palitans* Muesebeck, an internal parasite of the spotted alfalfa aphid, *Therioaphis maculata*

(Buckton) (Hymenoptera: Braconidae, Aphidiinae). *Ann. Entomol. Soc. Amer.* 53, 144–60.

Schlinger, E.I. and Hall, J.C. (1961). The biology, behavior, and morphology of *Trioxys* (*Trioxys*) *utilis*, an internal parasite of the spotted alfalfa aphid, *Therioaphis maculata* (Hymenoptera: Braconidae, Aphidiinae). *Ann. Entomol. Soc. Amer.* 54, 34–45.

Schmidt, G.H. (1974). Steuerung der Kastenbildung und Geschlechtsregulation im Waldameisenstaat. In *Sozialpolymorphismus bei Insekten* (ed. G.H. Schmidt) pp. 404–512. Wissenschaftliche Verlagsgesellschaft MBH, Stuttgart.

Schmidt-Nielsen, K. (1975). *Animal physiology*. Cambridge University Press, Cambridge.

Schmieder, R.G. (1933). The polymorphic forms of *Melittobia chalybii* Ashmead and the determining factors involved in their production (Hymenoptera: Chalcidoidea, Eulophidae). *Biol. Bull.* 65, 338–54.

Schmieder, R.G. (1939a). On the dimorphism of cocoons of *Sphecophaga burra* (Cresson). (Hymenoptera: Ichneumonidae). *Entomol. News* 50, 91–97.

Schmieder, R.G. (1939b). The significance of the two types of larvae in *Sphecophaga burra* (Cresson) and the factors conditioning them (Hymenoptera: Ichneumonidae). *Entomol. News* 50, 125–31.

Schneider, F. (1950). Die Entwicklung des Syrphidenparasiten *Diplazon fissorius* Grav. (Hym., Ichneum.). *Mitt. Schweiz. Entomol. Ges.* 23, 155–94.

Schneider, F. (1951). Einige physiologische Beziehungen zwischen Syrphidenlarven und ihren Parasiten. *Z. Ang. Entomol.* 33, 150–62.

Schneiderman, H.A. (1972). Insect hormones and insect control. In *Insect juvenile hormones* (eds. J.J. Menn and M. Beroza) pp. 3–27. Academic Press, New York.

Schneiderman, H.A. and Horwitz, J. (1958). The induction and termination of facultative diapause in the chalcid wasps *Mormoniella vitripennis* (Walker) and *Tritneptis klugii* (Ratzeburg). *J. Exp. Biol.* 35, 520–51.

Schneiderman, H.A. and Williams, C.M. (1953). The physiology of insect diapause. VII. The respiratory metabolism of the cecropia silkworm during diapause and development. *Biol. Bull.* 105, 320–24.

Schneirla, T.C. (1945). The army-ant behavior pattern: Nomad-statary relations in the swarmers and the problem of migration. *Biol. Bull.* 88, 166–93.

Schneirla, T.C. (1957). A comparison of species and genera in the ant subfamily Dorylinae with respect to functional pattern. *Insectes Sociaux* 4, 259–98.

Schneirla, T.C. (1963). The behaviour and biology of certain nearctic army ants: Springtime resurgence of cyclic function—Southeastern Arizona. *Anim. Behav.* 11, 583–95.

Schneirla, T.C. and Brown, R.Z. (1950). Army-ant life and behavior under dry-season conditions. *Bull. Amer. Mus. Natur. Hist.* 95, 263–354.

Schneirla, T.C. and Brown, R.Z. (1952). Sexual broods and the production of young queens in two species of army ants. *Zoologica* 37, 5–32.

Schooneveld, H., Otazo Sanchez, A., and de Wilde, J. (1977). Juvenile hormone-induced break and termination of diapause in the Colorado potato beetle. *J. Insect Physiol.* 23, 689–96.

Schoonhoven, L.M. (1962). Diapause and the physiology of host-parasite synchronization in *Bupalus piniarius* L. (Geometridae) and *Eucarcelia rutilla* Vill. (Tachinidae). *Archs. Neerl. Zool.* 15, 111-74.

Schopf, A. (1980). Zur Diapause des Puppenparasiten *Pimpla turionellae* L. (Hym., Ichneumonidae). *Zool. Jb. Syst.* 107, 537–67.

Schumm, M., Stinner, R.E., and Bradley, J.R., Jr. (1983). Characteristics of diapause in the bean leaf beetle, *Cerotoma trifurcata* (Forster) (Coleoptera: Chrysomelidae). *Environ. Entomol.* 12, 475–77.

Searle, J.B. and Mittler, T.E. (1981). Embryogenesis and the production of males by apterous viviparae of the green peach aphid *Myzus persicae* in relation to photoperiod. *J. Insect Physiol.* 27, 145–53.

Seeley, T.D. (1978). Life history strategy of the honey bee, *Apis mellifera*. *Oecologia* 32, 109–18.

Seeley, T.D. (1983). The ecology of temperate and tropical honeybee societies. *Amer. Sci.* 71, 264–72.

Seeley, T.D. and Heinrich, B. (1981). Regulation of temperature in the nests of social insects. In *Insect thermoregulation* (ed. B. Heinrich) pp. 160–234. Wiley-Interscience, New York.

Seger, J. (1983). Partial bivoltinism may cause alternating sex-ratio biases that favour eusociality. *Nature* 301, 59–62.

Selander, R.B. and Weddle, R.C. (1972). The ontogeny of blister beetles. III. Diapause termination in coarctate larvae of *Epicauta segmenta. Ann. Entomol. Soc. Amer.* 65, 1-17.

Sempala, S.D.K. (1983). Seasonal population dynamics of the immature stages of *Aedes africanus* (Theobald) (Diptera: Culicidae) in Zika Forest, Uganda. *Bull. Entomol. Res.* 73, 11–18.

Sevacherian, V., Stern, V.M., and Mueller, A.J. (1977). Heat accumulation for timing *Lygus* control measures in a safflower-cotton complex. *J. Econ. Entomol.* 70, 399–402.

Shaffer, P.L. and Rock, G.C. (1983). Tufted apple budworm (Lepidoptera: Tortricidae): effects of constant daylengths and temperatures on larval growth rate and determination of larval-pupal ecdysis. *Environ. Entomol.* 12, 76–80.

Shapiro, A.M. (1976). Seasonal polyphenism. In *Evolutionary biology*, Vol. 9 (eds. M.K. Hecht, W.C. Steere, and B. Wallace) pp. 259–333. Plenum Press, New York.

Shapiro, A.M. (1977). Phenotypic induction in *Pieris napi* L.: Role of temperature and photoperiod in a coastal California population. *Ecol. Entomol.* 2, 217–24.

Shapiro, A.M. (1978). The evolutionary significance of redundancy and variability in phenotypic-induction mechanisms of pierid butterflies (Lepidoptera). *Psyche* 85, 275–83.

Shapiro, A.M. (1979). The phenology of *Pieris napi microstriata* (Lepidoptera: Pieridae) during and after the 1975–77 California drought, and its evolutionary significance. *Psyche* 86, 1–10.

Shapiro, A.M. (1980a). Convergence in pierine polyphenisms (Lepidoptera). *J. Natur. Hist.* 14, 781–802.

Shapiro, A.M. (1980b). Egg-load assessment and carryover diapause in *Anthocharis* (Pieridae). *J. Lepid. Soc.* 34, 307–15.

Shapiro, A.M. (1980c). Physiological and developmental responses to photoperiod and temperature as data in phylogenetic and biogeographic inference. *Syst. Zool.* 29, 335–41.

Shapiro, A.M. (1982). Redundancy in pierid polyphenisms: Pupal chilling induces vernal phenotype in *Pieris occidentalis* (Pieridae). *J. Lepid. Soc.* 36, 174–77.

Sharpe, E.S. and Detroy, R.W. (1979). Susceptibility of Japanese beetle larvae to *Bacillus thuringiensis*: Associated effects of diapause, midgut pH, and milky disease. *J. Invert. Pathol.* 34, 90–91.

Sheldon, J.K. and MacLeod, E.G. (1971). Studies on the biology of the Chrysopidae. II. The feeding behavior of the adult of *Chrysopa carnea* (Neuroptera). *Psyche* 78, 107–21.

Sheldon, J.K. and MacLeod, E.G. (1974). Studies of the biology of the Chrysopidae. IV. A field and laboratory study of the seasonal cycle of *Chrysopa carnea* Stephens in Central Illinois (Neuroptera: Chrysopidae). *Trans. Amer. Entomol. Soc.* 100, 437–512.

Shepard, L.J. and Lutz, P.E. (1976). Larval responses of *Plathemis lydia* Drury to experimental photoperiods and temperatures (Odonata: Anisoptera). *Amer. Midl. Natur.* 95, 120–30.

Shimada, K. (1982). Glycerol accumulation in developmentally arrested pupae of *Papilio machaon* obtained by brain removal. *J. Insect Physiol.* 28, 975–78.

Shimada, K., Sakagami, S.F., Honma, K., and Tsutsui, H. (1984). Seasonal changes of glycogen/trehalose contents, supercooling points and survival rate in mature larvae of the overwintering soybean pod borer *Leguminivora glycinivorella. J. Insect Physiol.* 30, 369–73.

Shimizu, I. (1982). Photoperiodic induction in the silkworm, *Bombyx mori*, reared on artificial diet: Evidence for extraretinal photoreception. *J. Insect Physiol.* 28, 841–46.

Shiotsu, Y. (1977). Effects of temperature and photoperiod on the seasonal life cycle of *Hestina japonica* C. et R. Felder in Fukuoka City. *Japan. J. Ecol.* 27, 5–12.

Shiotsu, Y. and Arakawa, R. (1982). One host—one parasitoid system: Seasonal life cycles of *Pryeria sinica* (Lepidoptera) and *Agrothereutes minousubae* (Hymenoptera). *Res. Popul. Ecol.* 24, 43–57.

Shirozu, T. and Hara, A. (1962). *Early stages of Japanese butterflies in colour.* Vol. 1; Vol. 2. Hoikusha, Osaka (in Japanese).

Shorey, H.H. (1976). *Animal communication by pheromones.* Academic Press, New York.

Showers, W.B. (1981). Geographic variation of the diapause response in the European corn borer. In *Insect life history patterns* (eds. R.F. Denno and H. Dingle) pp. 97–111. Springer-Verlag, New York.

Showers, W.B., Reed, G.L., and Brindley, T.A. (1971). An adaptation of the European corn borer in the Gulf South. *Ann. Entomol. Soc. Amer.* 64, 1369–73.

Showers, W.B., Brindley, T.A., and Reed, G.L. (1972). Survival and diapause characteristics of hybrids of three geographical races of the European corn borer. *Ann. Entomol. Soc. Amer.* 65, 450–57.

Showers, W.B., Chiang, H.C., Keaster, A.J., Hill, R.E., Reed, G.L., Sparks, A.N., and Musick, G.J. (1975). Ecotypes of the European corn borer in North America. *Environ. Entomol.* 4, 753–60.

Shroyer, D.A. and Craig, G.B., Jr. (1980). Egg hatchability and diapause in *Aedes triseriatus* (Diptera: Culicidae): Temperature- and photoperiod-induced latencies. *Ann. Entomol. Soc. Amer.* 73, 39–43.

Shroyer, D.A. and Craig, G.B., Jr. (1981). Seasonal variation in sex ratio of *Aedes triseriatus* (Diptera: Culicidae) and its dependence on egg hatching behavior. *Environ. Entomol.* 10, 147–52.

Shroyer, D.A. and Craig, G.B., Jr. (1983). Egg diapause in *Aedes triseriatus* (Diptera: Culicidae): Geographic variation in photoperiodic response and factors influencing diapause termination. *J. Med. Entomol.* 20, 601–7.

Shuja-Uddin. (1977). Observations on normal and diapausing cocoons of the genus *Lipolexis* Foerster (Hymenoptera: Aphidiidae) from India. *Bull. Lab. Entomol. Agr.* 34, 51–54.

Sieber, R. and Benz, G. (1977). Juvenile hormone in larval diapause of the codling moth, *Laspeyresia pomonella* L. (Lepidoptera, Tortricidae). *Experimentia* 33, 1598–99.

Sieber, R. and Benz, G. (1980a). Termination of the facultative diapause in the codling moth, *Laspeyresia pomonella* (Lepidoptera, Tortricidae). *Entomol. Exp. Appl.* 28, 204–12.

Sieber, R. and Benz, G. (1980b). The hormonal regulation of the larval diapause in the codling moth, *Laspeyresia pomonella* (Lep. Tortricidae). *J. Insect Physiol.* 26, 213–18.

Siew, Y.C. (1966). Some physiological aspects of adult reproductive diapause in *Galeruca tanaceti* (L.) (Coleoptera: Chrysomelidae). *Trans. Roy. Entomol. Soc. Lond.* 118, 359–74.

Simmonds, F.J. (1948). The influence of maternal physiology on the incidence of diapause. *Phil. Trans. R. Soc. Ser. B.*, No. 603, 233, 385–414.

Simmonds, F.J. (1963). Genetics and biological control. *Can. Entomol.* 95, 561–67.

Simonenko, N.P. (1978). The special nature of seasonal changes in the development of the cabbage moth *Barathra brassicae* (Lepidoptera, Noctuidae) continuously reared in the laboratory. *Entomol. Rev.* 57, 323–27.

Sims, S.R. (1980). Diapause dynamics and host plant suitability of *Papilio zelicaon* (Lepidoptera: Papilionidae). *Amer. Midl. Natur.* 103, 375–84.

Sims, S.R. (1982). Larval diapause in the eastern tree-hole mosquito, *Aedes triseriatus*: Latitudinal variation in induction and intensity. *Ann. Entomol. Soc. Amer.* 75, 195–200.

Sims, S.R. (1983a). Inheritance of diapause induction and intensity in *Papilio zelicaon*. *Heredity* 51, 495–500.

Sims, S.R. (1983b). Prolonged diapause and pupal survival of *Papilio zelicaon* Lucas (Lepidoptera: Papilionidae). *J. Lepid. Soc.* 37, 29–37.

Sims, S.R. and Munstermann, L.E. (1983). Egg and larval diapause in two populations of *Aedes geniculatus* (Diptera: Culicidae). *J. Med. Entomol.* 20, 263–71.

Sims, S.R. and Shapiro, A.M. (1983a). Pupal colour dimorphism in California *Battus philenor*: Pupation sites, environmental control, and diapause linkage. *Ecol. Entomol.* 8, 95–104.

Sims, S.R. and Shapiro, A.M. (1983b). Pupal diapause in *Battus philenor* (Lepidoptera: Papilionidae). *Ann. Entomol. Soc. Amer.* 76, 407–12.

Skopik, S.D. and Bowen, M.F. (1976). Insect photoperiodism: An hourglass measures photoperiodic time in *Ostrinia nubilalis*. *J. Comp. Physiol.* 111, 249–59.

Slaff, M. and Crans, W.J. (1981). The activity and physiological status of pre-and posthibernating *Culex salinarius* (Diptera: Culicidae) populations. *J. Med. Entomol.* 18, 65–68.

Sláma, K. (1964). Hormonal control of respiratory metabolism during growth, reproduction, and diapause in female adults of *Pyrrhocoris apterus* L. (Hemiptera). *J. Insect Physiol.* 10, 283–303.

Slansky, F., Jr. (1974). Relationship of larval food-plants and voltinism patterns in temperate butterflies. *Psyche* 81, 243–53.

Slifer, E.H. (1949a). Variations, during development, in the resistance of the grasshopper egg to a toxic substance. *Ann. Entomol. Soc. Amer.* 42, 134–40.

Slifer, E.H. (1949b). Changes in certain of the grasshopper egg coverings during development as indicated by fast green and other dyes. *J. Exp. Zool.* 110, 183–203.

Siifer, E.H. and King, R.L. (1961). The inheritance of diapause in grasshopper eggs. *J. Hered.* 52, 39–44.

Smith, N.G. (1982). Population irruptions and periodic migrations in the day-flying moth *Urania fulgens*. In *The ecology of a tropical forest* (eds. E.G. Leigh, Jr., A.S. Rand, and D.M. Windsor) pp. 331–44. Smithsonian Institution Press, Washington, DC.

Smith, R.F. and van den Bosch, R. (1967). Integrated control. In *Pest control: Biological, phys-*

ical, and selected chemical methods (eds. W.W. Kilgore and R.L. Doutt) pp. 295–340. Academic Press, New York.

Smith, S.M. and Brust, R.A. (1971). Photoperiodic control of the maintenance and termination of larval diapause in *Wyeomyia smithii* (Coq.) with notes on oogenesis in the adult female. *Can. J. Zool.* 49, 1065–73.

Smith-Gill, S.J. (1983). Developmental plasticity: Developmental conversion *versus* phenotypic modulation. *Amer. Zool.* 23, 47–55.

Smythe, N. (1982). The seasonal abundance of night-flying insects in a neotropical forest. In *The ecology of a tropical forest* (eds. E.G. Leigh, Jr., A.S. Rand, and D.M. Windsor) pp. 309–18. Smithsonian Institution Press, Washington, DC.

Solbreck, C. (1974). Maturation of post-hibernation flight behaviour in the coccinellid *Coleomegilla maculata* (DeGeer). *Oecologia* 17, 265–75.

Solbreck, C. (1978). Migration, diapause, and direct development as alternative life histories in a seed bug, *Neacoryphus bicrucis*. In *Evolution of insect migration and diapause* (ed. H. Dingle) pp. 195–217. Springer-Verlag, New York.

Solbreck, C. (1979). Induction of diapause in a migratory seed bug, *Neacoryphus bicrucis* (Say) (Heteroptera: Lygaeidae). *Oecologia* 43, 41–49.

Solbreck, C. and Pehrson, I. (1979). Relations between environment, migration and reproduction in a seed bug, *Neacoryphus bicrucis* (Say) (Heteroptera: Lygaeidae). *Oecologia* 43, 51–62.

Solbreck, C. and Sillén-Tullberg, B. (1981). Control of diapause in a "monovoltine" insect, *Lygaeus equestris* (Heteroptera). *Oikos* 36, 68–74.

Sømme, L. (1964). Effects of glycerol on cold-hardiness in insects. *Can. J. Zool.* 42, 87–101.

Sømme, L. (1965). Further observations on glycerol and cold-hardiness in insects. *Can. J. Zool.* 43, 765–70.

Sømme, L. (1979). Overwintering ecology of alpine Collembola and oribatid mites from the Austrian Alps. *Ecol. Entomol.* 4, 175–80.

Sømme, L. (1982). Supercooling and winter survival in terrestrial arthropods. *Comp. Biochem. Physiol.* 73A, 519–43.

Sømme, L. and Block, W. (1982). Cold hardiness of Collembola at Signy Island, maritime Antarctic. *Oikos* 38, 168–76.

Sømme, L. and Conradi-Larsen, E.M. (1977a). Cold-hardiness of collembolans and oribatid mites from windswept mountain ridges. *Oikos* 29, 118–26.

Sømme, L. and Conradi-Larsen, E.M. (1977b). Anaerobiosis in overwintering collembolans and oribatid mites from windswept mountain ridges. *Oikos* 29, 127–32.

Sømme, L. and Zachariassen, K.E. (1981). Adaptations to low temperature in high altitude insects from Mount Kenya. *Ecol. Entomol.* 6, 199–204.

Soni, S.K. (1976). Effect of temperature and photoperiod on diapause induction in *Erioischia brassicae* (Bch.) (Diptera, Anthomyiidae) under controlled conditions. *Bull. Entomol. Res.* 66, 125–31.

Sonobe, H. and Ohnishi, E. (1971). Silkworm *Bombyx mori.* L.: Nature of diapause factor. *Science* 174, 835–38.

Sonobe, H., Ikeda, M., and Kaizuma, H. (1979). Oxygen permeability of the chorion in relation to diapause termination in *Bombyx* eggs. *Experientia* 35, 1650–51.

Southwick, E.E. (1983). The honey bee cluster as a homeothermic superorganism. *Comp. Biochem. Physiol.* 75A, 641–45.

Southwood, T.R.E. (1962). Migration of terrestrial arthropods in relation to habitat. *Biol. Rev.* 37, 171-214.

Southwood, T.R.E. (1964). Migration and population change in *Oscinella frit* L. (Diptera) on the oatcrop. *Proc. XIIth International Congress of Entomology* (ed. P. Freeman) pp. 420–21. Royal Entomological Society London, London.

Southwood, T.R.E. (1977). Habitat, the templet for ecological strategies? *J. Anim. Ecol.* 46, 337–65.

Southwood, T.R.E. (1978). The components of diversity. In *Diversity of insect faunas* (eds. L.A. Mound and N. Waloff) pp. 19–40. Blackwell, Oxford, London, Edinburgh, and Melbourne.

Southwood, T.R.E., Jepson, W.F., and van Emden, H.F. (1961). Studies on the behaviour of *Oscinella frit* L. (Diptera) adults of the panicle generation. *Entomol. Exp. Appl.* 4, 196–210.

Southwood, T.R.E., Brown, V.K., and Reader, P.M. (1983). Continuity of vegetation in space and time: A comparison of insects' habitat templet in different successional stages. *Res. Popul. Ecol.*, Suppl. 3, 61–74.

Spadoni, R.D., Nelson, R.L., and Reeves, W.C. (1974). Seasonal occurrence, egg production, and blood-feeding activity of autogenous *Culex tarsalis*. *Ann. Entomol. Soc. Amer.* 67, 895–902.

Sparks, A.N., Chiang, H.C., Keaster, A.J., Fairchild, M.L., and Brindley, T.A. (1966). Field studies of European corn borer biotypes in the Midwest. *J. Econ. Entomol.* 59, 922–28.

Spielman, A. (1957). The inheritance of autogeny in the *Culex pipiens* complex of mosquitoes. *Amer. J. Hyg.* 65, 404–25.

Spielman, A. (1962). The influence of rainfall upon the abundance of Cuban *Hippelates* (Diptera: Chloropidae). *Ann. Entomol. Soc. Amer.* 55, 39–42.

Spielman, A. (1971). Studies on autogeny in natural populations of *Culex pipiens*. II. Seasonal abundance of autogenous and anautogenous populations. *J. Med. Entomol.* 8, 555–61.

Spielman, A. (1979). Autogeny in *Culex pipiens* populations in nature: Effects of inbreeding. *Ann. Entomol. Soc. Amer.* 72, 826–28.

Spielman, A. and Wong, J. (1973a). Environmental control of ovarian diapause in *Culex pipiens*. *Ann. Entomol. Soc. Amer.* 66, 905–7.

Spielman, A. and Wong, J. (1973b). Studies on autogeny in natural populations of *Culex pipiens*. III. Midsummer preparation for hibernation in anautogenous populations. *J. Med. Entomol.* 10, 319–24.

Spradbery, J.P. (1971). Seasonal changes in the population structure of wasp colonies (Hymenoptera: Vespidae). *J. Anim. Ecol.* 40, 501–23.

Sprenkel, R.K. and Rabb, R.L. (1981). Effects of micrometeorological conditions on survival and fecundity of the Mexican bean beetle in soybean fields. *Environ. Entomol.* 10, 219–21.

Staal, G.B. (1975). Insect growth regulators with juvenile hormone activity. *Annu. Rev. Entomol.* 20, 417–60.

Stadelbacher, E.A. and Martin, D.F. (1981). Fall diapause and spring emergence of *Heliothis virescens*, *H. subflexa*, and backcrosses of their hybrid. *Environ. Entomol.* 10, 139–42.

Städler, E. (1970). Beitrag zur Kenntnis der Diapause bei der Möhrenfliege (*Psila rosae* Fabr., Diptera: Psilidae). *Bull. Soc. Entomol. Suisse* 43, 17–37.

Stanton, M.L. and Cook, R.E. (1983). Sources of intraspecific variation in the hostplant seeking behavior of *Colias* butterflies. *Oecologia* 60, 365–70.

Stark, S.B. and AliNiazee, M.T. (1982). Model of postdiapause development in the western cherry fruit fly. *Environ. Entomol.* 11, 471–74.

Stark, W.S. and Tan, K.E.W.P. (1982). Ultraviolet light: Photosensitivity and other effects on the visual system. *Photochem. Photobiol.* 36, 371–80.

Starý, P. (1972). Host life-cycle and adaptation of parasites of *Periphyllus*-aphids (Homoptera, Chaitophoridae; Hymenoptera, Aphidiidae). *Acta. Entomol. Bohemoslov.* 69, 89–96.

Stearns, S.C. (1976). Life-history tactics: A review of the ideas. *Q. Rev. Biol.* 51, 3–47.

Stearns, S.C. (1980). A new view of life-history evolution. *Oikos* 35, 266–81.

Stearns, S.C. (1981). On measuring fluctuating environments: Predictability, constancy, and contingency. *Ecology* 62, 185–99.

Stearns, S.C. (1983). Introduction to the symposium: The inter-face of life-history evolution, whole-organism ontogeny, and quantitative genetics. *Amer. Zool.* 23, 3–4.

Steel, C.G.H. (1976). Neurosecretory control of polymorphism in aphids. In *Phase and caste determination in insects* (ed. M. Lüscher) pp. 117–30. Pergamon Press, Oxford.

Steel, C.G.H. (1977). The neurosecretory system in the aphid *Megoura viciae*, with reference to unusual features associated with long distance transport of neurosecretion. *Gen. Comp. Endocrinol.* 31, 307–22.

Steel, C.G.H. and Lees, A.D. (1977). The role of neurosecretion in the photoperiodic control of polymorphism in the aphid *Megoura viciae*. *J. Exp. Biol.* 67, 117–35.

Steele, H.V. (1941). Some observations on the embryonic development of *Austroicetes cruciata* Sauss. (Acrididae) in the field. *Trans. Roy. Soc. S. Aust.* 65, 329–32.

Steele, J.E. (1981). The role of carbohydrate metabolism in physiological function. In *Energy metabolism in insects* (ed. R.G.H. Downer) pp. 101–33. Plenum Press, New York.

Stekol'nikov, A.A., Geyspits, K.A., Krusanova, Y.V., and Medvedeva, G.A. (1977). Seasonal variability in the photoperiodic reaction of the turnip moth, *Agrotis segetum* (Lepidoptera, Noctuidae). *Entomol. Rev.* 56, 8–14.

Sterling, W. (1972). Photoperiodic sensitivity in the ontogeny of the boll weevil. *Environ. Entomol.* 1, 568–71.

Sterling, W.L. and Adkisson, P.L. (1966). Differences in the diapause response of boll weevils from the High Plains and Central Texas and the significance of this phenomenon in revising present fall insecticidal control programs. *Texas A&M Univ. Bull.* 1047, 1–7.

Stern, V.M., Adkisson, P.L., Beingolea, G., and Viktorov, G.A. (1976). Cultural controls. In *Theory and practice of biological control* (eds. C.B. Huffaker and P.S. Messenger) pp. 593–613. Academic Press, New York.

Sternburg, J.G. and Waldbauer, G.P. (1978). Phenological adaptations in diapause termination by cecropia from different latitudes. *Entomol. Exp. Appl.* 23, 48–54.

Stewart, J.W., Whitcomb, W.H., and Bell, K.O. (1967). Estivation studies of the convergent lady beetle in Arkansas. *J. Econ. Entomol.* 60, 1730–5.

Stinner, R.E., Gutierrez, A.P., and Butler, G.D., Jr. (1974). An algorithm for temperature-dependent growth rate simulation. *Can. Entomol.* 106, 519–24.

Stinner, R.E., Butler, G.D., Jr., Bacheler, J.S., and Tuttle, C. (1975). Simulation of temperature-dependent development in population dynamics models. *Can. Entomol.* 107, 1167–74.

Stoffolano, J.G., Jr. (1968). The effect of diapause and age on the tarsal acceptance threshold of the fly *Musca autumnalis*. *J. Insect Physiol.* 14, 1205–14.

Stoffolano, J.G., Jr. (1974). Control of feeding and drinking in diapausing insects. In *Experimental analysis of insect behavior* (ed. L. Barton Browne) pp. 32–47. Springer-Verlag, Berlin.

Strand, M.R., Ratner, S., and Vinson, S.B. (1980). Maternally induced host regulation by the egg parasitoid *Telenomus heliothidis*. *Physiol. Entomol.* 8, 469–75.

Strassman, J.E. (1979). Honey caches help female paper wasps (*Polistes annularis*) survive Texas winters. *Science* 204, 207–9.

Strong, D.R., Jr., and Stiling, P.D. (1983). Wing dimorphism changed by experimental density manipulation in a planthopper (*Prokelisia marginata*, Homoptera, Delphacidae). *Ecology* 64, 206–9.

Strong, F.E. and Apple, J.W. (1958). Studies on the thermal constants and seasonal occurrence of the seed-corn maggot in Wisconsin. *J. Econ. Entomol.* 51, 704–7.

Stross, R.G. (1969a). Photoperiod control of diapause in *Daphnia*. II. Induction of winter diapause in the Arctic. *Biol. Bull.* 136, 264–73.

Stross, R.G. (1969b). Photoperiod control of diapause in *Daphnia*. III. Two-stimulus control of long-day, short-day induction. *Biol. Bull.* 137, 359–74.

Stross, R.G. and Hill, J.C. (1965). Diapause induction in *Daphnia* requires two stimuli. *Science* 150, 1462–64.

Stross, R.G. and Hill, J.C. (1968). Photoperiod control of winter diapause in the fresh-water crustacean, *Daphnia*. *Biol. Bull.* 134, 176–98.

Stross, R.G. and Kangas, D.A. (1969). The reproductive cycle of *Daphnia* in an arctic pool. *Ecology* 50, 457–60.

Sugg, P., Edwards, J.S., and Baust, J. (1983). Phenology and life history of *Belgica antarctica*, an Antarctic midge (Diptera: Chironomidae). *Ecol. Entomol.* 8, 105–13.

Sugiki, T. and Masaki, S. (1972). Photoperiodic control of larval and pupal development in *Spilactia imparilis* Butler (Lepidoptera: Arctiidae). *Kontyû* 40, 269–78.

Sugimoto, A. and Kobayashi, T. (1978). On the seasonal prevalence and possibility of seasonal migration of the black cutworm, *Agrotis ipsilon* (Hufn.), on Ishigaki Is., Okinawa. *Japan. J. Appl. Entomol. Zool.* 22, 40–43 (in Japanese).

Sullivan, C.R. and Wallace, D.R. (1967). Interaction of temperature and photoperiod in the induction of prolonged diapause in *Neodiprion sertifer*. *Can. Entomol.* 99, 834–50.

Sullivan, W.N., Oliver, M.Z., Hayes, D.K., and Schechter, M.S. (1970). Photoperiod manipulation to control diapause in the pink bollworm, *Pectinophora gossypiella*. *Experientia* 26, 1101–2.

Summy, K.R. and Gilstrap, F.E. (1983). Facultative production of alates by greenbug and corn leaf aphid, and implications in aphid population dynamics (Homoptera: Aphididae). *J. Kans. Entomol. Soc.* 56, 434–40.

Sunose, T. (1978). Studies on extended diapause in *Hasegawaia sasacola* Monzen (Diptera, Cecidomyiidae) and its parasites. *Kontyû* 46, 400–15.

Surgeoner, G.A. and Wallner, W.E. (1978). Evidence of prolonged diapause in prepupae of the variable oakleaf caterpillar, *Heterocampa manteo*. *Environ. Entomol.* 7, 186–88.

Sutherland, O.R.W. (1968). Dormancy and lipid storage in the Pemphigine aphid *Thecabius affinis*. *Entomol. Exp. Appl.* 11, 348–54.

Sutherst, R.W., Kerr, J.D., Maywald, G.F., and Stegeman, D.A. (1983). The effect of season and nutrition on the resistance of cattle to the tick *Boophilus microplus*. *Aust. J. Agric. Res.* 34, 329–39.

Sutton, R.D. (1983). Seasonal colour changes, sexual maturation and oviposition in *Psylla peregrina* (Homoptera: Psylloidea). *Ecol. Entomol.* 8, 195–201.

Suzuki, T. (1981). Effect of photoperiod on male egg production by foundresses of *Polistes chinensis antennalis* Pérez (Hymenoptera, Vespidae). *Japan. J. Ecol.* 31, 347–51.

Suzuki, T. (1982). Cessation and resumption of laying of female-producing eggs by foundresses of a polistine wasp *Polistes chinensis antennalis* (Hymenoptera, Vespidae) under experimental conditions. *Kontyû* 50, 652–55.

Suzuki, T., Suzuki, K., and Miya, K. (1980). Polyols in diapause eggs of the false melon beetle, *Atrachya menetriesi. Appl. Entomol. Zool.* 15, 492–4.

Syme, P.D. (1972). The influence of constant temperature on the non-diapause development of *Hyssopus thymus. Can. Entomol.* 104, 113–20.

Tabachnick, W.J., Munstermann, E., and Powell, J.R. (1979). Genetic distinctness of sympatric forms of *Aedes aegypti* in East Africa. *Evolution* 33, 287–95.

Takada, H. (1982a). Influence of photoperiod and temperature on the production of sexual morphs in a green and a red form of *Myzus persicae* (Sulzer) (Homptera, Aphididae). I. Experiments in the laboratory. *Kontyû* 50, 233–45.

Takada, H. (1982b). Influence of photoperiod and temperature on the production of sexual morphs in a green and a red form of *Myzus persicae* (Sulzer) (Homoptera, Aphididae). II. Rearing under natural conditions. *Kontyû* 50, 353–64.

Takafuji, A. and Kamezaki, H. (1984). Diapause incidence in eggs of the citrus red mite, *Panonychus citri* (M.) on pear twigs. *Appl. Entomol. Zool.* 19, 270–71.

Takafuji, A. and Morimoto, N. (1983). Diapause attributes and seasonal occurrences of two populations of the citrus red mite, *Panonychus citri* (McGregor) on pear (Acarina: Tetranychidae). *Appl. Entomol. Zool.* 18, 525–32.

Takahashi, F. (1977). Generation carryover of a fraction of population members as an animal adaptation to unstable environmental conditions. *Res. Popul. Ecol.* 18, 235–42.

Takeda, M. and Chippendale, G.M. (1982a). Environmental and genetic control of the larval diapause of the southwestern corn borer, *Diatraea grandiosella. Physiol. Entomol.* 77, 99–110.

Takeda, M. and Chippendale, G.M. (1982b). Phenological adaptations of a colonizing insect: The southwestern corn borer, *Diatraea grandiosella. Oecologia* 53, 386–93.

Takeda, M. and Masaki, S. (1976). Photoperiodic control of larval development in *Plodia interpunctella.* In *Proc. U.S.-Japan Seminar Stored Product Entomol.*, pp. 186–201. Manhattan, KS.

Takeda, M. and Masaki, S. (1979). Asymmetric perception of twilight affecting diapause induction by the fall webworm, *Hyphantria cunea. Entomol. Exp. Appl.* 25, 317–27.

Takeda, N. (1972). Activation of neurosecretory cells in *Monema flavescens* (Lepidoptera) during diapause break. *Gen. Comp. Endocrinol.* 18, 417–27.

Takeda, N. (1977). Brain hormone carrier haemocytes in the moth *Monema flavescens. J. Insect Physiol.* 23, 1245–54.

Takeda, N. (1978a). Glycerol uptake in the neuroendocrine system of two slug moth prepupae (Lepidoptera). *Comp. Biochem. Physiol.* 59A, 73–78.

Takeda, N. (1978b). Hormonal control of prepupal diapause in *Monema flavescens* (Lepidoptera). *Gen. Comp. Endocrinol.* 34, 123–31.

Takeda, S. (1977). Induction of egg diapause in *Bombyx mori* by some cephalo-thoracic organs of the cockroach, *Periplaneta americana. J. Insect Physiol.* 23, 813–16.

Takehara, I. (1966). Natural occurrence of glycerol in the slug caterpillar, *Monema flavescens. Contr. Inst. Low Temp. Sci.*, Series B, 14, 1–34.

Takehara, I. and Asahina, E. (1961). Glycerol in a slug caterpillar. I. Glycerol formation, diapause and frost resistance in insects reared at various graded temperatures. *Low Temp. Sci.*, Series B, 19, 29–36.

Tamaki, G., Annis, B., Fox, L., Gupta, R.K., and Meszleny, A. (1982a). Comparison of yellow holocyclic and green anholocyclic strains of *Myzus persicae* (Sulzer): Low temperature adaptability. *Environ. Entomol.* 11, 231–33.

Tamaki, G., Weiss, M.A., and Long, G.E. (1982b). Effective growth units in population dynamics of the green peach aphid (Homoptera: Aphididae). *Environ. Entomol.* 11, 1134–36.

Tamura, I., Iwata, T., and Kishino, K. (1959). Geographical races in the rice stem maggot, *Chlorops oryzae* Matsumura. *Japan. J. Appl. Entomol. Zool.* 3, 243–49 (in Japanese, with English summary).

Tanaka, S. (1976). Wing polymorphism, egg production and adult longevity in *Pteronemobius taprobanensis* Walker (Orthoptera, Gryllidae). *Kontyû* 44, 327–33.

Tanaka, S. (1978a). Effects of changing photoperiod on nymphal development in *Pteronemobius nitidus* Bolivar (Orthoptera, Gryllidae). *Kontyû* 46, 135–51.

Tanaka, S. (1978b). Photoperiodic determination of wing form in *Pteronemobius nitidus* Bolivar (Orthoptera, Gryllidae). *Kontyû* 46, 207–17.

Tanaka, S. (1979). Multiple photoperiodic control of the seasonal life cycle in *Pteronemobius nitidus* Bolivar (Orthoptera: Gryllidae). *Kontyû* 47, 465–75.

Tanaka, S. (1983). Seasonal control of nymphal diapause in the spring ground cricket, *Pteronemobius nitidus* (Orthoptera: Gryllidae). In *Diapause and life cycle strategies in insects* (eds. V.K. Brown and I. Hodek) pp. 35–53. Junk, The Hague, The Netherlands.

Tanaka, S. (1984). Seasonal variation in embryonic diapause of the striped ground cricket, *Allonemobius fasciatus. Physiol. Entomol.* 9, 97–105.

Tanaka, S., Matsuka, M., and Sakai, T. (1976). Effect of change in photoperiod on wing form in *Pteronemobius taprobanensis* Walker (Orthoptera: Gryllidae). *Appl. Entomol. Zool.* 11, 27–32.

Tanaka, Y. (1951). Studies on hibernation with special reference to photoperiodicity and breeding of the Chinese tussar silkworm (V). Nippon Sanshigaku Zasshi (*J. Sericult. Sci. Japan*) 20, 132–38.

Tanno, K. (1965). Frost-resistance in the poplar sawfly, *Trichiocampus popli* Okamoto. III. Frost-resistance and sugar content. *Low Temp. Sci.*, Series B, 23, 55–64 (in Japanese).

Tanno, K. (1967). Immediate termination of prepupal diapause in poplar sawfly by body freezing. *Low Temp. Sci.*, Series B, 25, 97–103 (in Japanese).

Tanno, K. (1977). Ecological observation and frost-resistance in overwintering prepupae, *Sciara* sp. (Sciaridae). *Low Temp. Sci.*, Series B, 35, 63–74.

Tanno, K. and Asahina, E. (1964). Frost resistance in the poplar sawfly, *Trichiocampus populi. Low Temp. Sci.* Series B, 22, 59–70 (in Japanese, English summary).

Tauber, M.J. and Tauber, C.A. (1969). Diapause in *Chrysopa carnea* (Neuroptera: Chrysopidae). I. Effect of photoperiod on reproductively active adults. *Can. Entomol.* 101, 364–70.

Tauber, M.J. and Tauber, C.A. (1970a). Photoperiodic induction and termination of diapause in an insect: Response to changing day lengths. *Science* 167, 170.

Tauber, M.J. and Tauber, C.A. (1970b). Adult diapause in *Chrysopa carnea*: Stages sensitive to photoperiodic induction. *J. Insect Physiol.* 16, 2075–80.

Tauber, M.J. and Tauber, C.A. (1972a). Geographic variation in critical photoperiod and in diapause intensity of *Chrysopa carnea* (Neuroptera). *J. Insect Physiol.* 18, 25–29.

Tauber, M.J. and Tauber, C.A. (1972b). Larval diapause in *Chrysopa nigricornis*: Sensitive stages, critical photoperiod, and termination (Neuroptera: Chrysopidae). *Entomol. Exp. Appl.* 15, 105–11.

Tauber, C.A. and Tauber, M.J. (1973a). Diversification and secondary intergradation of two *Chrysopa carnea* strains (Neuroptera: Chrysopidae). *Can. Entomol.* 105, 1153–67.

Tauber, M.J. and Tauber, C.A. (1973b). Insect phenology: Criteria for analyzing dormancy and for forecasting postdiapause development and reproduction in the field. *Search* (*Agr.*) *Cornell Univ. Agr. Exp. Stn.*, Ithaca, NY, 3, 1-16.

Tauber, M.J. and Tauber, C.A. (1973c). Nutritional and photoperiodic control of the seasonal reproductive cycle in *Chrysopa mohave* (Neuroptera). *J. Insect Physiol.* 19, 729–36.

Tauber, M.J. and Tauber, C.A. (1973d). Seasonal regulation of dormancy in *Chrysopa carnea* (Neuroptera). *J. Insect Physiol.* 19, 1455–63.

Tauber, M.J. and Tauber, C.A. (1973e). Quantitative response to daylength during diapause in insects. *Nature* 244, 296–97.

Tauber, M.J. and Tauber, C.A. (1974). Thermal accumulations, diapause, and oviposition in a conifer-inhabiting predator, *Chrysopa harrisii* (Neuroptera). *Can. Entomol.* 106, 969–78.

Tauber, M.J. and Tauber, C.A. (1975a). Natural daylengths regulate insect seasonality by two mechanisms. *Nature* 258, 711-12.

Tauber, M.J. and Tauber, C.A. (1975b). Criteria for selecting *Chrysopa carnea* biotypes for biological control: Adult dietary requirements. *Can. Entomol.* 107, 589–95.

Tauber, M.J. and Tauber, C.A. (1976a). Insect seasonality: Diapause maintenance, termination, and postdiapause development. *Annu. Rev. Entomol.* 21, 81–107.

Tauber, M.J. and Tauber, C.A. (1976b). Developmental requirements of the univoltine species *Chrysopa downesi*: Photoperiodic stimuli and sensitive stages. *J. Insect Physiol.* 22, 331–35.

Tauber, M.J. and Tauber, C.A. (1976c). Environmental control of univoltinism and its evolution in an insect species. *Can. J. Zool.* 54, 260–66.

Tauber, M.J. and Tauber, C.A. (1976d). Physiological responses underlying the timing of vernal activities in insects. *Intl. J. Biometeorol.* 20, 218–22.

Tauber, C.A. and Tauber, M.J. (1977a). Sympatric speciation based on allelic changes at three loci: Evidence from natural populations in two habitats. *Science* 197, 1298–99.

Tauber, C.A. and Tauber, M.J. (1977b). A genetic model for sympatric speciation through habitat diversification and seasonal isolation. *Nature* 268, 702–5.

Tauber, M.J. and Tauber, C.A. (1977c). Adaptive mechanisms underlying univoltinism in insects. *Proc. XII Intl. Conf. Intl. Soc. Chronobiol.*, pp. 639–42. Publishing House Il Ponte, Milan.

Tauber, M.J. and Tauber, C.A. (1978). Evolution of phenological strategies in insects: A comparative approach with eco-physiological and genetic considerations. In *Evolution of insect migration and diapause* (ed. H. Dingle) pp. 53–71. Springer-Verlag, New York.

Tauber, M.J. and Tauber, C.A. (1979). Inheritance of photoperiodic responses controlling diapause. *Bull. Entomol. Soc. Amer.* 25, 125–28.

Tauber, C.A. and Tauber, M.J. (1981a). Insect seasonal cycles: Genetics and evolution. *Annu. Rev. Ecol. Syst.* 12, 281–308.

Tauber, M.J. and Tauber, C.A. (1981b). Seasonal responses and their geographic variation in *Chrysopa downesi*: Ecological and evolutionary considerations. *Can. J. Zool.* 59, 370–76.

Tauber, C.A. and Tauber, M.J. (1982). Evolution of seasonal adaptations and life history traits in *Chrysopa*: Response to diverse selective pressures. In *Evolution and genetics of life histories* (eds. H. Dingle and J.P. Hegmann) pp. 51–72. Springer-Verlag, New York.

Tauber, M.J. and Tauber, C.A. (1983). Life history traits of *Chrysopa carnea* and *Chrysopa rufilabris* (Neuroptera: Chrysopidae): influence of humidity. *Ann. Entomol. Soc. Amer.* 76, 282–85.

Tauber, C.A. and Tauber, M.J. (1985). Host specificity in insect predators: An unaddressed evolutionary problem. (submitted).

Tauber, M.J., Tauber, C.A., and Denys, C.J. (1970a). Diapause in *Chrysopa carnea* (Neuroptera: Chrysopidae). II. Maintenance by photoperiod. *Can. Entomol.* 102, 474–78.

Tauber, M.J., Tauber, C.A., and Denys, C.J. (1970b). Adult diapause in *Chrysopa carnea*: Photoperiodic control of duration and colour. *J. Insect Physiol.* 16, 949–55.

Tauber, C.A., Tauber, M.J., and Nechols, J.R. (1977). Two genes control seasonal isolation in sibling species. *Science* 197, 592–93.

Tauber, M.J., Tauber, C.A., Nechols, J.R., and Helgesen, R.G. (1982). A new role for temperature in insect dormancy: Cold maintains diapause in temperate zone diptera. *Science* 218, 690–91.

Tauber, M.J., Tauber, C.A., Nechols, J.R., and Obrycki, J.J. (1983). Seasonal activity of parasitoids: Control by external, internal, and genetic factors. In *Diapause and life cycle strategies in insects* (eds. V.K. Brown and I. Hodek) pp. 87–108. Junk, The Hague, The Netherlands.

Tauber, M.J., Tauber, C.A., and Masaki, S. (1984). Adaptations to hazardous seasonal conditions: Dormancy, migration, and polyphenism. In *Ecological entomology* (eds. C.B. Huffaker and R.L. Rabb) pp. 149–83. John Wiley & Sons, New York.

Tauthong, P. and Brust, R.A. (1977). The effect of photoperiod on diapause induction, and temperature on diapause termination in embryos of *Aedes campestris* Dyar and Knab (Diptera:Culicidae). *Can. J. Zool.* 55, 129–34.

Taylor, F. (1980a). Timing in the life histories of insects. *Theor. Popul. Biol.* 18, 112–24.

Taylor, F. (1980b). Optimal switching in diapause in relation to the onset of winter. *Theor. Popul. Biol.* 18, 125–33.

Taylor, F. (1981). Ecology and evolution of physiological time in insects. *Amer. Natur.* 117, 1–23.

Taylor, F. (1982). Sensitivity of physiological time in arthropods to variation of its parameters. *Environ. Entomol.* 11, 573–77.

Taylor, F. (1984). Mexican bean beetles mate successfully in diapause. *Int. J. Invert. Reprod.* 7, 297–302.

Taylor, F. and Schrader, R. (1984). Transient effects of photoperiod on reproduction in the Mexican bean beetle. *Physiol. Entomol.* 9, 459–64.

Taylor, V.A. (1981). The adaptive and evolutionary significance of wing polymorphism and parthenogenesis in *Ptinella* Motschulsky (Coleoptera: Ptilidae). *Ecol. Entomol.* 6, 89–98.

Tazima, Y. (1964). *The genetics of the silkworm.* Logos Press, London.

Templeton, A.R. (1980). Modes of speciation and inferences based on genetic differences. *Evolution* 34, 719–29.

Ternovoy, V.I. (1978). A study of the diapause in *Wohlfahrtia magnifica* (Diptera, Sarcophagidae). *Entomol. Rev.* 57, 328–32.

Thiele, H.U. (1966). Einflüsse der Photoperiode auf die Diapause von Carabiden. *Z. Ang. Entomol.* 58, 143–49.

Thiele, H.U. (1968). Formen der Diapausesteuerung bei Carabiden. *Verh. Dtsch. Zool. Ges. Heidelberg 1967* (*Zool. Anz. Suppl.*) 31, 358–64.

Thiele, H.U. (1969). The control of larval hibernation and of adult aestivation in the carabid beetles *Nebria brevicollis* F. and *Patrobus atrorufus* Stroem. *Oecologia* 2, 347–61.

Thiele, H.U. (1971). Die Steuerung der Jahresrhythmik von Carabiden durch exogene und endogene Faktoren. *Zool. Jb. Syst. Bd.* 98, 341–71.

Thiele, H.U. (1973). Remarks about Mansingh's and Müller's classifications of dormancies in insects. *Can. Entomol.* 105, 925–28.

Thiele, H.U. (1977a). Measurement of day-length as a basis for photoperiodism and annual periodicity in the carabid beetle *Pterostichus nigrita* F. *Oecologia* 30, 331–48.

Thiele, H.U. (1977b). *Carabid beetles in their environments.* Springer-Verlag, Berlin.

Thiele, H.U. (1979a). Relationships between annual and daily rhythms, climatic demands and habitat selection in carabid beetles. In *Carabid beetles: Their evolution, natural history, and classification* (eds. T.L. Erwin, G.E. Ball, and D.R. Whitehead) pp. 449–70. Junk, The Hague, The Netherlands.

Thiele, H.U. (1979b). Intraspecific differences in photoperiodism and measurement of day length in *Pterostichus nigrita* Paykull (Coleoptera, Carabidae). In *On the evolution of behaviour in carabid beetles* (eds. P.J. den Boer, H.U. Thiele and F. Weber) pp. 53–62. Misc. Pap. No. 18; Agricultural University of Wageningen, H. Veerman & Zonen, B.V., Wageningen, The Netherlands.

Thiele, H.U. and Krehan, I. (1969). Experimentelle Untersuchungen zur Larvaldiapause des Carabiden *Pterostichus vulgaris. Entomol. Exp. Appl.* 12, 67–73.

Thompson, J.N. and Price, P.W. (1977). Plant plasticity, phenology, and herbivore dispersion: Wild parsnip and the parsnip webworm. *Ecology* 58, 1112–19.

Tingle, F.C. and Lloyd, E.P. (1969). Influence of temperature and diet on attainment of firm diapause in the boll weevil. *J. Econ. Entomol.* 62, 596–99.

Tingle, F.C., Lane, H.C., King, E.E., and Lloyd, E.P. (1971). Influence of nutrients in the adult diet on diapause in the boll weevil. *J. Econ. Entomol.* 64, 812–14.

Tischler, W. (1967). Zur Biologie und Ökologie des Opilioniden *Mitopus morio* F. *Biol. Zentralblatt* 4, 473–84.

Tobe, S.S. and Chapman, C.S. (1979). The effects of starvation and subsequent feeding on juvenile hormone synthesis and oöcyte growth in *Schistocerca americana gregaria. J. Insect Physiol.* 25, 701–8.

Toda, M.J. and Kimura, M.T. (1978). Bionomics of Drosophilidae (Diptera) in Hokkaido. I. *Scaptomyza pallida* and *Drosophila nipponica. Kontyû* 46, 83–98.

Tombes, A.S. (1964a). Respiratory and compositional study on the aestivating insect, *Hypera postica* (Gyll.) (Curculionidae). *J. Insect Physiol.* 10, 997–1003.

Tombes, A.S. (1964b). Seasonal changes in the reproductive organs of the alfalfa weevil, *Hypera postica* (Coleoptera: Curculionidae), in South Carolina. *Ann. Entomol. Soc. Amer.* 57, 422–26.

Tombes, A.S. (1970). *An introduction to invertebrate endocrinology.* Academic Press, New York.

Tombes, A.S. and Marganian, L. (1967). Aestivation (summer diapause) in *Hypera postica* (Coleoptera: Curculionidae). II. Morphological and histological studies of the alimentary canal. *Ann. Entomol. Soc. Amer.* 60, 1–8.

Torchio, P.F. (1975). The biology of *Perdita nuda* and descriptions of its immature forms and those of its *Sphecodes* parasite (Hymenoptera: Apoidea). *J. Kans. Entomol. Soc.* 48, 257–79.

Torchio, P.F. and Tepedino, V.J. (1982). Parsivoltinism in three species of *Osmia* bees. *Psyche* 89, 221–38.

Toshima, A., Homma, K., and Masaki, S. (1961). Factors influencing the seasonal incidence and breaking of diapause in *Carposina niponensis* Walsingham. *Japan. J. Appl. Entomol. Zool.* 5, 260–69.

Trimble, R.M. and Smith, S.M. (1978). Geographic variation in developmental time and predation in the tree-hole mosquito, *Toxorhynchites rutilus septentrionalis* (Diptera: Culicidae). *Can. J. Zool.* 56, 2156–65.

Trimble, R.M. and Smith, S.M. (1979). Geographic variation in the effects of temperature and photoperiod on dormancy induction, development time, and predation in the tree-hole mosquito, *Toxorhynchites rutilus septentrionalis* Diptera: Culicidae). *Can. J. Zool.* 57, 1612–18.

Troester, S.J., Ruesink, W.G., and Rings, R.W. (1982). A model of black cutworm (*Agrotis ipsilon*) development: Description, uses, and implications. *Univ. Illinois Agr. Exp. Sta. Bull.* 774, 1–33.

Trottier, R. (1971). Effect of temperature on the life-cycle of *Anax junius* in Canada. *Can. Entomol.* 103, 1671–83.

Trpis, M. (1978). Genetics of hematophagy and autogeny in the *Aedes scutellaris* complex (Diptera: Culicidae). *J. Med. Entomol.* 15, 73–80.

Trpis, M. and Horsfall, W.R. (1967). Eggs of floodwater mosquitoes (Diptera: Culicidae). XI. Effect of medium on hatching of *Aedes sticticus*. *Ann. Entomol. Soc. Amer.* 60, 1150–52.

Truman, J.W. (1973). Temperature sensitive programming of the silkmoth flight clock: A mechanism for adapting to the seasons. *Science* 182, 727–29.

Truman, J.W. and Riddiford, L.M. (1974). Hormonal mechanisms underlying insect behaviour. *Adv. Insect Physiol.* 10, 297–352.

Tsuchida, K. and Yoshitake, N. (1983a). Effect of different artificial diets on diapause induction under controlled temperature and photoperiod in the silkworm, *Bombyx mori* L. *Physiol. Entomol.* 8, 333–38.

Tsuchida, K. and Yoshitake, N. (1983b). Relationship between photoperiod and secretion of the diapause hormone during larval stages of the silkworm, *Bombyx mori* L., reared on an artificial diet. *J. Insect Physiol.* 29, 755–59.

Tsuda, Y. (1982). Reproductive strategy of insects as adaptation to temporally varying environments. *Res. Popul. Ecol.* 24, 388–404.

Tsuji, H. (1958). Studies on diapause in the Indian meal moth, *Plodia interpunctella* Hübner. I. The influence of temperature on the diapause and the type of diapause. *Japan. J. Appl. Entomol. Zool.* 2, 17–23 (in Japanese).

Tsuji, H. (1960). Studies on the ecological life history of the Indian meal moth, *Plodia interpunctella* Hübner. I. Comparative studies of the three stocks with special reference to the onset of diapause. *Japan. J. Appl. Entomol. Zool.* 4, 173–81.

Tsuji, H. (1963). Experimental studies on the larval diapause of the Indian-meal moth, *Plodia interpunctella* Hübner (Lepidoptera: Pyralidae). Ph.D. thesis, Kyushu University, Fukuoka, Kokodo Ltd., Tokyo.

Tsuji, H. (1966). Rice bran extracts effective in terminating diapause in *Plodia interpunctella* Hübner (Lepidoptera: Pyralidae). *Appl. Entomol. Zool.* 1, 51.

Tsumuki, H. and Kanehisa, K. (1979). Enzymes associated with glycogen metabolism in larvae of the rice stem borer, *Chilo suppressalis* Walker: Some properties and changes in activities during hibernation. *Appl. Entomol. Zool.* 14, 270–77.

Tsumuki, H. and Kanehisa, K. (1980a). Effect of low temperature on glycerol and trehalose concentration in haemolymph of the rice stem borer, *Chilo suppressalis* Walker. *Japan. J. Appl. Entomol. Zool.* 24, 189–93.

Tsumuki, H. and Kanehisa, K. (1980b). Changes in enzyme activities related to glycerol synthesis in hibernating larvae of the rice stem borer, *Chilo suppressalis* Walker. *Appl. Entomol. Zool.* 15, 285–92.

Tsumuki, H. and Kanehisa, K. (1980c). Metabolism of ^{14}C-glucose and *UDP*-^{14}C-G- in hibernating larvae of the rice stem borer, *Chilo suppressalis* Walker. *Berichte des Ohara Inst. für landwirtsch. Biol. Okayama Univ.* 18, 43–53.

Tsumuki, H. and Kanehisa, K. (1981a). Effect of JH and ecdysone on glycerol and carbohydrate contents in diapausing larvae of the rice stem borer, *Chilo suppressalis* Walker (Lepidoptera: Pyralidae). *Appl. Entomol. Zool.* 16, 7–15.

Tsumuki, H. and Kanehisa, K. (1981b). The fate of C-glycerol in the rice stem borer, *Chilo suppressalis* Walker (Lepidoptera: Pyralidae). *Appl. Entomol. Zool.* 16, 200–8.

Tummala, R.L., Haynes, D.L., and Croft, B.A. (1976). *Modeling for pest management*. Michigan State University Press, East Lansing.

Tuomi, J., Halaka, T., and Haukioja, E. (1983). Alternative concepts of reproductive effort, costs of reproduction, and selection in life-history evolution. *Amer. Zool.* 23, 25–34.

Turnock, W.J. (1973). Factors influencing the fall emergence of *Bessa harveyi* (Tachinidae: Diptera). *Can. Entomol.* 105, 399–409.

Turnock, W.J., Lamb, R.J., and Bodnaryk, R.P. (1983). Effects of cold stress during pupal diapause on the survival and development of *Mamestra configurata* (Lepidoptera: Noctuidae). *Oecologia* 56, 185–92.

Turunen, S. and Chippendale, G.M. (1979). Possible function of juvenile hormone-dependent protein in larval insect diapause. *Nature* 280, 836–38.

Turunen, S. and Chippendale, G.M. (1980). Proteins of the fat body of non-diapausing and diapausing larvae of the southwestern corn borer, *Diatraea grandiosella*: Effect of juvenile hormone. *J. Insect Physiol.* 26, 163–69.

Tuskes, P.M. and Brower, L.P. (1978). Overwintering ecology of the monarch butterfly, *Danaus plexippus* L., in California. *Ecol. Entomol.* 3, 141-53.

Tyler, B.M.J. and Jones, P.A. (1974). Hibernation study with *Lysiphlebus testaceipes*, parasite of the greenbug. *Environ. Entomol.* 3, 412–14.

Tyshchenko, V.P. (1977). Physiology of insect photoperiodism. *Trud. Vses. Entomol. Obshches.* 59, 1–155 (in Russian).

Tyshchenko, V.P. (1980). The physiological mechanism of the photoperiodic reaction controlling the onset of pupal diapause in Lepidoptera. *Entomol. Rev.* 59, 1–8.

Tyshchenko, V.P. and Tyshchenko, G.F. (1982). New evidence for the participation of two circadian processes in the physiological mechanism of photoperiodic reaction in insects. *Dokl. Akad. Nauk. SSSR* 270, 494–96.

Tyshchenko, V.P., Volkovich, T.A., and Lanevich, V.P. (1983). The intermediate photoperiodic reaction and its role in regulating seasonal development in *Axylia putris* L. (Lepidoptera, Noctuidae). *Entomol. Rev.* 62, 1–10.

Ubukata, H. (1980). Life history and behavior of a corduliid dragonfly, *Cordulia aenea amurensis* Selys. III. Aquatic period, with special reference to larval growth. *Kontyû* 48, 414–27.

Uchida, M. and Shinkaji, N. (1980). Oviposition and hatching time of the winter eggs of the citrus red mite, *Panonychus citri* (McGregor), on pear in Tottori Prefecture (Acarina: Tetranychidae). *Japan. J. Appl. Entomol. Zool.* 24, 18–23.

Ujiye, T. (1983). Studies on the diapause of the apple leaf miner, *Phyllonorycter ringoniella* Matsumura (Lepidoptera: Gracillariidae). I. Effect of photoperiod on the induction of diapause in the Morioka population. *Japan. J. Appl. Entomol. Zool.* 27, 117–23 (in Japanese, English abstract).

Umeya, K. and Miyata, T. (1979). Effects of photoperiod and temperature on the maculation of the fall webworm moth, *Hyphantria cunea* Drury (Lepidoptera: Arctiidae). *Japan. J. Appl. Entomol. Zool.* 23, 17–21 (in Japanese, English summary).

Urquhart, F.A. and Urquhart, N.R. (1976a). A study of the peninsular Florida populations of the monarch butterfly (*Danaus p. plexippus*; Danaidae). *J. Lepid. Soc.* 30, 73–87.

Urquhart, F.A. and Urquhart, N.R. (1976b). Monarch butterfly (*Danaus plexippus* L.) overwintering population in Mexico. *Atlanta* 7, 51–60.

Urquhart, F.A. and Urquhart, N.R. (1976c). The overwintering site of the eastern population of the monarch butterfly (*Danaus p. plexippus*; Danaidae) in southern Mexico. *J. Lepid. Soc.* 30, 153–58.

Urquhart, F.A. and Urquhart, N.R. (1977). Overwintering areas and migratory routes of the monarch butterfly (*Danaus p. plexippus*, Lepidoptera: Danaidae) in North America, with special reference to the western population. *Can. Entomol.* 109, 1583–89.

Urquhart, F.A. and Urquhart, N.R. (1978). Autumnal migration routes of the eastern population of the monarch butterfly (*Danaus p. plexippus* L.; Danaidae; Lepidoptera) in North America to the overwintering site in the Neovolcanic Plateau of Mexico. *Can. J. Zool.* 56, 1759–64.

Urquhart, F.A. and Urquhart, N.R. (1979a). Vernal migration of the monarch butterfly (*Danaus p. plexippus*, Lepidoptera: Danaidae) in North America from the overwintering site in the Neo-volcanic Plateau of Mexico. *Can. Entomol.* 111, 15–18.

Urquhart, F.A. and Urquhart, N.R. (1979b). Breeding areas and overnight roosting locations in the northern range of the monarch butterfly (*Danaus plexippus plexippus*) with a summary of associated migratory routes. *Can. Field- Natur.* 93, 41–47.

Urquhart, F.A. and Urquhart, N.R. (1979c). Aberrant autumnal migration of the eastern population of the monarch butterfly, *Danaus p. plexippus* (Lepidoptera: Danaidae) as it relates to the occurrence of strong westerly winds. *Can. Entomol.* 111, 1281–86.

Ushatinskaya, R.S. (1966). Prolonged diapause in Colorado beetle and conditions of its formation. In *Ecology and physiology of diapause in the Colorado beetle* (ed. K.V. Arnoldi) pp. 168–200. English translation (1976) by Indian National Scientific Documentation Centre, New Delhi.

Ushatinskaya, R.S. (1976a). Insect dormancy and its classification. *Zool. Jb. Syst. Bd.* 103, 76–97.

Ushatinskaya, R.S. (1976b). The lability of diapause and its modifications in the Colorado beetle, *Leptinotarsa decemlineata* (Coleoptera, Chrysomelidae). *Entomol. Rev.* 55, 18–21.

Usua, E.J. (1973). Induction of diapause in the maize stemborer, *Busseola fusca. Entomol. Exp. Appl.* 16, 322–28.

Uvarov, B. (1966). *Grasshoppers and locusts*, Vol. 1. Cambridge University Press, London.

Vaartaja, O. (1959). Evidence of photoperiodic ecotypes in trees. *Ecol. Monogr.* 29, 91–111.

van den Berg, M.A. (1971). Studies on the induction and termination of diapause in *Mesocomys pulchriceps* Cam. (Hymenoptera: Eupelmidae) an egg parasite of Saturniidae (Lepidoptera). *Phytophylactica* 3, 85–88.

van den Bosch, R. (1968). Comments on population dynamics of exotic insects. *Bull. Entomol. Soc. Amer.* 14, 112–15.

van den Bosch, R. and Messenger, P.S. (1973). *Biological control.* Intext Press, New York.

van den Bosch, R., Schlinger, E.I., Dietrick, E.J., and Hall, I.M. (1959). The role of imported parasites in the biological control of spotted alfalfa aphid in southern California in 1957. *J. Econ. Entomol.* 52, 142–57.

van den Bosch, R., Schlinger, E.I., Dietrick, E.J., Hall, J.C., and Puttler, B. (1964). Studies on succession, distribution, and phenology of imported parasites of *Therioaphis trifolii* (Monell) in southern California. *Ecology* 45, 602–21.

van den Bosch, R., Hom, R., Matteson, P., Frazer, B.D., Messenger, P.S., and Davis, C.S. (1979). Biological control of the walnut aphid in California: Impact of the parasite, *Trioxys pallidus. Hilgardia* 47, 1–13.

van der Laak, S. (1982). Physiological adaptations to low temperature in freezing-tolerant *Phyllodecta laticollis* beetles. *Comp. Biochem. Physiol.* 73A, 613–20.

van Dinther, J.B.M. (1962). Flight periods of the white rice borer *Rupela albinella* (Cr.) in Wageningen, Surinam (South America). *Meded. LandbHoogesch. OpzoekStns. Gent.* 27, 829–36.

Van Kirk, J.R. and AliNiazee, M.T. (1981). Determining low-temperature threshold for pupal development of the western cherry fruit fly for use in phenology models. *Environ. Entomol.* 10, 968–71.

Van Kirk, J.R. and AliNiazee, M.T. (1982). Diapause development in the western cherry fruit fly, *Rhagoletis indifferens* Curran (Diptera, Tephritidae). *Z. Ang. Entomol.* 93, 440–45.

Vaz Nunes, M. and Veerman, A. (1982). Photoperiodic time measurement in the spider mite *Tetranychus urticae*: A novel concept. *J. Insect Physiol.* 28, 1041–53.

Veerman, A. (1977a). Aspects of the induction of diapause in a laboratory strain of the mite *Tetranychus urticae. J. Insect Physiol.* 23, 703–11.

Veerman, A. (1977b). Photoperiodic termination of diapause in spider mites. *Nature* 266, 526–27.

Veerman, A. (1980). Functional involvement of carotenoids in photoperiodic induction of diapause in the spider mite, *Tetranychus urticae. Physiol. Entomol.* 5, 291–300.

Veerman, A. and Herrebout, W.M. (1982). Photoperiodic response curve for *Yponomeuta vigintipunctatus* (Retz.) (Lepidoptera: Yponomeutidae). *Netherlands J. Zool.* 32, 117–22.

Veerman, A., Overmeer, W.P.J., van Zon, A.Q., de Boer, J.M., de Waard, E.R., and Huisman, H.O. (1983). Vitamin A is essential for photoperiodic induction of diapause in an eyeless mite. *Nature* 302, 248–49.

Vepsäläinen, K. (1971). The role of gradually changing daylength in determination of wing length, alary dimorphism and diapause in a *Gerris odontogaster* (Zett.) population (Gerridae, Heteroptera) in South Finland. *Ann. Acad. Sci. Fenn. A, IV*, 183, 1–25.

Vepsäläinen, K. (1974a). Determination of wing length and diapause in water-striders (*Gerris* Fabr., Heteroptera). *Hereditas* 77, 163–75.

Vepsäläinen, K. (1974b). Lengthening of illumination period is a factor in averting diapause. *Nature* 247, 385–86.

Vepsäläinen, K. (1978). Wing dimorphism and diapause in *Gerris*: Determination and adaptive significance. In *Evolution of insect migration and diapause* (ed. H. Dingle) pp. 218–53. Springer-Verlag, New York.

Villavaso, E.J. and Earle, N.W. (1974). Attraction of female boll weevils to diapausing and reproducing males. *J. Econ. Entomol.* 67, 171–72.

Vinogradova, E.B. (1958). The photoperiodic reaction in the malaria mosquito (*Anopheles maculipennis messeae* Fall.). *Sci. Mem. Leningrad State Univ.* 240, 52–60 (in Russian) (through Danilevsky 1965).

Vinogradova, E.B. (1960). An experimental investigation of the ecological factors inducing imaginal diapause in bloodsucking mosquitoes (Diptera, Culicidae). *Entomol. Rev.* 39, 210–19.

Vinogradova, E.B. (1961). The biological isolation of subspecies of *Culex pipiens* L. (Diptera, Culicidae). *Entomol. Rev.* 40, 29–35.

Vinogradova, E.B. (1974). The pattern of reactivation of diapausing larvae in the blowfly, *Calliphora vicina*. *J. Insect Physiol.* 20, 2487–96.

Vinogradova, E.B. (1975a). Photoperiodic and temperature induction of egg diapause in *Aedes caspius caspius* Pall. (Diptera, Culicidae). *Akad. Nauk. SSSR, Parazitologiia* 9, 385–91 (in Russian, English summary).

Vinogradova, E.B. (1975b). Intraspecific variability of reactions controlling the larval diapause in *Calliphora vicina* R.-D. (Diptera, Calliphoridae). *Rev. d' Entomol. de l' URSS* 4, 720–35 (in Russian, English summary).

Vinogradova, E.B. (1976a). The effect of changes of the photoperiodic regime during the life of adult blow flies, *Calliphora vicina* (Diptera, Calliphoridae), on the induction of larval diapause in their progeny. *Entomol. Rev.* 55, 35–40.

Vinogradova, E.B. (1976b). Embryonic photoperiodic sensitivity in two species of fleshflies, *Parasarcophaga similis* and *Boettcherisca septentrionalis*. *J. Insect Physiol.* 22, 819–22.

Vinogradova, E.B. (1978). Effect of gradual change of daylength on induction of pupal diapause in onion fly *Hylemyia antigua* MG. (Diptera, Anthomyidae). *Dokl. Akad. Nauk. SSSR* 242, 423–25 (English translation).

Vinogradova, E.B. and Bogdanova, T.P. (1980). Endogenous cyclic changes in the incidence of diapause in blow-flies and flesh-flies (Diptera) under continuous laboratory rearing at constant conditions. *Rev. d'Entomol. l'URSS* 59, 26–38 (in Russian, English summary).

Vinogradova, E.B. and Tsutskova, I.P. (1978). The inheritance of larval diapause in the crossing of intraspecific forms of *Calliphora vicina* (Diptera, Calliphoridae). *Entomol. Rev.* 57, 168–76.

Vinogradova, E.B. and Zinovyeva, K.B. (1972a). Experimental investigation of the seasonal aspect of the relationship between blowflies and their parasites. *J. Insect Physiol.* 18, 1629–38.

Vinogradova, E.B. and Zinovyeva, K.B. (1972b). The control of seasonal development in parasites of blow flies. I. Ecological control of pupal diapause in Sarcophagidae (Diptera). In *Host-parasite relations in insects* (ed. V.A. Zaslavsky) pp. 77–89. Publishing House Nauka, Leningrad (in Russian, English summary).

Vinogradova, E.B. and Zinovyeva, K.B. (1972c). The control of seasonal development in parasites of blow flies. IV. Patterns of photoperiodic reaction in *Alysia manducator* Panz. (Hymenoptera, Braconidae). In *Host-parasite relations in insects* (ed. V.A. Zaslavsky) pp. 112–17. Publishing House Nauka, Leningrad (in Russian, English summary).

Vinogradova, E.B. and Zinovyeva, K.B. (1972d). Maternal induction of larval diapause in the blowfly, *Calliphora vicina*. *J. Insect Physiol.* 18, 2401–9.

Vinson, S.B. and Iwantsch, G.F. (1980). Host regulation by insect parasitoids. *Q. Rev. Biol.* 55, 143–65.

Visscher, S.N. (1976). The embryonic diapause of *Aulocara elliotti* (Orthoptera, Acrididae). *Cell. Tiss. Res.* 174, 433–52.

Visscher, S.N. (1980). Regulation of grasshopper fecundity, longevity and egg viability by plant growth hormones. *Experientia* 36, 130–31.

Visscher, S.N., Lund, R., and Whitmore, W. (1979). Host plant growth temperatures and insect rearing temperatures influence reproduction and longevity in the grasshopper, *Aulocara elliotti* (Orthoptera: Acrididae). *Environ. Entomol.* 8, 253–58.

Volney, W.J.A., Waters, W.E., Akers, R.P., and Liebhold, A.M. (1983). Variation in spring emergence patterns among western *Choristoneura* spp. (Lepidoptera: Tortricidae) populations in southern Oregon. *Can. Entomol.* 115, 199–209.

von Kaster, L. and Showers, W.B. (1982). Evidence of spring immigration and autumn reproductive diapause of the adult black cutworm in Iowa. *Environ. Entomol.* 11, 306–12.

Wada, T., Kobayashi, M., and Shimazu, M. (1980). Seasonal changes of the proportions of mated females in the field population of the rice leaf roller, *Cnaphalocrocis medinalis* Guenée (Lepidoptera: Pyralidae). *Appl. Entomol. Zool.* 15, 81–89.

Wagner, T.L., Wu, H.-I., Sharpe, P.J.H., Schoolfield, R.M., and Coulson, R.N. (1984). Modeling insect development rates: A literature review and application of a biophysical model. *Ann. Entomol. Soc. Amer.* 77, 208–25.

Waldbauer, G.P. (1978). Phenological adaptation and the polymodal emergence patterns of insects. In *Evolution of insect migration and diapause* (ed. H. Dingle) pp. 127–44. Springer-Verlag, New York.

Waldbauer, G.P. and Kogan, M. (1976). Bean leaf beetle: Phenological relationship with soybean in Illinois. *Environ. Entomol.* 5, 35–44.

Waldbauer, G.P. and Sheldon, J.K. (1971). Phenological relationship of some aculeate Hymenoptera, their dipteran mimics, and insectivorous birds. *Evolution* 25, 371–82.

Waldbauer, G.P. and Sternburg, J.G. (1973). Polymorphic termination of diapause by cecropia: Genetic and geographical aspects. *Biol. Bull.* 145, 627–41.

Waldbauer, G.P. and Sternburg, J.G. (1978). The bimodal termination of diapause in the laboratory by *Hyalophora cecropia. Entomol. Exp. Appl.* 23, 121–30.

Waldbauer, G.P., Sternburg, J.G., and Maier, C.T. (1977). Phenological relationships of wasps, bumblebees, their mimics, and insectivorous birds in an Illinois sand area. *Ecology* 58, 583–91.

Waldbauer, G.P., Sternburg, J.G., and Wilson, G.R. (1978). The effect of injections of β-ecdysone on the bimodal emergence of *Hyalophora cecropia. J. Insect Physiol.* 24, 623–27.

Walker, G.P. and Denlinger, D.L. (1980). Juvenile hormone and moulting hormone titres in diapause- and non-diapause destined flesh flies. *J. Insect Physiol.* 26, 661–64.

Walker, T.J. (1974). *Gryllus ovisopis* n. sp.: A taciturn cricket with a life cycle suggesting allochronic speciation. *Fla. Entomol.* 57, 13–22.

Walker, T.J. (1980). Mixed oviposition in individual females of *Gryllus firmus*: graded proportions of fast-developing and diapause eggs. *Oecologia* 47, 291–98.

Walker, T.J., Reinert, J.A., and Schuster, D.J. (1983). Geographical variation in flights of the mole cricket, *Scapteriscus* spp. (Orthoptera: Gryllotalpidae). *Ann. Entomol. Soc. Amer.* 76, 507–17.

Wall, C. (1974). Effect of temperature on embryonic development and diapause in *Chesias legatella. J. Zool. Lond.* 172, 147–68.

Wallace, B. (1985). Reflections on the still-"hopeful monster." *Quart. Rev. Biol.* 60, 31–42.

Wallace, D.R. and Sullivan, C.R. (1963). Laboratory and field investigations on the effect of temperature on the development of *Neodiprion sertifer* (Geoff.) in the cocoon. *Can. Entomol.* 95, 1051–66.

Wallace, D.R. and Sullivan, C.R. (1974). Photoperiodism in the early balsam strain of the *Neodiprion abietis* complex (Hymenoptera: Diprionidae). *Can. J. Zool.* 52, 507–13.

Wallace, M.M.H. (1968). The ecology of *Sminthurus viridus* (Collembola) II. Diapause in the aestivating egg. *Aust. J. Zool.* 16, 871–83.

Wallace, M.M.H. (1970a). Diapause in the aestivating egg of *Halotydeus destructor* (Acari: Eupodidae). *Aust. J. Zool.* 18, 295–313.

Wallace, M.M.H. (1970b). The influence of temperature on post-diapause development and survival in the aestivating eggs of *Halotydeus destructor* (Acari: Eupodidae). *Aust. J. Zool.* 18, 315–29.

Wallace, M.M.H. (1971). The influence of temperature and moisture on diapause development in the eggs of *Bdellodes lapidaria. J. Aust. Entomol. Soc.* 10, 276–80.

Wallner, W.E. (1979). Induction of diapause in *Rogas indiscretus*, a larval parasite of the gypsy moth, *Lymantria dispar. Ann. Entomol. Soc. Amer.* 72, 358–60.

Waloff, N. (1949). Observations on larvae of *Ephestia elutella* Hübner (Lep. Phycitidae) during diapause. *Trans. R. Entomol. Soc. Lond.* 100, 147–59.

Waloff, Z. (1966). The upsurges and recessions of the desert locust plague: An historical survey. *Anti-locust Memoir* 8, 1–111.

Ward, S.A., Leather, S.R., and Dixon, A.F.G. (1984). Temperature prediction and the timing of sex in aphids. *Oecologia* 62, 230–33.

Wardhaugh, K.G. (1970). The development of eggs of the Australian plague locust, *Chortoicetes terminifera* (Walk.), in relation to temperature and moisture. *Proc. Intl. Study Conf. Current and Future Problems of Acridology* (eds. C.F. Hemming and T.H.C. Taylor) pp. 261–74. Centre for Overseas Pest Research, London.

Wardhaugh, K.G. (1977). The effects of temperature and photoperiod on the morphology of the egg-pod of the Australian plague locust (*Chortoicetes terminifera* Walker, Orthoptera: Acrididae). *Aust. J. Ecol.* 2, 81–88.

Wardhaugh, K.G. (1980a). The effects of temperature and photoperiod on the induction of diapause in eggs of the Australian plague locust, *Chortoicetes terminifera* (Walker) (Orthoptera: Acrididae). *Bull. Entomol. Res.* 70, 635–47.

Wardhaugh, K.G. (1980b). The effects of temperature and moisture on the inception of diapause in eggs of the Australian plague locust, *Chortoicetes terminifera* Walker (Orthoptera: Acrididae). *Aust. J. Ecol.* 5, 187–91.

Washino, R.K. (1970). Physiological condition of overwintering female *Anopheles freeborni* in California (Diptera: Culicidae). *Ann. Entomol. Soc. Amer.* 63, 210–16.

Washino, R.K. (1977). The physiological ecology of gonotrophic dissociation and related phenomena in mosquitoes. *J. Med. Entomol.* 13, 381–88.

Washino, R.K. and Bailey, S.F. (1970). Overwintering of *Anopheles punctipennis* (Diptera: Culicidae) in California. *J. Med. Entomol.* 7, 95–98.

Washino, R.K., Gieke, P.A., and Schaefer, C.H. (1971). Physiological changes in the overwintering females of *Anopheles freeborni* (Diptera: Culicidae) in California. *J. Med. Entomol.* 8, 279–82.

Watanabe, N. (1978). An improved method for computing heat accumulation from daily maximum and minimum temperatures. *Appl. Entomol. Zool.* 13, 44–46.

Waterhouse, D.F. and Norris, K.R. (1980). Insects and insect physiology in the scheme of things. In *Insect biology in the future* (eds. M. Locke and D.S. Smith), pp. 19–37. Academic Press, New York.

Watson, J.A.L. (1974). Caste development and its seasonal cycle in the Australian harvester termine, *Drepanotermes perniger* (Froggatt) (Isoptera: Termitinae). *Aust. J. Zool.* 22, 471–87.

Watson, T.F., Crowder, L.A., and Langston, D.T. (1974). Geographical variation of diapause termination of the pink bollworm. *Environ. Entomol.* 3, 933–34.

Way, M.J. (1960). The effects of freezing temperatures on the developing egg of *Leptohylemyia coarctata* Fall. (Diptera, Muscidae) with special reference to diapause development. *J. Insect Physiol.* 4, 92–101.

Weeda, E. (1981). Hormonal regulation of proline synthesis and glucose release in the fat body of the Colorado potato beetle, *Leptinotarsa decemlineata. J. Insect Physiol.* 27, 411–17.

Weeda, E., de Kort, C.A.D., and Beenakkers, A.M.T. (1979). Fuels for energy metabolism in the Colorado potato beetle, *Leptinotarsa decemlineata* Say. *J. Insect Physiol.* 25, 951–55.

Weeda, E., Koopmanschap, A.B., de Kort, C.A.D., and Beenakkers, A.M.T. (1980). Proline synthesis in fat body of *Leptinotarsa decemlineata. Insect Biochem.* 10, 631–36.

Wellings, P.W., Leather, S.R., and Dixon, A.F.G. (1980). Seasonal variation in reproductive potential: A programmed feature of aphid life cycles. *J. Anim. Ecol.* 49, 975–85.

Wellington, W.G. (1983). Biometeorology of dispersal. *Bull. Entomol. Soc. Amer.* 29, 24–29.

Wellington, W.G. and Trimble, R.M. (1984). Weather. In *Ecological entomology* (eds. C.B. Huffaker and R.L. Rabb) pp. 399–425. John Wiley & Sons, New York.

Wellso, S.G. and Adkisson, P.L. (1964). Photoperiod and moisture as factors involved in the termination of diapause in the pink bollworm, *Pectinophora gossypiella. Ann. Entomol. Soc. Amer.* 57, 170–73.

Wellso, S.G. and Adkisson, P.L. (1966). A long-day short-day effect in the photoperiodic control of the pupal diapause of the bollworm, *Heliothis zea* (Boddie). *J. Insect Physiol.* 12, 1455–65.

Wellso, S.G. and Hoxie, R.P. (1981). Diapause and nondiapause behavior of the cereal leaf beetle. *Entomol. Exp. Appl.* 30, 19–25.

Wellso, S.G., Connin, R.V., Hoxie, R.P., and Cobb, D.L. (1970). Storage and behavior of plant and diet-fed adult cereal leaf beetle, *Oulema melanoplus* (Coleoptera: Chrysomelidae). *Mich. Entomol.* 3, 101–7.

Weseloh, R.M. (1973). Termination and induction of diapause in the gypsy moth larval parasitoid, *Apanteles melanoscelus. J. Insect Physiol.* 19, 2025–33.

Weseloh, R.M. (1982). Comparison of U.S. and Indian strains of the gypsy moth (Lepidoptera: Lymantriidae) parasitoid *Apanteles melanoscelus* (Hymenoptera: Braconidae). *Ann. Entomol. Soc. Amer.* 75, 563–67.

Weseloh, R.M. (1984). Effect of size, stress, and ligation of gypsy moth (Lepidoptera: Lymantriidae) larvae on development of the tachinid parasite *Compsilura concinnata* Meigen (Diptera: Tachinidae). *Ann. Entomol. Soc. Amer.* 77, 423–28.

West, D.A., Snellings, W.M. and Herbek, T.A. (1972). Pupal color dimorphism and its environmental control in *Papilio polyxenes asterius* Stoll (Lepidoptera: Papilionidae). *J. N. Y. Entomol. Soc.* 80, 205–11.

West-Eberhard, M.J. (1978). Polygyny and the evolution of social behavior in wasps. *J. Kans. Entomol. Soc.* 51, 832–56.

West-Eberhard, M.J. (1982). The nature and evolution of swarming in tropical social wasps (Vespidae, Polistinae, Polybiini). In *Social insects in the tropics*, Vol. 1 (ed. P. Jaisson) pp. 97–128. Université Paris-Nord.

Wheatley, G.A. and Dunn, J.A. (1962). The influence of diapause on the time of emergence of the pea moth *Laspeyresia nigricana* (Steph.). *Ann. Appl. Biol.* 50, 609–11.

Wheeler, W.M. (1893). Contribution to insect embryology. *J. Morph.* 8, 141–60.

Whitcomb, R.F., Kramer, J.P., and Coan, M.E. (1972). *Stirellus bicolor* and *S. obtutus* (Homoptera: Cicadellidae): Winter and summer forms of a single species. *Ann. Entomol. Soc. Amer.* 65, 797–98.

White, E.B., DeBach, P., and Garber, M.J. (1970). Artificial selection for genetic adaptation to temperature extremes in *Aphytis lingnanensis* Compere (Hymenoptera: Aphelinidae). *Hilgardia* 40, 161–92.

White, J.A. and Lloyd, M. (1975). Growth rates of 17– and 13–year periodical cicadas. *Amer. Midl. Natur.* 94, 127–43.

White, M.J.D. (1978). *Modes of speciation.* Freeman, San Francisco.

Whitten, M.J. and Foster, G.G. (1975). Genetical methods of pest control. *Annu. Rev. Entomol.* 20, 461–76.

Wielgolaski, F.E. (1974). Phenological studies in tundra. In *Phenology and seasonality modeling* (ed. H. Lieth) pp. 209–14. Springer-Verlag, Berlin.

Wigglesworth, V.B. (1970). *Insect hormones.* Oliver & Boyd, Edinburgh.

Wigglesworth, V.B. (1972). *The principles of insect physiology* (7th ed.) Methuen, London.

Wiklund, C. (1975). Pupal colour polymorphism in *Papilio machaon* L. and the survival in the field of cryptic versus non-cryptic pupae. *Trans. Roy. Entomol. Soc. Lond.* 127, 73–84.

Wiklund, G., Persson, A., and Wickman, P.O. (1983). Larval aestivation and direct development as alternative strategies in the speckled wood butterfly, *Pararge aegeria*, in Sweden. *Ecol. Entomol.* 8, 233–38.

Wildbolz, T. and Riggenbach, W. (1969). Untersuchungen über die Induktion und die Beendigung de Diapause bei Apfelwicklern aus der Zentral- und Ostschweiz. *Mitt. Schweiz. Entomol. Ges.* 42, 58–78.

Wilkinson, P.R. (1968). Phenology, behavior, and host-relations of *Dermacentor andersoni* Stiles in an outdoor "rodentaria" and in nature. *Can. J. Zool.* 46, 677–89.

Willey, R.L., Bowen, W.R., and Durban, E. (1970). Symbiosis between *Euglena* and damselfly nymphs is seasonal. *Science* 170, 80–81.

Williams, C.M. (1952). Physiology of insect diapause. IV. The brain and prothoracic glands as an endocrine system in the Cecropia silkworm. *Biol. Bull.* 103, 120–38.

Williams, C.M. (1956). Physiology of insect diapause. X. An endocrine mechanism for the influence of temperature on the diapausing pupa of the Cecropia silkworm. *Biol. Bull.* 110, 201-18.

Williams, C.M. and Adkisson, P.L. (1964). Physiology of insect diapause. XIV. An endocrine mechanism for the photoperiodic control of pupal diapause in the oak silkworm, *Antheraea pernyi. Biol. Bull.* 127, 511-25.

Willmer, P.G. (1982). Microclimate and the environmental physiology of insects. In *Advances in insect physiology*, Vol. 16 (eds. M.J. Berridge, J.E. Treherne, and V.B. Wigglesworth) pp. 1–57. Academic Press, London.

Wilson, A.G.L., Lewis, T., and Cunningham, R.B. (1979). Overwintering and spring emergence of *Heliothis armigera* (Hübner) (Lepidoptera: Noctuidae) in the Namoi Valley, New South Wales. *Bull. Entomol. Res.* 69, 97–109.

Wilson, E.O. (1971). *The insect societies.* Belknap Press, Cambridge, MA.

Wilson, G.R. and Horsfall, W.R. (1970). Eggs of floodwater mosquitoes XII. Installment hatching of *Aedes vexans* (Diptera: Culicidae). *Ann. Entomol. Soc. Amer.* 63, 1644–47.

Wise, E.J. (1980). Seasonal distribution and life histories of Ephemeroptera in a Northumbrian river. *Freshwat. Biol.* 10, 101–11.

Wolda, H. (1980). Seasonality of tropical insects. I. Leafhoppers (Homoptera) in Las Cumbres, Panama. *J. Anim. Ecol.* 49, 277–90.

Wolda, H. (1982). Seasonality of Homoptera on Barro Colorado. In *The ecology of a tropical forest* (eds. E.G. Leigh, Jr., A.S. Rand, and D.M. Windsor) pp. 319–30. Smithsonian Institution Press, Washington, DC.

Wolda, H. (1983). Spatial and temporal variation in abundance in tropical animals. In *The tropical rainforest: Ecology and management* (eds. S.L. Sutton, T.C. Whitmore, and A.C. Chadwick) pp. 93–105. Blackwell Scientific Publ., Oxford, U.K.

Wolda, H. and Denlinger, D.L. (1984). Diapause in a large aggregation of a tropical beetle. *Ecol. Entomol.* 9, 217–30.

Wolda, H. and Galindo, P. (1981). Population fluctuations of mosquitoes in the non-seasonal tropics. *Ecol. Entomol.* 6, 99–106.

Wolfenbarger, D.A., Graham, H.M., Parker, R.D., and Davis, J.W. (1976). Boll weevil: Seasonal patterns of response to traps baited with grandlure in the lower Rio Grande Valley. *Environ. Entomol.* 5, 403–8.

Wood, T.K. (1980). Divergence in the *Enchenopa binotata* Say complex (Homoptera: Membracidae) effected by host plant adaptation. *Evolution* 34, 147–60.

Wood, T.K. and Guttman, S.I. (1981). The role of host plants in the speciation of treehoppers: An example from the *Enchenopa binotata* complex. In *Insect life history patterns* (eds. R.F. Denno and H. Dingle) pp. 39–54. Springer-Verlag, New York.

Wood, T.K. and Guttman, S.I. (1982). Ecological and behavioral basis for reproductive isolation in the sympatric *Enchenopa binotata* complex (Homoptera. Membracidae). *Evolution* 36, 233–42.

Wood, T.K. and Olmstead, K.L. (1984). Latitudinal effects on treehopper species richness (Homoptera: Membracidae). *Ecol. Entomol.* 9, 109–15.

Wu, C.-I. (1985). A stochastic simulation study on speciation by sexual selection. *Evolution* 39, 66–82.

Wylie, H.G. (1977). Preventing and terminating pupal diapause in *Athrycia cinerea* (Diptera: Tachinidae). *Can. Entomol.* 109, 1083–90.

Wylie, H.G. (1980). Factors affecting facultative diapause of *Microctonus vittatae* (Hymenoptera: Braconidae). *Can. Entomol.* 112, 747–49.

Wylie, H.G. (1982). An effect of parasitism by *Microctonus vittatae* (Hymenoptera: Braconidae) on emergence of *Phyllotreta cruciferae* and *Phyllotreta striolata* (Coleoptera: Chrysomelidae) from overwintering sites. *Can. Entomol.* 114, 727–32.

Yagi, S. (1975). Endocrinological studies on diapause in some lepidopterous insects. *Mem. Fac. Agric. Tokyo Univ. Educ.* 21, 1–50 (in Japanese, English summary).

Yagi, S. (1981). Physiological aspects of diapause in rice stem borers and the effect of juvenile hormone (Lepidoptera: Pyralidae). *Entomol. Gen.* 7, 213–21.

Yagi, S. and Fukaya, M. (1974). Juvenile hormone as a key factor regulating larval diapause of the rice stem borer, *Chilo suppressalis* (Lepidoptera: Pyralidae). *Appl. Entomol. Zool.* 9, 247–55.

Yaginuma, T. and Yamashita, O. (1978). Polyol metabolism related to diapause in *Bombyx* eggs: Different behaviour of sorbitol from glycerol during diapause and post-diapause. *J. Insect Physiol.* 24, 347–54.

Yaginuma, T. and Yamashita, O. (1979). NAD-dependent sorbitol dehydrogenase activity in relation to the termination of diapause in eggs of *Bombyx mori. Insect Biochem.* 9, 547–53.

Yaginuma, T. and Yamashita, O. (1980). The origin of free glycerol accumulated in diapause eggs of *Bombyx mori. Physiol. Entomol.* 5, 93–97.

Yamashita, O. (1983). Egg diapause. In *Endocrinology of insects* (eds. R.G.H. Downer and H. Laufer) pp. 337–42. Alan R. Liss, New York.

Yamashita, O. and Hasegawa, K. (1970). Oöcyte age sensitive to the diapause hormone from the standpoint of glycogen synthesis in the silkworm, *Bombyx mori. J. Insect Physiol.* 16, 2377–83.

Yamashita, O. and Hasegawa, K. (1976). Diapause hormone action in silkworm ovaries incubated *in vitro*; ^{14}C-trehalose incorporation into glycogen. *J. Insect Physiol.* 22, 409–14.

Yamashita, O., Suzuki, K., and Hasegawa, K. (1975). Glycogen phosphorylase activity in relation to diapause initiation in *Bombyx* eggs. *Insect Biochem.* 5, 707–18.

Yamashita, O., Isobe, M., Imai, K., Kondo, N., and Goto, T. (1980). Serum albumin as an effective carrier for diapause hormone of the silkworm, *Bombyx mori* L. (Lepidoptera: Bombycidae). *Appl. Entomol. Zool.* 15, 90–95.

Yamashita, O., Yaginuma, T., and Hasegawa, K. (1981). Hormonal and metabolic control of egg diapause of the silkworm, *Bombyx mori* (Lepidoptera: Bombycidae). *Entomol. Gen.* 7, 195–211.

Yin, C.-M. and Chippendale, G.M. (1973). Juvenile hormone regulation of the larval diapause of the southwestern corn borer *Diatraea grandiosella. J. Insect Physiol.* 19, 2403–20.

Yin, C.-M. and Chippendale, G.M. (1974). Juvenile hormone and the induction of larval polymorphism and diapause of the southwestern corn borer, *Diatraea grandiosella. J. Insect Physiol.* 20, 1833–47.

Yin, C.-M. and Chippendale, G.M. (1976). Hormonal control of larval diapause and metamorphosis of the southwestern corn borer *Diatraea grandiosella. J. Exp. Biol.* 64, 303–10.

Yin, C.-M. and Chippendale, G.M. (1979a). Diapause of the southwestern corn borer, *Diatraea grandiosella*: Further evidence showing juvenile hormone to be the regulator. *J. Insect Physiol.* 25, 513–23.

Yin, C.-M. and Chippendale, G.M. (1979b). Ultrastructural characteristics of insect corpora allata in relation to larval diapause. *Cell Tissue Res.* 197, 453–61.

Young, A.M. (1982). *Population biology of tropical insects.* Plenum Press, New York.

Young, A.M. and Thomason, J.H. (1974). The demography of a confined population of the butterfly *Morpho peleides* during a tropical dry season. *Studies Neotrop. Fauna* 9, 1–34.

Young (Krieger), D.L. (1972). Photoperiod and wing production by the aphid *Dactynotus ambrosiae* on the short day plant *Xanthium pensylvanicum. Physiol. Zool.* 45, 60–67.

Young, S.R. and Block, W. (1980). Experimental studies on the cold tolerance of *Alaskozetes antarcticus. J. Insect Physiol.* 26, 189–200.

Zachariassen, K.E. (1979). The mechanism of the cryoprotective effect of glycerol in beetles tolerant to freezing. *J. Insect Physiol.* 25, 29–32.

Zachariassen, K.E. (1982). Nucleating agents in cold-hardy insects. *Comp. Biochem. Physiol.* 73A, 557–62.

Zachariassen, K.E. and Husby, J.A. (1982). Antifreeze effect of thermal hysteresis agents protects highly supercooled insects. *Nature* 298, 865–67.

Zar, J.H. (1968). The fatty acid composition of the ladybird beetle, *Coleomegilla maculata* (DeGeer) during hibernation. *Comp. Biochem. Physiol.* 26, 1127–29.

Zaslavsky, V.A. (1972). Two-stage photoperiodic reactions as a starting point for the production of a model of the photoperiodic control of arthropod development. *Entomol. Rev.* 51, 133–45.

Zaslavsky, V.A. (1975). The occurrence of changing photoperiodic reactions among insects, mites and ticks. *Entomol. Rev.* 54, 36–45.

Zaslavsky, V.A. and Fomenko, R.B. (1973). Possibility of a common nature of the physiological mechanism of group effect and photoperiodic responses in *Chilocorus bipustulatus* L. (Coleoptera, Coccinelidae). *Dokl. Akad. Nauk. SSSR* 212, 1472–74 (English translation).

Zaslavsky, V.A. and Umarova, T.Y. (1981). Photoperiodic and temperature control of diapause in *Trichogramma evanescens* Westw. (Hymenoptera, Trichogrammatidae). *Entomol. Rev.* 60, 1–12.

Zeleny, J. (1961). Contribution to the knowledge of diapause in insects. 6. Influence of the photoperiod and temperature on the induction of diapause in host and parasite. *Acta Soc. Zool. Bohemoslov.* 25, 258–70.

Zinovyeva, K.B. (1972). The effects of ecological conditions and host's physiological state on diapause induction in *Perilissus lutescens* Hlmgr. (Hymenoptera, Ichneumonidae) and in *Meigenia bisignata* Meig. (Diptera, Tachinidae) parasites of the cabbage sawfly *Athalia rosae* L. (Hymenoptera, Tenthredinidae). In *Host-parasite relations in insects* (ed. V.A. Zaslavsky) pp. 118–27. Publishing House Nauka, Leningrad.

Zinovyeva, K.B. (1974). Stimulation of pupation of larvae of *Calliphora vicina* R.-D. (Diptera, Calliphoridae) by the parasite *Aphaceta minuta* Nees (Hymenoptera: Braconidae). *Dokl. Akad. Nauk. SSSR* 216, 306–7 (English translation).

Zinovyeva, K.B. (1976). The role of light and temperature rhythms in diapause induction in *Alysia manducator* Panz. (Hymenoptera, Braconidae). *Entomol. Rev.* 55, 6–11.

Zinovyeva, K.B. (1980). Inheritance of larval diapause in the crossing of two geographic forms of *Calliphora vicina* R.-D. (Diptera, Calliphoridae). *Entomol. Rev.* 59, 9–20.

Zucker, I., Johnston, P.G., and Frost, D. (1980). Comparative, physiological and biochronometric analyses of rodent seasonal reproductive cycles. *Prog. Reprod. Biol.* 5, 102–33.

Author Index

Abdinbekova, A. A., 125
Abdul Nabi, A. A., 74
Abo-Ghalia, A., 63
Abou-Elela, R., 41
Ackerman, J. D., 172
Ackerman, T. L., 187
Adedokun, T. A., 49, 96, 101–2
Adesiyun, A. A., 13, 268
Adkisson, P. L., 54, 73, 80, 117, 119, 131, 139, 155, 159, 180, 222, 234, 237, 241–42, 298–301
Adler, V. E., 114, 302
Ae, S. A., 180, 189
Aeschlimann, J., 237
Aida, S., 108
Akers, R. P., 149, 216
Akhmedov, R. M., 125
Akingbohungbe, A. E., 299
Akinlosotu, T. A., 180
Akkawi, M. M., 267
Albrecht, F. O., 17
Alden, B. M., 18, 180, 182, 203, 252, 273
Alexander, R. D., 248, 258, 283–85
Ali, M., 155
Ali, M. A., 95
AliNiazee, M. T., 142, 149, 290
Allemand, R., 215
Almeida, L. C., 306
Alrouechdi, K., 253
Alves, S. B., 306
Amir, M., 171
Anderson, J. F., 44, 131, 164, 166–67
Anderson, T. E., 149, 243–44, 289
Ando, Y., 19, 50, 54–56, 105, 143, 145, 147, 158–59, 180, 194, 203, 208, 232, 236–37
Andres, L. A., 305
Andrewartha, H. G., 4, 21, 26, 41, 56, 77, 103–4, 142, 154, 163, 203, 289
Andries, J. C., 75
Ankersmit, G. W., 180, 222, 241–42, 281, 302
Annecke, D. P., 162–63, 165–66, 305
Annila, E., 198
Annis, B., 24, 36, 199
Apple, J. W., 5, 60, 203, 244, 289–90, 296
Arai, T., 63
Arakawa, R., 89, 165, 197
Arbuthnot, K. D., 211, 233, 244
Armbrust, E. J., 89
Arnold, C. Y., 290
Arntfield, P. W., 84
Arora, G., 18, 23, 182
Asahina, E., 60, 96–99, 101
Ascerno, M. E., 39, 70, 85, 94, 169
Askew, R. R., 162, 165–66
Atwood, C. E., 283, 285
Awiti, L. R. S., 19
Azaryan, A. G., 65, 203, 232

Baard, G., 68, 70, 81
Bacheler, J. S., 290
Baehr, J.-C., 75
Baerwald, R. J., 235
Bährmann, R., 232
Bailey, C. L., 84
Bailey, S. F., 83–85
Bajusz, B. A., 290
Baker, C. R. B., 149, 203, 290
Baker, E. W., 39
Baker, G. L., 172
Baker, H. G., 178
Baker, J. E., 136, 241
Baldwin, D., 180, 182
Bale, J. S., 94, 132, 194
Balling, S. S., 184
Bamberg, S. A., 187
Banks, W. A., 172
Barbosa, P., 105
Barker, J. F., 88, 96
Barnard, D. R., 84, 95
Barnes, H. F., 198
Barnes, J. K., 39, 85, 93
Barnes, M. M., 240, 283
Baronio, P., 168–69

Barry, B. D., 211, 234, 237
Bartell, D. P., 301
Bartelt, R. J., 143
Bartholomew, G. A., 248
Bartlett, B. R., 306
Barton, N. H., 280
Basedow, T., 159
Baskerville, G. L., 290
Batten, A., 16
Baust, J. G., 96–98, 100, 102, 186–87
Baxendale, F. P., 160, 198
Beach, R. F., 44, 203, 302
Bean, D. W., 75–77
Beards, G. W., 57
Beatley, J. C., 187
Beck, S. D., 4–5, 39, 55–57, 75–77, 80, 94, 96, 112–16, 127, 133, 138, 140, 143, 159, 166, 203, 210, 220, 222–26, 243–44, 250, 289–90, 296
Beckage, N. E., 162, 164, 168–69
Beenakkers, A. M. T., 93
Begnaud, J. E., 149, 289
Begon, M., 23, 82, 215
Behrendt, K., 102
Beingolea, G., 298
Bell, C. H., 102, 136, 156, 180, 241
Bell, K. O., 95, 152
Bell, R. A., 54, 104, 155, 210–11, 242, 303
Bell, W. J., 171
Bellamy, R. E., 84–85, 89
Belozerov, V. N., 41, 83, 85, 112–13
Bengston, M., 114
Bennett, F. A., 305
Benschoter, C. A., 233
Benz, G., 54, 75–76, 142
Beppu, K., 215
Berg, C. O., 159, 261
Bergot, B. J., 70
Bergmann, E. D., 174
Berlinger, M. J., 302
Berry, R. E., 24
Berry, S. J., 80
Berryman, A. A., 157
Bevan, D., 24
Bewley, J. D., 21
Bier, K., 176
Bigelow, R. S., 258, 283–85
Birch, L. C., 4, 26, 41, 56, 59, 82, 103–4, 145, 167
Black, M., 21
Blackman, R. L., 15, 24, 29, 35–36, 43, 90, 109, 180, 199, 203, 213
Blake, G. M., 66, 124
Blakley, N., 180, 182–83, 201, 203–4, 252, 270, 273
Blau, W. S., 209, 266, 273, 275
Bleken, E., 197, 262
Blickenstaff, C. C., 12, 70, 233
Block, W., 24–25, 96–102
Boctor, I. Z., 93
Bodegas V., P. R., 180
Bodnaryk, R. P., 50, 90, 97, 102, 142
Boethel, D. J., 125
Bogdanova, T. P., 66
Bogyo, T. P., 157
Bohle, H. W., 149
Bohm, M. K., 173
Boiteau, G., 39, 94
Boller, E. F., 203, 283, 303
Bollenbacher, W. E., 67, 69, 73, 80
Bombosch, S., 306
Bonet, A., 200
Bongers, W., 154, 240
Bonnemaison, L., 15, 36, 65
Booth, C. O., 15–16, 27, 36
Borden, J. H., 86
Borg, T. K., 104
Borror, D. J., 143
Bottrell, D. G., 82, 301
Boudreaux, H. B., 125
Boush, G. M., 235
Bowden, J., 183
Bowen, M. F., 69, 73, 96, 225
Bowers, W. S., 70, 302
Bowley, C. R., 136, 180, 241
Boyland, P. B., 290
Bradfield, J. Y., IV, 50, 54, 73, 120, 196
Bradley, B. P., 274
Bradley, J. R., Jr., 39, 94–95, 204, 211, 233, 237
Bradshaw, W. E., 39, 42, 55, 114, 125, 154, 180, 197–98, 202–3, 205–6, 220, 222, 252–53, 261–63, 273, 289–90
Brakefield, P. M., 179, 184
Branson, T. F., 180–81, 203–4, 211, 220, 233, 241, 287, 298
Brandenburg, R. K., 288
Braune, H. J., 32
Brazzel, J. R., 95
Breed, M. D., 171, 173
Brelje, N. N., 12
Brian, M. V., 171–76
Briers, T., 93
Brindley, T. A., 210–11, 243–44
Brittain, J. E., 100
Brockmann, H. J., 171–72, 270–71
Broodryk, S. W., 163
Brower, J. H., 50
Brower, L. P., 86–88
Brown, B. W., Jr., 244

Brown, C. K., 196, 201, 209, 211, 215, 231–32, 236–37, 266, 273, 275
Brown, G. C., 157
Brown, R. Z., 177, 180
Brown, V. K., 5, 9, 39, 219, 258
Brown, W. L., Jr., 281
Browning, T. O., 51, 73, 120, 139, 141, 159
Broza, M., 113, 134
Brunel, E., 136
Brunetti, R., 245
Brunnarius, J., 140, 222, 225
Brussard, P. F., 245
Brust, R. A., 79, 83–84, 89, 119, 129, 131, 143, 269, 305
Bryan, G., 162
Bues, R., 115, 139, 203, 233
Buechi, R., 245
Buffington, J. D., 89
Bünning, E., 80
Burbutis, P. P., 165
Burn, A. J., 144
Burns, M. D., 15
Bush, G. L., 203, 279–80, 283–84, 303, 306
Butler, G. D., Jr., 290
Butler, M. G., 39, 186–87, 257, 261
Butterfield, J., 129, 132–33, 193

Calcote, V. R., 149, 289, 293
Callan, E. McC., 297
Calvert, W. H., 87–88
Campbell, A., 203, 289
Campbell, L. E., 302
Campbell, R. K., 289
Campbell, R. W., 61
Canard, M., 196, 202, 253
Canfield, N. L., 289
Cantelo, W. W., 180
Capinera, J. L., 236
Cardé, R. T., 30, 108, 245
Cardwell, D. L., 104
Carlisle, D. B., 153
Carmona, A. S., 105
Carson, H. L., 93, 215
Carter, C. I., 24
Carter, N., 288
Carton, Y., 136, 144, 166–67, 193
Casagrande, R. A., 89
Casanova, D., 17
Case, T. J., 70
Cassier, P., 17
Catt, C. L., 140
Cave, R. D., 289
Cawley, B. M., 302
Čerkasov, J., 95
Ceusters, R., 177, 180
Chambers, R. J., 157
Chantarasa-ard, S., 162
Chaplin, S. B., 95
Chaplin, S. J., 4, 23, 183
Chapman, C. S., 17
Chapman, R. F., 159
Chapman, R. K., 289
Chappell, M. A., 188
Charlesworth, B., 280
Charlesworth, P., 82, 116, 215
Cherednikov, A. V., 174
Chernysh, S. I., 74, 169
Chetverikov, S. S., 233
Chiang, H. C., 211, 244, 299
Chino, H., 93, 96, 101
Chippendale, G. M., 34, 45, 61, 68, 75–76, 80–81, 93, 98, 106–7, 139–40, 143, 159, 180, 195, 211, 233, 237, 242, 291, 298–300, 302
Chowdhury, A. B., 219
Christian, J., 87–88
Church, N. S., 55, 113, 130, 143, 149
Claret, J., 80, 114, 136, 144, 164–67, 169, 193
Claridge, M. F., 27, 90
Clay, M. E., 206
Clayton, T. E., 159
Cleere, J. S., 302
Clements, A. N., 12, 57, 83–84, 94, 184
Clench, H. K., 30, 108
Cloutier, E. J., 94
Coaker, T. H., 144
Coan, M. E., 31
Cobb, D. L., 81
Cogan, P. M., 136, 180, 241
Cohen, A. C., 188
Cohen, J. A., 87
Cohen, J. L., 188
Cohet, Y., 11
Collier, R. H., 143, 197, 203
Connin, R. V., 81, 85
Conradi-Larsen, E. M., 99, 120
Cook, R. E., 28
Cooper, J. N., 149, 289
Corbet, P. S., 25, 41, 145, 149, 198, 248–49
Cothran, W. R., 89
Coulson, J. C., 132, 193
Coulson, R. N., 290
Cousin, G., 231–32
Cox, D. F., 211
Cox, P. D., 241
Craig, G. B., Jr., 43, 195, 202, 206, 214, 261, 302
Cranham, J. E., 145, 195
Crans, W. J., 82

Crawford, C. S., 189
Creech, J. F., 303
Croft, B. A., 203, 239, 288, 291
Croset, H., 49
Crowder, L. A., 96, 155, 203–4
Crozier, A. J. G., 96
Cui, Z., 101
Cullen, J. M., 120, 139
Cunningham, R. B., 55, 148
Curl, G. D., 165
Curtis, G. A., 13
Cymborowski, B., 74

Dadd, R. H., 154
Dallmann, S. H., 72
Danilevsky, A. S., 4, 39, 94, 114, 133, 138, 154, 163, 166, 193–94, 202–4, 208–11, 245, 273, 295
Danks, H. V., 96, 185–87, 274
Davey, K. G., 50
David, J. R., 11
Davidson, J., 289
Davies, D. E., 16
Davis, C. P., 165
Davis, C. S., 295, 307
Davis, D. W., 165–66
Davis, J. R., 86
Davis, J. W., 82
Dawe, A. R., 21
Daxl, R., 287
Day, K., 283
Dean, C. G., 219
Dean, J. M., 121
Dean, R. L., 39, 150, 199, 293
DeBach, P., 297, 305, 307–8
de Boer, J. A., 68, 70, 80–81
deCoss, F., M. E., 180
Defossez, A., 75
de Kort, C. A. D., 68, 70–71, 80–81, 93, 233
Delong, D. M., 143
de Loof, A., 68–70, 81, 93, 169
den Boer, P. J., 274
Denlinger, D. L., 44, 49, 50, 54, 57, 63, 66, 73–74, 80, 95–96, 101–2, 120, 125, 127, 136, 143–44, 179–81, 183, 195–96, 220, 222, 231, 235, 251–52, 268–69
Denno, R. F., 14–15, 19, 204, 215
Denys, C. J., 30, 34, 50, 63, 107–8, 120, 133
Depner, K. R., 44
De Reggi, M., 79
Derr, J. A., 18, 179–80, 182
Desai, A. K., 233
Deseö, K. V., 63, 125, 267
Dethier, B. E., 289
Detroy, R. W., 96, 305
Deura, K., 137, 148
de Waard, E. R., 80
de Wilde, J., 4, 21–22, 43, 67–70, 80–81, 89, 95, 133, 143, 154, 233, 240, 268, 289, 296
Dharma, T. R., 273, 278
Diaz Castro, G., 180, 203–4, 220, 241
Dickson, R. C., 118, 139
Dietrick, E. J., 305
Dillwith, J. W., 76
Dimock, R. V., Jr., 100
Dingle, H., 5, 9, 13–14, 17–18, 22–23, 26, 39, 53, 86, 88, 133, 180, 182, 195–96, 201, 203–4, 208–11, 214–15, 219–22, 231–32, 236–37, 252, 265–66, 270, 273, 275, 287
Dixon, A. F. G., 13, 15–16, 26–27, 32, 35–36, 65, 90, 109, 154, 157, 267, 270, 273, 277–78, 288
Dobzhansky, T., 82, 215
Dortland, J. F., 93
Douglas, E. E., 83
Douglas, M. M., 14, 33, 227
Doutt, R. L., 162–66, 305–6
Downer, R. G. H., 68
Downes, J. A., 96, 185–87
Dray, F., 169
Drummond, F. H., 79
Druzhelyubova, T. S., 29
Dubynina, T. S., 65–66, 113
Duintjer, C. S., 43, 80, 133
Duman, J. G., 96–101
Dumortier, B., 140, 222, 225
Dunn, J. A., 293
Dunn, R. L., 70
Durban, E., 96
Dutrieu, J., 96
Dyer, E. D. A., 136

Earle, N. W., 86, 155
Eckenrode, C. J., 289
Edmunds, M., 23, 182
Edney, E., 103, 158
Edwards, J. S., 100, 186–87
Eertmoed, G. E., 25, 202–3
Eghtedar, E., 121
Ehler, L. E., 295, 305–6
Ehrlich, P. R., 273, 281
Ehrman, L., 186, 214
Eichhorn, O., 164–65, 169, 197–99, 202–3, 211, 309
Eichler, W., 12
Eickwort, G. C., 171
Eidt, D. C., 298
Eisenbach, J., 72
Elbert, A., 157
Eldridge, B. F., 44, 79, 84–85, 141

El-Hariri, G., 93, 95
Ellis, P. E., 17, 153
El-Saedy, A. H. A., 95
Emberson, R. M., 243
Embree, D. G., 149
Emin, P., 290
Emme, A. M., 219
Emmel, J. F., 189
Emmel, T. C., 189
Endler, J. A., 205, 280–82
Endo, K., 34, 85–86, 106
Englemann, F., 67–69, 130
Epling, C., 82, 215
Erpenbeck, A., 172
Eskafi, F. M., 12, 141, 166
Etchegaray, J. B., 88
Evans, A. A. F., 219
Evans, F. D. S., 63, 269
Evans, H. T., 63, 269
Evans, K. W., 83
Everett, T. R., 299
Ewen, A. B., 77
Ewer, D. W., 158

Fairchild, M. L., 211, 244
Falkovich, M. I., 39, 63, 187
Farley, R. D., 188
Farner, D. S., 66, 246–47
Farris, S. H., 70, 293
Farrow, R. A., 17, 23
Fasulati, S. R., 34
Faure, J. C., 19
Faustini, D. L., 73, 96
Featherston, N. H., 233
Feeny, P., 259
Ferenz, H.-J., 39, 71, 80, 95, 121, 132, 194
Ferket, P., 296
Ferrari, D. C., 186, 203
Ferris, G. F., 104
Fetter-Lasko, J. L., 306
Finch, S., 143, 197, 203
Fisher, P. D., 288
Fisher, R. C., 162, 166, 168–70
Flaherty, D. L., 39, 57
Flanagan, T. R., 74
Flanders, S. E., 162
Flint, A. P. F., 246, 248
Flint, M. L., 166, 295, 309
Flint, W. P., 296, 299
Flitters, N. E., 290
Flynn, M. S., 149, 289
Follett, B. K., 246–48
Follett, D. E., 246, 248
Fomenko, R. B., 157, 222
Fontana, P. G., 273
Force, D. C., 289, 306
Fortescue-Foulkes, J., 16
Foster, D. R., 96, 155
Foster, G. G., 303
Foster, W. A., 15
Fourche, J., 102
Fowler, H. W., Jr., 159
Fox, L., 24, 36, 199
Fraenkel, G., 21, 96, 143, 252
Frank, J. H., 13
Frankie, G. W., 178
Fraser, A., 210, 214
Frazer, B. D., 203, 289, 295, 307
Free, J. B., 174–76
Freeman, B. E., 165–66, 169
Friedlander, M., 76
Frisbie, R. E., 300
Frost, D., 248
Fuchs, T. W., 149, 289
Fujiie, A., 63, 267
Fujisaki, Y., 299
Fujiyama, S., 136
Fukatami, A., 215
Fukaya, M., 61, 75
Fukuda, S., 77, 79, 106
Fullard, J. H., 264
Fulton, W. C., 149, 196, 211, 290
Furunishi, S., 26, 41, 44, 113, 115, 121, 123, 132, 154, 196, 260
Fuseini, B. A., 18, 182
Futuyma, D. J., 277
Fuzeau-Braesch, S., 75, 125, 138, 199, 215, 232, 236

Gage, S. H., 289
Galindo, P., 178
Gallaway, W. J., 84, 89
Galloway, T. D., 305
Galyal'murad, M., 41
Gambaro-Ivancich, P., 155
Garber, M. J., 308
Garcia, R. F., 180
Garthe, W. A., 86
Gast, R. T., 232
Gatehouse, A. G., 141, 144, 148
Geier, P. W., 295
Gelman, D. B., 77, 243
Georgian, T., 262
Gerber, G. H., 85, 149
Geyspits, K. F., 26, 65–66, 83, 113, 116, 133, 235, 237
Gharib, B., 79
Gibbons, J. R. H., 281, 283
Gibbs, D., 139
Giebultowicz, J. M., 73

Gieke, P. A., 83
Giese, R. L., 288
Giesel, J. T., 273
Gilbert, L. I., 69, 73
Gilbert, N., 203, 289
Gillett, J. D., 159, 272
Gillett, S. D., 17
Gillott, C., 143, 149
Gilstrap, F. E., 305
Girardie, A., 17, 70, 79
Glancey, B. M., 172
Glass, E. H., 231, 233
Glen, D. M., 15, 27, 35–36, 65, 90, 109
Glinyanaya, E. I., 65–66, 113, 235, 237
Goettel, M. S., 101–2, 143
Golden, J. W., 156
Goldschmidt, R., 203–4, 208
Goldson, S. L., 243
Gollands, B., 118, 129, 289, 295, 300
Goodman, W. G., 76
Goodner, S. R., 183
Gordon, H. T., 287, 289
Goryshin, N. I., 4, 66, 114, 133
Gösswald, K., 176
Goto, T., 78
Gottlieb, R., 174
Gourdoux, L., 96
Graham, H. M., 82
Granger, N. A., 67, 70, 80
Granier, S., 70
Grant, V., 281
Gray, R., 73
Greany, P., 307
Grechka, E. O., 173
Greenbank, D. O., 289
Gregg, P. C., 141
Griffiths, E., 101
Griffiths, K. J., 44, 149–50, 162, 166–67, 305
Grigo, F., 95
Grinfeld, E. K., 173–74
Grissell, E. E., 14–15, 19, 204, 215
Gross, G., 140
Grula, J. W., 14, 227
Grzelak, K., 74
Guerra, A. A., 180
Gunn, D. L., 21
Gupta, A. P., 68
Gupta, R. K., 24, 36, 199
Gurjanova, T. M., 169
Guss, P. L., 180
Gustin, R. D., 241
Guthrie, W. D., 211, 243–44
Gutierrez, A. P., 203, 287, 289–90, 300
Guttman, S. I., 279, 283–84
Gyrisco, G. G., 89
Habu, N., 200
Hackett, D. S., 141, 144, 148
Hadley, N. F., 19
Haeger, J. S., 208, 269
Haga, Y., 78
Hagedorn, H. H., 74
Hagen, K. S., 87, 89, 152, 306–7
Hagen, R. H., 197
Haglund, B. M., 153
Hagstrum, D. W., 156
Hain, F. P., 149
Hairston, N. G., Jr., 201, 263
Halaka, T., 258, 265
Halberg, F., 96
Hall, I. M., 305
Hall, J. C., 108, 165
Hall, P. M., 136
Hall, R. W., 295, 305–6
Hamamura, T., 114, 121
Hammond, A. M., Jr., 83
Hand (née Smeeton), L., 24
Hand, S. C., 176–77
Hanec, W., 243
Hans, H., 57
Hansen, T. E., 102
Hanski, I., 259, 262
Hara, A., 259
Harbo, J. R., 169
Hardee, D. D., 82, 301
Hardie, J., 16, 26–27, 36, 72, 90
Hare, J. D., 154, 240, 277, 296
Harper, A. M., 102
Harrewijn, P., 15, 154
Harris, M. K., 149, 289, 293
Harrison, R. G., 214, 245, 270, 285
Hartley, J. C., 19, 39, 137, 148, 150, 199, 293
Hartman, M. J., 39
Hartstack, A. W., Jr., 57
Harvey, A. W., 16
Harvey, G. T., 64, 108, 114, 194, 200, 210–11, 233
Harwood, R. F., 44, 269
Hasegawa, K., 77–78, 93, 101
Hashimoto, Y., 115, 121, 196
Hassell, M. P., 162
Haukioja, E., 258, 265
Havelka, J., 118, 198
Hawke, S. D., 188
Hayakawa, Y., 101
Hayashida, K., 172
Hayes, D. K., 73, 79, 89, 96, 114, 243, 288, 302
Hayes, J. L., 39, 86, 194–95, 211, 216, 261
Hays, S. B., 233
Haynes, D. L., 289

Hazel, W. N., 19, 33–34, 51, 108, 204
Hebert, P. D. N., 203
Hedin, P. A., 293
Hedlin, A. F., 198
Hedrick, L. E., 87–88
Heed, W. B., 70, 215
Hegdekar, B. M., 45, 104
Hegmann, J. P., 5, 196, 201, 209, 211, 215, 231–32, 236–37, 265–66, 273, 275
Heinrich, B., 103, 173, 175
Helfert, B., 40
Helgesen, R. G., 50, 53–54, 61, 73, 129, 143–45, 149, 167, 193, 203, 271, 289, 296
Helle, W., 194, 216, 235
Henderson, C. F., 167
Henneberry, T. J., 159
Henrich, V. C., 66, 231, 235
Hensle, R., 283
Hensley, S. D., 166
Henson, J. L., 149, 289
Herbek, T. A., 19, 108
Herbert, H. J., 149, 186
Herfst, M., 234
Herman, W. S., 26, 71–72, 88, 96
Heron, R. J., 149, 203
Herrebout, W. M., 89
Herron, J. C., 85
Hew, C. L., 98
Hewing, A. N., 302
Herzog, G. A., 233
Hibbs, E. T., 299
Hidaka, T., 19, 33, 40, 63, 65, 73, 80, 93–94, 108, 113–14, 117, 124, 133
Highnam, K. C., 67–68, 77
Hill, H. F., Jr., 88
Hill, J. C., 34, 157
Hill, L., 67–68, 77
Hill, R. E., 244
Hille Ris Lambers, D., 15, 24, 27, 35–36, 90, 104, 109
Hillman, W. S., 225
Hilmy, N., 41
Hinton, H. E., 12, 183
Hiruma, K., 74
Hobbs, G. A., 235
Hodek, I., 5, 21, 39, 46, 57, 65, 89, 93–95, 113, 127, 129–30, 133–34, 153, 200, 203, 219, 232, 306
Hodgson, C. J., 65
Hodková, M., 71, 95, 200
Hoffmann, H. J., 55, 71, 132
Hoffmann, J., 17
Hoffmann, R. J., 228
Hogan, T. W., 43–44, 51, 79, 104, 141, 143, 273, 303
Hokama, Y., 159
Hokkanen, T., 260
Hokyo, N., 149, 200
Holling, C. S., 288
Hollingsworth, C. S., 236
Holtzer, T. O., 204, 211, 233, 237
Holzapfel, C. M., 39, 42, 55, 125, 202, 206, 261–63, 273
Hom, R., 295, 307
Homma, K., 40, 108
Honda, K., 19
Honek, A., 50, 125, 142, 200, 232, 273
Hong, J. W., 210
Honma, K., 100
Hontela, A., 248
Hoopingarner, R. A., 85
Hopp, R. J., 289
Horobin, J. C., 132, 193
Horsfall, W. R., 159
Horton, J., 96
Horwarth, K. L., 96, 98, 100–101
Horwitz, J., 142–43, 164, 167
House, H. L., 112, 231, 235, 309
Howe, R. W., 136, 241
Howell, F., 114
Hower, A. A., Jr., 29, 70, 85, 94, 169
Hoxie, R. P., 39, 81–82, 85, 94
Hoy, M. A., 39, 45, 57, 165, 194, 196, 203, 208–11, 214, 230, 234–36, 273, 303–6, 308–9
Hoyt, S. C., 289–90
Hsiao, T., 143, 210, 240, 252, 268, 296–97
Hubbell, T. H., 258
Huber, R. T., 149, 288
Hudson, A., 107
Hudson, J. E., 84, 120, 139
Huff, B. L., Jr., 41
Huffaker, C. B., 287
Huggans, J. L., 233
Huisman, H. O., 80
Hummelen, P. J., 153, 169, 180, 222
Hunter, D. M., 29, 141
Husby, J. A., 98
Hussey, N. W., 30, 39, 65, 108, 145
Hynes, C. D., 39

Ibaraki, A., 70, 293
Ichijo, N., 117, 215
Ichinosé, T., 181, 203
Iheagwam, E. U., 89, 198
Ikan, R., 174
Ikeda, M., 78
Ilyinskaya, N. B., 50, 102
Imai, K., 78
Ingram, B. R., 131

Injeyan, H. S., 17
Iny, Y., 173
Iperti, G., 134, 153
Ishay, J., 173–74
Ishii, M., 33, 40, 114, 133
Ishikawa, K., 163
Ishikura, H., 299
Ishitani, M., 148
Ishizuka, Y., 40
Ismail, S., 75, 125, 138, 199, 215, 232, 236
Isobe, M., 78
Istock, C. A., 13, 154, 196, 201, 211–12, 215–16, 235–37, 253, 256–57, 265, 269, 273, 276, 280, 283
Ito, Y., 290
Ittyeipe, K., 165–66, 169
Ivanovic, J., 68
Ives, W. G. H., 149
Iwantsch, G. F., 170
Iwao, S., 14, 17, 19, 157
Iwao, Y., 215
Iwasaki, N., 181
Iwata, T., 273
Izumiyama, S., 100–101

Jackman, J. A., 149, 289
Jackson, D. J., 164–65, 167
Jacobson, J. W., 240, 297
Jacquemard, P., 168, 180
James, B. D., 100, 102
James, D. G., 88, 96, 243
Jankovic-Hladni, M., 68
Janzen, D. H., 179–80, 184
Järvinen, O., 82
Jarvis, R. A., 167
Jenner, C. E., 40, 127, 129
Jeppson, L. R., 39
Jepson, W. F., 13
Jermy, T., 125
Jervis, M. S., 169
Johansson, M. P., 175
Johansson, T. S. K., 175
Johki, Y., 114
Johnson, C. G., 13–14, 26, 28, 86
Johnson, G. D., 71
Johnson, H. G., 202
Johnston, P. G., 248
Joiner, R. L., 204, 241
Joly, L., 17
Joly, P., 17
Jones, P. A., 167
Jones, R. E., 289
Jones, R. L., 143
Jones, S. L., 307
Jordan, R. G., 39, 202, 273
Jorgensen, C. D., 289
Jorgensen, J., 200
Judson, C. J., 159

Kai, H., 78
Kaizuma, H., 78
Kalmes, R., 166
Kalpage, K. S. P., 143, 269
Kambysellis, M. P., 70, 215
Kamezaki, H., 199, 203
Kamm, J. A., 57, 125
Kanehisa, K., 75, 93, 101–2
Kaneto, A., 78
Kangas, D. A., 186
Kao, M. H., 98
Kappus, K. D., 43, 79
Katakura, H., 264
Kato, Y., 131
Kats, T. S., 72
Katsu, Y., 131
Kawai, T., 78
Kawaski, R., 155, 292, 295
Kaya, H. K., 44, 164, 166–67
Keaster, A. J., 211, 244
Keeley, L. L., 204, 241
Kefuss, J. A., 175
Keifer, H. H., 39
Kelleher, J. S., 307
Kelly, A. F., 176
Kenagy, G. J., 248
Kennedy, G. G., 14, 149, 243–44, 289, 293
Kennedy, J. S., 13, 15–17, 22, 26–27, 36, 71, 82, 86
Keränen, L., 215
Kerr, J. D., 300
Khaldey, Y. L., 113
Khalil, G. M., 113
Khan, M. A., 70
Khelevin, N. V., 44
Khomyakova, V. O., 245
Khoo, S. G., 199, 211
Kidd, N. A. C., 32
Kidokoro, T., 43, 89, 205, 271, 273, 281–82, 293, 299
Kikukawa, S., 45, 89, 140, 180, 225, 231, 233–34, 237, 241–42
Kim, K. C., 244
Kimura, M., 27, 33
Kimura, M. T., 16, 51, 85, 117, 203, 208, 210, 215
Kimura, T., 34, 39–40, 43, 54, 65, 90, 106, 109, 118, 148, 196
Kind, T. V., 43, 73, 78
King, A. B. S., 208, 211, 234
King, E. E., 155

King, J. R., 66
King, R. L., 232
Kingsolver, J. G., 201
Kipyatkov, V. E., 173, 177
Kirchner, W., 172
Kirkland, R. L., 86
Kishino, K., 203, 205–6, 239, 273
Kisimoto, R., 19
Kitto, G. B., 306
Klassen, W., 210–11, 242, 303
Kliewer, J. W., 159
Klun, J. A., 245
Knerer, G., 283, 285
Knipling, E. F., 303
Knop, N. F., 236, 303
Knutson, L., 159, 261
Kobayashi, M., 80, 82, 257
Kobayashi, T., 29
Kochar, V. K., 219
Koga, K., 78
Kogan, M., 298–99
Kogure, M., 4, 43, 115
Koidsumi, K., 83
Kolesova, D. A., 199
Koller, D., 187
Komazaki, S., 24
Kondo, N., 78
Kono, Y., 43, 45, 49, 51, 69, 71, 80, 92–93, 108, 155, 208
Koopmanschap, A. B., 93
Kopec, D., 180, 203, 252, 273
Koref-Santibañez, S., 186, 214
Koskela, H., 262
Kozhantshikov, I. W., 137
Kraft, K. J., 169
Kramer, J. P., 31
Krasnick, G. J., 208
Krehan, I., 121, 132, 136
Krówczyńska, A., 74
Krüll, F., 186
Krusanova, Y. V., 66
Krysan, J. L., 57, 158–59, 180, 203–4, 211, 220, 233, 241, 287, 298
Kulman, H. M., 143
Kumar, R., 18, 182
Kurahashi, H., 180, 183, 210, 252
Kurata, S., 78
Kurihara, K., 128–29, 246
Kuznetsova, I. A., 66, 139
Kuznetsova, V. G., 199
Kvitko, N. V., 66, 113

LaChance, L. E., 211
Lakovaara, S., 82, 186, 214–15
Lamb, R. J., 90, 97, 149
Landa, V., 39, 94–95
Lande, R., 280–82
Landis, B. J., 24
Lane, H. C., 155
Lanevich, V. P., 275
Langston, D. T., 203–4, 211, 234, 237
Lankinen, P., 202
Larsen, J. R., 84, 141, 159
Larsen, T. B., 179, 184
Larson, G. J., 260
Lassota, Z., 74
Laudien, H., 154, 290
Laufer, H., 68
Lauga, J., 17
Laugé, G., 17
Launois, M., 17
Lazarovici, P., 29
Lea, A. O., 208
Leather, S. R., 267, 273, 278
LeBerre, J.-R., 232
Lederhouse, R. C., 197
Lee, R. E., Jr., 97, 101–2
Lees, A. D., 4, 15–16, 21, 26–27, 30, 34–36, 39, 43–44, 65–66, 71–72, 80, 90, 93, 104, 108–9, 114, 118, 138, 142, 154, 158, 163, 165, 208, 210–11, 222–23, 225
Lees, E., 42, 296
Legner, E. F., 12, 141, 166, 169
Leibee, G. L., 148
Leigh, E. G., Jr., 178
Leigh, T. F., 289
Leinaas, H. P., 197, 262
Lepage, M. G., 172
Lessman, C. A., 71–72, 88
Levings, S. C., 180
Levins, R., 220, 228, 295, 305
Lewis, R. A., 246–47
Lewis, T., 55, 148
Liebherr, J., 245
Liebhold, A. M., 149, 216
Lieth, H., 288
Lilly, C. E., 102
Lim, K.-P., 98
Lincoln, C., 39
Linley, J. R., 63, 269
Litsinger, J. A., 60
Little, C. H. A., 298
Lloyd, E. P., 155, 232
Lloyd, J. E., 285
Lloyd, M., 39
Loeb, M. J., 73, 79
Lofgren, C. S., 172
Logan, J. A., 290
Logan, S. H., 290
Lokki, J., 214–15

Long, G. E., 101, 289
Longfield, C., 248
Lopez, J. D., Jr., 57, 165, 198
Lounibos, L. P., 13, 39, 42, 202–3, 205, 220, 222, 252–53, 273
Lowman, M. D., 180
Lucas, B. A., 233
Ludwig, D., 137
Luff, M. L., 100, 102
Lumme, J., 82, 186, 198, 202, 208, 214–15, 234, 248, 282
Lund, R., 153
Lutz, P. E., 40–41, 127, 129, 139
Luu, T. A., 102
Luyk, A. K., 102
Lyman, C. P., 21
Lynch, C. B., 208, 211, 304
Lynch, R. E., 244
Lynn, D., 204, 241
Lyon, R. L., 233

MacFarlane, J. R., 79
Machida, J., 163
Mack, T. P., 290
Mackauer, M., 203, 289, 306, 308
MacKay, P. A., 65
MacLeod, E. G., 30, 39, 52, 83
Maddrell, S., 103, 158
Madrid, F. J., 97
Maeda, N., 197–99, 263
Magnarelli, L. A., 84
Mahr, D. L., 202
Maier, C. T., 263
Makino, K., 83
Malan, A., 21
Mane, S. D., 76
Mani, G. S., 281, 283
Mansingh, A., 101, 220
Marcovitch, S., 4
Marcus, N. H., 219, 246, 248
Marganian, L., 93
Marion, K. R., 248
Marsella, P. A., 302
Marshall, H. V., Jr., 300
Martin, D. F., 211
Maruyama, T., 280
Masaki, S., 4, 23, 26, 33–35, 39–41, 43–44, 49, 54–55, 58, 65–66, 82, 89–90, 103, 106, 108–9, 113–16, 118, 121, 123–25, 131–33, 137, 139–40, 142–45, 147, 152, 154–56, 180–81, 196, 200, 202–5, 208–11, 220, 225, 228, 232, 237–38, 241–42, 248, 260, 263, 271, 273, 281–83, 285, 290, 292, 295–96
Maslennikova, V. A., 74, 165–66, 168–69, 203, 309
Matsuka, M., 36, 73, 125
Matsumoto, K., 301
Matsuura, I., 258
Matteson, P., 295, 307
Matthée, J. J., 29–30, 41, 104, 136, 158–59
Maynard Smith, J., 283
Mayr, E., 280–81, 284
Maywald, G. F., 300
McCaffery, A. R., 32
McCafferty, W. P., 41
McCoy, J. R., 86
McGuire, J. U., Jr., 303
McHaffey, D. G., 44
McIntyre, R. C., 301
McKelvey, J. J., Jr., 306
McLachlan, A., 183–84
McLeod, D. G. R., 56–57, 113, 127, 133, 143, 211, 243–45, 273
McMullen, R. D., 115, 133
McMurtry, J. A., 202, 306
McNeil, J. N., 44, 164, 166–68
McPherson, R. M., 166
McQueen, D. J., 246
McRae, K. B., 149
Meats, A., 23, 101
Medvedeva, G. A., 66
Meidell, E. M., 99
Mellini, E., 163–64, 168–69
Memmi, M., 41
Menaker, M., 140
Meola, R. W., 73
Merkl, M. E., 86
Merritt, R. W., 260
Messenger, P. S., 166, 289–90, 294–95, 298, 305–7
Messina, F. J., 14
Meszleny, A., 24, 36, 199
Metcalf, C. L., 296, 299
Metcalf, R. L., 296, 299
Meyer, A., 19
Meyer, R. P., 82
Mfon, M., 241
Michel, R., 17
Michener, C. D., 171
Michieli, S., 30, 51, 108
Micinski, S., 125
Miethke, P. M., 77
Milanovic, M., 68
Miller, E. R., 23, 180, 182, 203–4, 252, 270, 273, 275
Miller, G. E., 198
Miller, G. W., 203
Miller, K., 98–99
Miller, L. K., 100, 102
Mills, N. J., 95
Minami, N., 51, 117, 203, 215

Minder, I. F., 89, 133, 275
Minis, D. H., 234
Missonnier, J., 136
Mitchell, C. J., 84
Mitchell, E. B., 82, 301
Mitsuhashi, J., 75
Mitter, C., 277
Mittler, T. E., 15, 27, 36, 72, 154
Miya, K., 100–2, 208, 232, 236
Miyata, T., 64, 108
Mochida, O., 75
Modder, W. W. D., 159
Moeur, J. E., 13, 269
Mogi, M., 39, 41, 206
Möller, F. W., 24, 36
Montgomery, M. E., 236
Moody, D. S., 204, 241
Mook, L., 43, 80, 133
Moore, C. G., 269
Moore, N. W., 248
Morallo-Rejesus, B., 302
Morant, V., 16
Morden, R. D., 143, 146, 196
Moreau, R., 96
Moretti, L. J., 159
Morgan, T. D., 93, 98
Moriarty, F., 159
Morimoto, N., 203
Morohoshi, S., 77
Morris, R. F., 149, 155, 196, 211, 290
Morrison, R. K., 165
Morrissey, R. E., 98
Mouze, M., 75
Mueller, A. J., 289
Mukerji, M. K., 289
Mulla, M. S., 84, 95
Müller, F. P., 24, 36
Müller, H. J., 22, 24, 34, 121–22, 132, 199, 220
Muller, W. H., 189
Munns, W. R., Jr., 201
Munroe, E., 245
Munstermann, E., 278
Munstermann, L. E., 39, 195
Münster-Swendsen, M., 162
Muona, O., 82, 198
Murai, M., 200
Muroga, H., 233
Musick, G. J., 244
Mustafa, T. M., 65
Mustafayeva, T. M., 203, 309
Mutuura, A., 245

Nachman, G., 162
Nagai, T., 244–45, 273
Nagell, B., 100
Nair, K. S. S., 75, 233
Nakamura, I., 189
Nakasuji, F., 27, 33
Nakata, J., 162, 306
Nakatsuka, K., 299
Nassar, S. G., 72
Natuhara, Y., 13
Nawalinski, T. A., 219
Nayar, J. K., 208, 269
Nealis, V. G., 289
Nechols, J. R., 50, 53–54, 61, 73, 108, 129, 143–45, 149, 163, 165–68, 193, 197, 203, 209–10, 212, 271, 284, 296, 307
Neck, R. W., 306
Neeb, C. W., 149, 289
Negishi, H., 181, 203
Nei, M., 280
Neill, G. B., 85
Nelson, D. R., 104
Nelson, R. L., 82, 84, 269
Nemec, S. J., 302
Nenadovic, V., 68
Neudecker, C., 132
Neunzig, H. H., 300
Neuvonen, S., 260
Newsom, L. D., 95, 155
Niehaus, M., 194, 233
Niemelä, P., 260
Nijhout, H. F., 19, 68, 71, 74–75
Nilsen, E. T., 189
Nishi, K., 78
Nishida, T., 88
Noda, T., 114
Nolan, E. S., 290
Nolte, D. J., 16–17, 29, 173
Nopp, H., 34
Nopp-Pammer, E., 34. *See also* Pammer
Nordin, J. H., 101
Norling, U., 41
Norris, K. H., 114
Norris, K. R., 218
Norris, M. J., 17, 29, 39, 113, 120, 180
Norton, G. A., 288
Norton, R. M., 258
Numata, H., 65, 73, 80, 93–94, 113, 117, 124, 133

Oatman, E. R., 165
Obrycki, J. J., 86, 89, 108, 118, 129, 165–70, 203, 289, 295, 300, 307
Ochando, D., 215
Odesser, D. B., 302
Odiyo, P. O., 29
Ogata, T., 83
Ogura, N., 19, 77
Ohmachi, F., 210, 248, 258

Ohnishi, E., 78–79
Ohta, T., 302
Ohtaki, T., 19, 73, 79, 180, 183, 210, 252
Oikarinen, A., 214–15, 234, 282
Okada, M., 104
Oku, T., 29, 108, 159
Okuda, T., 203
Oldfield, G. N., 109
Olds, E. J., 263
Oliver, D. R., 186
Oliver, M. Z., 302
Olivier, D., 96
Olmstead, K. L., 4
Olszak, R., 149, 289
Olton, G. A., 12
Omata, K., 164
O'Meara, G. F., 13, 208, 214
Omura, S., 90
O'Neill, A., 23
Opler, P. A., 178
Orell, M., 82
Orshan, L., 39, 65, 85, 94, 113, 118, 133–34, 140, 193–94, 271, 281–82
Osborne, D. J., 153
Oshiki, T., 77
Otazo Sanchez, A., 70
Overmeer, W. P. J., 80
Owen, D. F., 107, 178, 180, 268
Owen, W. L., Jr., 301
Oyama, N., 131

Paarmann, W., 23, 103, 123, 136, 139, 180, 188, 220, 222, 250–51
Pace, A. E., 285
Page, T. L., 80
Page, W. W., 32, 159
Palm, C. E., 39
Palmer, J. O., 266
Pammer, E., 34–35
Papaj, D. R., 28
Parker, F. D., 211
Parker, R. D., 82, 149, 289
Parkin, S., 241
Parr, W. J., 30, 39, 108
Parrish, D. S., 165–66
Parsons, P. A., 294, 297
Pasqualini, E., 41
Pass, B. C., 148
Passera, L., 178
Patterson, J. L., 96, 98, 100–101
Peart, R. M., 288
Pedgley, D., 293
Peet, F. G., 70, 293
Peferoen, M., 93
Pehrson, I., 18
Pener, M. P., 17, 29, 34, 39, 65, 68–69, 85, 94, 106, 113, 118, 133–34, 140, 173, 193–94, 271, 281–82
Perry, R. N., 219
Persson, A., 267
Peter, R. E., 248
Pfaender, S. L., 93–95
Phillips, D. L., 114
Phillips, J. R., 82, 233, 300
Phillips, P. A., 240, 283–84
Phillips, V., 159
Philogène, B. J. R., 83, 101–2, 143
Phipps, J., 178–79
Pickford, R., 232
Pidzhakova, T. V., 66, 113
Pienkowski, R. L., 81
Pierce, P. A., 208, 269
Pinger, R. R., 44, 79
Pittendrigh, C. S., 223–25, 234
Platner, G. R., 165
Platt, A. P., 210
Pohjola, L., 234
Poitout, S., 115, 139, 203, 233
Polis, G. A., 190
Poras, M., 40, 63, 69, 71, 127, 268
Porcheron, P., 75, 169
Potter, M. F., 149, 243
Powell, D. M., 24
Powell, J. A., 53, 188, 198
Powell, J. R., 278
Prebble, M. L., 216, 235
Price, P. W., 4, 162, 275, 287, 307
Priesner, E., 245
Principi, M. M., 41
Prokopy, R. J., 89
Propp, G. D., 45, 60, 64, 115, 127
Proshold, F. I., 211
Pschorn-Walcher, H., 165
Puttler, B., 305

Raabe, M., 68
Rabb, R. L., 14, 44–45, 54, 57, 93–95, 149, 164, 166–68, 180, 203–4, 210–11, 233, 237, 287, 289, 293, 300
Rabbinge, R., 288
Raignier, A., 177, 180
Raina, A. K., 210–11, 242
Rainey, R. C., 13, 16–17
Rains, T. D., 100
Rakshpal, R., 141
Rand, A. S., 178
Randell, R. L., 232
Rankin, M. A., 13, 18, 26, 68, 71, 86, 180, 182, 227
Rankin, S. M., 71, 86

Ratner, S., 162
Rausher, M. D., 28
Raven, P. H., 281
Razumova, A. P., 66, 113
Reader, P. M., 9, 258
Readio, P. A., 137
Ready, P. D., 49
Reagan, T. E., 300
Reddy, A. S., 107, 139–40, 143, 159
Reed, G. L., 210–11, 244
Reeves, W. C., 82, 84–85, 89, 269
Régnière, J., 149
Reinert, J. A., 241
Remington, C. L., 294
Remmert, H., 185–87
Renfer, A., 167
Renfree, M. B., 246, 248
Resh, V. H., 184, 259, 283
Rettenmeyer, C. W., 17, 177, 180
Reyes, R. J., 180–81, 220
Reynolds, H. T., 300
Rice, R. E., 289
Richards, K. W., 235
Richmond, C. E., 233
Riddiford, L. M., 67, 69, 71
Riddle, D. L., 156
Ridgway, R. L., 307
Riedl, H., 239–40, 283, 290–92
Riggenbach, W., 211, 233
Ring, D. R., 149, 289, 293
Ring, R. A., 44, 96–99, 101, 136, 195, 235
Rings, R. W., 29, 87
Risco, S. H., 306
Ritchot, C., 244–45, 273
Rivnay, E., 198
Roach, S. H., 119, 139
Roberts, B., 252
Robertson, J. L., 233
Robin, J. C., 245
Robinson, A. S., 234
Rock, G. C., 40, 45, 54, 57, 131, 140, 289
Roe, R. M., 83
Roelofs, W. L., 245
Roff, D., 205
Rojas, R. R., 102
Rolley, F., 134, 153
Roman, E. A., 159
Root, R. B., 4, 23, 182
Rose, M. R., 271, 273
Röseler, P. F., 174
Rosenzweig, E., 173
Rosenzweig, M. L., 281–83
Ross, D. H., 260
Roush, R. T., 308
Rowell, C. H. F., 19
Ruesink, W. G., 29, 87, 89
Rummel, D. R., 82, 301
Ruth, D. S., 70, 198, 293
Růžička, 133, 200
Ryan, R. B., 44, 164, 166

Saeki, H., 32
Safranek, L., 74
Sahota, T. S., 70, 293
Saito, T., 77
Sakagami, S. F., 100, 172
Sakagami, Y., 40
Sakaguchi, B., 78
Sakai, T., 115–16, 125, 139, 144
Sakate, S., 131
Salt, G., 169
Salt, R. W., 96, 100, 143, 149
Sanborn, J. R., 301
Sanburg, L. L., 84, 141
Sapozhnikova, F. D., 65–66, 113, 235, 237
Sardesai, J. B., 301
Sargent, T. D., 263
Sáringer, G., 63, 105, 125, 155, 267, 302
Sasakawa, M., 83
Sato, N., 148
Sato, S., 197–99, 263
Sauer, K. P., 283
Saulich, A. K., 26
Saunders, D. S., 4, 39, 44, 50, 73, 80, 114, 117, 124–25, 130, 138, 140, 154, 164, 166–67, 208, 222–25, 228, 252
Saura, A., 186, 214
Sawall, E. F., Jr., 306–7
Sawchyn, W. W., 55, 113, 130, 149
Schad, G. A., 219
Schaefer, C. H., 83, 96
Schaller, F., 61, 75, 145
Schechter, M. S., 96, 114, 302
Scheltes, P., 75, 153–54, 180, 222
Schipper, A. L., 104
Schlinger, E. I., 108, 165, 305
Schmidt, G. H., 176
Schmidt-Nielsen, K., 103
Schmieder, R. G., 108, 165–66, 169
Schneider, F., 169
Schneider, J. C., 277
Schneiderman, H. A., 50, 104, 142–43, 164, 167, 302
Schneirla, T. C., 17, 177, 180
Schooley, D. A., 70
Schoolfield, R. M., 290
Schooneveld, H., 70, 154, 233, 240
Schoonhoven, L. M., 163, 168–69
Schopf, A., 164, 166
Schrader, R., 118

Schumm, M., 95
Schuster, D. J., 241
Scott, D. R., 267
Seaman, J. E., 241
Searle, J. B., 36, 72
Seeley, T. D., 173, 175
Seger, J., 171–72, 270–71
Sehnal, F., 168–69
Sekiguchi, Y., 155, 292, 295
Selander, R. B., 143, 150
Sempala, S. D. K., 180, 184
Sevacherian, V., 289
Shaffer, P. L., 40, 45, 131
Shaltout, A. D., 45
Shapiro, A. M., 22, 30–31, 33–35, 51, 64, 68, 71, 108–9, 145, 179, 184, 188, 195, 197–98, 204, 209, 220, 224–25, 227–28, 260, 273
Shaposhnikov, G. K., 199
Shappirio, D. G., 130
Sharma, S., 136, 180, 241
Sharpe, E. S., 96, 305
Sharpe, P. J. H., 290
Sheldon, J. K., 52, 83, 263
Shepard, L. J., 139
Shimada, K., 100–1
Shimazu, M., 82
Shimizu, I., 45
Shinkaji, N., 199
Shiotsu, Y., 89, 137, 165, 197
Shirozu, T., 259
Shorey, H. H., 85
Shorrocks, B., 82, 116, 215
Showers, W. B., 29, 87, 210–11, 243–44
Shroyer, D. A., 43, 195, 202, 206, 214, 261
Shuja-Uddin, 165
Shukla, M., 73, 96, 183
Sieber, R., 54, 75–76, 142
Siew, Y. C., 50, 123, 133
Silhacek, D. L., 156
Sillén-Tullberg, B., 112
Simmonds, F. J., 164, 166, 308
Simonenko, N. P., 65–66, 113, 116, 133
Sims, S. R., 39, 42, 145, 195, 198, 204, 206, 231, 233, 240–41, 260, 273
Singer, M. C., 13, 26, 71
Skopik, S. D., 225
Slaff, M., 82
Sláma, K., 71
Slansky, F., Jr., 248, 260–61
Slifer, E. H., 104, 232
Smallman, B. N., 101
Smilowitz, Z., 39, 70, 85, 94, 169, 290
Smith, G. R. J., 132, 193
Smith, J. S., 100
Smith, N. G., 183
Smith, R. F., 287, 300
Smith, S. M., 119, 129, 131, 202–3
Smith, W. F., 210, 214
Smith-Gill, S. J., 273, 281
Smythe, N., 180
Snellings, W. M., 19, 108
So, Y. P., 98
Sobrio, G., 198
Solbreck, C., 18, 89, 112, 120
Sømme, L., 96–101, 120
Soni, S. K., 63
Sonobe, H., 78
Sorjonen, J., 260
Southwick, E. E., 175
Southwood, T. R. E., 9, 13–14, 22–23, 201, 258, 268, 283–84
Spadoni, R. D., 82, 84, 269
Sparks, A. N., 211, 243–44
Spielman, A., 12, 85, 89, 208, 213, 269, 278, 284
Spradbery, J. P., 174
Sprenkel, R. K., 93–95, 289
Springett, B. P., 295
Staal, G. B., 68, 70, 72, 81, 302
Stadelbacher, E. A., 211
Städler, E., 42, 296
Stalker, H. D., 93, 215
Stanic, V., 68
Stanton, M. L., 28
Stark, S. B., 149, 290
Stark, W. S., 114
Starý, P., 162
Stearns, S. C., 5, 201, 248, 265
Steel, C. G. H., 72, 80, 246
Steele, H. V., 26, 41
Steele, J. E., 100–1
Stegeman, D. A., 300
Stekol'nikov, A. A., 66
Sterling, W., 155
Sterling, W. L., 301
Stern, V. M., 289, 298
Sternburg, J. G., 197, 203, 233, 263
Stewart, J. W., 95, 152
Stewart, R. K., 97
Stiling, P. D., 19
Stinner, R. E., 14, 95, 149, 243–44, 289–90, 300
Stoffolano, J. G., Jr., 82, 84–85
Strand, M. R., 162
Strassman, J. E., 83, 174
Strong, D. R., Jr., 19
Strong, F. E., 57, 289
Stross, R. G., 34, 55, 157, 186
Sugg, P., 100, 186–87

Sugiki, T., 40, 137
Sugimoto, A., 29
Sullivan, C. R., 141, 149, 198
Sullivan, W. N., 114, 302
Summy, K. R., 305
Sunose, T., 198, 275, 309
Surgeoner, G. A., 198, 301
Sutherland, O. R. W., 24, 36, 101, 154, 199
Sutherst, R. W., 300
Sutton, D., 167
Sutton, R. D., 32
Suzuki, H., 200
Suzuki, K., 93, 100–102
Suzuki, M., 115, 121, 196
Suzuki, T., 102, 174
Syme, P. D., 289
Szczesna, E., 74
Szentkirályi, F., 63, 105

Tabachnick, W. J., 278
Tahvanainen, J., 260
Takada, H., 36
Takafuji, A., 199, 203, 210
Takahashi, F., 136, 197, 274
Takahashi, H., 108
Takahashi, M., 73
Takano, H., 54, 65, 118, 196
Takano, T., 299
Takeda, M., 114, 156, 195, 211, 233, 242, 291
Takeda, N., 74, 101
Takeda, S., 77
Takehara, I., 96, 100–101
Takeuchi, S., 77
Tamaki, G., 24, 36, 101, 199, 289
Tamura, I., 273
Tan, K. E. W. P., 114
Tanaka, S., 34, 40, 49, 88, 122, 125, 129, 131, 137, 270
Tanaka, Y., 233
Tanigoshi, L. K., 290
Tanno, K., 89, 98–100
Taranets, M. N., 65–66, 113
Tassan, R. L., 306–7
Tauber, C. A., 4, 23, 26, 30, 34, 41, 43, 45, 50, 53–58, 60–61, 63–64, 73, 83, 85, 94, 107–8, 112–15, 117–20, 122, 124, 126–29, 131–33, 142–44, 149, 152–54, 159, 165–68, 189, 193, 196, 200–204, 208–10, 212, 219–20, 228, 231–32, 246–48, 253–54, 256, 260, 264, 271–73, 280–81, 283–84, 289, 293–96, 300, 306–7, 309
Tauber, M. J., 4, 23, 26, 30, 34, 41, 43, 45, 50, 53–58, 60–61, 63–64, 73, 83, 85–86, 89, 94, 107–8, 112–15, 117–20, 122, 124, 126–29, 131–33, 142–45, 149, 152–54, 159, 163, 165–68, 170, 189, 193, 196, 200–204, 208–10, 212, 219–20, 228, 231–32, 246–48, 253–54, 256, 260, 264, 271–73, 280–81, 283–84, 289, 293–96, 300, 306–7, 309
Tauber, P. J., 197
Tauthong, P., 79
Taylor, F., 46, 85, 118, 258, 289, 290
Taylor, L. R., 13
Taylor, V. A., 35
Tazima, Y., 5, 208, 211
Teetes, G. L., 160, 198
Templeton, A. R., 280
Tepedino, V. J., 40, 198–90, 211
Ternovoy, V. I., 112, 145
Tesar, D., 97–99
Thibout, E., 63
Thiele, H.-U., 39, 55, 71, 89, 95, 121, 123, 132, 136, 194, 225, 250–51
Thomas, J., 219
Thomason, J. H., 180
Thompson, J. N., 4
Thurston, R., 164
Tilley, R. J. D., 42, 296
Tingey, W. M., 300
Tingle, F. C., 155, 232
Tischler, W., 136
Tobe, S. S., 17
Toda, M. J., 16, 85, 215
Tombes, A. S., 50, 60, 67, 93
Tomchaney, A., 96, 98, 100–101
Tonascia, J. A., 219
Topp, W., 95
Torchio, P. F., 40, 169, 198–99
Toshima, A., 108
Treherne, J. E., 15
Tremblay, E., 162–63, 165–66, 305
Trimble, R. M., 8, 202–3, 289
Triplehorn, C. A., 143
Tripp, H. A., 275
Troester, S. J., 29, 87
Trottier, R., 149
Trpis, M., 159, 208, 213, 269
Truman, J. W., 62, 67, 69, 71
Tsuchida, K., 45, 208
Tsuda, Y., 201
Tsuji, H., 89, 154, 156, 200–201, 231, 234, 241, 301
Tsumuki, H., 75, 93, 101–2
Tsutskova, I. P., 208–9, 213
Tsutsui, H., 100
Tummala, R. L., 288
Tuomi, J., 258, 265
Turunen, S., 76, 80, 93

Turnock, W. J., 90, 97, 309
Tuskes, P. M., 86, 88
Tuttle, C., 290, 300
Tyler, B. M. J., 167
Tyshchenko, G. F., 66, 225
Tyshchenko, V. P., 4, 66, 80, 114, 133, 139, 219, 221–25, 228, 275

Ubukata, H., 39
Uchida, M., 199
Ujiye, T., 125
Umarova, T. Y., 66, 164–65
Umeya, K., 64, 155, 292, 295
Urquhart, F. A., 86–88
Urquhart, N. R., 86–88
Ushatinskaya, R. S., 220, 296
Usua, E. J., 153, 180, 222
Uvarov, B., 17, 258

Vaartaja, O., 186
Valdes, M. H., 180–81, 220
van Boven, J., 177, 180
van den Berg, M. A., 168, 198
van den Bosch, R., 287, 289, 294–95, 305–7
van der Laak, S., 99
Vanderroost, C., 169
van Dinther, J. B. M., 153, 180, 222
van Dover, C., 13
Van Duyn, J. W., 39, 94
van Emden, H. F., 13
Van Kirk, J. R., 142, 290
Van Loon, J., 169
van Zon, A. Q., 80
Vaz Nunes, M., 225
Vavra, K. J., 211–12, 235–37, 273
Vawter, A. T., 245
Veerman, A., 57, 65–66, 80, 89, 140, 225
Venard, C. E., 43, 79, 206
Vepsäläinen, K., 19, 35, 109, 122, 200–201, 204, 214
Viktorov, G. A., 298
Villavaso, E. J., 86
Vinogradova, E. B., 44–45, 54, 66, 83, 114, 118, 124, 139–40, 167, 194, 202–3, 208–9, 212–13, 234, 237, 252, 269
Vinson, S. B., 162, 170, 204, 241
Visscher, S. N., 26, 41, 79, 153
Vittum, M. T., 289
Viyk, M. O., 102
Volkovich, T. A., 275
Volney, W. J. A., 149, 216
von Kaster, L., 29, 87
Vosselman, L., 234
Vouidibio, J., 11

Wada, T., 82
Wagner, T. L., 290
Waldbauer, G. P., 52, 143, 196–98, 203, 233, 263, 299
Walker, G. P., 73–74
Walker, T. J., 201, 241, 285
Wall, C., 145
Wallace, B., 218
Wallace, D. R., 141, 198
Wallace, J. B., 262
Wallace, M. M. H., 104, 144, 148–49, 153
Wallner, W. E., 149, 166, 198, 301
Waloff, N., 233
Waloff, Z., 16
Wang, L. C. H., 21
Wang, Y., 287, 289
Wangboonkong, S., 15
Ward, S. A., 267
Wardhaugh, K. G., 29, 34, 90, 141, 158–59, 189
Warne, A. C., 39
Warren, M. A., 252
Washino, R. K., 70, 83–85, 96, 306
Wasserman, S. S., 13, 154, 253, 269, 273
Watabe, H., 215
Watanabe, A., 55, 143, 147
Watanabe, N., 290
Waterhouse, D. F., 218
Waterman, P. G., 179
Waters, W. E., 149, 216
Watson, J. A. L., 172
Watson, T. F., 149, 203–4, 211, 234, 237, 243
Way, M. J., 55
Weddle, R. C., 143, 150
Weeda, E., 93
Weir, B. J., 246, 248
Weiss, M. A., 101, 289
Wellings, P. W., 273, 278
Wells, A., 77
Wells, P. H., 88, 95
Wellso, S. G., 39, 81–82, 85, 94, 119, 155, 159
Wellington, W. G., 8, 289, 293
Wenner, A. M., 88
Weseloh, R. M., 165, 167–69, 210, 214, 309
West, D. A., 19, 33–34, 51, 108, 204
Westdal, P. H., 85
West-Eberhard, M. J., 171–72
Westigard, P. H., 289
Wheatley, G. A., 293
Wheeler, D. E., 19, 68, 71
Wheeler, W. M., 21
Whitcomb, R. F., 31

Whitcomb, W. H., 95, 152
White, E. B., 308
White, J. A., 39
White, M. J. D., 280
White, R. R., 273
Whitmore, W., 153
Whitten, M. J., 303
Wickman, P. O., 267
Wielgolaski, F. E., 185
Wigglesworth, V. B., 19, 50, 67–68, 77, 95, 290
Wiklund, C., 107–8
Wiklund, G., 267
Wildbolz, T., 211, 233
Wilkinson, P. R., 85
Willey, R. L., 96
Williams, C. M., 50, 57, 68, 73–75, 80, 104, 117, 131, 146
Williams, I. H., 176
Willis, J. H., 96
Willis, J. S., 21
Willmer, P. G., 103–4, 158, 188
Wilson, A. G. L., 55, 148
Wilson, E. O., 171–73, 281
Wilson, G. R., 159, 197
Wilson, M. R., 27, 90
Windsor, D. M., 178, 180
Wise, E. J., 197–98, 274
Wisniewski, W., 187
Witz, J. A., 57
Wolda, H., 178–80
Wolfenbarger, D. A., 82
Wollkind, D. J., 290
Wong, J., 85, 89, 269
Wood, K. A., 301
Wood, T. K., 4, 283–84
Woods, C. W., 77
Wratten, S. D., 101
Wu, C.-I., 280
Wu, H.-I., 290
Wylie, H. G., 164, 166–67, 170

Yagi, S., 61, 75, 121
Yaginuma, T., 78, 93
Yamashita, O., 77–78, 93, 101
Yamauchi, M., 131
Yeargan, D. R., 45, 54, 57, 289
Yeargan, K. V., 148
Yin, C.-M., 34, 61, 75–76, 101, 106
Yoshimeki, M., 75
Yoshitake, N., 45, 108
Young, A. M., 179–80
Young (Krieger), D. L., 36
Young, S. R., 24–25, 100–102
Yukawa, J., 197–99, 263

Zachariassen, K. E., 96, 98–100
Zakharova, L. V., 174
Zar, J. H., 95
Zarankina, A. I., 26, 83
Zaslavsky, V. A., 56, 66, 118, 157, 164–65, 222
Zeleny, J., 166
Žener, B., 30, 51, 108
Zimmer, H., 13, 154, 253, 269, 273
Zinovyeva, K. B., 44, 139–40, 167–70, 208–9, 213
Zisfein, J., 211–12, 235–37, 273
Zucker, I., 248

Species Index

Abraxas miranda, 116
Acheta commodus. *See Teleogryllus (= Acheta) commodus*
Acraea quirina, 282
Acrobasis nuxvorella, 149, 293
Acronycta rumicis, 66, 73, 138, 202, 210–11, 295
Acrotylus insubricus, 41
Adalia bipunctata, 118, 129
Adelphocoris lineolatus, 77
Adoxophyes orana, 40, 302
Aedes aegypti, 272, 278
Aedes atropalpus, 44, 143, 213, 302
Aedes geniculatus, 39, 195
Aedes sierrensis, 39, 202
Aedes solicitans, 131
Aedes taeniorhynchus, 12
Aedes togoi, 39–40, 206
Aedes triseriatus, 39, 42–43, 55–56, 195, 202, 206, 214, 261, 272
Aedes vexans, 159
Aelia acuminata, 65, 113, 133
Aelia rostrata, 232
Aeshna mixta, 61, 145
Agelena limbata, 128–29
Aglais urticae, 194, 233
Agromyza frontella, 61, 143–44, 193, 271, 296
Agrothereutus minousubae, 165
Agrotis ipsilon, 29–30, 87
Alaskozetes antarcticus, 24–25, 101
Aleyrodes asari, 232
Aleyrodes brassicae, 88
Allonemobius fasciatus, 137
Alsophila pometaria, 277
Altalia rosae, 168
Alysia manducator, 167
Amathes c-nigrum, 125
Anacridium aegyptium, 70
Anaphothrips obscurus, 57
Anax imperator, 198
Anomala cuprea, 136
Anopheles freeborni, 83, 85–86
Anopheles labranchiae atroparvus, 83
Anopheles lyrcanus, 83
Anopheles maculipennis messeae, 12, 83, 114, 194, 202
Anopheles punctipennis, 83–85
Anopheles sacharovi, 83
Anopheles superpictus, 83
Antheraea pernyi, 62, 73, 77, 80, 117, 131, 302
Antheraea yamamai, 77, 131, 233
Atheta fungi, 95
Anthonomus grandis, 82, 155, 180, 232, 300
Anthrenus verbasci, 66, 124
Aonidiella aurantii, 308
Apanteles congretatus, 168
Apanteles glomeratus, 168
Apanteles melanoscelus, 211, 214, 235, 309
Aphaereta minuta, 167, 170
Aphidoletes aphidimyza, 118
Aphis fabae, 15–16, 72
Aphodius, 258–59
Aphytis lingnanensis, 308
Apis mellifera, 83, 172, 174–77
Aporia crataegi, 168
Araschnia levana, 64
Arenivaga, 188
Argas arboreus, 113
Athalia rosae, 302
Atrachya menetriesi, 105, 145, 158–59, 194, 232, 236–37
Aulocara elliotti, 41, 79
Austracris proxima, 23
Austroicetes cruciata, 41, 59, 104, 142, 145

Barathra brassicae. *See Mamestra (= Barathra) brassicae*
Barbara colfaxiana, 293
Bathyplectes curculionis, 165–66
Battus philenor, 28, 145, 194–95, 204
Bdellodes lapidaria, 144, 148
Bembidion andreae, 23

Bessa harveyi, 309
Boettcherisca septentrionalis, 66
Bombus, 174, 186
Bombyx mandarina, 90
Bombyx mori, 43–45, 77–80, 93, 101, 104, 115, 158, 233, 257
Boophilus microplus, 300
Broscus laevigatus, 123, 251
Bucculatrix pyrivorella, 267
Bupalus piniarius, 169
Busseola fusca, 153

Caenorhabditis elegans, 156
Calliphora, 209
Calliphora vicina, 45, 66, 118, 139, 212–15, 234, 237
Callophrys macfarlandi, 189
Caraphractus cinctus, 165
Carposina niponensis, 108
Casignetella polynella, 63
Cataloaccus aeneoviridus, 167–68
Catocala, 263
Cavelerius saccharivorus, 200
Cephus cinctus, 149
Ceraclea transversa, 283
Cerotoma trifurcata, 95, 299
Ceuthorrhynchus pleurostigma, 281
Chaoborus americanus, 154, 197
Chelonus curvimaculatus, 163
Chesias legatella, 145
Chilo, 80
Chilo argyrolepia, 153
Chilo orichalcociliella, 153
Chilo partellus, 153
Chilo suppressalis, 75–76, 83, 93, 101–2, 205, 239, 260, 299
Chilo zonellus, 153
Chilocorus bipustulatus, 157, 222
Chironomus, 39, 187
Chironomus imicola, 183
Chironomus tentans, 130
Choristoneura, 149, 211, 216
Choristoneura fumiferana, 64, 114, 194, 233
Choristoneura orae, 211
Chorthippus brunneus, 159
Chortoicetes terminifera, 29, 90, 141, 158–59, 189
Chortophila brassicae, 135
Chrysopa, 26, 209, 285
Chrysopa carnea, 30, 34, 45, 53, 55, 57, 83, 85, 94, 107–8, 113–15, 118, 120, 123–24, 127–29, 132–33, 142–43, 149, 152–53, 189, 196, 200, 202–3, 212, 215, 231–32, 253–54, 260, 272, 284, 293, 307, 309
Chrysopa coloradensis, 45
Chrysopa downesi, 55, 108, 122, 129, 131–32, 212, 246–47, 254, 264, 272, 284
Chrysopa flavifrons, 41
Chrysopa harrisii, 57, 60, 118, 127, 129
Chrysopa nigricornis, 45, 60, 133
Chrysopa oculata, 45, 60, 64, 113–15, 127, 133
Cirphus unipunctata, 26
Cloeon, 100
Cloeon dipterum, 99
Cnaphalocrocis medinalis, 14, 82
Coccinella novemnotata, 115–16, 133
Coccinella septempunctata, 57, 65, 134, 203
Coleomegilla maculata, 86, 89, 129, 167–68, 170
Colias alexandra, 194, 216
Colias philodice eriphyle, 27
Compsilura concinnata, 169
Conotrachelus nenuphar, 233
Contarinia tritici, 159
Contarmia sorghicola, 160
Cordulia aenea amurensis, 39
Crambus tutillus, 125
Cryptoglossa verrucosa, 19
Cucullia boryphora, 63
Culex, 209
Culex nigripalpus, 12
Culex pipiens, 84–85, 141, 278, 284
Culex pipiens molestus, 213
Culex pipiens pipiens, 213
Culex tarsalis, 84–85, 96
Culex tritaeniorhychus, 84
Culicoides furens, 269
Culiseta inornata, 84, 120, 139
Cydia (= Laspeyresia) pomonella, 54, 76, 80, 142, 155–56, 211, 233, 239, 267, 291, 302

Dacus tryoni, 23, 101
Danaus plexippus, 71, 87, 95, 243
Daphnephila machilicola, 262–63
Daphnia, 186
Daphnia middendorffiana, 157
Daphnia pulex, 55
Dasychira pudibunda, 26, 83
Delia antiqua, 234
Delia radicum, 197
Dendroides canadensis, 101
Dendroides concolor, 101
Dendrolimus spectabilis, 200
Dermacentor andersoni, 85
Dermacentor marginalis, 113
Diabrotica virgifera, 158–59, 180–81, 204, 211, 220, 233, 241

Diaptomus sanguineus, 263
Diatraea, 80
Diatraea grandiosella, 75–76, 81, 107, 139–40, 143, 159, 180, 195, 211, 233, 242, 291, 299
Diatraea saccharalis, 83
Diplazon, 169
Diprion pini, 197, 202, 211
Diura bicaudata, 211
Dolycoris baccarum, 120, 133
Drepanosiphum dixoni, 270
Drepanosiphum platanoides, 16, 32, 157, 276–77
Drepanotermes perniger, 172
Drosophila, 11, 82, 136, 144, 193, 215, 308
Drosophila auraria, 116–17
Drosophila littoralis, 202, 214–15, 234
Drosophila lummei, 215
Drosophila nipponica, 85
Drosophila phalerata, 66, 116, 133
Drosophila subobscura, 23
Drosophila transversa, 66, 139
Dysaphis anthrisci, 232
Dyscia malatyana, 63
Dysdercus, 23, 182
Dysdercus fasciatus, 18
Dysdercus nigrofasciatus, 18
Dysdercus superstitiosus, 18

Eciton burchelli, 17
Eciton hamatum, 17
Ectobius lapponicus, 39
Edwardsiana rosae, 27, 90
Empoasca fabae, 14
Enallagma aspersum, 131
Enallagma hageni, 131
Enchenopa, 278–79, 285
Entomoscelis americana, 149
Eotetranychus hicoriae, 125
Ephemerella ignita, 149
Ephestia cautella, 136, 156
Ephestia elutella, 233
Ephestia kuehniella, 241
Ephippiger cruciger, 39, 150, 198, 293
Epilachna varivestis, 85, 118
Epilachna vigintioctopunctata, 71, 155
Euboettcheria trejosi, 183
Eucallipterus tilliae, 32
Eucarcelia, 169
Eucarcelia rutilla, 169
Euryeloma oleracea, 34
Euscelis incisus, 24, 199
Euxoa sibirica, 159
Exoteleia nepheos, 149

Forficula tomis, 113
Formica, 172, 176
Formica polyctena, 176

Galeruca tanaceti, 123, 133
Gerris, 35, 204, 214
Gerris odontogaster, 122, 200
Gilpinia polytoma, 216, 235
Gonia cinerascens, 169
Grapholitha funebrana, 105, 267
Grapholitha molesta. See Laspeyresia (= Grapholitha) molesta
Gryllodes sigillatus, 23
Gryllus bimaculatus, 101
Gryllus campestris, 125, 137–38, 215, 232, 236

Hagenomyia micans, 41, 121, 123, 260
Halotydeus destructor, 104, 144, 148–49, 153
Hasegawia sassacola, 274, 309
Heliothis, 141
Heliothis armigera, 55, 144, 148
Heliothis punctigera, 73, 119, 139
Heliothis virescens, 57, 73, 149, 211, 233, 243, 302
Heliothis zea, 73, 80, 119, 139, 233, 267, 299
Henosepilachna pustulosa, 264
Henosepilachna vigintioctomaculata, 264
Henosepilachna vigintioctopunctata, 45
Hestina japonica, 137
Hexacola, 141, 166
Hippelates, 12, 166
Hippodamia convergens, 71, 86–87, 101, 152
Homoeosoma electellum, 45
Hyalomma anatolicum, 41
Hyalophora (= Platysamia) cecropia, 57, 72, 80, 142, 197, 233
Hydrellia griseola, 299
Hylemyia antiqua, 124
Hylemyia floralis, 148
Hypera postica, 28, 60, 81, 85, 233
Hyperodes bonariensis, 243
Hyphantria cunea, 35, 60, 62, 64, 82, 89, 142, 149, 155, 211, 290, 292, 295
Hypopteramalus tabacum, 168
Hysterosia subfumida, 63

Ips pini, 82
Isia isabella, 89

Knutsonia lineata, 159

Laphygma exigua, 26
Lapidomera clivicollis, 266

Laspeyresia (= Grapholitha) molesta, 139, 231, 233, 267
Laspeyresia nigricana, 293
Laspeyresia pomonella. See Cydia (= Laspeyresia) pomonella
Lepidocyrtus lignorum, 262
Lepidosaphes, 79
Leptinotarsa decemlineata, 65, 70, 72, 80, 89, 93, 133, 143, 154, 233, 240, 268, 296
Leptohylemyia coarctata, 55
Leptophlebia verpertina, 100
Leptophyes punctatissima, 137, 148
Leptopiliana boulardi, 136, 144, 193
Leptopterna dolobrata, 32
Leptothorax, 172
Lestes disjunctus, 55, 113, 130
Lestes eurinus, 41, 127
Lestes sponsa, 145, 249
Lestes unguiculatus, 55, 113, 130
Lethe, 30, 108
Leucania separata, 77
Leucoma salicis, 211
Leuhdorfia japonica, 40, 133
Leuhdorfia puziloi inexpecta, 39, 148
Lindbergina aurovittata, 27, 90
Lipolexis, 165
Locusta migratoria gallica, 232
Locusta migratoria migratoria, 29
Locustana pardalina, 29, 41, 104, 136, 158–59
Lucilia caesar, 195, 214, 235
Lucilia illustris, 214
Lucilia sericata, 101
Lygus hesperus, 57
Lymantria dispar. See Porthetria (= Lymantria) dispar

Malacosoma americanum, 105
Malacosoma disstria, 149
Mallada perfecta, 41, 83
Mamestra (= Barathra) brassicae, 66, 73–74, 115–16, 121, 139, 144, 202, 233
Mamestra configurata, 45, 89, 104, 142
Mamestra genistae, 125
Mamestra oleracea, 139, 233
Manduca sexta, 45, 54, 73, 89, 104, 120–21, 211, 299–300
Margarodes vitium, 104
Mayetiola (= Phytophaga) destructor, 299
Megachile pacifica, 235
Megachile rotundata, 211
Megastigmus spermotrophus, 145
Megoura, 44
Megoura viciae, 35, 43, 65, 72
Meigenia bisignata, 168
Melanoplus bivittatus, 143, 149
Melanoplus differentalis, 104, 232
Melanoplus sanguinipes, 12, 121, 232, 236
Melasoma collaris, 99
Meleoma emuncta, 45
Meleoma signoretti, 114–15, 117–19, 129, 131, 246–47
Melittobia, 165–66
Melittobia chalybii, 169
Mesocomys pulchriceps, 168, 198
Metaseiulus occidentalis, 45, 57
Metatetranychus ulmi. See Panonychus (= Metatetranychus) ulmi
Metrioptera hime, 63
Microctonus vittatae, 164, 170
Mitopus morio, 136
Mocydia crocea, 121
Molophilus ater, 192
Monema flavescens, 74, 101
Mormoniella vitripennis. See Nasonia (= Mormoniella) vitripennis
Musca autumnalis, 85
Myrmecophilus sapporoensis, 23
Myrmeleon, 26
Myrmeleon formicarius, 41, 44, 113, 154, 260
Myrmica, 172
Myrmica rubra, 176–77
Mythimna separata, 14
Myzus, 44
Myzus persicae, 43, 72, 199, 213, 215

Naranga senescens, 157
Nasonia, 169
Nasonia (= Mormoniella) vitripennis, 140, 142, 164, 167
Nathalis iole, 14, 33, 227
Neacoryphus bicrucis, 18, 120
Nebria brevicollis, 123
Neivamyrmex nigrescens, 177
Nemobius yezoensis. See Pteronemobius nitidus
Neodiprion abietis, 141
Neodiprion sertifer, 133, 141, 149
Neodiprion swainei, 275
Neoseiulus fallacis, 45, 57
Nephotettix cincticeps, 149
Nexara viridula, 30, 50, 108
Nilaparvata lugens, 14
Nineta flava, 195
Nomadacris septemfasciata, 29, 120

Oedipoda miniata, 65, 85, 113, 118, 133–34, 140, 193, 271

Oncopeltus, 23, 182–83, 252
Oncopeltus cingulifer, 182
Oncopeltus fasciatus, 18, 71–72, 86, 88, 133, 182, 195, 203, 211, 231–32, 237, 252, 270, 275
Oncopeltus sandarachatus, 182
Oncopeltus unifasciatellus, 182
Operophtera brumata, 149
Orgyia, 44
Orgyia antiqua, 43, 77–78, 80
Orgyia gonostigma, 137
Orgyia thyellina, 43, 90 106, 109
Orthomus barbarus atlanticus, 123, 139, 251
Orthoporus, 189
Oscinella frit, 13, 268
Osmia, 40
Osmia californica, 199
Osmia iridis, 199
Osmia montana, 199
Ostrinia (= Pyrausta) nubilalis, 28, 57, 76, 80, 101, 113, 127, 133, 140, 143, 149, 155, 159, 211, 222, 233, 243–45, 290, 299, 302
Otiorhynchus ligustici, 39
Oulema melanopus, 85
Oulema oryzae, 293

Palagiolepis, 172
Panonychus citri, 199
Panonychus (= Metatetranychus) ulmi, 66, 90, 118, 145, 149, 154, 195
Papilio, 19
Papilio alexanor, 188–89
Papilio glaucus, 197
Papilio machaon, 99, 108
Papilio polyxenes, 34, 51, 108, 273
Papilio xuthus, 73, 99
Papilio zelicaon, 231, 233, 240, 260
Pararge aegeria, 42, 267, 296
Parnara guttata guttata, 27, 33
Paruroctonus mesaenis, 190
Patrobus atrorufus, 123
Pattonella intermutans, 183
Paulinia acuminata, 19
Pectinophora gossypiella, 96, 140, 155, 159, 211, 234, 237, 241, 299–300, 302
Pelophila borealis, 99
Pemphigus treherni, 15
Perilitus coccinellae, 129, 167–68, 170
Periphyllus, 104
Periplaneta americana, 77
Peripsocus quadrifasciatus, 25, 202
Petrobia latens, 104
Phalaenoides glycinae, 77
Pheidole pallidula, 177
Philodromus subaureolus, 114, 121
Philosamia cynthia, 35, 101
Phyllonorycter ringoniella, 125
Phyllotreta cruciferae, 170
Phyllotreta striolata, 170
Phytophaga destructor. See Mayetiola (= Phytophaga) destructor
Pieris brassicae, 96, 138, 140, 168, 202, 222
Pieris napi, 33
Pieris napi microstriata, 188–89
Pieris virginiensis, 33
Pimpla instigator, 166–67
Pimpla turionellae, 166
Pionea forficalis, 211, 234
Plathemis lydia, 139
Platygaster, 309
Platynota idaeusalis, 40, 44, 131, 140
Platysamia cecropia. See Hyalophora (= Platysamia) cecropia
Pleolophus basizonus, 149–50
Plodia interpunctella, 89, 102, 136, 140, 154, 156, 200, 231, 234, 241
Poecilometopa punctipennis, 136
Poecilometopa spilogaster, 235
Pogonus chalceus, 23, 136
Polistes, 172–73
Polistes annularis, 83, 174
Polistes chinensis antennalis, 174
Polistes metricus, 173
Polygonia c-aureum, 86, 108
Polypedilum, 103
Polypedilum vanderplanki, 12, 183
Popillia japonica, 96, 137, 149
Porthetria (= Lymantria) dispar, 28, 77, 79, 145, 158, 194, 204, 211, 234, 236, 304
Praon palitans, 165
Pristiphora erichisonii, 149
Prokelisia, 204
Prokelisia marginata, 14–15
Protoparce quinquemaculata, 107
Pseudosarcophaga affinis, 231, 235, 309
Psila rosae, 42, 136, 144, 296
Psylla peregrina, 32
Pteromalus puparum, 168–69
Pteronemobius, 44, 237
Pteronemobius fascipes, 34, 43, 181, 205, 232, 237, 248
Pteronemobius mikado, 205, 242. *See also Pteronemobius taprobanensis*
Pteronemobius nigrofasciatus, 205. *See also Pteronemobius fascipes*
Pteronemobius nitidus (= Nemobius yezoensis), 40, 122, 125, 129, 131, 137
Pteronemobius taprobanensis, 34, 125, 205, 237, 242, 248, 270

Pterostichus nigrita, 55, 71, 94, 132, 194
Pterostichus vulgaris, 136
Ptinella, 35
Pyrausta nubilalis. *See Ostrinia (= Pyrausta) nubilalis*
Pyrrhocoris apterus, 57, 71, 117, 125, 127, 129–30, 232
Pyrrhalta humeralis, 83
Pyrrharctia isabella, 101, 155

Reduvius personatus, 137
Rhagoletis, 278–79, 285
Rhagoletis indifferens, 142, 149
Rhagoletis pomonella, 235
Rhopalosiphum padi, 35
Rhynchaenus fagi, 132
Riptortus clavatus, 113, 117, 124, 133
Ripula subnordicornis, 192
Rothschildia lebeau, 184
Rupela albinella, 153

Sarcophaga, 73
Sarcophaga argyrostoma, 125, 138–39, 143, 222
Sarcophaga bullata, 57, 65, 101–2, 127, 143, 195, 231, 235, 268
Sarcophaga crassipalpis, 73, 95, 101–2
Sarcophaga spilogaster, 144
Scapsipedus asperus, 32
Scapteriscus acletus, 241
Scaptomyza pallida, 16, 85
Schistocerca gregaria, 29
Schizotetranychus schizopus, 66
Sciara, 89, 99
Scirpophage incertulas, 299
Scolytus ratzeburgi, 101
Semiadalia 11-notata, 95
Semiadalia undecimnotata, 133, 153
Sitodiplosis mosellana, 159
Sitona cylindricollis, 57, 85–86
Sitona hispidulus, 148
Sminthurus viridis, 104, 153
Sogatella furcifera, 14
Solenopsis invicta, 172
Solenopsis richteri, 172
Sphecophaga burra, 165–66
Spilarctia imparilis, 40, 64, 118, 137
Spilosoma lubricipeda, 149
Spilosoma menthastri, 194
Spodoptera exigua, 14
Spodoptera frugiperda, 14, 299
Stenacron interpunctatum, 41
Stenocranus minutus, 121
Stirellus bicolor, 31
Sturmia sericariae, 163
Subcoccinella 24-punctata, 155

Teleogryllus, 44, 143, 248
Teleogryllus (= Acheta) commodus, 43, 51, 55, 79, 104, 141, 143, 146–47, 159, 231–32
Teleogryllus emma, 142, 204–5, 210
Teleogryllus oceanicus, 79, 104
Teleogryllus yezoemma, 205
Tetragoneuria cynosura, 40, 127, 129
Tetramorium, 172
Tetranychus crataegi, 66
Tetranychus telarius. *See Tetranychus urticae (= T. telarius)*
Tetranychus urticae (= T. telarius), 30, 57, 65–66, 108, 114, 216, 235
Tetrastichus julis, 53–54, 129, 145, 149, 167
Tetrix undulata, 40, 71, 127, 268
Thecabius affinis, 24
Thermophilum sexmaculatum, 188
Thyridopteryx ephemeraeformis, 146
Tipula pagana, 129, 132
Tipula subnodicornis, 129, 132
Toxorhynchites rutilus, 125
Toxorhynchites rutilus rutilus, 13
Toxorhynchites rutilus septentionalis, 202
Trichiocampus populi, 98
Trichogramma, 164–65
Trichogramma evanescens, 66, 164
Trioxys complanatus, 308
Trioxys pallidus, 307
Trioxys utilus, 165
Trirhabda borealis, 14
Trirhabda virgata, 14
Trogoderma granarium, 233
Trogoderma variabile, 157
Trogus mactator, 164

Urania fulgens, 183

Vespula, 165

Wohlfahrtia magnifica, 145
Wyeomyia medioalbipes, 12
Wyeomyia smithii, 12, 39, 41, 83, 117, 119, 129, 131, 154, 202, 211–12, 216, 235–37, 252, 269, 275–76, 282
Wyeomyia vanduzeei, 12

Yponomeuta vigintipunctatus, 89

Zonocerus variegatus, 32, 159, 198
Zonotrichia leucophrys gambelii, 246–47
Zygogramma exclamationis, 85

Subject Index

Anoxia tolerance, 99–100
Arctic insects, 161, 185–87
Autogeny, 83–85, 269
Autoparasitism, 162

Biological control, 238, 293–94, 305–9. *See also* Pest management
Biotypes, 243, 287, 297, 306

Character displacement, 281
Chemical control, 300–301. *See also* Pest management
Clock, biological, 80
Cold-hardiness, 21, 25, 49, 58–60, 75, 96–103, 108, 113, 140, 143, 172, 186–87, 219, 226, 257
  and feeding, 25
  and diapause, 50–51, 97, 100–103
  freezing tolerance, 97–98, 186–87
  supercooling, 98–99, 101, 186–87
  without diapause, 24–25, 100
Colonization, 8, 229, 238–45, 292–97
Coloration. *See* Polyphenism, coloration
Comparative studies, 229, 245–54
Competition, 262
Crowding: and diapause, 154, 156–68, 166, 177, 180, 200–201, 231, 241
  and migration, 15, 17, 30
  and polyphenism, 19–20, 32, 34, 36, 154
  and wing length, 19–20, 154
Cultural control, 298–300. *See also* Pest management

Desert insects, 19, 103–4, 161, 187–90
Diapause: aestival, 39–40, 58, 60, 63–64, 74, 84, 87, 103, 115–16, 121, 123, 127, 131–32, 136, 139, 144, 148, 152–53, 157, 195, 200, 203, 260, 262–63, 267
  behavioral changes, 81, 85–86, 89, 92, 103, 180
  and cold-hardiness. *See* Cold-hardiness
  and crowding. *See* Crowding and diapause
  definition, 21
  density-dependent. *See* Crowding and diapause
  diapause development, 52, 56, 73, 121, 130, 142, 145, 166–67, 203, 210
  and disease, 96, 305
  and dormancy. *See* Dormancy, diapause-mediated
  and drought-hardiness. *See* Drought-hardiness
  duration, 45, 53, 118, 120, 127, 181, 195–96, 203–4, 210, 241
  endocrine control, 68, 93, 169, 222, 226, 252, 257. *See also* Endocrine control.
  evolution of. *See* Evolution of diapause
  facultative, 111
  feeding during, 26, 50, 81–83, 173–74
  and food. *See* Food and diapause
  genetic variation. *See* Variation, genetic
  genetics of. *See* Genetics of diapause
  geographic variation. *See* Variation, geographic
  growth during, 26, 41, 50
  and heat-hardiness. *See* Heat-hardiness
  hormone, 69, 77–78
  induction, 76, 111–12, 126, 135–38, 164–65, 173–74, 194, 290
  intensification, 34, 50, 107, 126, 141–42
  life-cycle timer, 258–64
  maintenance, 52, 54, 77, 126–30, 142–46, 166
  in males, 39, 86, 94–95, 194–95, 268
  mating during, 85
  and migration. *See* Migration, diapause-mediated
  and moisture. *See* Moisture
  obligatory, 111–12
  in parasitoids. *See* Parasitoid diapause
  phases, 47, 54–55, 58, 71, 86, 113, 121, 137, 174

and photoperiod. *See* Photoperiod
physiological changes during, 50, 91–92, 180
polymorphism. *See* Polymorphism, diapause
and polyphenism 30, 34–35. *See also* Polyphenism, diapause-mediated
and population density. *See* Crowding and diapause
prediapause changes, 47, 70, 76, 81–82, 89, 92, 203
prolonged, 53, 141, 165, 188, 198, 214, 274, 296
postdormancy effects, 62
sensitive stage, 43, 64–66, 77, 79, 122, 164, 173, 203, 208, 223, 290–91
stage, 38, 77, 164, 172, 186, 257–61, 271
syndrome, 20–21, 26, 28, 31, 34, 47, 50–51, 68, 70, 81, 86, 90, 107, 111, 135, 140, 142, 164, 174, 227
and temperature. *See* Temperature
termination, 52, 56, 85–86, 130–33, 146–48, 166
variation. *See* Variation
Dormancy, 8–11, 22–23, 257
aestivation, 22. *See also* Diapause, aestival
classification of, 220–23
definition, 22
and diapause, 22
diapause-mediated, 23, 25, 220
evolution of. *See* Evolution of dormancy
genetics of. *See* Genetics of dormancy
hibernation, 22
nondiapause, 23–25
periods of, 48, 74, 174
postdiapause period, 59, 61–62
and quiescence, 23, 186, 219
site, 86, 88–90, 258
survival during, 257, 262, 267
Drought: and quiescence, 10, 12
Drought-hardiness, 21, 103–5, 183, 188, 200, 226, 258

Endocrine system, 9, 17–19, 26, 31, 34, 43, 67–68, 77, 79, 85. *See also* Diapause, endocrine control
adult, 69–72
egg, 77–79
larva, 74–77
pupa, 72–74
Evolution: of diapause, 29, 69, 90, 219–29, 238, 242, 248, 250, 257, 259, 269
of dormancy, 257
of life history. *See* Life history, evolution of
of photoperiodism, 221
of sociality, 171

Fat body, 71, 76, 82–83, 92–93, 95, 103, 175
Food: and color change, 19. *See also* Food and polyphenism
and diapause, 41, 45, 49, 55, 57, 82, 152–55, 166, 176, 198, 201, 203, 240, 253
and dormancy, 23, 174
and migration, 15, 17–18
and polyphenism, 19–20, 32, 36
and postdiapause development, 60, 94, 155, 259, 268
and quiescence, 10, 12
and wing length, 19–20
postdormancy, 63
Foraging behavior, 16, 26–28, 31, 36, 82–84, 90, 174

Genetic analysis, 208–10
Genetic control, 303. *See also* Pest management
Genetic improvement, 308–9
Genetic switch mechanism, 203, 213–14
Genetics: of diapause, 79, 181, 203, 210, 230–38, 253, 272, 309
of dormancy, 203, 210–16, 276
of seasonal cycles, 210–17
sex linkage, 208, 210, 214–15, 236
Gonotrophic dissociation, 83–84, 94
Growth, nondiapause, 289

Heat-hardiness, 21, 49, 103–5, 108, 226, 258
Heritability: of diapause, 214
of life-history traits, 273
Hormonal control, 301–2. *See also* Pest management
Host resistance, 300. *See also* Pest management
Humidity. *See* Moisture
Hybridization, 209, 271, 276
Hyperparasitism, 162, 168

Insect pest management. *See* Pest management

Life history: evolution of, 230, 265–79
and natural enemies, 262, 275
and seasonal food resources, 259–60, 270–71, 276–77
and seasonal habitats, 261
and seasonal temperature, 260

Life history (*continued*)
timing of events, 259
trade-offs, 256, 261, 266–69, 271, 274
Life-history traits, 236, 243, 257–65
coevolution, 266–73
sex differences, 261, 267–69

Metabolism, 91–95, 103, 175, 257
Migration, 9–11, 226
aseasonal, 10, 13–14, 17
and crowding. *See* Crowding and migration
diapause-mediated, 26, 28–29, 49, 82, 87–89, 203–4, 226, 270, 293
and food. *See* Food and migration
and humidity. *See* Moisture and migration
in locusts, 16–17, 29
nondiapause, 26–27, 29, 87, 182
and photoperiod, *See* Photoperiod and migration
"Pied Piper," 14
seasonal, 8, 21–22, 86, 176
in social insects, 17
and temperature. *See* Temperature and wing length
variation in, 17, 88, 268, 270, 275. *See also* Variation
Moisture, 189. *See also* Drought-hardiness
and diapause, 57, 103–5, 153–54, 158–60, 166, 177, 180
and egg development, 158–59
and migration, 17
and quiescence, 158

Neuroendocrine system. *See* Endocrine system

Parasitoid, 92, 112, 136, 140–42, 144, 161–71, 187, 193, 197–98, 214, 262, 275, 295, 305–9
development, 163, 168–69
diapause, 44, 53–54, 57, 66
effect on host development, 170
polyphenism, 165
Pest management, 238, 287–88
prediction, 288–97
Phenological models, 288–93
Phenology, 300, 306
definition, 20
host plant, 240, 260
phenophase, 22
Pheromones, 82, 85, 173, 245, 289, 301
Photoperiod, 4, 112, 135
changing day length, 118–24, 127, 166, 246
critical photoperiod, 34, 41, 50, 114, 117, 127, 138, 194–95, 201–2, 206, 210, 214–15, 231, 239–40, 243, 271, 290–92, 295, 309
diapause induction, 40, 45, 70, 126, 164–67, 175, 177, 186, 231, 251, 291
diapause maintenance, 54, 71, 117, 126–30, 167
diapause termination, 56–57, 113, 117, 128, 130–33, 198
and growth, 113, 124, 226
and migration, 17, 88–89
and polyphenism, 31, 33–36, 107, 122, 125
postdiapause development, 133–34
postdormancy, 63, 113
and prediapause changes, 49, 113, 124–25
response curve, 114–18, 194, 239
sensitive stage, 44–45, 54, 64–66
and sex ratio, 174
spectral sensitivity, 114
Photoperiodic control, 301–2. *See also* Pest management
Photoperiodic counter, 223–25
Photoperiodic receptors, 80, 165–66, 223
Photothermograph, 290–91
Physiological time, 289
Phylogeny, 251
Polymorphism, 8, 180, 196, 203–4
diapause, 197, 199, 272, 278
dormancy, 9, 184
emergence times, 197
genetics of, 215–16
life histories, 274, 278
migration, 9, 270, 275
morphology, 197, 277–78
wing length, 9, 200
Polyphenism, 10–11, 176–77
and crowding. *See* Crowding
and food. *See* Food
and photoperiod *See* Photoperiod
and temperature. *See* Temperature
aseasonal, 10, 15–16, 18
cocoon, 33, 97, 104, 108, 165–66
coloration, 19, 31–34, 50–51, 59–60, 106–7, 120, 142, 184, 204, 263
definition, 30
diapause-mediated, 31, 33, 50–51, 106–7, 203–4
hormonal control, 71–72, 106
nondiapause, 31–32
postdormancy, 64
seasonal, 8, 21–22, 31, 106, 165, 179, 184, 213, 227
waxy secretion, 19, 33, 97, 104

wing length, 19, 32–35, 50, 106–7, 109, 204, 214
Population density. *See* Crowding
Postdiapause: development, 59–61, 74–75, 133–34, 149–50, 168, 203, 292
and food. *See* Food and postdiapause development
and moisture. *See* Moisture
period. *See* Dormancy, postdiapause period
and photoperiod. *See* Photoperiod, postdiapause development
postdormancy, 62. *See also* Postdormancy
quiescence, 24, 58–59, 196
and temperature. *See* Temperature and postdiapause development
Postdormancy, 62
behavior, 62
life-history traits, 63, 94, 259, 266–69
phenotypes, 64, 106, 109
Prediction. *See* Pest management, prediction

Quiescence, 9–10, 22–24. *See also* Dormancy and quiescence
aseasonal, 8, 11
definition, 11
and drought. *See* Drought and quiescence
and food. *See* Food and quiescence
postdiapause. *See* Postdiapause quiescence
seasonal, 24
and temperature. *See* Temperature and quiescence

Reproductive isolation, 242, 264, 272, 276, 278–82
Resource partitioning, 262
Rhythmicity, annual, 66, 124, 209, 225–26, 236

Seasonal migration. *See* Migration, seasonal
Seasonal polyphenism. *See* Polyphenism, seasonal
Selection: artificial, 194, 196, 209, 214, 216, 229–38, 240, 242, 245, 273, 297
density-dependent, 281, 283
disruptive, 277, 280, 283
natural, 196, 212, 238, 240–41, 244, 254, 258, 270, 273, 275, 281
seasonally variable, 275, 277–78
Sex ratio, 174–75, 195, 268, 270
Social insects, 83, 94, 112, 161, 171–78, 186, 270–71
Speciation, 251, 276, 280–85, 297
allochronic, 284–85
allopatric, 280–82
climatic, 282
parapatric, 282–83
sympatric, 283–84
Spectral sensitivity, 114
Swarming, 171, 174–75

Temperature. *See also* Cold-hardiness
"chilling" and diapause termination, 69, 73, 78, 143, 146
and critical photoperiod, 138–39
and diapause development, 73, 146
diapause induction, 135–38, 164, 166–67, 176, 180, 231
diapause intensification, 141–42
diapause maintenance, 54, 73, 142–46, 166–67
diapause termination, 56–57, 73, 146–48
and migration, 17, 88
and polyphenism, 19–20, 35–36, 184
postdiapause development, 60–62, 142, 145, 149–50, 241, 292
and quiescence, 10–12, 41, 83
and wing length, 19–20
Thermoperiod, 140–41, 166, 225, 250
Thermoregulation, 32–33, 103, 106, 109, 173, 175, 186, 188, 227, 273
Token stimulus, 21, 25, 27–28, 34, 50, 52, 59, 90, 163–64, 220, 250. *See also* Crowding, Food, Moisture, Photoperiod, Temperature
food, 22, 260
moisture, 189, 222
parasitoid host, 164–65, 167, 169
temperature, 135, 222
Torpor. *See* Quiescence
Tropical insects, 11, 23, 32, 136, 144, 153, 161, 171, 177–84, 205, 237, 241, 248, 251–52, 268
diapause, 179–81, 220–22
nondiapause seasonal adaptations, 17, 179, 181–84
Twilight. *See* Spectral sensitivity

Variation: concealed, 196, 204
disjunct. *See* Polymorphism
genetic, 42, 44, 53, 82, 164, 180–81, 193, 216, 230, 238, 247, 276, 287, 296, 307
geographical, 41, 192, 201–6, 214–15, 230, 238–45, 259, 270–71, 276, 309
intrapopulation, 193–201, 260, 274–75, 277–78, 296
life-history, 274, 276, 278
maintenance of, 212, 273–79
in migration. *See* Migration, variation in
phenotypic, 192, 242, 271, 296
sex-related, 194